Molecular Nano Dynamics

*Edited by
Hiroshi Fukumura,
Masahiro Irie,
Yasuhiro Iwasawa,
Hiroshi Masuhara,
and Kohei Uosaki*

Related Titles

Matta, C. F. (ed.)

Quantum Biochemistry

2010
ISBN: 978-3-527-32322-7

Meyer, H.-D., Gatti, F., Worth, G. A. (eds.)

Multidimensional Quantum Dynamics

MCTDH Theory and Applications

2009
ISBN: 978-3-527-32018-9

Reiher, M., Wolf, A.

Relativistic Quantum Chemistry

The Fundamental Theory of Molecular Science

2009
ISBN: 978-3-527-31292-4

Höltje, H.-D., Sippl, W., Rognan, D., Folkers, G.

Molecular Modeling

Basic Principles and Applications
Third, Revised and Expanded Edition

2008
ISBN: 978-3-527-31568-0

Matta, C. F., Boyd, R. J. (eds.)

The Quantum Theory of Atoms in Molecules

From Solid State to DNA and Drug Design

2007
ISBN: 978-3-527-30748-7

Rode, B. M., Hofer, T., Kugler, M.

The Basics of Theoretical and Computational Chemistry

2007
ISBN: 978-3-527-31773-8

Molecular Nano Dynamics

Volume II: Active Surfaces, Single Crystals and Single Biocells

Edited by
Hiroshi Fukumura, Masahiro Irie, Yasuhiro Iwasawa,
Hiroshi Masuhara, and Kohei Uosaki

WILEY-VCH Verlag GmbH & Co. KGaA

The Editors

Prof. Dr. Hiroshi Fukumura
Tohoku University
Graduate School of Science
6-3 Aoba Aramaki, Aoba-ku
Sendai 980-8578
Japan

Prof. Dr. Masahiro Irie
Rikkyo University
Department of Chemistry
Nishi-Ikebukuro 3-34-1
Toshima-ku
Tokyo 171-8501
Japan

Prof. Dr. Yasuhiro Iwasawa
University of Electro-Communications
Department of Applied Physics and Chemistry
1-5-1 Chofu
Tokyo 182-8585

and

Emeritus Professor
University of Tokyo
7-3-1 Hongo, Bunkyo-ku
Tokyo 113-0033
Japan

Dr. Hiroshi Masuhara
Nara Institute of Science and Technology
Graduate School of Material Science
8916-5 Takayama, Ikoma
Nara, 630-0192
Japan

and

National Chiao Tung University
Department of Applied Chemistry and
Institute of Molecular Science
1001 Ta Hsueh Road
Hsinchu 30010
Taiwan

Prof. Dr. Kohei Uosaki
Hokkaido University
Graduate School of Science
N 10, W 8 , Kita-ku
Sapporo 060-0810
Japan

All books published by **Wiley-VCH** are carefully produced. Nevertheless, authors, editors, and publisher do not warrant the information contained in these books, including this book, to be free of errors. Readers are advised to keep in mind that statements, data, illustrations, procedural details or other items may inadvertently be inaccurate.

Library of Congress Card No.: applied for

British Library Cataloguing-in-Publication Data
A catalogue record for this book is available from the British Library.

Bibliographic information published by the Deutsche Nationalbibliothek
The Deutsche Nationalbibliothek lists this publication in the Deutsche Nationalbibliografie; detailed bibliographic data are available on the Internet at http://dnb.d-nb.de.

© 2009 WILEY-VCH Verlag GmbH & Co. KGaA, Weinheim

All rights reserved (including those of translation into other languages). No part of this book may be reproduced in any form – by photoprinting, microfilm, or any other means – nor transmitted or translated into a machine language without written permission from the publishers. Registered names, trademarks, etc. used in this book, even when not specifically marked as such, are not to be considered unprotected by law.

Cover Design   Adam-Design, Weinheim
Typesetting   Thomson Digital, Noida, India
Printing and Binding   betz-druck GmbH, Darmstadt

Printed in the Federal Republic of Germany
Printed on acid-free paper

ISBN: 978-3-527-32017-2

Contents to Volume 2

Contents to Volume 1 *XV*
Preface *XVII*
About the Editors *XIX*
List of Contributors for Both Volumes *XXIII*

Part Three Active Surfaces *315*

18   **The Genesis and Principle of Catalysis at Oxide Surfaces: Surface-Mediated Dynamic Aspects of Catalytic Dehydration and Dehydrogenation on TiO_2(110) by STM and DFT** *317*
         Yohei Uemura, Toshiaki Taniike, Takehiko Sasaki, Mizuki Tada, and Yasuhiro Iwasawa
18.1   Introduction *317*
18.2   Experimental *318*
18.2.1   STM Measurements of TiO_2(110) *318*
18.2.2   Computational Methods *318*
18.3   Results and Discussion *319*
18.3.1   Dynamic Mechanism for Catalytic Dehydration of Formic Acid on a TiO_2(110) Surface, Much Different from the Traditional Static Acid Catalysis *319*
18.3.2   Dynamic Catalytic Dehydrogenation of Formic Acid on a TiO_2(110) Surface *327*
18.3.2.1 Mechanism of the Switchover of Reaction Paths *331*
18.4   Conclusion and Perspective *332*
         References *333*

19   **Nuclear Wavepacket Dynamics at Surfaces** *337*
         Kazuya Watanabe
19.1   Introduction *337*
19.2   Experimental Techniques *338*

19.2.1	Time-Resolved Two-Photon Photoemission with Femtosecond Laser Pulses	*338*
19.2.1.1	Principles	*338*
19.2.1.2	Experimental Set-Up	*339*
19.2.2	Time-Resolved Second Harmonic Generation	*340*
19.2.2.1	Principles and Brief History	*340*
19.2.2.2	Experimental Set-Up	*341*
19.3	Nuclear Wavepacket Motions of Adsorbate Probed by Time-Resolved 2PPE	*343*
19.3.1	Alkali Atom Desorption from a Metal Surface	*343*
19.3.2	Solvation Dynamics at Metal Surfaces	*344*
19.3.3	Ultrafast Proton-Coupled Electron Transfer at Interfaces	*345*
19.4	Nuclear Wavepacket Motion at Surfaces Probed by Time-Resolved SHG	*345*
19.4.1	Vibrational Coherence and Coherent Phonons at Alkali-Covered Metal Surfaces	*345*
19.4.2	Dephasing of the Vibrational Coherence: Excitation Fluence Dependence	*347*
19.4.3	Excitation Mechanisms	*349*
19.4.4	Mode Selective Excitation of Coherent Surface Phonons	*351*
19.5	Concluding Remarks	*352*
	References	*353*
20	**Theoretical Aspects of Charge Transfer/Transport at Interfaces and Reaction Dynamics**	*357*
	Hisao Nakamura and Koichi Yamashita	
20.1	Introduction and Theoretical Concepts	*357*
20.1.1	Introduction	*357*
20.1.2	Molecular Orbital Theory and Band Theory	*358*
20.1.3	Charge Transfer vs. Charge Transport	*359*
20.1.4	Electronic Excitation	*361*
20.1.5	Reaction Dynamics	*363*
20.2	Electrode–Molecule–Electrode Junctions	*365*
20.2.1	Nonequilibrium Green's Function Formalism	*365*
20.2.2	Efficient MO Approach	*367*
20.2.3	*Ab Initio* Calculations: Single Molecular Conductance and Waveguide Effects	*370*
20.2.4	Inelastic Transport and Inelastic Electron Tunneling Spectroscopy	*375*
20.3	Photochemistry on Surfaces	*381*
20.3.1	Theoretical Model of Hot Electron Transport and Reaction Probability	*381*
20.3.2	Photodesorption Mechanism of Nitric Oxide on an Ag(111) Surface	*384*
20.4	Summary and Outlook	*392*
	References	*394*

Contents to Volume 2 | VII

21 **Dynamic Behavior of Active Ag Species in NOx Reduction on Ag/Al$_2$O$_3$** 401
Atsushi Satsuma and Ken-ichi Shimizu
21.1 Introduction 401
21.1.1 NOx Reduction Technologies for Diesel and Lean-Burn Gasoline Engines 401
21.1.2 Selective Catalytic Reduction of NOx by Hydrocarbons Over Ag/Al$_2$O$_3$ 402
21.2 Hydrogen Effect of HC-SCR Over Ag/Al$_2$O$_3$ 403
21.2.1 Boosting of HC-SCR Activity of Ag/Al$_2$O$_3$ by Addition of H$_2$ 403
21.2.2 Surface Dynamics of Ag Species Analyzed by *in situ* UV–Vis 405
21.3 The Role of Surface Adsorbed Species Analyzed by *in situ* FTIR 410
21.3.1 Reaction Scheme of HC-SCR Over Ag/Al$_2$O$_3$ 410
21.3.2 Effect of H$_2$ Addition on Reaction Pathways of HC-SCR Over Ag/Al$_2$O$_3$ 414
21.4 Relation Between Ag Cluster and Oxidative Activation of Hydrocarbons 416
21.4.1 Debates on Role of Ag Clusters 416
21.4.2 Reductive Activation of O$_2$ and Promoted HC-SCR on Ag Cluster 420
References 422

22 **Dynamic Structural Change of Pd Induced by Interaction with Zeolites Studied by Means of Dispersive and Quick XAFS** 427
Kazu Okumura
22.1 Introduction 427
22.2 Formation and Structure of Highly Dispersed PdO Interacted with Brønsted Acid Sites 428
22.3 Energy-Dispersive XAFS Studies on the Spontaneous Dispersion of PdO and Reversible Formation of Stable Pd Clusters in H-ZSM-5 and H-Mordenite 430
22.4 *In Situ* QXAFS Studies on the Dynamic Coalescence and Dispersion Processes of Pd in USY Zeolite 432
22.5 Time-Resolved EXAFS Measurement of the Stepwise Clustering Process of Pd Clusters at Room Temperature 435
22.6 Summary 438
References 439

Part Four Single Crystals 441

23 **Morphology Changes of Photochromic Single Crystals** 443
Seiya Kobatake and Masahiro Irie
23.1 Introduction 443
23.2 Photochromic Diarylethene Crystals 444
23.3 X-Ray Crystallographic Analysis 444
23.4 Reactivity in the Crystal 447

23.5	Photomechanical Effect *448*
23.6	Crystal Surface Changes *449*
23.7	Photoreversible Crystal Shape Changes *450*
	References *454*

24 Direct Observation of Change in Crystal Structures During Solid-State Reactions of 1,3-Diene Compounds *459*
Akikazu Matsumoto

24.1	Introduction *459*
24.1.1	Crystal Engineering Renaissance *459*
24.2	EZ-Photoisomerization *460*
24.2.1	Model of Photoisomerization *460*
24.2.2	Photoisomerization of Benzyl Muconate *462*
24.2.3	Change in Crystal Structures During Photoisomerization *463*
24.3	[2 + 2] Photodimerization *465*
24.3.1	[2 + 2] Photodimerization of 1,3-Dienes *465*
24.3.2	[2 + 2] Photodimerization of Benzyl Muconates *465*
24.4	Topochemical Polymerization *469*
24.4.1	Features of Topochemical Polymerization *469*
24.4.2	Monomer Stacking Structure and Polymerization Reactivity *470*
24.4.3	Shrinking and Expanding Crystals *473*
24.4.4	Accumulation and Release of Strain During Polymerization *474*
24.4.5	Homogeneous and Heterogeneous Polymerizations *476*
24.5	Conclusion *480*
	References *481*

25 Reaction Dynamics Studies on Crystalline-State Photochromism of Rhodium Dithionite Complexes *487*
Hidetaka Nakai and Kiyoshi Isobe

25.1	Introduction *487*
25.2	Photochromism of Rhodium Dithionite Complexes *488*
25.3	Reaction Dynamics of Crystalline-State Photochromism *490*
25.3.1	Dynamics of Molecular Structural Changes in Single Crystals *490*
25.3.2	Dynamics of Reaction Cavities in a Crystalline-State Reaction *495*
25.3.3	Dynamics of Surface Morphology Changes of Photochromic Single Crystals *498*
25.4	Summary *499*
	References *500*

26 Dynamics in Organic Inclusion Crystals of Steroids and Primary Ammonium Salts *505*
Mikiji Miyata, Norimitsu Tohnai, and Ichiro Hisaki

26.1	Introduction *505*
26.2	Dynamics of Steroidal Inclusion Crystals *506*
26.2.1	Guest-Responsive Molecular Assemblies *506*

26.2.2	Intercalation in Steroidal Bilayer Crystals	508
26.2.3	Guest Fit Through Weak Non-Covalent Bonds	510
26.3	Dynamics of Organic Crystals of Primary Ammonium Salts	512
26.3.1	Solid-State Fluorescence Emission	512
26.3.2	Hydrogen Bond Clusters	514
26.4	Dynamical Expression of Molecular Information in Organic Crystals	516
26.4.1	Hierarchical Structures with Supramolecular Chirality	516
26.4.2	Expression of Supramolecular Chirality in Hierarchical Assemblies	517
26.4.2.1	Three-Axial Chirality	517
26.4.2.2	Tilt Chirality	518
26.4.2.3	Helical and Bundle Chirality in a 2_1 Assembly	519
26.4.3	Supramolecular Chirality of Hydrogen Bonding Networks	520
26.4.4	Expression of Molecular Information	522
26.5	Conclusion and Perspectives	523
	References 523	

27 Morphology Changes of Organic Crystals by Single-Crystal-to-Single-Crystal Photocyclization *527*
Hideko Koshima

27.1	Introduction	527
27.2	Surface Morphology Changes in the Salt Crystals of a Diisopropylbenzophenone Derivative with Amines via Single-Crystal-to-Single-Crystal Photocyclization	528
27.2.1	Solid-State Photocylization	528
27.2.2	Crystal Structures and the Reaction Mechanism	529
27.2.3	Morphology Changes in Bulk Crystals	531
27.2.4	Morphology Changes in Microcrystals	532
27.2.5	Correlation between the Morphology Changes and the Crystal Strucural Changes	535
27.3	Morphology Changes in Triisobenzophenone Crystals via Diastereospecific Single-Crystal-to-Single-Crystal Photocyclization	537
27.3.1	Solid-State Photocyclization and the Crystal Structures	537
27.3.2	Morphology Changes	539
27.4	Concluding Remarks	541
	References 541	

Part Five Single Biocells *545*

28 Femtosecond Laser Tsunami Processing and Light Scattering Spectroscopic Imaging of Single Animal Cells *547*
Hiroshi Masuhara, Yoichiroh Hosokawa, Takayuki Uwada, Guillaume Louit, and Tsuyoshi Asahi

28.1	Introduction	547

28.2	Femtosecond Laser Ablation and Generated Impulsive Force in Water: Laser Tsunami *548*	
28.2.1	Manipulation of a Single Polymer Bead by Laser Tsunami *551*	
28.2.2	Manipulation of Single Animal Cells by Laser Tsunami *554*	
28.2.3	Modification and Regeneration Process in Single Animal Cells by Laser Tsunami *556*	
28.2.4	Injection of Nanoparticles into Single Animal Cells by the Laser Tsunami *558*	
28.3	Development of Rayleigh Light Scattering Spectroscopy/Imaging System and its Application to Single Animal Cells *561*	
28.4	Summary *565*	
	References *566*	

29 **Super-Resolution Infrared Microspectroscopy for Single Cells** *571*
Makoto Sakai, Keiichi Inoue, and Masaaki Fujii

29.1	Introduction *571*	
29.1.1	Infrared Microscopy *571*	
29.1.2	Super-Resolution Microscopy by Two-Color Double Resonance Spectroscopy *571*	
29.1.3	Transient Fluorescence Detected IR Spectroscopy *572*	
29.1.4	Application to Super-Resolution Infrared Microscopy *573*	
29.2	Experimental Set-Up for Super-Resolution Infrared Microscopy *574*	
29.2.1	Picosecond Laser System *574*	
29.2.2	Fluorescence Detection System *574*	
29.2.2.1	Optical Layout for the Solution and Fluorescent Beads *574*	
29.2.2.2	Optical Layout for Biological Samples *575*	
29.2.3	Sample *576*	
29.3	Results and Discussion *576*	
29.3.1	Transient Fluorescence Image with IR Super-Resolution in Solution *576*	
29.3.2	Picosecond Time-Resolved Measurement *578*	
29.3.3	Application to Fluorescent Beads *579*	
29.3.4	Application to Whole Cells *581*	
29.3.4.1	Super-Resolution IR Imaging of *Arabidopsis thaliana* Roots *581*	
29.3.4.2	Vibrational Relaxation Dynamics in the Cells *582*	
29.4	Summary *584*	
	References *585*	

30 **Three-Dimensional High-Resolution Microspectroscopic Study of Environment-Sensitive Photosynthetic Membranes** *589*
Shigeichi Kumazaki, Makotoh Hasegawa, Mohammad Ghoneim, Takahiko Yoshida, Masahide Terazima, Takashi Shiina, and Isamu Ikegami

30.1	Introduction *589*	
30.1.1	Thylakoid Membranes of Oxygenic Photosynthesis *589*	

30.1.2	Thylakoid Membranes in Chloroplasts	590
30.1.3	Thylakoid Membrane of Cyanobacteria	590
30.1.4	Applications of Fluorescence Microscopy to a Thylakoid Membrane	590
30.1.5	Simultaneous Spectral Imaging and its Merits	591
30.2	Spectral Fluorescence Imaging of Thylakoid Membrane	592
30.2.1	Realization of Fast Broadband Spectral Acquisition in Two-Photon Excitation Fluorescence Imaging	592
30.2.2	Spectral Imaging of a Filamentous Cyanobacterium, *Anabaena*	594
30.2.2.1	Thylakoid Membrane of Cyanobacterium	594
30.2.2.2	Stability of the *Anabaena* Fluorescence Spectra Under Photoautotrophic Conditions	594
30.2.2.3	Change of the *Anabaena* Fluorescence Spectra by Dark Some Conditions	595
30.2.2.4	Intracellular Spectral Gradient in *Anabaena* Cells	596
30.2.3	Spectral Imaging of Chloroplasts	598
30.2.3.1	Chloroplasts from a Plant, *Zea mays*	598
30.2.3.2	Chloroplast from the Green Alga, *Chlorella*	600
30.3	Technical Verification and Perspective	601
30.4	Summary	602
	References	604

31   **Fluorescence Lifetime Imaging Study on Living Cells with Particular Regard to Electric Field Effects and pH Dependence**   607
Nobuhiro Ohta and Takakazu Nakabayashi

31.1	Introduction	607
31.2	Experimental	608
31.2.1	FLIM Measurement System	608
31.2.2	Preparation of *Hb. salinarum* Loaded with BCECF	610
31.2.3	Measurements of External Electric Field Effects	610
31.3	Results and Discussion	611
31.3.1	FLIM of *Hb. salinarum*	611
31.3.2	pH Dependence of the Fluorescence Lifetime in Solution and in Living Cells	614
31.3.3	External Electric Field Effect on Fluorescence of BCECF	616
31.3.4	Electric-Field-Induced Aggregate Formation in *Hb. salinarum*	617
31.4	Summary	619
	References	619

32   **Multidimensional Fluorescence Imaging for Non-Invasive Tracking of Cell Responses**   623
Ryosuke Nakamura and Yasuo Kanematsu

32.1	Introduction	623
32.2	Materials and Methods	625
32.2.1	Time-Gated Excitation–Emission Matrix Spectroscopy	625

32.2.2	Time- and Spectrally-Resolved Fluorescence Imaging	*626*
32.2.3	PARAFAC Model	*628*
32.2.4	Sample Preparation	*630*
32.3	Time-Gated Excitation–Emission Matrix Spectroscopy	*630*
32.3.1	The 3D Fluorescence Properties of Dye Solutions	*630*
32.3.2	The 3D Fluorescence Property of a Mixed Solution	*631*
32.3.3	PARAFAC Decomposition Without any Prior Knowledge of Constituents	*633*
32.4	Time- and Spectrally-Resolved Fluorescence Imaging	*635*
32.4.1	Characterization of γ–Em Maps	*635*
32.4.2	Spatial Localization of Fluorescent Components	*637*
32.4.3	PARAFAC Decomposition	*637*
32.4.4	Possible Assignments of Fluorescent Components	*639*
32.5	Concluding Remarks	*640*
	References	*642*

33 Fluorescence Correlation Spectroscopy on Molecular Diffusion Inside and Outside a Single Living Cell *645*

Kiminori Ushida and Masataka Kinjo

33.1	Introduction	*645*
33.1.1	Investigation on Biological System Based on Molecular Identification and Visualization	*645*
33.1.2	Technical Restrictions and Regulations in Real-Time Visualization of Material Transport in Biological System	*647*
33.1.2.1	Spatial Resolution	*647*
33.1.2.2	Time Resolution	*648*
33.1.2.3	Sensitivity	*648*
33.1.3	Time and Space Resolution Required to Observe Anomalous Diffusion of a Single Molecule in Biological Tissues	*648*
33.1.4	General Importance of Anomalous Diffusion in a Signaling Reaction	*652*
33.2	Use of Fluorescence Correlation Spectroscopy (FCS) for Investigation of Biological Systems	*655*
33.2.1	Use of FCS for Biological Systems	*655*
33.2.2	Experimental Example of Anomalous Diffusion Observed in a Model System for Extracellular Matrices	*656*
33.2.3	Quantitative Estimation of Reaction Volume in Signaling Reaction	*661*
33.3	A Short Review of Recent Literature Concerning FCS Inside and Outside a Single Cell	*662*
33.3.1	FCS Measurement Inside Single Cells	*662*
33.3.2	FCS Measurement Outside Cells	*664*
33.4	Summary	*664*
	References	*665*

34	**Spectroscopy and Photoreactions of Gold Nanorods in Living Cells and Organisms** 669	
	Yasuro Niidome and Takuro Niidome	
34.1	Introduction 669	
34.1.1	Spectroscopic Properties of Gold Nanorods 669	
34.1.2	Biocompatible Gold Nanorods 670	
34.2	Spectroscopy of Gold Nanorods in Living Cells 674	
34.2.1	Gold Nanorods Targeting Tumor Cells 674	
34.2.2	Spectroscopy of Gold Nanorods *In Vivo* 675	
34.3	Photoreactions of Gold Nanorods for Biochemical Applications 680	
34.4	Conclusions and Future Outlook 682	
	References 683	
35	**Dynamic Motion of Single Cells and its Relation to Cellular Properties** 689	
	Hideki Matsune, Daisuke Sakurai, Akitomo Hirukawa, Sakae Takenaka, and Masahiro Kishida	
35.1	Introduction 689	
35.1.1	Single Cell Analysis 689	
35.1.2	Dynamic Motion of Murine Embryonic Stem Cell 690	
35.2	Laser Trapping of Biological Cells 691	
35.2.1	Optical Tweezers 691	
35.2.2	Set-up for Optical Trapping of a Living Cell 692	
35.2.3	Murine Embryonic Stem Cell Trapped with Optical Tweezers 693	
35.3	Relationship Between Cellular Motion and Proliferation 694	
35.3.1	Dynamic Motion of a Murine Embryonic Stem Cell 694	
35.3.2	Experimental Procedure 695	
35.4	Cell Separation by Specific Gravity 699	
35.4.1	Cell Separation 699	
35.5	Summary 700	
	References 701	

Index 703

Contents to Volume 1

Part One Spectroscopic Methods for Nano Interfaces *1*

1 **Raman and Fluorescence Spectroscopy Coupled with Scanning Tunneling Microscopy** *3*
   Noriko Nishizawa Horimoto and Hiroshi Fukumura

2 **Vibrational Nanospectroscopy for Biomolecules and Nanomaterials** *19*
   Yasushi Inouye, Atsushi Taguchi, and Taro Ichimura

3 **Near-Field Optical Imaging of Localized Plasmon Resonances in Metal Nanoparticles** *39*
   Hiromi Okamoto and Kohei Imura

4 **Structure and Dynamics of a Confined Polymer Chain Studied by Spatially and Temporally Resolved Fluorescence Techniques** *55*
   Hiroyuki Aoki

5 **Real Time Monitoring of Molecular Structure at Solid/Liquid Interfaces by Non-Linear Spectroscopy** *71*
   Hidenori Noguchi, Katsuyoshi Ikeda, and Kohei Uosaki

6 **Fourth-Order Coherent Raman Scattering at Buried Interfaces** *103*
   Hiroshi Onishi

7 **Dynamic Analysis Using Photon Force Measurement** *117*
   Hideki Fujiwara and Keiji Sasaki

8 **Construction of Micro-Spectroscopic Systems and their Application to the Detection of Molecular Dynamics in a Small Domain** *133*
   Syoji Ito, Hirohisa Matsuda, Takashi Sugiyama, Naoki Toitani, Yutaka Nagasawa, and Hiroshi Miyasaka

Molecular Nano Dynamics, Volume II: Active Surfaces, Single Crystals and Single Biocells
Edited by H. Fukumura, M. Irie, Y. Iwasawa, H. Masuhara, and K. Uosaki
Copyright © 2009 WILEY-VCH Verlag GmbH & Co. KGaA, Weinheim
ISBN: 978-3-527-32017-2

9   Nonlinear Optical Properties and Single Particle Spectroscopy of CdTe Quantum Dots   *155*
Lingyun Pan, Yoichi Kobayashi, and Naoto Tamai

Part Two   Nanostructure Characteristics and Dynamics   *171*

10   Morphosynthesis in Polymeric Systems Using Photochemical Reactions   *173*
Hideyuki Nakanishi, Tomohisa Norisuye, and Qui Tran-Cong-Miyata

11   Self-Organization of Materials Into Microscale Patterns by Using Dissipative Structures   *187*
Olaf Karthaus

12   Formation of Nanosize Morphology of Dye-Doped Copolymer Films and Evaluation of Organic Dye Nanocrystals Using a Laser   *203*
Akira Itaya, Shinjiro Machida, and Sadahiro Masuo

13   Molecular Segregation at Periodic Metal Nano-Architectures on a Solid Surface   *225*
Hideki Nabika and Kei Murakoshi

14   Microspectroscopic Study of Self-Organization in Oscillatory Electrodeposition   *239*
Shuji Nakanishi

15   Construction of Nanostructures by use of Magnetic Fields and Spin Chemistry in Solid/Liquid Interfaces   *259*
Hiroaki Yonemura

16   Controlling Surface Wetting by Electrochemical Reactions of Monolayers and Applications for Droplet Manipulation   *279*
Ryo Yamada

17   Photoluminescence of CdSe Quantum Dots: Shifting, Enhancement and Blinking   *293*
Vasudevanpillai Biju and Mitsuru Ishikawa

Preface

Over the past two decades, studies of chemical reaction dynamics have shifted from ideal systems of isolated molecules in the gas phase, of molecular clusters in jet beams, on ultra-clean surfaces, in homogeneous and in dilute molecular solutions, and in bulk crystals, towards nanosystems of supramolecules, colloids, and ultra-small materials, following the contemporary trends in nanoscience and nanotechnology. The preparation, characterization, and functionalization of supramolecules, molecular assemblies, nanoparticles, nanodots, nanocrystals, nanotubes, nanowires, and so on, have been conducted extensively, and their chemical reactions and dynamic processes are now being elucidated. The systematic investigation of molecular nanosystems gives us a platform from which we can understand the nature of the dynamic behavior and chemical reactions occurring in complex systems such as molecular devices, catalysts, living cells, and so on. Thus we have conducted the KAKENHI (The Grant-in-Aid for Scientific Research) Project on Priority Area "Molecular Nanodynamics" (Project Leader: Hiroshi Masuhara) for the period from 2004 April to 2007 March, involving 86 laboratories in Japan.

For the investigation of such complex systems new methodologies which enable us to analyze dynamics and mechanisms in terms of space and time are indispensable. Methods for simultaneous direct dynamic measurements in both time and real space domains needed to be devised and applied. Spectroscopy with novel space-resolution and ultrafast spectroscopy with high sensitivity have been developed, the manipulation and fabrication of single molecules, nanoparticles, and single living cells have been realized, molecules and nanoparticles for probing chemical reactions spectroscopically and by imaging have been synthesized, new catalyses for cleaning air and new reactions have been found, and the way in which a reaction in a single molecular crystal leads to its morphological change has been elucidated under the umbrella of this research program. The recent development of these new methods and the advances in understanding chemical reaction dynamics in nanosystems are summarized in the present two volumes.

The presented results are based on our activities over three years, including 1146 published papers and 1112 presentations at international conferences. We hope readers will understand the present status and new movement in Molecular

Molecular Nano Dynamics, Volume II: Active Surfaces, Single Crystals and Single Biocells
Edited by H. Fukumura, M. Irie, Y. Iwasawa, H. Masuhara, and K. Uosaki
Copyright © 2009 WILEY-VCH Verlag GmbH & Co. KGaA, Weinheim
ISBN: 978-3-527-32017-2

Nano Dynamics and its relevant research fields. The editors thank the contributors and the Ministry of Education, Culture, Sport, Science, and Technology (MEXT), Japan for their support of the project. We would also like to thank our publishers for their constant support.

Sendai, Tokyo, Nara
and Sapporo
August 2009

Hiroshi Fukumura
Masahiro Irie
Yasuhiro Iwasawa
Hiroshi Masuhara
Kohei Uosaki

About the Editors

Hiroshi Fukumura received his M.Sc and Ph.D. degrees from Tohoku University, Japan. He studied biocompatibility of polymers in the Government Industrial Research Institute of Osaka from 1983 to 1988. He became an assistant professor at Kyoto Institute of Technology in 1988, and then moved to the Department of Applied Physics, Osaka University in 1991, where he worked on the mechanism of laser ablation and laser molecular implantation. Since 1998, he is a professor in the Department of Chemistry at Tohoku University. He received the Award of the Japanese Photochemistry Association in 2000, and the Award for Creative Work from The Chemical Society Japan in 2005. His main research interest is the physical chemistry of organic molecules including polymeric materials studied with various kinds of time-resolved techniques and scanning probe microscopes.

Masahiro Irie received his B.S. and M.S. degrees from Kyoto University and his Ph.D. in radiation chemistry from Osaka University. He joined Hokkaido University as a research associate in 1968 and started his research on photochemistry. In 1973 he moved to Osaka University and developed various types of photoresponsive polymers. In 1988 he was appointed Professor at Kyushu University. In the middle of the 1980's he invented a new class of photochromic molecules – diarylethenes - which undergo thermally irreversible and fatigue resistant photochromic reactions. He is currently interested in developing single-crystalline photochromism of the diarylethene derivatives.

Molecular Nano Dynamics, Volume II: Active Surfaces, Single Crystals and Single Biocells
Edited by H. Fukumura, M. Irie, Y. Iwasawa, H. Masuhara, and K. Uosaki
Copyright © 2009 WILEY-VCH Verlag GmbH & Co. KGaA, Weinheim
ISBN: 978-3-527-32017-2

About the Editors

Yasuhiro Iwasawa received his B.S., M.S. and Ph.D. degrees in chemistry from The University of Tokyo. His main research interests come under the general term "Catalytic Chemistry" and "Surface Chemistry", but more specifically, catalyst surface design, new catalytic materials, reaction mechanism, in situ characterization, oxide surfaces by SPM, time-resolved XAFS, etc. His honors include the Progress Award for Young Chemists in The Chemical Society of Japan (1979), The Japan IBM Science Award (1990), Inoue Prize for Science (1996), Catalysis Society of Japan Award (1999), The Surface Science Society of Japan Award (2000), Medal with Purple Ribbon (2003), and The Chemical Society of Japan Award (2004). The research reported by Yasuhiro Iwasawa represents a pioneering integration of modern surface science and organometallic chemistry into surface chemistry and catalysis in an atomic/ molecular scale. Iwasawa is a leader in the creation of the new filed of catalysis and surface chemistry at oxide surfaces by XAFS and SPM techniques.

Hiroshi Masuhara received his B.S. and M.S. degrees from Tohoku University and Ph.D. from Osaka University. He started his research in photochemistry and was the first to use nanosecond laser spectroscopy in Japan. He studied electronic states, electron transfer, ionic photodissociation of molecular complexes, polymers, films, and powders by developing various time-resolved absorption, fluorescence, reflection, and grating spectroscopies until the mid 1990s. The Masuhara Group combined microscope with laser and created a new field on Microchemistry, which has now developed to Laser Nano Chemistry. After retiring from Osaka University he shifted to Hamano Foundation and is now extending his exploratory research on femtosecond laser crystallization and laser trapping crystallization in National Chiao Tung University in Taiwan and Nara Institute of Science and Technology. He is a foreign member of Royal Flemish Academy of Belgium for Science and the Arts and his honors include The Purple Ribbon Medal, Doctor Honoris Causa de Ecole Normale Superier de Cachan, Porter Medal, the Chemical Society of Japan Award, Osaka Science Prize, and Moet Hennessy Louis Vuitton International Prize "Science for Art" Excellence de Da Vinci.

**Kohei Uosaki** received his B.Eng. and M.Eng. degrees from Osaka University and his Ph.D. in Physical Chemistry from Flinders University of South Australia. He was a Research Chemist at Mitsubishi Petrochemical Co. Ltd. From 1971 to 1978 and a Research Officer at Inorganic Chemistry Laboratory, Oxford University, U.K. between 1978 and 1980 before joining Hokkaido University in 1980 as Assistant Professor in the Department of Chemistry. He was promoted to Associate Professor in 1981 and Professor in 1990. He is also a Principal Investigator of International Center for Materials Nanoarchitectonics (MANA) Satellite, National Institute for Materials Science (NIMS) since 2008. His scientific interests include photoelectrochemistry of semiconductor electrodes, surface electrochemistry of single crystalline metal electrodes, electrocatalysis, modification of solid surfaces by molecular layers, and non-linear optical spectroscopy at interfaces.

List of Contributors for Both Volumes

Hiroyuki Aoki
Kyoto University
Department of Polymer Chemistry
Katsura, Nishikyo
Kyoto 615-8510
Japan

Tsuyoshi Asahi
Osaka University
Department of Applied Physics
Suita 565-0871
Japan

Vasudevanpillai Biju
National Institute of Advanced
Industrial Science and Technology
(AIST)
Health Technology Research Center
Nano-bioanalysis Team
2217-14 Hayashi-cho, Takamatsu
Kagawa 761-0395
Japan

Masaaki Fujii
Tokyo Institute of Technology
Chemical Resources Laboratory
4259 Nagatsuta-cho, Midori-ku
Yokohama 226-8503
Japan

Hideki Fujiwara
Hokkaido University
Research Institute for Electronic
Science
Kita-12, Nishi-6, Sapporo
Hokkaido 060-0812
Japan

Hiroshi Fukumura
Tohoku University
Graduate School of Science
6-3 Aramaki Aoba
Sendai 980-8578
Japan

Mohammad Ghoneim
Kyoto University
Graduate School of Science
Department of Chemistry
Kyoto 606-8502
Japan

Makotoh Hasegawa
Kyoto University
Graduate School of Science
Department of Chemistry
Kyoto 606-8502
Japan

Molecular Nano Dynamics, Volume II: Active Surfaces, Single Crystals and Single Biocells
Edited by H. Fukumura, M. Irie, Y. Iwasawa, H. Masuhara, and K. Uosaki
Copyright © 2009 WILEY-VCH Verlag GmbH & Co. KGaA, Weinheim
ISBN: 978-3-527-32017-2

Akitomo Hirukawa
Kyushu University
Faculty of Engineering
Department of Chemical Engineering
Moto-Oka, Nishi Ku
Fukuoka 819-0395
Japan

Ichiro Hisaki
Osaka University
Graduate School of Engineering
2-1 Yamadaoka, Suita
Osaka 565-0871
Japan

Noriko Nishizawa Horimoto
Tohoku University
Graduate School of Science
6-3 Aramaki Aoba
Sendai 980-8578
Japan

Yoichiroh Hosokawa
Nara Institute of Science and Technology
Graduate School of Materials Science
Takayama 8916-5
Ikoma 630-0192
Japan

Taro Ichimura
Osaka University
Graduate School of Frontier Biosciences
& Graduate School of Engineering
Suita, Osaka
Japan

Katsuyoshi Ikeda
Hokkaido University
Graduate School of Science
Division of Chemistry
Sapporo 060-0810
Japan

Isamu Ikegami
Kyoto University
Graduate School of Science
Department of Chemistry
Kyoto 606-8502
Japan

Kohei Imura
The Graduate University for Advanced Studies
Institute for Molecular Science
Myodaiji, Okazaki
Aichi 444-8585
Japan

Keiichi Inoue
Tokyo Institute of Technology
Chemical Resources Laboratory
4259 Nagatsuta-cho, Midori-ku
Yokohama 226-8503
Japan

Yasushi Inouye
Osaka University
Graduate School of Frontier Biosciences
& Graduate School of Engineering
Suita, Osaka
Japan

Masahiro Irie
Rikkyo University
Department of Chemistry
Nishi-Ikebukuro 3-34-1, Toshima-ku
Tokyo 171-8501
Japan

Mitsuru Ishikawa
National Institute of Advanced Industrial Science and Technology (AIST)
Health Technology Research Center
Nano-bioanalysis Team
2217-14 Hayashi-cho, Takamatsu
Kagawa 761-0395
Japan

Kiyoshi Isobe
Kanazawa University
Graduate School of Natural Science and Technology
Department of Chemistry
Kakuma-machi
Kanazawa 920-1192
Japan

Akira Itaya
Kyoto Institute of Technology
Department of Polymer Science and Engineering
Matsugasaki, Sakyo-ku
Kyoto 606-8585
Japan

Syoji Ito
Osaka University
Graduate School of Engineering Science
Center for Quantum Science and Technology under Extreme Conditions
Division of Frontier Materials Science
Toyonaka
Osaka 560-8531
Japan

Yasuhiro Iwasawa
The University of Tokyo
Graduate School of Science
Department of Chemistry
Hongo, Bunkyo-ku
Tokyo 113-0033
Japan

Yasuo Kanematsu
Osaka University
Center for Advanced Science and Innovation
Venture Business Laboratory
JST-CREST
Suita
Osaka 565-0871
Japan

Olaf Karthaus
Chitose Institute of Science and Technology
758-65 Bibi, Chitose
Hokkaido 066-8655
Japan

Masataka Kinjo
Riken
Hirosawa 2-1, Wako
Saitama 351-0198
Japan

Masahiro Kishida
Kyushu University
Faculty of Engineering
Department of Chemical Engineering
Moto-Oka, Nishi Ku
Fukuoka 819-0395
Japan

Yoichi Kobayashi
Kwansei Gakuin University
School of Science and Technology
Department of Chemistry
2-1 Gakuen
Sanda 669-1337
Japan

Seiya Kobatake
Osaka City University
Graduate School of Engineering
Department of Applied Chemistry
Sugimoto 3-3-138, Sumiyoshi-ku
Osaka 558-8585
Japan

Hideko Koshima
Ehime University
Graduate School of Science and Engineering
Department of Materials Science and Biotechnology
Matsuyama 790-8577
Japan

Shigeichi Kumazaki
Kyoto University
Graduate School of Science
Department of Chemistry
Kyoto 606-8502
Japan

Guillaume Louit
Osaka University
Department of Applied Physics
Suita 565-0871
Japan

Shinjiro Machida
Kyoto Institute of Technology
Department of Polymer Science and Engineering
Matsugasaki, Sakyo-ku
Kyoto 606-8585
Japan

Hiroshi Masuhara
Nara Institute of Science and Technology
Graduate School of Materials Science
Takayama 8916-5
Ikoma 630-0192
Japan

and

Osaka University
Department of Applied Physics
Suita 565-0871
Japan

Sadahiro Masuo
Kyoto Institute of Technology
Department of Polymer Science and Engineering
Matsugasaki, Sakyo-ku
Kyoto 606-8585
Japan

Hirohisa Matsuda
Osaka University
Graduate School of Engineering Science
Center for Quantum Science and Technology under Extreme Conditions
Division of Frontier Materials Science
Toyonaka
Osaka 560-8531
Japan

Akikazu Matsumoto
Osaka City University
Graduate School of Engineering
Department of Applied Chemistry
3-3-138 Sugimoto, Sumiyoshi-ku
Osaka 558-8585
Japan

Hideki Matsune
Kyushu University
Faculty of Engineering
Department of Chemical Engineering
Moto-Oka, Nishi Ku
Fukuoka 819-0395
Japan

Hiroshi Miyasaka
Osaka University
Graduate School of Engineering Science
Center for Quantum Science and Technology under Extreme Conditions
Division of Frontier Materials Science
Toyonaka
Osaka 560-8531
Japan

Mikiji Miyata
Osaka University
Graduate School of Engineering
2-1 Yamadaoka, Suita
Osaka 565-0871
Japan

Kei Murakoshi
Hokkaido University
Graduate School of Science
Department of Chemistry
Sapporo
Hokkaido 060-0810
Japan

Takakazu Nakabayashi
Hokkaido University
Research Institute for Electronic
Science (RIES)
Sapporo 001-0020
Japan

Hideki Nabika
Hokkaido University
Graduate School of Science
Department of Chemistry
Sapporo
Hokkaido 060-0810
Japan

Yutaka Nagasawa
Osaka University
Graduate School of Engineering Science
Center for Quantum Science and
Technology under Extreme Conditions
Division of Frontier Materials Science
Toyonaka
Osaka 560-8531
Japan

Hidetaka Nakai
Kanazawa University
Graduate School of Natural Science and
Technology
Department of Chemistry
Kakuma-machi
Kanazawa 920-1192
Japan

Hisao Nakamura
The University of Tokyo
Graduate School of Engineering
Department of Chemical System
Engineering
Tokyo 113-8656
Japan

Ryosuke Nakamura
Osaka University
Center for Advanced Science and
Innovation
Venture Business Laboratory
JST-CREST
Suita
Osaka 565-0871
Japan

Hideyuki Nakanishi
Graduate School of Science and
Technology
Kyoto Institute of Technology
Department of Macromolecular Science
and Engineering
Matsugasaki
Kyoto 606-8585
Japan

Shuji Nakanishi
Osaka University
Graduate School of Engineering Science
Division of Chemistry
Toyonaka
Osaka 560-8531
Japan

Takuro Niidome
Kyushu University
Department of Applied Chemistry
744 Moto-Oka, Nishi Ku
Fukuoka 819-0395
Japan

Yasuro Niidome
Kyushu University
Department of Applied Chemistry
744 Moto-Oka, Nishi Ku
Fukuoka 819-0395
Japan

Hidenori Noguchi
Hokkaido University
Graduate School of Science
Division of Chemistry
Sapporo 060-0810
Japan

Tomohisa Norisuye
Graduate School of Science and
Technology
Kyoto Institute of Technology
Department of Macromolecular Science
and Engineering
Matsugasaki
Kyoto 606-8585
Japan

Nobuhiro Ohta
Hokkaido University
Research Institute for Electronic
Science (RIES)
Sapporo 001-0020
Japan

Hiromi Okamoto
The Graduate University for Advanced
Studies
Institute for Molecular Science
Myodaiji, Okazaki
Aichi 444-8585
Japan

Kazu Okumura
Tottori University
Faculty of Engineering
Department of Materials Science
Koyama-cho, Minami
Tottori 680-8552
Japan

Hiroshi Onishi
Kobe University
Faculty of Science
Department of Chemistry
Rokko-dai, Nada, Kobe
Hyogo 657-8501
Japan

Lingyun Pan
Kwansei Gakuin University
School of Science and Technology
Department of Chemistry
2-1 Gakuen
Sanda 669-1337
Japan

Makoto Sakai
Tokyo Institute of Technology
Chemical Resources Laboratory
4259 Nagatsuta-cho, Midori-ku
Yokohama 226-8503
Japan

Daisuke Sakurai
Kyushu University
Faculty of Engineering
Department of Chemical Engineering
Moto-Oka, Nishi Ku
Fukuoka 819-0395
Japan

Keiji Sasaki
Hokkaido University
Research Institute for Electronic Science
Kita-12, Nishi-6, Sapporo
Hokkaido 060-0812
Japan

Takehiko Sasaki
The University of Tokyo
Graduate School of Frontier Science
Department of Chemistry
Kashiwanoha, Kashiwa
Chiba 277-8561
Japan

Atsushi Satsuma
Nagoya University
Graduate School of Engineering
Department of Molecular Design and Engineering
Chikusa
Nagoya 464-8603
Japan

Takashi Shiina
Kyoto University
Graduate School of Science
Department of Chemistry
Kyoto 606-8502
Japan

Ken-ichi Shimizu
Nagoya University
Graduate School of Engineering
Department of Molecular Design and Engineering
Chikusa
Nagoya 464-8603
Japan

Takashi Sugiyama
Osaka University
Graduate School of Engineering Science
Center for Quantum Science and Technology under Extreme Conditions
Division of Frontier Materials Science
Toyonaka
Osaka 560-8531
Japan

Mizuki Tada
The University of Tokyo
Graduate School of Frontier Science
Department of Chemistry
Kashiwanoha, Kashiwa
Chiba 277-8561
Japan

Atsushi Taguchi
Osaka University
Graduate School of Frontier Biosciences
& Graduate School of Engineering
Suita, Osaka
Japan

Sakae Takenaka
Kyushu University
Faculty of Engineering
Department of Chemical Engineering
Moto-Oka, Nishi Ku
Fukuoka 819-0395
Japan

Naoto Tamai
Kwansei Gakuin University
School of Science and Technology
Department of Chemistry
2-1 Gakuen
Sanda 669-1337
Japan

Toshiaki Taniike
The University of Tokyo
Graduate School of Science
Department of Chemistry
Hongo, Bunkyo-ku
Tokyo 113-0033
Japan

Masahide Terazima
Kyoto University
Graduate School of Science
Department of Chemistry
Kyoto 606-8502
Japan

Norimitsu Tohnai
Osaka University
Graduate School of Engineering
2-1 Yamadaoka, Suita
Osaka 565-0871
Japan

Naoki Toitani
Osaka University
Graduate School of Engineering Science
Center for Quantum Science and Technology under Extreme Conditions
Division of Frontier Materials Science
Toyonaka
Osaka 560-8531
Japan

Qui Tran-Cong-Miyata
Graduate School of Science and Technology
Kyoto Institute of Technology
Department of Macromolecular Science and Engineering
Matsugasaki
Kyoto 606-8585
Japan

Yohei Uemura
The University of Tokyo
Graduate School of Science
Department of Chemistry
Hongo, Bunkyo-ku
Tokyo 113-0033
Japan

Kohei Uosaki
Hokkaido University
Graduate School of Science
Division of Chemistry
Sapporo 060-0810
Japan

Kiminori Ushida
Riken
Hirosawa 2-1, Wako
Saitama 351-0198
Japan

Takayuki Uwada
Nara Institute of Science and Technology
Graduate School of Materials Science
Takayama 8916-5
Ikoma 630-0192
Japan

and

National Chiao Tung University
Institute of Molecular Science
Department of Applied Chemistry
1001 Ta Hsueh Road
Hsinchu 30010
Taiwan

Kazuya Watanabe
Kyoto University
Graduate School of Science
Department of Chemistry
Kyoto 606-8502
Japan

and

PRESTO, JST
4-1-8 Honcho Kawaguchi
Saitama
Japan

Ryo Yamada
Osaka University
Graduate School of Engineering Science
Division of Materials Physics
Department of Materials Engineering Science
Toyonaka, Osaka
Japan

Hiroaki Yonemura
Kyushu University
Department of Applied Chemistry
6-10-1 Hakozaki, Higashi-ku
Fukuoka 812-8581
Japan

Takahiko Yoshida
Kyoto University
Graduate School of Science
Department of Chemistry
Kyoto 606-8502
Japan

Part Three
Active Surfaces

18
The Genesis and Principle of Catalysis at Oxide Surfaces: Surface-Mediated Dynamic Aspects of Catalytic Dehydration and Dehydrogenation on TiO$_2$(110) by STM and DFT

Yohei Uemura, Toshiaki Taniike, Takehiko Sasaki, Mizuki Tada, and Yasuhiro Iwasawa

18.1
Introduction

An important catalytic property of metal oxides is the surface acid–base property, which regulates the performance of catalysts for the dehydrogenation and dehydration of alcohols, formic acid, ethanolamine, and so on [1]. The acid–base property of heterogeneous oxide catalysts has been characterized by the decomposition reaction of formic acid as a probe reaction, because the products are regulated by the acid–base property of the oxide surface [2–8]. Typically, the dehydration reaction (HCOOH → CO + H$_2$O) occurs on acidic oxides such as Al$_2$O$_3$ and TiO$_2$, while the dehydrogenation reaction (HCOOH → CO$_2$ + H$_2$) occurs on basic oxides such as MgO and ZnO. Both the reactions proceed via the formate anion intermediate (HCOO$^-$). We found that the catalytic reaction of HCOOH on a TiO$_2$(110) surface violated the uniqueness rule of the acidity–basicity of the catalysts [9–11]. The HCOOH dehydration dominantly occurred under a low-pressure atmosphere of HCOOH and at high temperatures, above 500 K, while the dehydrogenation of HCOOH occurred under a relatively high-pressure atmosphere and at low temperatures, below 450 K [12–15]. In the latter case an additional acidic HCOOH molecule promotes the dehydrogenation categorized as basic catalysis. Note that TiO$_2$ powder predominantly catalyzes the dehydration. We proposed the mechanism for the dehydrogenation of HCOOH on a TiO$_2$(110) surface by means of density functional theory (DFT) calculations and scanning tunneling microscopy (STM) [14].

A key issue of developing catalytic technologies is to understand site specific surface dynamic processes from the atomic-scale view point. The inherent compositional and structural inhomogeneity of oxide surfaces makes the problem of identifying the essential issues for their functions extremely difficult. STM has particularly great potential to overcome the difficulty of heterogeneity of oxide surfaces, discriminating specific sites from the other sites on an atomic scale. It is well known that the images obtained by STM do not naturally reflect the physical geometry of surfaces because the tunneling current used for regulating the tip–sample separation for STM reflects the local electronic density of states of the sample and tip in addition to physical geometry. Thus additional experimental evidence may be demanded for identification of the STM image. Chemical identification of Ti atoms at a $TiO_2(110)$–(1×1) surface has been performed by the use of formic acid as a probe molecule that adsorbs selectively on Ti^{4+} cations of the surface, which can be imaged by STM [16, 17]. Thus, formic acid is regarded not only as a probe reactant molecule for catalysis research but also a useful molecule for visualization of chemical events on oxide single crystal surfaces.

The present *in situ* STM and DFT studies on the dehydration and dehydrogenation of formic acid on a $TiO_2(110)$ surface as typical probe catalytic reactions document new dynamic aspects of acid–base catalysis. Active sites for the dehydration are oxygen vacancies that are produced *in situ* under the catalytic reactions and the formate intermediates migrate on the surface to find the oxygen vacancies where they decompose to CO and OH. The pre-adsorbed formates modify the acidic surface to a new surface with basic character on which acidic HCOOH adsorbs with Coulombic attractive interaction to promote a bimolecular dehydrogenation. These findings provide a new concept of the genesis of acid–base catalysis on oxide surfaces and also a new implication to oxide catalyst design on a molecular scale.

18.2
Experimental

18.2.1
STM Measurements of $TiO_2(110)$

All experiments were performed in an ultrahigh vacuum (UHV) STM (JEOL JSTM 4500VT) with ion guns and low-energy electron diffraction (LEED) optics. The base pressure of the chamber was less than 1×10^{-8} Pa. Electrically etched tungsten tips were used in the STM observation, and all STM images were taken in constant current mode.

A polished $TiO_2(110)$ wafer of $6.5 \times 1 \times 0.25$ mm^3 (Earth Chemical) was used after deposition of Ni film on the rear side of the sample to resistively heat the sample on a sample holder. The surface was cleaned by cycles of Ar^+ ion sputtering (3 keV for 3 min) and annealing under UHV at 900 K for 30 s until a clear 1×1 LEED pattern was obtained. Deuterated formic acid (DCOOD, Wako, 98% purity, most of the contaminant is water) was purified by repeated freeze–pump–thaw cycles and introduced into the chamber by backfilling. The surface temperature of the crystal was monitored by an infrared radiation thermometer.

18.2.2
Computational Methods

Density functional calculations with the GGA-PBE [46] functional were performed with Materials Studio Dmol3 from Accelrys. The basis sets were DND and effective core potentials (DND is as accurate as 6-31G*) [18]. A transition state (TS) between two immediate stable structures was first identified by linear synchronous transit (LST) [19], and then cyclically refined by quadratic synchronous transit and conjugate gradient methods. Each TS was converged within 0.1 Ha nm^{-1}. The slab method was employed to model a TiO$_2$(110) surface, where the thickness of a vacuum layer was 1 nm. The slab thickness was decided to be three layers (one layer involves three atomic layers, i.e., O−Ti−O) for the reason given later. The atomic arrangement of a bottom layer was fixed at the bulk arrangement. In the calculations of the adsorption energies of HCOOH on a defective surface, an oxygen defect (O-defect) was formed by removing a bridging oxygen from a stoichiometric p(1 × 2) surface. For the transition state search, p(1 × 4) or p(2 × 2)-HCOOH surfaces were used, where the k-point meshes were 3 × 2 × 1 and 2 × 3 × 1, respectively. In the calculations for the transition state search on the defective surface, an O-defect was formed by removing 1/4 bridging oxygen from a stoichiometric p(2 × 2) surface, and thermal smearing of moderate strength [20] was imposed to improve the SCF convergence. The chosen adsorption states were the most stable bridging formate and a quasi-stable monodentate formate produced by the dissociative adsorption of HCOOH, and a less stable molecular form. We employed the most efficient three-layer slab throughout the calculations involving TS search. Independently, very recently, Perron et al. have also reported that the surface energy on TiO$_2$(110) was almost convergent at four-layers slab [21]. At the cost of the thin slab in deciding plausible reaction paths for HCOOH dehydrogenation, the difference of 0.5 eV in the activation barrier was regarded as a margin, and all the reaction paths with activation energies less than 0.5 eV larger than the smallest activation energy were considered as candidates.

18.3
Results and Discussion

18.3.1
Dynamic Mechanism for Catalytic Dehydration of Formic Acid on a TiO$_2$(110) Surface, Much Different from the Traditional Static Acid Catalysis

The rutile-TiO$_2$(110)–(1 × 1) surface has been studied extensively and well characterized experimentally and theoretically to be close to that of the bulk-truncated structure [22, 23]. It consists of alternating rows of fivefold coordinated Ti^{4+} rows at the troughs and bridging O (denoted O$_B$ hereafter) rows at the ridges that locate 0.11 nm above the underlying Ti−O plane (Figure 18.1(b)). The O-ridges and the Ti rows are aligned with a 0.649-nm separation. Figure 18.1(a) shows a typical STM image of the TiO$_2$(110)–(1 × 1) surface at positive sample bias voltages [23, 24]. Bright rows consisting of bright spots along the [001] direction are imaged with a constant separation of 0.649 nm. The periodic bright spots form a rectangle of 0.649 × 0.296 nm^2 which coincides with the (1 × 1) unit cell. At a positive sample bias voltage, the major path for tunneling electrons is from the Fermi energy of the tip to the unoccupied states of the sample surface [25]. Rutile TiO$_2$ bulk has a filled valence band of predominantly O 2p character and an empty conduction band of

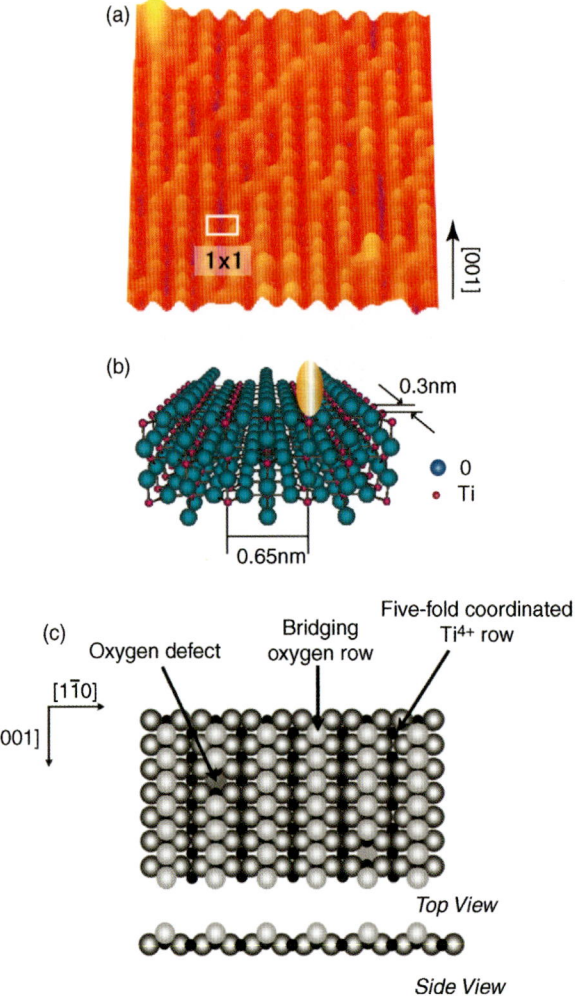

Figure 18.1 Surface structure of $TiO_2(110)$ characterized by STM and non-contact atomic force microscope (NC-AFM). (a) An empty state STM image (7.3×7.3 nm^2, Vs = 1.2 V, It = 0.15 nA) of a $TiO_2(110)$ surface at RT. (b) A structure model of a clean $TiO_2(110)$ surface. The fivefold coordinated Ti atoms are visualized. (c) A perspective view of a $TiO_2(110)$ surface with oxygen defect.

predominantly Ti 3d character, separated by a band gap of 3.1 eV [26]. However, it becomes an n-type semiconductor due to a slight deficiency of oxygen atoms from the stoichiometry caused by Ar^+ ion sputtering and annealing at 900–1000 K under UHV, which are typical procedures used for cleaning the surface. The Fermi level is close to the conduction-band minimum and electrons can tunnel to the states at positive sample bias voltages. Thus, the Ti atom is imaged as the "bright spot" by STM through its 3d empty state although the topmost atoms are the oxygen atoms.

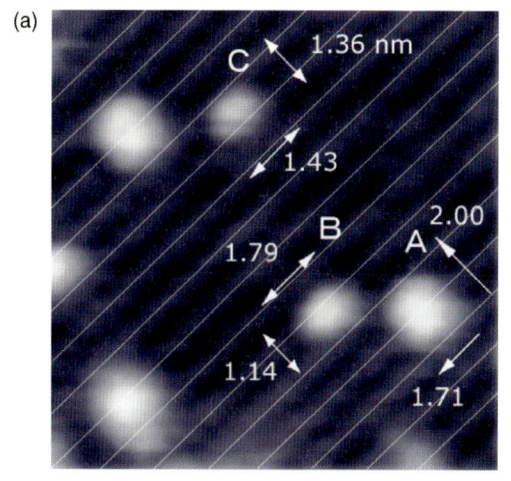

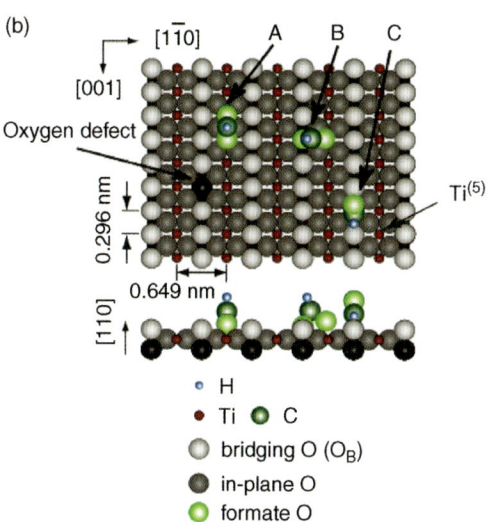

(a)

[STM image with labels: C 1.36 nm, 1.43, B 1.79, 2.00 A, 1.14, 1.71]

(b)

[Structural model with labels: [1̄10], [001], A B C, Oxygen defect, 0.296 nm, Ti(5), 0.649 nm, [110]]

- H
- Ti ○ C
- ○ bridging O (O$_B$)
- ● in-plane O
- ○ formate O

Figure 18.2 (a) An STM image (7.1 × 7.1 nm^2, Vs = 1.2 V, It = 0.1 nA) of TiO$_2$(110) exposed to 1 L formic acid at room temperature followed by annealing at 350 K. Three types of formate species are observed. The lengths of each formate along the [001] and [1̄10] directions are indicated in nanometers. (b) Models of the three formate configurations.

Figure 18.2(a) shows a typical STM image of a TiO$_2$(110) surface exposed to 1 L (1 L = 1.33 × 10^{-4} Pa s) of formic acid at room temperature followed by annealing at 350 K. In this STM image, titanium rows are observed as bright lines in spite of the poor atomic resolution, and traced with white lines on which formate species are observed as white protrusions with different shapes in three different configurations, labeled "A", "B", and "C", as shown in our recent report [12]. Species A in the "A" configuration is located on a Ti row, species B in the "B" configuration is located between Ti and O rows, and species C in the "C" configuration is located on an O row. STM can discriminate the three different kinds of formates by the locations at the surface.

The directions of O–C–O planes of species "A" and "B" can be determined using the shape of the STM image in Figure 18.2(a) because the shapes of the observed formates reflect the shape of their lowest unoccupied molecular orbitals (LUMOs), which expand to the direction perpendicular to the O–C–O plane. The observed species A and B have ellipsoidal shapes, and the major axes of them are along $[1\bar{1}0]$ and [001], respectively. Thus, the O–C–O planes of species A and B are along the [001] and $[1\bar{1}0]$ directions, respectively. This result was also confirmed through our theoretical calculation, and the obtained configurations are shown in Figure 18.2(b). The detail of the "C" configuration was also determined through the same calculation. Formate species A in the "A" configuration is on two fivefold coordinated Ti^{4+} ions (denoted by $Ti^{(5)}$, hereafter) in a bridge configuration, and is the same configuration as that previously reported by STM observations at room temperature [16, 27–29]. Species B is located on an oxygen vacancy site and a $Ti^{(5)}$ ion, respectively, at an oxygen atom and another oxygen atom of formate species in the "B" configuration. When the $Ti^{(5)}$–O bond is broken to transform to a monodentate type, the STM image is observed as a bright contrast of a round shape on an oxygen vacancy site. This formate species is assigned as species C in the "C" configuration, as shown in Figure 18.2(b). The "B" and "C" configurations were not observed at room temperature though the B type was observed on a highly defective surface or by much greater exposure of formic acid than 1 L. Among the three formate species the monodentate formate species in the "C" configuration is imaged as a circular protrusion in Figure 18.2(a) although the O–C–O plane is illustrated along the [001] direction in Figure 18.2(b). This is because the monodentate species is expected to rotate thermally around the C–O bonding at 350 K. The thermal effect yields a circular shape.

It is believed that formate ions are reaction intermediates in catalytic formic acid decomposition. However, the three formates A, B, and C were found at a catalytic reaction temperature of 350 K [13]. The dynamic behavior of these formates was found through successive STM observation at 350 K. Figure 18.3 shows successive STM images of the same area on a TiO_2(110) surface. Although many successive STM images were taken at intervals of 80 s, only five selected ones are shown in Figure 18.3. One finds two formate species in the yellow rectangular area in each image. In the image (a), both of them are seen as ellipses along the $[1\bar{1}0]$ direction on Ti rows. Thus, they are in the "A" configuration at $t = 0$ s. However, in the image (b), one of them, which is indicated by a white arrow in the lower part, changes into an ellipse along the [001] direction. It also shifts slightly to the right side of the Ti row with a white center line. Since these features are characteristic of the "B" configuration, we can safely say that the initial "A" configuration, which is the major configuration at RT, has changed into the "B" configuration at $t = 80$ s. Since this kind of dynamics was not observed at RT, it must be due to the elevated temperature effect, and hence can be considered as a process of the catalytic reaction. We superimposed white solid and broken lines along the $[1\bar{1}0]$ direction in each STM image. The

(a)

on Ti row

(b)

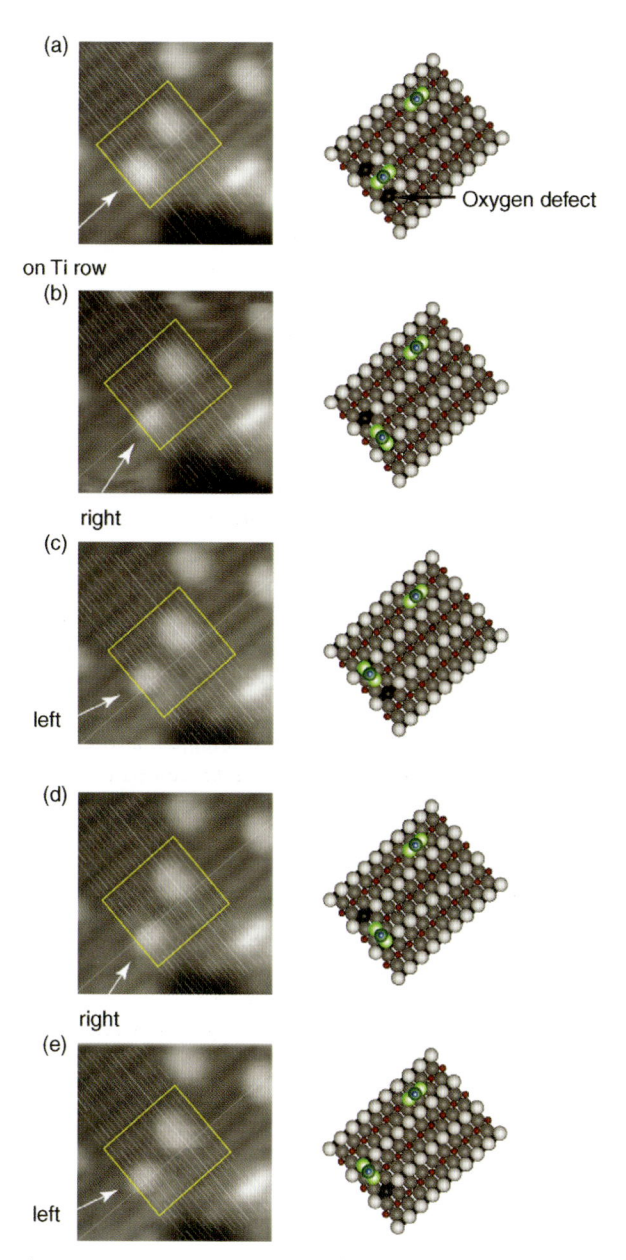

right

(c)

left

(d)

right

(e)

left

Oxygen defect

Figure 18.3 Successive STM images of formates on TiO$_2$(110) at 350 K (7.5 × 7.5 nm^2, Vs = 1.2 V, It = 0.1 nA). (a) $t = 0$, (b) $t = 80$ s, (c) $t = 320$ s (d) $t = 640$ s, (e) $t = 880$ s. The structure model inside the yellow rectangle is shown in the right panel of each STM image. The interval of solid and dotted lines in each figure is 0.296 nm, which corresponds to half of the distance between two adjacent Ti atoms on the Ti row. The solid lines are drawn so that one of them passes over the center of the upper formate species.

interval between a solid line and the adjacent broken line is 0.148 nm, which corresponds to half of the Ti–Ti distance along the [001] axis. Since one of the solid lines is drawn so as to pass through the center of the upper formate species, each solid line is on the second layer O atoms, and the broken line is on the Ti atoms in the same atomic layer. The center of the lower formate species is crossed also by a solid line in the image (a), but it shifts to the lower side along the Ti row and is crossed by a broken line in the image (b). This situation is illustrated schematically illustrated at the right side of the STM images. The change from "A" to "B" yields a shift of the species by 0.148 nm along the Ti row.

We found another dynamics of the species through further observation. As mentioned above, the species is in the "B" configuration on the right side of a Ti row in (b). The species remains at the same site until (c), recorded at $t = 240$ s. However, the species shifts to the left side of the same Ti row in the image (b), which corresponds to the mirror site of the "B" configuration in (b) against the Ti row. Since a formate species is stable in the "B" configuration only at the vacancy of O_B, two O defects are needed on both sides of the corresponding Ti row for this kind of hopping behavior. It is known that hydroxy groups on $TiO_2(110)$ generate H_2O and make O vacancy defects at 350 K. When this kind of H_2O generation occurs on the neighboring O site to a formate in the "A" configuration, the formate is thermally activated into the "B" configuration. Furthermore, when another O vacancy is generated on the other side of the formate on the Ti row, the species hops between the two "B" configurations. The hopping is repeated unless the species changes into a configuration other than "B". The present formate species hops again to the initial "B" configuration in the image (d) at $t = 640$ s, then to the mirror "B" configuration again in the image (e) taken at 880 s. The hopping event occurred three times in the present observation of 800 s from (b) to (e). Assuming a typical value of the vibration frequency of the formate adsorbate as 10^{13} s^{-1}, the activation energy E (kJ mol^{-1}) for a formate to hop between the two "B" configurations is simply estimated by the following relation:

$$v_0 \exp(-E/RT) = 3/800$$

This relation leads to $E \sim 100$ kJ mol^{-1}, which is of the order of magnitude for a chemical binding energy. The value is also similar to the value for the binding energy estimated by temperature programmed desorption (TPD) [10]. Although we could not observe other dynamics than hopping for the formates in the "B" configuration, the observed hopping event indicates that the formate species B can change its configuration easily, just like the species A.

Our successive STM observation in another sample area revealed the final step of the formate dehydration process. Figure 18.4a and b are STM images of the same sample area, but image (b) was taken at 80 s after image (a). The formate species in the top part of (a) is located on the O_B row, and hence it is species C in the "C" configuration. However, after 80 s in (b), the species changed into a rather faint protrusion, and shifted a little upward. This indicates some kind of reaction for the species. The line-profiles along 1–2 in (a) and 3–4 in (b) are shown in Figure 18.4(c) and (d). The formate in the "C" configuration is known as a protrusion of 0.20 nm

18.3 Results and Discussion | 325

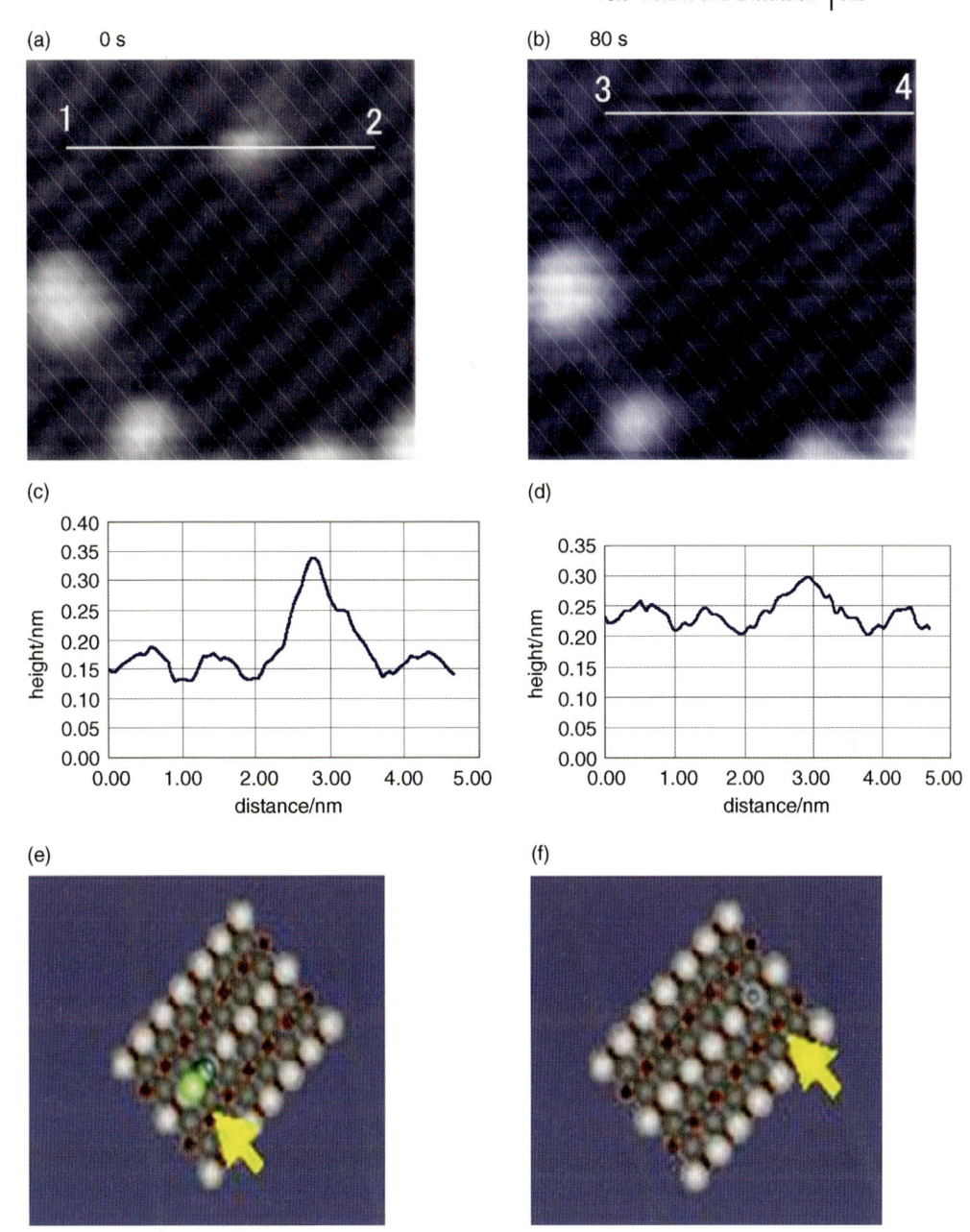

Figure 18.4 Successive STM images (7.5 × 7.5 nm²) taken under the same conditions as Figure 18.3. A decomposition process of species C and a formation process of an OH group are visualized. The OH species diffused along the oxygen row by a distance of three oxygen atoms. Image (b) was taken 80 s after image (a). Line profiles of species C in (a) and an OH group in (b) are shown in (c) and (d), respectively. The location models of the two species are shown in (e) and (f), respectively.

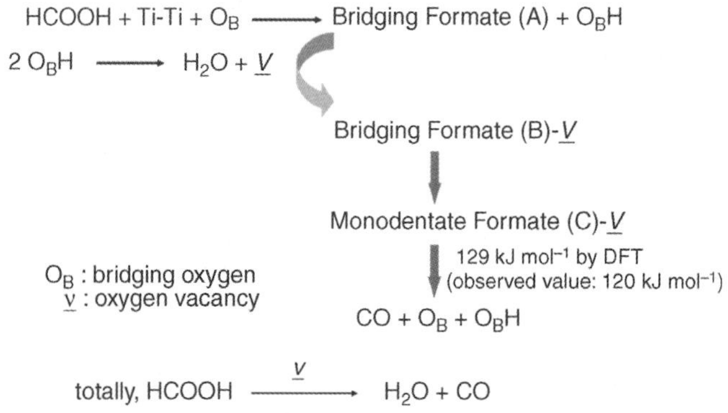

$$\text{HCOOH} + \text{Ti-Ti} + \text{O}_\text{B} \longrightarrow \text{Bridging Formate (A)} + \text{O}_\text{B}\text{H}$$

$$2\,\text{O}_\text{B}\text{H} \longrightarrow \text{H}_2\text{O} + \underline{V}$$

$$\text{Bridging Formate (B)-}\underline{V}$$

$$\downarrow$$

$$\text{Monodentate Formate (C)-}\underline{V}$$

$$\downarrow \text{129 kJ mol}^{-1} \text{ by DFT (observed value: 120 kJ mol}^{-1})$$

$$\text{CO} + \text{O}_\text{B} + \text{O}_\text{B}\text{H}$$

O_B : bridging oxygen
$\underline{V}$: oxygen vacancy

$$\text{totally, HCOOH} \xrightarrow{V} \text{H}_2\text{O} + \text{CO}$$

Scheme 18.1 A proposed reaction mechanism for the HCOOH dehydration at oxygen vacancy on a TiO$_2$(110) surface.

height measured from the top of a Ti row in (a). However, it changes to a 0.05 nm height protrusion in (b). Considering the previous study [17], the protrusion is assigned to a surface OH group adsorbed on a bridge oxygen row. Thus, it is suggested that an OH group and a CO molecule were produced from the formate species C in the monodentate "C" configuration. A proposed reaction mechanism based on the STM images and DFT calculations is shown in Scheme 18.1.

Now we succeeded in observing the change in formate configuration from "A" to "B", and the dehydration process from "C". As mentioned above, these formates A, B, and C can be considered as reaction intermediates because they are only observed at elevated temperature. Thus we elucidated that the formate on the TiO$_2$(110) surface changed its configuration from "A" to "C" through "B" in the dehydration process. This scenario has actually been confirmed by our previous theoretical calculations [12]. A bridging formate species A first changes into a bridging formate species B, then into a monodentate formate species C, and the monodentate formate C finally decomposes. The first and the last processes were shown directly in the present STM observations. Although the "B" to "C" transition was not observed in the present study, it is inevitable for the decomposition process. However, the observed dynamics of the species B strongly indicates such a transition. In this mechanism, formate species A migrates to find an oxygen vacancy on which one of the two oxygen atoms in formate A is trapped with the molecular axis rotated by 90° (formate B). It is to be noted that the active site for the dehydration is the oxygen vacancy but it is not necessary for the oxygen vacancies to be present at the surface. Even if there are no active oxygen vacancies at the catalyst surface, the active site can be produced *in situ* under the catalytic reaction conditions. This indicates the importance of *in situ* characterization of the catalyst surface under the working conditions. The formate species has long been regarded to be the reaction intermediate for the formic acid dehydration, but the dehydration of formic acid involves four dynamic processes including dynamic migration of formic acids and the production of different formate species in three configurations.

18.3.2
Dynamic Catalytic Dehydrogenation of Formic Acid on a TiO_2(110) Surface

Various adsorption configurations of HCOOH on a TiO_2(110) surface (Figure 18.5) were first investigated to seek possible reactants and intermediates for the HCOOH dehydrogenation by DFT calculations, considering previous experimental and theoretical studies on adsorption structures on TiO_2 (110) surfaces [12–14]. Table 18.1 summarizes their adsorption energies and the Mulliken charges of the hydrogen atoms of associative and dissociative adsorbates [14]. R1 and R2 are molecularly adsorbed species, where R1 adsorbs on a fivefold coordinated Ti^{4+} with a carbonyl group of formic acid, while R2 adsorbs with a hydroxy group. The adsorption energies of the molecular adsorption states were similar and small, and they may be regarded as precursor states for the dissociative adsorption of formic acid. For the dissociative adsorption on a stoichiometric surface, there were two configurations of R3 and R4. R3 is monodentate formate which binds to a fivefold coordinated Ti^{4+} with an O atom of the carbonyl group. R4 is a bridging formate, where two O atoms of a formate bind to the neighboring surface Ti^{4+} ions, and the bridging formate (R4) was most

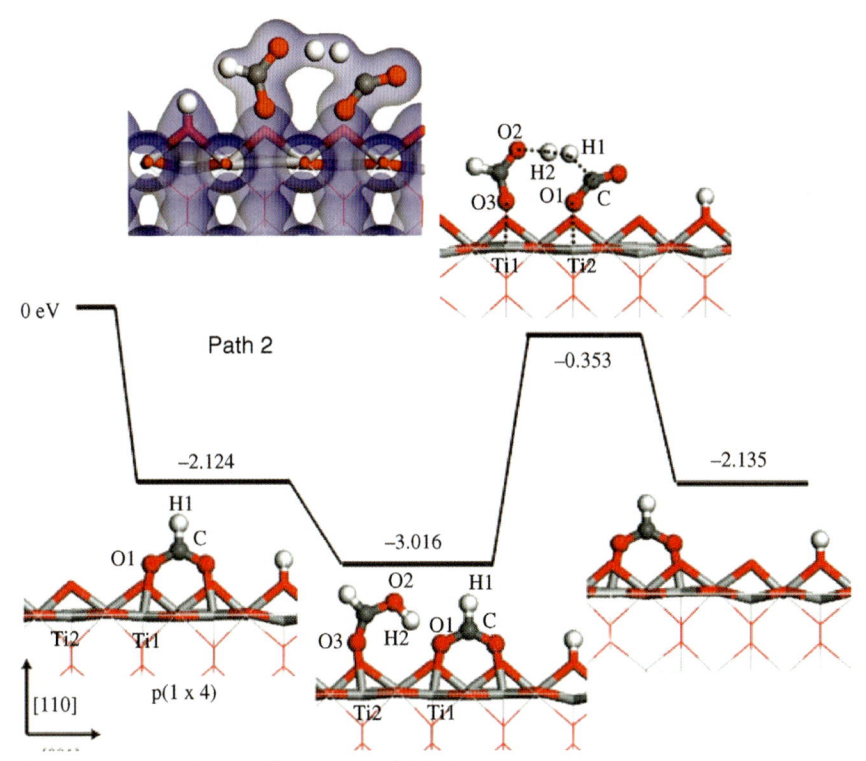

Figure 18.5 Energy diagrams for a proposed dehydrogenation pathway on a stoichiometric TiO_2(110) surface. The zero-level energy (E_0) as reference is defined as the sum of the energies of the clean surface and of two HCOOH(g).

Table 18.1 Adsorption energies of HCOOH on a TiO$_2$ (110) surface and Mulliken charges of the hydrogen atoms of C—H and O—H. R1 and R2: associative adsorption, R3-R8: dissociative adsorption.

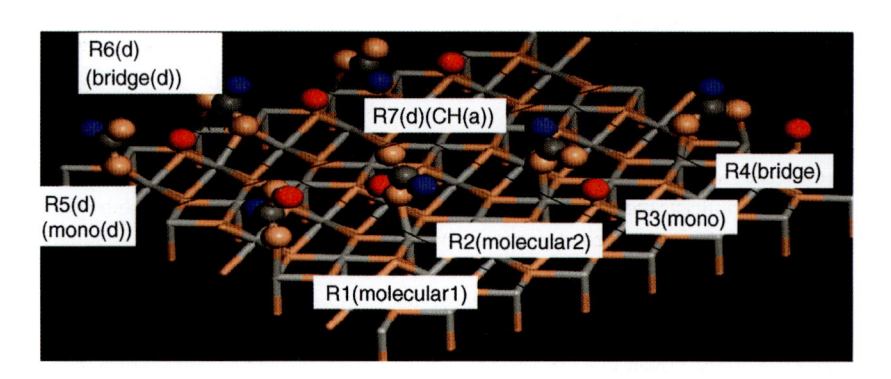

Adsorbate conformations	Adsorption energy/eV	Charge of H1[a]	Charge of H2[b]
R1	−0.559	0.330	0.437
R2	−0.562	0.241	0.552
R3	−1.059	0.215	0.552
R4	−1.872	0.219	0.472
R5(d)[c]	−1.584	0.181	0.490
R6(d)[c]	−1.992	0.221	0.480
R7(d)[c]	−0.292	0.047	0.476
R8(d)[c]	−0.181	0.006	0.468
Gas phase	—	0.140	0.421

[a] H1 is the hydrogen atom of C—H.
[b] H2 is the hydrogen atom of O—H groups. Note that there are two types of OH groups for molecularly adsorbed HCOOH and bridging OH at the surface.
[c] Adsorption on oxgen defect sites.

stable. The adsorption energies of R1–R4 were very similar to the values found by previous DFT calculations, which confirms the quality of the present calculation method. On the other hand, four dissociative configurations (R5(d)–R8(d)) were found to be specific to a defective surface. R5(d) offers an O atom to a defect site and another O atom directs to the surface normal, while R6(d) binds to a defect site with an O atom and another O atom binds to a fivefold coordinated Ti^{4+} in front of the defect site. R7(d) adsorbs on a fivefold coordinated Ti^{4+} with an O atom and on a defect site with a C—H hydrogen. R8(d) is a formate species, where the C—H bond of R7(d) is nearly broken (not shown). The most stable species on the defective surface was R6(d), while unstable R7(d) and R8(d) may be regarded as transient configurations. This is due to an electron transfer from the neighboring Ti^{3+} ions at the defect site, which weakens the Lewis acidity of the five-coordinated Ti^{4+} site in front of the defect site. Experimentally, only R4 and R6(d) were observed at RT on a TiO$_2$(110)

surface by means of Fourier transform reflection-absorption infrared (FT-RAIR) spectroscopy [15] and X-ray photoelectron diffraction (XPD) [30, 31]. The majority was R4 species, as imaged by STM [16] at elevated temperatures. STM revealed an increasing population of R6(d) and the existence of R5(d) as intermediate species for the unimolecular dehydration, which was suggested to proceed by the steps R4 → R6 (d) → R5(d) [12]. The unimolecular decomposition of R5(d) to CO + OH is rate determining, as suggested by STM [16] and DFT calculation [12].

The electron density at the transition state in the most plausible dehydrogenation pathway (Path 2) is also shown. Note that both the reactant molecules at the transition state interact with the surface.

In all the configurations, the C–H hydrogen atoms denoted as H1 are electron richer than the OH hydrogen atoms denoted as H2 (Table 18.1) and, particularly, H1 of R7(d) and R8(d) is richest due to direct electron donation from the two Ti^{3+} ions. Except for these species, the adsorption on the Ti^{4+} Lewis acid site decreased the electron negativity of H1 compared to that of the gas-phase molecule. On the other hand, the most positively charged H2 was observed with R2 and R3, which hydrogen bonded with the neighboring surface oxygen atoms. The other H2 had a similar charge to that of a gas-phase molecule. The modification of the H charge through adsorption was much larger for electron-negative H1 than for H2. These charges on H1 and H2 were strongly related to the genesis of a bimolecular reaction path, as described below.

The dehydrogenation of HCOOH on a TiO_2 (110) surface is given by the second-order rate equation which is proportional to the formate coverage and the gas-phase HCOOH pressure [10]. We performed transition state searches for the bimolecular dehydrogenation reaction between a HCOOH(g) or molecularly weakly adsorbed HCOOH and dissociatively adsorbed species on the basis of the adsorption states mentioned above. The energy diagrams of the obtained pathways on the stoichiometric and defective surfaces have been reported previously by us [14]. Two kinds of activation energies for each path are listed in Table 18.2; ΔE_{el} is the activation energy of the rate-limiting elementary step, which corresponds to the height of the transition state from the neighboring stable or quasi-stable state, and ΔE_{ap} is the apparent activation energy, which is the height of the transition state from the zero-level reference energy, that is, the energy of a clean surface and two HCOOH(g). Note that the experimentally observed activation barrier 0.16 eV corresponds to ΔE_{ap}.

As a result of the series of transition state searches, we found two important factors to decide the height of the transition state. One factor is the charges of the reacting H atoms to produce H_2(g). Since H1 and H2 are electro-negative and positive, respectively, the dehydrogenation reaction tends to be a heterolytic reaction. This fact suggests that the polarization between the two H atoms was important. The relation between the activation energies and the charges of the reacting H atoms in Table 18.2 indicates that the H1 charge is critical for ΔE_{el}. Concretely, the defect-trapped H1 atom gave very small ΔE_{el} in Paths 5 and 7 (Path 6 was an exceptional case, where the large distance between the reacting H atoms raises ΔE_{el}).

The other factor is the amount of the adsorption energy, which affects the energy level of the transition states. ΔE_{ap} of Path 2 was comparable to or even smaller than those of Paths 5 and 7, in spite of the much larger value of ΔE_{el}. This was because the stabilization of the transition state by the adsorption energies was most effective in Path 2, as shown in Figure 18.5. At the transition state in Path 2 the two reactant molecules have a monodentate-like configuration, whose adsorption energies should be about 1 eV, as listed in Table 18.1 and Figure 18.5. As seen in the electron density distribution of the transition state in Path 2 (Figure 18.5), both the leaving and adsorbing reactants interact with surface fivefold coordinated Ti^{4+} ions. On the other hand, at the rate-limiting transition states of Paths 5 and 7, the adsorption energy of only one monodentate-like reactant is available because of the instability of R7(d) and R8(d). Furthermore, the transition state in Path 2 seems not to have any notable steric repulsion or distortion because of the moderate distance between the two neighboring Ti^{4+} ions (Figure 18.5). On the contrary, the elongated distance between the two reacting H atoms in Paths 5 and 7 made the use of the adsorption energy at the transition states ineffective.

The results of Table 18.2 lead to the conclusion that Paths 2, 5 and 7 are plausible pathways with sufficiently low activation energies compared to the experimental value (0.16 eV). Path 2 on the stoichiometric surface is advantageous for the available adsorption energies at the transition state, whereas Paths 5 and 7 on the defective surface are stable due to the electronegativity of the defect-trapped H atom [14].

Which pathway is the most plausible under the actual catalytic reaction conditions? While Path 2 proceeds on the three neighboring fivefold coordinated Ti^{4+} ions at a perfect surface, Paths 5 and 7 require a priori an oxygen-defect site. It is known that the concentration of oxygen defects on a TiO_2(110) surface is typically only less than 5% under the dehydrogenation reaction conditions [32]. In addition, the defect formation by H_2O removal from two OH groups observed at elevated temperature, >500 K, is not probable at a dehydrogenation temperature below 450 K [10, 12, 32]. Considering that the R6(d) species was the most stable one at the oxygen defects, most of the defects must be occupied by R6(d), particularly at the higher pressures of HCOOH under the dehydrogenation reaction conditions. Namely, Paths 5 and 7 cannot be reaction paths for the HCOOH. The R6(d) species is a reaction intermediate for the HCOOH dehydration, as reported previously [33]. Thus, Paths 5 and 7 are excluded, based on the lack of the reactants R7 and R8 under the reaction conditions [14].

On the other hand, the concentrations of the reactants, bridging formate R4 and weakly adsorbed formic acid, in Path 2 were sufficient at the surface because no special sites are required for this path and one of the reactants, R4, is the most stable species. Furthermore, high HCOOH coverage by increasing HCOOH pressure increases the coadsorption state in such a manner that a weakly adsorbed HCOOH is located adjacent to a R4 species. Thus Path 2 via the transition state activated in a concerted manner by three Ti^{4+} ions (Figure 18.5) should be the most plausible dehydrogenation pathway under the reaction conditions [14].

Table 18.2 Relationship between the activation energies and the Mulliken charges of the H atoms of two reactants at the transition state.

Path	H1	H2	$\Delta E_{el}/eV^a$	$\Delta E_{ap}/eV^b$
1	0.14	0.552	2.708	0.585
2	0.219	0.437	2.662	−0.353
3	0.221	0.437	2.424	1.508
3	0.221	0.552	2.173	1.257
5	0.047	0.437	0.952	0.212
6	−0.046	0.552	1.782	1.007
7	−0.046	0.552	0.907	−0.041

$^a\Delta E_{el}$ is the intrinsic activation energy between the transition state (TS) and the adsorption state, as shown below.
$^b\Delta E_{ap}$ is the apparent activation energy which is the height of the TS from the initial state before adsorption, as shown below.

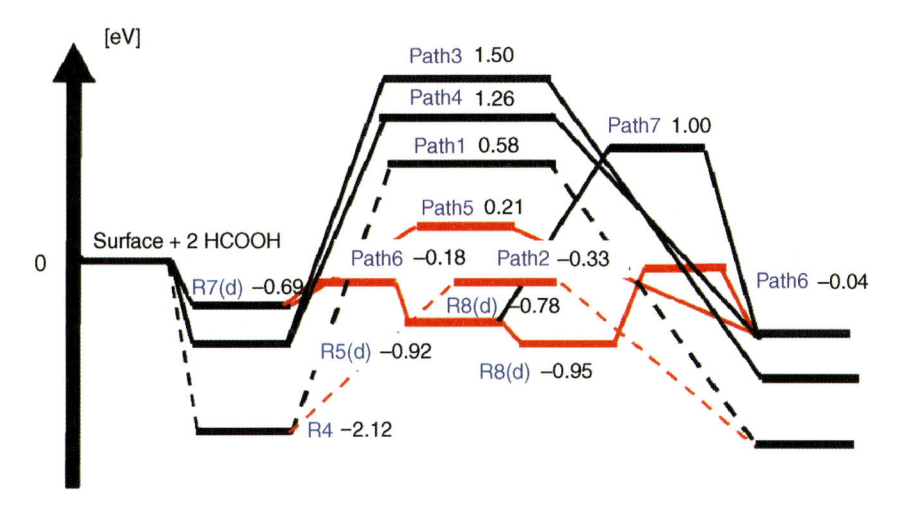

18.3.2.1 Mechanism of the Switchover of Reaction Paths

We suggested the bimolecular dehydrogenation mechanism, where the reaction proceeds on the three neighboring Ti^{4+} sites, with the aid of the large adsorption energy of the reactants. As this mechanism requires neither special sites nor unstable reactants, the concentration of the reactants as well as the apparent activation barrier are quite reasonable. The dehydrogenation pathways on the oxygen defect sites were discarded because the concentration of the active site was regarded to be too small at the dehydrogenation temperature below 450 K. Henderson and Bowker et al. stated that there was no evidence for the dehydrogenation of HCOOH on $TiO_2(110)$ [34, 35]. This discrepancy with our results should be attributed to the difference in the employed reaction atmospheres. We performed the decomposition of HCOOH under

HCOOH atmospheres, whereas they examined it under ultra-high vacuum. According to our results, the bimolecular dehydrogenation never proceeds in the absence of the gas-phase HCOOH. Kecskés et al. [36] reported that HCOOH was converted to HCHO on a defective TiO_2 (110) surface at 300 K, and that the produced HCHO reacted with HCOOH(g), leading to CO and H_2 at 473 K. We could not find any evidence for the previous STM work [7–10]. Wang et al. studied the decomposition of DCOOD at 500 K on TiO_2(110) surfaces with different initial concentrations of the oxygen defects [34]. They observed the production of $CO_2 + D_2$ on a perfect TiO_2 (110) surface, similar to our result [9–11], while the product changed to $CO + D_2$ as the concentration of the defect sites increased. These results on the defective TiO_2(110) surface reproduce the dehydration reactivity of TiO_2 power catalysts [37].

We found a denuded zone of 1.4 nm on the terrace from the edge of an atomic-height step on a TiO_2(110) surface. This denuded zone may be explained by the assumption that formate species near a step are more reactive for decomposition or more easily desorbed compared to other formate species. Alternatively, formate species are much less stable near a step, which leads to almost no population of formate species near steps because anions like chloride ions do not adsorb near a step. Anyhow, it means that no dehydrogenation reaction proceeds unless there are flat areas larger than 2.8 nm dimension without defects and steps on the surface of TiO_2 powders [11]. Regarding the formic acid dehydration, oxygen vacancies which are formed *in situ* under the catalytic reaction conditions are indispensable for catalysis at higher temperatures than the dehydrogenation reaction. Thus the reaction pathways of HCOOH decomposition depend on the defect concentration as well as the geometric arrangement of the TiO_2 surface.

18.4
Conclusion and Perspective

We investigated the dehydration of formic acid on a TiO_2(110) surface. Contrary to the conventional knowledge of acid–base catalysis, which is regulated by the intrinsic acid–base property of a catalyst surface, we showed experimental and theoretical evidence for a surface-mediated catalytic reaction mechanism on TiO_2(001), where the existence of oxygen vacancies as active sites at the beginning of the catalysis is not necessary and the active oxygen vacancies are produced *in situ* under the catalytic reaction. We propose a dynamic acid–base concept for catalysis at oxide surfaces.

The mechanism of the bimolecular dehydrogenation reaction of HCOOH on a TiO_2(110) surface was also investigated by DFT calculations. The most plausible reaction pathway was that between a bridging formate adsorbed on two fivefold coordinated Ti^{4+} ions and HCOOH molecule weakly adsorbed at the adjacent Ti^{4+} ion. The dehydrogenation occurs by Coulombic interaction between the two adsorbates. The intrinsic acidic TiO_2 surface is modified by adsorbed formate anions, and the new basic surface reacts with the second formic acid. The difference in the reactivity between the TiO_2(110) single crystal surface and TiO_2 powder catalysts should be due to the concentration of oxygen defects and the dimensions of the flat terrace area without steps. The classical acid–base concept for oxide catalysis is not appropriate for explanation of the oxide catalysis and should be modified.

The present finding shows the critical importance of the atomic-scale design of the surface of a material. The arrangements of Ti^{4+} ions and O^{2-} ions create new catalytic functions for desired chemical processes. The active sites can be produced *in situ* under the catalytic reaction conditions, even if there are no active sites at the surface before the catalysis. The surface is also modified by adsorption of a reactant to form a new surface with a different acid–base character from the intrinsic property. The dynamic acid–base aspect at a catalyst surface is the key issue to regulate the acid–base catalysis, which may provide a new strategy for creation of acid–base catalysts.

References

1 Tanabe, K., Hattori, H., Yamaguchi, T. and Tanaka, T. (1989) *Acid-Base Catalysis*, Kodansha.
2 Mars, P., Scholten, J.F. and Zwietering, P. (1963) The catalytic decomposition of formic acid. *Adv. Catal.*, **14**, 35–113.
3 Aramedia, M.A., Borau, V., Garcia, I.M., Jimenez, C., Marinas, A., Marinas, J.M., Porras, A. and Urbano, F.J. (1999) Comparison of different organic test reactions over acid–base catalysts. *Appl. Catal. A-General*, **184**, 115–125.
4 Patermarakis, G. (2003) The parallel dehydrative and dehydrogenative catalytic action of γ-Al_2O_3 pure and doped by MgO: kinetics, selectivity, time dependence of catalytic behaviour, mechanisms and interpretations. *Appl. Catal. A-General*, **252**, 231–241.
5 Poulston, S., Rowbotham, E., Stone, P., Parlett, P. and Bowker, M. (1998) Temperature-programmed desorption studies of methanol and formic acid decomposition on copper oxide surfaces. *Catal. Lett.*, **52**, 63–67.
6 Halawy, S.A., Al-Shihry, S.S. and Mohamed, M.A. (1997) Gas-phase decomposition of formic acid over Fe_2O_3 catalysts. *Catal. Lett.*, **48**, 247–251.
7 Haffad, D., Chambellan, A. and Lavalley, J.C. (2001) Propan-2-ol transformation on simple metal oxides TiO_2, ZrO_2 and CeO_2. *J. Mol. Catal. A-Chem.*, **168**, 153–164.
8 Munuera, G. (1970) A study of the mechanisms of formic acid dehydration on TiO_2. *J. Catal.*, **18**, 19–29.
9 Onishi, H., Aruga, T. and Iwasawa, Y. (1993) Catalytic reactions on a metal oxide single crystal: switchover of the reaction paths in formic acid decomposition on titanium dioxide TiO_2(110). *J. Am. Chem. Soc.*, **115**, 10460–10461.
10 Onishi, H., Aruga, T. and Iwasawa, Y. (1994) Switchover of reaction paths in the catalytic decomposition of formic acid on TiO_2(110) surface. *J. Catal.*, **146**, 557–567.
11 Iwasawa, Y., Onishi, H., Fukui, K., Suzuki, S. and Sasaki, T. (1999) The selective adsorption and kinetic behaviour of molecules on TiO_2(110) observed by STM and NC-AFM. *Faraday Discuss.*, **114**, 259–266.
12 Morikawa, Y., Takahashi, I., Aizawa, M., Namai, Y., Sasaki, T. and Iwasawa, Y. (2004) First-principles theoretical study and scanning tunneling microscopic observation of dehydration process of formic acid on a TiO_2(110) surface. *J. Phys. Chem. B*, **108**, 14446–14451.
13 Aizawa, M., Morikawa, Y., Namai, Y., Morikawa, H. and Iwasawa, Y. (2005)

Oxygen vacancy promoting catalytic dehydration of formic acid on TiO_2(110) by in situ scanning tunneling microscopic observation. *J. Phys. Chem. B*, **109**, 18831–18838.

14 Uemura, Y., Taniike, T., Tada, M., Morikawa, Y. and Iwasawa, Y. (2007) Switchover of reaction mechanism for the catalytic decomposition of HCOOH on a TiO_2(110) surface. *J. Phys. Chem. C*, **111**, 16379–16386.

15 Hayden, B.E., King, A. and Newton, M.A. (1999) Fourier transform reflection absorption IR spectroscopy study of formate adsorption on TiO_2(110). *J. Phys. Chem. B*, **103**, 203–208.

16 Onishi, H. and Iwasawa, Y. (1994) STM-imaging of formate intermediates adsorbed on a TiO_2(110) surface. *Chem. Phys. Lett.*, **226**, 111–114.

17 Onishi, H., Fukui, K. and Iwasawa, Y. (1995) Atomic-scale surface structures of TiO_2(110) determined by scanning tunneling microscopy: a new surface-limited phase of titanium oxide. *Bull. Chem. Soc. Jpn.*, **68**, 2447–2458.

18 Delley, B. (1990) An all-electron numerical method for solving the local density functional for polyatomic molecules. *J. Chem. Phys.*, **92**, 508–517.

19 Halgern, T.A. and Lispscomb, W.N. (1977) The synchronous-transit method for determining reaction pathways and locating molecular transition states. *Chem. Phys. Lett.*, **49**, 225–232.

20 Delley, B. (1995) *Modern Density Functional Theory: A Tool for Chemistry, Theoretical and Computational Chemistry*, Vol. 2 (eds J.M. Seminario and P. Politzer), Elsevier, Amsterdam.

21 Perron, H., Domain, C., Rogers, J., Drot, R., Simon, E. and Xatalette, H. (2007) Optimisation of accurate rutile TiO_2 (110), (100), (101) and (001) surface models from periodic DFT calculations. *Theor. Chem. Acc.*, **117**, 565–574.

22 Lai, X., St Clair, T.P., Valden, M. and Goodman, D.W. (1998) Scanning tunneling microscopy studies of metal clusters supported on TiO_2 (110): morphology and electronic structure. *Prog. Surf. Sci.*, **59**, 25–52.

23 Diebold, U. (2003) The surface science of titanium dioxide. *Surf. Sci. Rep.*, **48**, 53–229.

24 Onishi, H. and Iwasawa, Y. (1994) Reconstruction of TiO_2(110) surface: STM study with atomic-scale resolution. *Surf. Sci.*, **313**, L783–L789.

25 Wiesendanger, R. (1994) *Scanning Probe Microscopy and Spectroscopy: Methods and Applications*, Cambridge University Press, New York.

26 Henrich, V.E. and Cox, P.A. (1994) *The Surface Science of Metal Oxides*, Cambridge University Press, Cambridge.

27 Onishi, H. and Iwasawa, Y. (1994) Observation of anisotropic migration of adsorbed organic species using nanoscale patchworks fabricated with a scanning tunneling microscope. *Langmuir*, **10**, 4414–4416.

28 Onishi, H., Fukui, K. and Iwasawa, Y. (1996) Molecularly resolved observation of anisotropic intermolecular force in a formate-ion monolayer on a TiO_2 (110) surface by scanning tunneling microscopy. *Colloids Surf., A: Physicochem. Eng., Aspects*, **109**, 335–343.

29 Iwasawa, Y. (1997) Recent progress in surface scientific approaches to oxide catalysis. *Catal. Surv. Jpn.*, **1**, 3–16.

30 Thevuthasan, S., Herman, G.S., Kim, Y.J., Chambers, S.A., Peden, C.H.F., Wang, Z., Ynzunza, R.X., Tober, E.D., Morais, J. and Fadley, C.S. (1998) The structure of formate on TiO_2(110) by scanned-energy and scanned-angle photoelectron diffraction. *Surf. Sci.*, **401**, 261–268.

31 Chambers, S.A., Thevuthasan, S., Kim, Y.J., Herman, G.S., Wang, Z., Tober, E., Ynzunza, R., Morais, J., Peden, C.H.F., Ferris, K. and Fadley, C.S. (1997) Chemisorption geometry of formate on TiO_2(110) by photoelectron diffraction. *Chem. Phys. Lett.*, **267**, 51–57.

32 Wang, Q., Biener, J., Guo, X.C., Farfran-Arribas, E. and Madix, R.J. (2003) Reactivity of stoichiometric and defective TiO_2(110) surfaces toward DCOOD decomposition. *J. Phys. Chem. B*, **107**, 11709–11720.

33 Bennett, R.A., Stone, P., Smith, R.D. and Bowker, M. (2000) Formic acid adsorption and decomposition on non-stoichiometric TiO_2(110). *Surf. Sci.*, **454**, 390–395.

34 Henderson, M.A. (1997) Complexity in the decomposition of formic acid on the TiO_2(110) surface. *J. Phys. Chem. B*, **101**, 221–229.

35 Bowker, M., Stone, P., Bennett, R. and Perkins, N. (2002) Formic acid adsorption and decomposition on TiO_2(110) and on Pd/TiO_2(110) model catalysts. *Surf. Sci.*, **511**, 435–448.

36 Keceskés, T., Németh, R., Rask, ó J. and Kiss, J. (2005) New reaction route of HCOOH catalytic decomposition. *Vacuum*, **80**, 64–68.

37 Anpo, M. (1997) Photocatalysis on titanium oxide catalysts: Approaches in achieving highly efficient reactions and realizing the use of visible light. *Catal. Surv. Jpn.*, **1**, 169–179.

19
Nuclear Wavepacket Dynamics at Surfaces

Kazuya Watanabe

19.1
Introduction

Since the development of ultrashort lasers, nuclear wavepacket dynamics of various matters have attracted continuing attention [1, 2]. The research targets extend from gas phase molecules [3, 4] to molecules in solution [5, 6], and solids [7]. In general, an excitation of matter by an ultrashort pulse with sufficient bandwidth leads to the creation of coherence between vibrational (or vibronic) eigenstates [1]. The induced nuclear wavepacket then starts to evolve on a certain potential energy surface and the dynamics is probed by a suitable pump–probe spectroscopy. The direct time-domain observation of the nuclear motion provides us with valuable information on photochemical reaction dynamics, vibrational excitation/relaxation mechanisms, electron-vibration (phonon) coupling, and so on.

Although there is an accumulated body of literature on nuclear wavepacket dynamics, those of surface adsorbates are less explored. This is mainly attributed to the fact that the experiments are demanding: one needs to combine surface science techniques and ultrafast spectroscopy and, generally, signals from monolayer adsorbates are far smaller than those from the bulk.

Nevertheless, detailed information on nuclear wavepacket dynamics of surface adsorbates is important both from fundamental and from practical points of view. As is evidenced from the huge success of catalysis, solid surfaces sustain various kinds of reactions [8]. In order to understand the elementary steps of these reactions, the electronic and vibrational dynamics of surface adsorbates should be investigated in depth. Photochemistry at surfaces involves the photoinduced nuclear dynamics of adsorbates, which needs to be elucidated by ultrafast spectroscopy. Furthermore, combining with recently developed pulse shaping technologies [9], elucidation of the wavepacket dynamics will open up a novel laser control scheme of surface photochemical reactions.

This chapter will first describe the principles of experimental techniques which enable us to study the nuclear wavepacket dynamics at surfaces. We focus on time-resolved two-photon photoemission (2PPE) and time-resolved second harmonic generation (SHG). After discussing these techniques, selected studies conducted by our group and others by using these methods are reviewed.

19.2
Experimental Techniques

19.2.1
Time-Resolved Two-Photon Photoemission with Femtosecond Laser Pulses

19.2.1.1 Principles

Two-photon photoemission spectroscopy is known for its capability to reveal not only occupied but also unoccupied electronic density of states [10]. In this scheme, one photon excites an electron below the Fermi level to an intermediate state. A second photon then excites the electron from the intermediate state to a final state above the vacuum level. The photoelectron yields are strongly enhanced if the excitation photon energy is tuned to the resonance conditions, and the photoelectron spectrum reflects the electron lifetime in the intermediate states as well as their density of states. It is necessary to keep the employed photon energy below the work function of the sample, otherwise one photon photoemission signal becomes excessive and buries the 2PPE signals.

2PPE signals from metal surfaces can be observed with nanosecond laser pulses. Early studies with tunable nanosecond dye lasers focused on image potential states formed at clean metal surfaces [11]. When the light is replaced by two ultrashort pulses and the 2PPE spectra are obtained as a function of the relative delay time of the pulses, the signal, which depends on the time delay, contains important physical information, that is, excited state lifetime, dephasing time, and so on. Time-resolved 2PPE has been proven to be a powerful technique to investigate ultrafast electron dynamics in solid states, bulk carrier dynamics in semiconductors and metals, electron dynamics at well-defined surfaces, and so on [12–14].

2PPE is sensitive to the electron dynamics in the adsorbate induced states, when molecules or atoms cover the surfaces. Upon electron injection into the unoccupied states with an ultrashort pulse, adsorbate molecules or atoms respond to the sudden change in their electronic states, and a nuclear wavepacket motion towards a new equilibrium starts. The adsorbate induced states change their energy and/or width as the wavepacket evolution, and it emerges as time-dependent changes of the 2PPE spectra as a function of the delay time.

2PPE signals give direct information on the density of states of the unoccupied states which is obtained only indirectly with other optical methods. One drawback is that since the excited electrons are detected, the observation time window is limited to the lifetime of the excited electrons. The excited state lifetimes at metal surfaces are typically less than a few hundreds of femtoseconds and much shorter than vibrational relaxation times. Hence the information is limited to that in the very beginning of the nuclear wavepacket motion, right after the photoexcitation.

19.2.1.2 Experimental Set-Up

For time-resolved 2PPE spectroscopy, a combined set-up of an ultrafast laser system and an ultrahigh-vacuum photoemission spectroscopic system is indispensable. Typical electron energy analyzers have been used as the spectrometer, such as a cylindrical mirror analyzer, a hemispherical analyzer and a time-of-flight (TOF) analyzer. The TOF analyzer is mainly used for low repetition rate (<1 kHz) laser sources, and the others are used for the lasers with multi-kHz or MHz repetition rates [11–14].

The pump–probe pulses are obtained by splitting a femtosecond pulse into two equal pulses for one-color experiments, or by frequency converting a part of the output to the ultraviolet region for bichromatic measurements. The relative time delay of the two pulses is adjusted by a computer-controlled stepping motor. Petek and coworkers have developed interferometric time-resolved 2PPE spectroscopy in which the delay time of the pulses is controlled by a piezo stage with a resolution of 50 attoseconds [14]. This set-up made it possible to probe decoherence times of electronic excitations at solid surfaces.

As an illustrative example, the 2PPE system developed by our group is depicted in Figure 19.1 [15, 30]. The UHV chamber was equipped with a home-made TOF electron energy analyzer, a hemispherical electrostatic electron energy analyzer, an

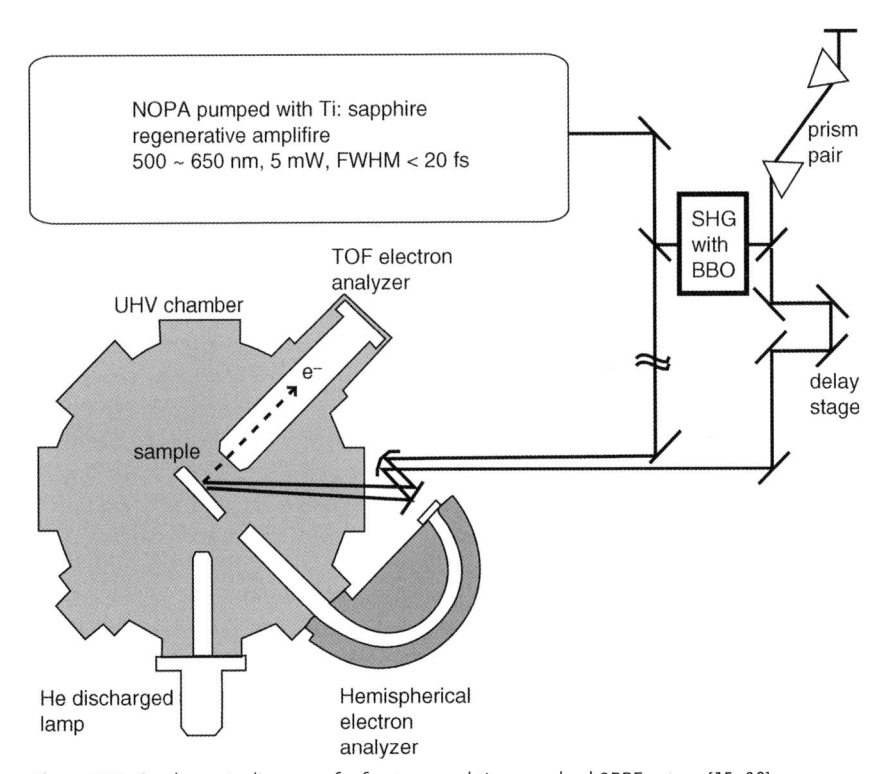

Figure 19.1 A schematic diagram of a femtosecond time-resolved 2PPE set-up [15, 30].

X-ray gun, and a He-discharge lamp. The light source for femtosecond two-color time-resolved 2PPE was a homemade double-pass noncollinear optical parametric amplification (NOPA) system pumped by the second harmonic output of a Ti: sapphire regenerative amplifier. The NOPA output beam was split into two by a 50% ultrathin beamsplitter. A split beam was focused on a 60-μm thick BBO crystal to generate UV pulses that compressed by a 45° prism. The typical pulse width of the NOPA output and its second harmonic are 20 and 35 fs, respectively. The pump and probe pulses were noncollinearly overlapped and focused by a spherical mirror onto the sample held in the UHV chamber. The relative time delay of the two pulses is varied by a delay stage and the kinetic energy distributions of photoelectrons in 2PPE experiments were measured by a home-made TOF electron energy analyzer as a function of the delay time.

19.2.2
Time-Resolved Second Harmonic Generation

19.2.2.1 Principles and Brief History

Second harmonic generation has been recognized as a powerful probe to study the electronic states at surfaces and interfaces [16]. Under the electric dipole approximation, second-order nonlinear processes are forbidden in centrosymmetric systems. This principle makes the phenomena surface-specific in many cases. Indeed, the capability of SHG spectroscopy to explore surface electronic states has been demonstrated on various systems, dye molecules at solid/liquid interfaces [17], organic molecules at liquid/air interfaces [18], semiconductor surface states [19], organic molecules at metal surfaces [20], and so on.

When an ultrashort laser pulse is used as the light source for the SHG spectroscopy, the signal contains dynamical information of the system. In time-resolved SHG spectroscopy, two ultrashort pulses with tunable relative time-delay irradiate the sample. The first pulse (pump pulse) excites the system, and the SHG intensity or its spectra induced by the second pulse (probe pulse) are measured as a function of the time delay.

The SHG intensity from interfaces is determined by the second-order nonlinear susceptibility and the Fresnel coefficients. The SHG spectra of the probe pulses change depending on the transient electronic population and the orientation of the chromophores through these physical quantities. Hohlfeld and coworkers have studied hot electron dynamics in thin metal films by this technique [21]. From the transient response of the SHG intensity, electronic temperature decay due to the electron–phonon coupling in the metal substrate is extracted. Eisenthal and coworkers have studied ultrafast excited state dynamics of dye molecules at liquid interfaces [22]. Particularly, the isomerization dynamics of an organic dye at the interfaces was found to become significantly slower than in the bulk.

In 1997, a seminal paper of the time-resolved SHG study on a GaAs surface appeared [23]. It was shown that the time-resolved SHG probes not only electronic dynamics but also lattice (phonon) dynamics. The detection scheme is as follows: The pump pulse impulsively excites the longitudinal optical (LO) phonon in the GaAs crystal, which results in coherent phonon oscillation at the surface as well as in the bulk. The phonon oscillation dynamically modulates the nonlinear susceptibility of the system as a function of the lattice displacement and that leads to modulation of the SHG intensity of the probe pulse which is linear to the lattice displacement. The experiment was conducted on a carefully cleaned surface under ultra-high vacuum, and several oscillatory components with frequencies distinct from that of the bulk LO phonon were observed. These were ascribed to surface phonon modes of GaAs. Although the bulk coherent phonons in solids had been well investigated by that time, no report on the dynamics of surface modes had appeared.

In 2002, our group extended this approach to a metal surface covered with adsorbates [24]. Time-resolved SHG was applied to a cesium-covered Pt(111) crystal surface under ultrahigh vacuum. The intensity variation of the SHG shows oscillatory components as a function of the delay time, due to coherent nuclear wavepacket dynamics of the Cs–Pt stretching mode. The nonlinear susceptibility of the system is considered to depend on, among other things, nuclear displacements of surface normal modes, that is,

$$\chi^{(2)} = \chi_0^{(2)} + \frac{\partial \chi^{(2)}}{\partial Q} \cdot \delta Q + \cdots, \tag{19.1}$$

where $\chi_0^{(2)}$ stands for the susceptibility under the equilibrium condition, δQ is the displacement of the vibrational (or phonon) coordinate, and the higher terms are omitted. The SHG intensity is proportional to the square of $\chi^{(2)}$, and the leading term, which is linear to the nuclear displacement, dominates the oscillatory part of the time-resolved SHG signals [25].

19.2.2.2 Experimental Set-Up

Here, the time-resolved SHG measurement scheme used in our studies are described [24, 25].

Light Sources A Ti:sapphire femtosecond laser system was used to generate the ultrashort pulses. The output of a commercial Ti:sapphire regenerative amplifier which delivers 130 fs pulses at 800 nm was used as the pump and probe pulses for low frequency modes such as the Cs–Pt stretching mode (77 cm^{-1}). For other alkali adsorbates which are lighter than Cs, the vibrational frequency becomes higher than 100 cm^{-1} and the surface electronic transition occurs in the visible region, so that one needs to compress the pulse width and to convert the wavelength. Both have been achieved by constructing a Ti:sapphire-based NOPA system. Figure 19.2 shows a schematic diagram of the NOPA built in our group, the details have been described in ref. [25]. Briefly, the output of the Ti:sapphire regenerative amplifier (800 μJ pulse^{-1}, 1 kHz) is converted to the second harmonic (400 nm) to be used to pump BBO crystals. A small portion of the fundamental output of the regenerative amplifier is focused onto a sapphire plate to generate a white continuum which is parametrically amplified at the BBO crystal. The white continuum and the 400 nm pump is mixed at the BBO with a non-collinear phase matching angle, and the center wavelength of the output is tuned from 620 to 490 nm by adjusting the relative delay of the two pulses.

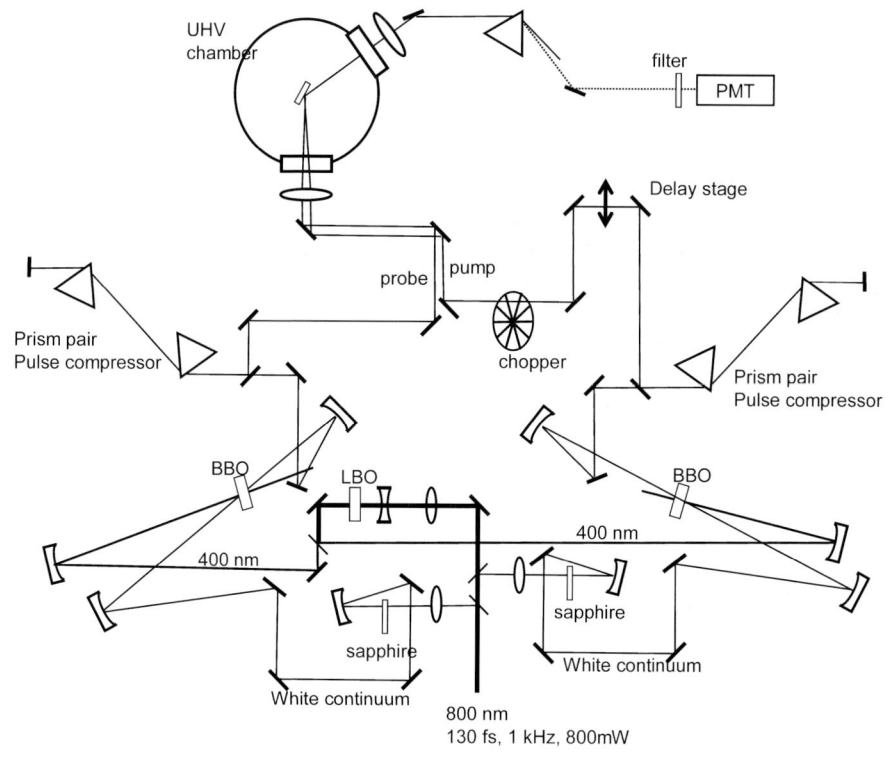

Figure 19.2 A schematic diagram of a time-resolved SHG spectroscopic system.

The amplified signal pulse is compressed with a pair of quartz prisms and the final pulse width is typically 20–30 fs. We built two NOPA set-ups in order to independently tune the pump and probe for the time-resolved SHG, and the NOPA pump pulse is delivered by dividing the 400 nm pump pulse into two. Typical output fluence of the NOPA is 4 µJ pulse^{-1} right after the BBO crystal.

Time-Resolved SHG Under an Ultrahigh Vacuum Figure 19.2 shows a schematic of the experimental set-up for the time-resolved SHG. The sample single crystal of metal (Pt, Cu) is held in an ultrahigh vacuum chamber with a base pressure better than 2×10^{-10} Torr. The chamber is equipped with a cylindrical mirror analyzer for Auger electron spectroscopy (or a hemispherical analyzer for X-ray photoelectron spectroscopy), a quadrupole mass spectrometer and a sputtering ion gun for sample cleaning and surface characterization. SAES getter alkali sources are used to deposit alkali atoms on the sample surface. The sample single crystal with 1 cm diameter is held with Ta wire and welded to a Ta rod, which is attached to a Cu cold-block cooled by liquid N_2. The sample heating is achieved by resistive heating of the Ta wires.

In the case of a Pt single crystal, the sample is cleaned by cycles of Ar^+ sputtering (500 eV energy, 15 min and 1 µA sample current) and annealing at 1000 K for 10 min, followed by oxygen treatment for 1 h (at 800 K, 1×10^{-7} Torr). For Cu, sputtering and annealing (650 K) are repeated until no contamination is detectable. In both cases, the sample is cooled to 110 K and the experiment is done at this temperature if not indicated explicitly.

The laser beams are introduced into the chamber from an inlet window of quartz plate with 1 mm thickness, and the incident angle at the surface is about 70°. The beams are focused onto the sample surface with a quartz lens with focusing length of 300–500 mm. The reflected beams are collimated by a quartz lens and are fed into a photomultiplier after passing through filters and a prism to reject unwanted fundamental light. In the time-resolved SHG measurement, the intensity of the second harmonic of the probe pulse generated coaxially with the probe beam is detected. An optical chopper is inserted in the pump beam path, enabling sensitive detection of the pump-induced intensity variation of the SHG by a lock-in amplifier. The pump–probe delay is varied by a stepping motor delay stage which is computer controlled, and the lock-in output is accumulated for a few seconds at each delay time.

19.3
Nuclear Wavepacket Motions of Adsorbate Probed by Time-Resolved 2PPE

19.3.1
Alkali Atom Desorption from a Metal Surface

Petek and coworkers have investigated the wavepacket dynamics of Cs atoms adsorbed on Cu(111) by the interferometric time-resolved 2PPE [26, 27]. Two femtosecond pulses, whose relative optical phases are locked, are irradiated onto the Cs-covered Cu(111). The pump pulse with photon energy resonant with a transition from an occupied surface state to a Cs–Cu antibonding unoccupied state creates coherent polarization. A delayed probe pulse interacts with the created coherent polarization of the system. The resulting interferometric time-resolved 2PPE trace shows a broader profile than the autocorrelation trace of the laser pulses, revealing the phase relaxation of the created coherent polarization and the population decay of the antibonding state. By analyzing the 2PPE time domain profiles, the phase relaxation time and the population decay time were deduced to 15 and 50 fs at 33 K, respectively.

When the photon energy is tuned to the resonance transition, the two-pulse correlation consists of a fast decay of the coherent polarization and a slower, strongly energy dependent, non-exponential decay. This non-exponential decay corresponds to the anti-bonding state population dynamics. This feature is attributed to the desorptive motion of Cs atom from the Cu surface. The time-resolved 2PPE spectra become broader and shift to lower energy over the delay time due to the evolution of the wavepacket of Cs on the excited state potential energy surface. The time-dependent shift of the antibonding state provides direct information on the mechanical forces acting on the Cs atom. Based on a classical model, elongation of the Cs–Cu bond is estimated to be about 0.35 Å within 160 fs of the excitation.

19.3.2
Solvation Dynamics at Metal Surfaces

Harris and coworkers have studied electron solvation and the resulting localization dynamics on acetonitrile/Ag(111) and butyronitrile/Ag(111) [28]. Time-resolved 2PPE was employed to populate the electron's image potential states and to observe their subsequent dynamics. Image potential states result from a confinement of an electron in a well built by the image potential and the crystal potential. Electrons in image potential states reside only a few angstroms outside the interface, making them particularly sensitive to the change in the electrostatic potential at the interface. When the pump pulse excites electrons from the substrate to the image potential states of the Ag(111) covered with nitriles, the electrostatic potentials experienced by the adsorbates are perturbed significantly. The adsorbates respond to the perturbation by reorientation of the molecular axis. The rotation of the molecule causes a reduction in the local work function, resulting in a stabilization of the electron energy at the interface. The time-dependent shift of the image potential state provides information on the molecular solvation dynamics at the metal surface.

The reorientation of the molecular adsorbates also induces a localization of the electron in the image state which is initially delocalized parallel to the surface. The localization phenomenon was proved by examining the emission angle dependence of the photoelectron spectra. It is concluded that the electrons in the delocalized state are trapped in the localized state at around 300 fs after the excitation, where the electron localization size is estimated to be 12 Å.

A similar solvation dynamics of electrons in an ice layer on Cu(111) has been reported by Gahl et al. [29]. In this case, a pump pulse excites electrons from the substrate into a conduction band of an ice film of four bilayer thickness. This initially delocalized electron is subsequently trapped by a localized state within 100 fs, which leads to a pronounced flattening of the dispersion in the 2PPE spectra. The localized electron further undergoes solvation on a picosecond time scale, which manifests itself experimentally as a shift in the binding energy.

Ino et al. have investigated electron injection dynamics at noble metal/tris-(8-hydroxyquinoline) aluminum (Alq_3) interfaces [30]. The electron dynamics at the electron affinity level (conduction band) of the Alq_3 layer adsorbed on Cu(111) and Au(111) were studied by time-resolved 2PPE. By examining time-dependent signals from the electron affinity level of the Alq_3 monolayer, the electron life-time in the state is estimated to be 31 ± 2 fs on Cu(111) and about three times shorter on Au(111). Contrary to the above two examples, the electronic wavefunction of the affinity state is localized within almost one Alq_3 molecule from the initial stage of the photoinduced charge transfer, which is proved by angle-resolved 2PPE. On Cu(111), the peak of the affinity level shifts to lower energy during its life-time with a slope of $-1.2\,\mathrm{eV\,ps^{-1}}$. This electronic energy lowering is attributed to nuclear wavepacket motion of the Alq_3 molecule in the anion state potential energy surface and/or solvation of the surrounding molecules in response to the sudden charging of their neighboring molecule.

19.3.3
Ultrafast Proton-Coupled Electron Transfer at Interfaces

Li et al. have reported time-resolved 2PPE studies on a $TiO_2(110)$ surface covered with methanol [31]. 10 fs pulses with 3.05 eV photon energy were employed to excite electrons trapped in oxygen vacancy sites to the acceptor sites of a CH_3OH overlayer at around 2.3 eV above the Fermi level. The subsequent decay dynamics of the "wet electron" solvated in the CH_3OH layer is probed by photoemission induced by time-delayed probe pulses. The two pulse correlation traces show ultrafast decay which depends on the CH_3OH coverage. Above 1-ML coverage, both the excited state population and its energy decay with fast (<30 fs) and slow (50–200 fs) components, and the slow population decay component shows a pronounced deuterium effect. This deuterium isotope effect cannot be explained by a purely electronic process but could be explained by a proton-coupled electron transfer. The population decay rate of the excited state at a fixed energy is successfully decomposed into two components; an isotope independent solvation term and a proton-coupled electron transfer term with a marked deuterium effect. The latter terms for the CH_3OH overlayer are found to be about twice those for the CH_3OD overlayer. Thus, with time-resolved 2PPE, the ultrafast dielectric response of a protic/solvent metal-oxide interface has been revealed.

19.4
Nuclear Wavepacket Motion at Surfaces Probed by Time-Resolved SHG

19.4.1
Vibrational Coherence and Coherent Phonons at Alkali-Covered Metal Surfaces [24, 25, 32–34]

Figure 19.3 shows typical traces of time-resolved SHG from alkali-covered Pt(111). In both cases, clear oscillatory components appear and they are ascribed to nuclear wavepacket motion of surface modes. There exist more than two components, which becomes clear by Fourier transforming the time-domain data (Figure 19.4). The Fourier spectra are obtained from the raw data with a delay time larger than 50 fs by subtracting background components whose frequencies are less than 1 THz. For Cs adsorbate, a peak at 2.3 THz is prominent and is due to the Cs–Pt stretching mode, while the corresponding stretching mode is observed at 4.8 THz for K adsorbate.

There appear some small peaks at 2.7 and 3.3 THz for K/Pt(111), and they are assigned to surface phonon modes of Pt substrate. These phonon modes are at the zone boundary on the clean surface and they are optically inactive without adsorbates. Since the K adsorbate forms a ($\sqrt{3} \times \sqrt{3}$) superstructure at the coverage, the Brillouin zone of a clean surface is reduced such that the zone boundary at the K point is folded back to the Γ point and so the zone boundary phonon modes become optically active.

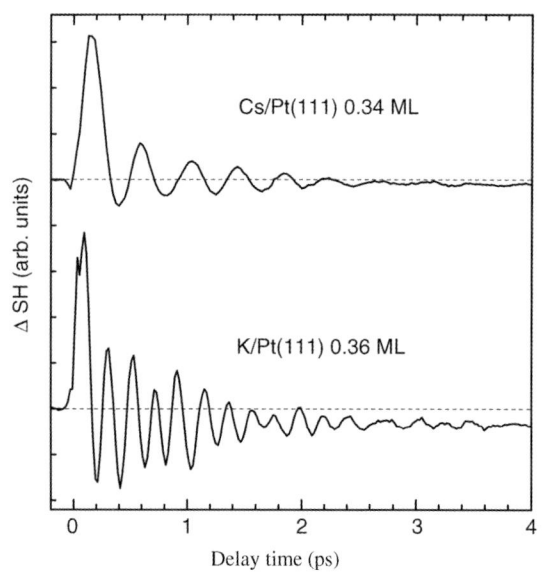

Figure 19.3 Time-resolved SHG traces taken from Pt(111) surfaces covered with (a) Cs and (b) K. The coverages of Cs and K are 0.34 and 0.36 ML, respectively. 25 fs laser pulses with center wavelength at 580 nm were used for the measurement.

A similar zone folding also occurs at Cs/Pt(111) and the phonon mode appears as a small dip in the Fourier spectrum in Figure 19.4. A detailed analysis of the time domain data by linear prediction singular value decomposition has been performed and a decomposition of the time-domain data to phonon modes and alkali–substrate stretching modes has been carried out. Coherent nuclear motions have been observed on substrates other than Pt. Figure 19.5a shows time-resolved SHG traces

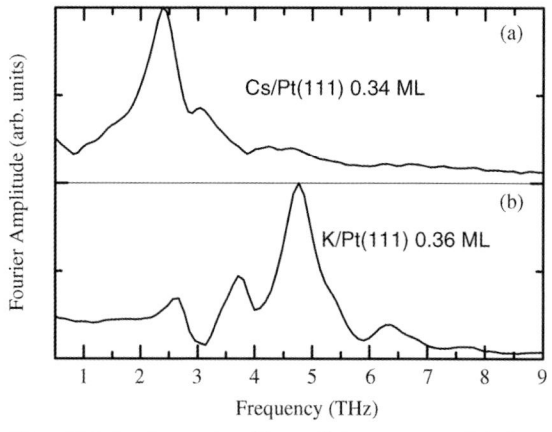

Figure 19.4 Fourier spectra of the oscillatory components in time-resolved SHG traces from Pt(111) covered with Cs (a) and K (b).

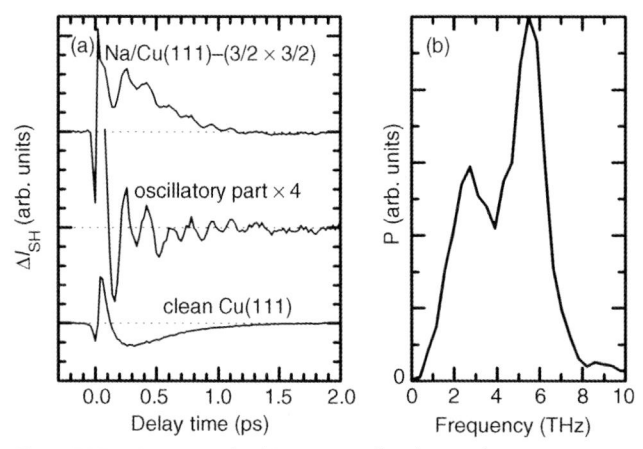

Figure 19.5 (a) time-resolved SHG traces for clean and full-monolayer Na-covered Cu(111) surfaces. The middle trace shows the oscillatory components in the top trace magnified by a factor of 4. (b) The Fourier power spectrum of the time-resolved SHG trace in (a) for Na-covered Cu(111) [34].

from clean and Na-covered Cu(111) surfaces. The response from the Cu clean surface is due to a transient temperature jump and subsequent cooling of the substrate electronic system, whereas some oscillatory components appear when the surface is covered with Na atoms. Figure 19.5b shows the Fourier-transformed spectrum of the time domain data from Na/Cu(111), which shows two prominent frequency components at 2.7 and 5.5 THz. According to a recent calculation, these peaks are ascribed to Na–Cu stretching modes and the stretching modes perturbed by a strong mixing with Cu surface phonon modes, respectively. A striking feature, which is different from the Pt substrate, is that a relative contribution of substrate hot electron dynamics to that of coherent nuclear motions is significant. This is due to the fact that the employed probe photon energy (2.1 eV) is close to the photoexcitation threshold from bulk d-bands, and the nonlinear susceptibility at the photon energy becomes sensitive to the electronic temperature change.

19.4.2
Dephasing of the Vibrational Coherence: Excitation Fluence Dependence [33, 35]

Upon considering a laser manipulation of surface dynamics, the information of the time scale of the coherence loss and their determining factor is important. To explore the effect of hot electrons in the substrate, pump fluence dependence of the time-resolved SHG has been investigated on Cs/Pt(111) and K/Pt(111). Figure 19.6 shows time-domain data for Cs/Pt(111) when varying the pump fluence from 1.7 to 13 mJ cm^{-2}. As the fluence of the pump pulse increases, the initial modulation amplitude due to the Cs–Pt stretching mode rapidly increases and decays much faster. The cause of the enhanced dephasing rate is due to the very effective excitation of the lateral modes by inelastic scattering of hot electrons in the substrate when the

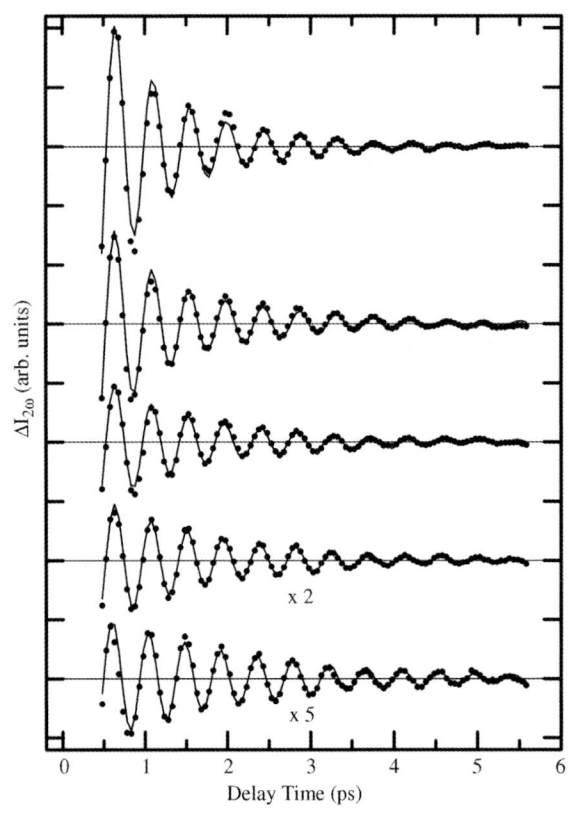

Figure 19.6 Pump fluence dependence of the time-resolved SHG traces taken from Cs/Pt(111). The pump fluences were 1.7, 3.4, 6.7, 10 and 13 mJ cm⁻ from the bottom to the top traces [35].

high laser fluence is utilized. Here, the lateral mode indicates the surface normal mode with atomic motions of adsorbate in parallel with the surface. As has been discussed above, the photon transition by the pump pulse occurs not only between adsorbate-induced electronic states but between bulk continuum bands. As a result, the electron temperature in the substrate increases in a short period of time. For an absorption fluence of 1.1 mJ cm^{-2}, a maximum electronic temperature is estimated to be 1600 K at about 30 fs and decays with a time constant of about 1 ps after the pump pulse irradiation. This hot electron is resonantly scattered with alkali-metal adsorbates, resulting in excitation of the lateral modes. As the fluence increases, multiple inelastic scattering populates the higher vibrational states of the lateral modes, as in the case of desorption induced by multiple electronic transitions.

A similar enhancement of the dephasing rate has been observed for K/Pt(111). Figure 19.7 shows Fourier spectra of the oscillatory components when varying the pump fluence. It is evident that K–Pt stretching mode at 4.8 THz shows a marked red shift and broadening. This feature can be ascribed to the incoherent excitation of the lateral modes, as in the case of Cs/Pt(111), but also notable in this system is that

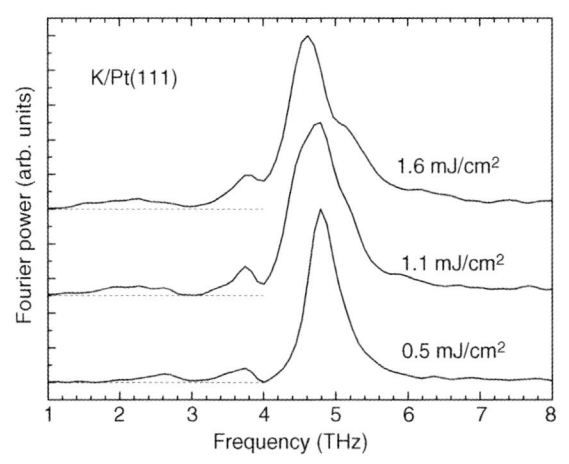

Figure 19.7 Fourier-transformed spectra of the oscillatory components in the time-resolved SHG taken from 0.38 ML K-covered Pt(111) surfaces as a function of the laser fluence absorbed by the Pt substrate.

new components appears at around 2 THz with higher fluences. The frequencies of the new components correspond to those of lateral modes, indicating that the lateral modes are excited coherently by the interaction with the substrate hot electrons.

19.4.3
Excitation Mechanisms [25, 34]

The excitation mechanism which generates coherent nuclear wavepackets has been discussed extensively [1, 36]. In transparent material, electronically off-resonant excitation leads to impulsive stimulated Raman scattering if the pulse duration is sufficiently shorter than the vibrational period. Vibrational wavepackets in the electronic ground state are formed and the expectation value of the normal coordinate starts to oscillate back and forth around the equilibrium position. This excitation process is substantially enhanced when the excitation wavelength is near an electronic absorption resonance. In this electronic resonance case, coherent vibrational motion can be initiated also in the electronically excited state [1]. In the molecular case, the distinction between the nuclear wavepacket in the ground and the electronically excited state is clear both conceptually and practically. That is, vibrational frequencies are shifted in the electronic excited states compared to those in the ground state and the excited state life-time is longer than the vibrational dephasing time in molecules. However, the deformation potentials for ions in solids and on solid surfaces hardly change upon excitation in the weak excitation limit where the perturbation treatment is valid. Therefore, the oscillation frequency is insensitive to the electronic excitation and distinction between the two cases is practically impossible [37]. Actually, in most of the pump–probe studies on opaque solids, the experimental observables are macroscopic polarizations modulated by coherent nuclear displacements, Q, and contributions from different electronic states are averaged out.

One of the key aspects concerning the excitation mechanism is which electronic transitions couple to the coherent nuclear motions. As for the surface adsorbate excitations, there are two extreme cases for the electronic transition which leads to surface dynamics. One is the adsorbate localized excitation and the other is the substrate-mediated excitation. In many cases, investigating the reaction yield by changing the characters of incident photons (polarization, energy, etc.) helps to confirm which mechanism operates. If a substrate-mediated process dominates, the reaction yield follows the features of bulk absorption, whereas a deviation from the bulk absorption property would be observed for the surface localized excitations.

In the case of coherent phonons at Cs/Pt(111), adsorbate localized excitation has been proposed as the driving force of the coherent nuclear motions [25, 32]. From a study of the coverage dependence of the initial amplitude of the Cs–Pt stretching mode, it is apparent that the coherent nuclear motions are enhanced at 0.25 ML < θ < 0.4 ML. It is known that the image potential states, like unoccupied states, are formed at alkali-covered metal surfaces, and their energies shift as a function of the coverage. Thus, with increase in coverage, the energy positions (from the Fermi level) of these unoccupied states at Cs/Pt become close to the excitation energy (1.55 eV) at 0.25 ML < θ < 0.4 ML, and the surface interband transitions occur effectively, that is, resonant electronic transitions involving adsorbate localized states likely to be responsible for the coherent nuclear motions.

A more direct evidence of the surface localized excitation mechanism has been obtained by a polarization dependence study. For K/Pt(111) at 0.36 ML, it has been demonstrated that the coherent excitation of the K–Pt stretching mode occurs with p-polarized excitation and not with s-polarized excitation. Since the s-polarization absorptance is about one fourth of that with p-polarization under the experimental conditions (2.19 eV photon energy, 70° angle of incidence), the coherent amplitude should be detectable with s-polarization if the substrate-mediated process operates. Therefore, the negligible oscillatory component with s-polarization is inconsistent with the substrate-mediated excitation model and it is indicated that some electronic transitions involving K-induced surface states are responsible for the coherent excitations.

Whereas adsorbate localized excitations play a crucial rule in some cases, an opposite trend was observed for Na/Cu(111). Figure 19.8 shows the action spectrum for the coherent amplitude of the Na–Cu stretching mode (5.5 THz). The carrier density in bulk Cu is estimated numerically and its variation with the photon energy is also depicted in Figure 19.8. Note that the photon energy dependence of the coherent amplitude coincides with that of the estimated carrier density within experimental error, which indicates initial electronic transitions in bulk drive the coherent nuclear motions. The laser excites electron–hole pairs near the surface by promoting electrons from the d band to the conduction band. The creation of d band holes can modulate the bonding charge density of the overlayer, either through the screening of the holes or by changing the state occupations. In addition, photoinduced hot electrons can fill the partially occupied Na-induced surface states just above

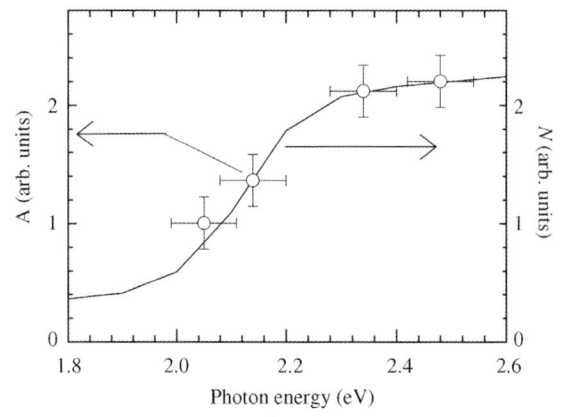

Figure 19.8 Action spectrum for the initial amplitude (A) of the 5.5 THz component. The curves drawn are the relative number of excited carriers (N) within a distance from a surface of 1000 nm normalized at 2.25 eV [34].

the Fermi level and the higher unoccupied surface states. Alternatively, electrons in the Na-induced surface states can undergo Auger recombination with photoinduced d-band holes. These changes in the occupations of the Na-induced surface states directly affect the Na–Cu bonding density and drive the coherent nuclear motions.

19.4.4
Mode Selective Excitation of Coherent Surface Phonons [37, 38]

For the coherent control of reactions at surfaces, the manipulation of adsorbate motion is essential. Cs/Pt(111) is a suitable system which provides us with a good opportunity to test whether or not we can excite preferentially one of the two modes whose frequencies are very close to each other by using tailored laser pulses. We have demonstrated the mode-selective excitation of coherent surface phonon modes on Cs/Pt(111) by synthesized femtosecond pulse trains.

Figure 19.9 shows the Fourier-transformed spectra of a time-resolved SHG trace taken from 0.27 ML Cs/Pt(111) by multiple pulse excitation with various pulse rates. As has been discussed in Section 19.4.1, the time-resolved SHG traces from Cs/Pt(111) contain two contributions of coherent surface phonons; the Cs–Pt stretching mode at 2.3 THz and the Pt surface phonon mode at 2.6 THz. According to the theoretical analysis of the multiple pulse excitation, the electric field of a femtosecond pulse train acts as a frequency-domain filter with a power spectrum of the pump field. In the case of the pulse train with the 2.3 THz rate, which coincides with the Cs–Pt stretching mode frequency, the corresponding Fourier spectrum shows a strong peak at the same frequency as that of the pulse envelope and the dip due to the surface phonon mode is absent. When the repetition rate is tuned to 2.9 THz, the contribution of the Cs–Pt stretching mode is negligible and only a peak at 2.7 THz appears, which corresponds to the Pt surface phonon mode. It has been

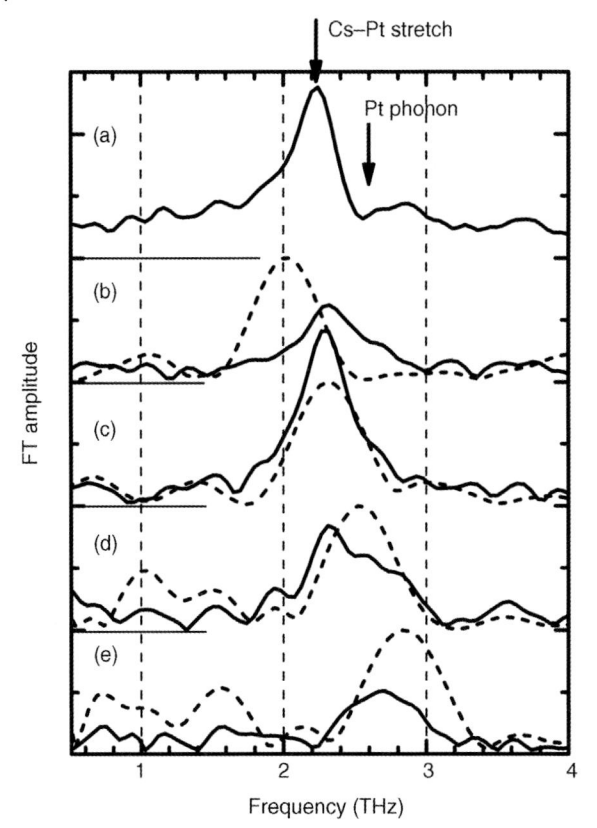

Figure 19.9 Fourier-transformed spectra of the oscillatory parts of time-resolved SHG traces obtained by varying the repetition rate (solid curves). Trace (a) is obtained by single pulse excitation. The repetition rate was tuned to (b) 2.0, (c) 2.3, (d) 2.6, and (e) 2.9 THz. The Fourier spectra of the excitation pulse trains are shown with dashed curves for each case. [38].

demonstrated that the relative amplitude of the Pt surface phonon modes can be enhanced by a factor of 4 as compared to the case of the single pulse excitation by using suitably tuned pulse trains [38].

19.5
Concluding Remarks

In this chapter we have surveyed recent experimental progress on the investigation of ultrafast nuclear wavepacket dynamics at surfaces. Nuclear (or vibrational) wavepackets of adsorbates are excited with ultrashort laser pulses, and subsequently their evolutions are probed with surface nonlinear spectroscopy such as 2PPE and SHG. These studies provide rich information on the initial stages of photoinduced processes at surfaces. So far, the application of these techniques has been limited to systems with reversible change. When photoinduced reactions with an accumulation of the products on the surface occurs, it is difficult to carry out a time-resolved measurement. It is necessary to develop experimental techniques which enable us to make repetitive measurements even under conditions with accumulating products on the surfaces.

Acknowledgments

The author would like to acknowledge coworkers who helped with the original work discussed here: Professor Y. Matsumoto, Professor N. Takagi, Dr. D. Ino, Dr. M. Fuyuki, and Professor H. Petek. I would also like to acknowledge Grants-in-Aid Scientific Research on Priority Areas "Molecular Nano Dynamics" (432).

References

1 Dhar, L., Rogers, J.A. and Nelson, K.A. (1994) Time-resolved vibrational spectroscopy in the impulsive limit. *Chem. Rev.*, **94**, 157–193, and references therein.

2 Polanyi, J.C. and Zewail, A.H. (1995) Direct observation of the transition-state. *Acc. Chem. Res.*, **28**, 119–132.

3 Scherer, N.F., Carlson, R.J., Matro, A., Du, M., Ruggiero, A.J., Romero-Rochin, V., Cina, J.A., Fleming, G.F. and Rice, S.A. (1991) Fluorescence-detected wave packet interferometry - time resolved molecular-spectroscopy with sequences of femtosecond phase-locked pulses. *J. Chem. Phys.*, **95**, 1487–1511.

4 Bowman, R.M., Dantus, M. and Zewail, A.H. (1989) Femtosecond transition-state spectroscopy of iodine – from strongly bound to repulsive surface dynamics. *Chem. Phys. Lett.*, **161**, 297–302.

5 Wise, F.M., Rosker, M.J. and Tang, C.L. (1987) Oscillatory femtosecond relaxation of photoexcited organic-molecules. *J. Chem. Phys.*, **86**, 2827–2832.

6 Chesnoy, J. and Mokhtari, A. (1988) Resonant impulsive-stimulated Raman-scattering on Malachite Green. *Phys. Rev. B*, **38**, 3566–3576.

7 Kuett, W., Albrecht, W. and Kurz, H. (1992) Generation of coherent phonons in condensed media. *IEEE J. Quantum. Electron.*, **42**, 2434–2444.

8 Somorjai, G.A. (1994) *Introduction to Surface Chemistry and Catalysis*, John Wiley & Sons, Inc., New York.

9 Dantus, M. and Lozovoy, V.V. (2004) Experimental coherent laser control of physicochemical processes. *Chem. Rev.*, **104**, 1813–1859, and references therein.

10 Steinmann, W. (1989) Spectroscopy of image-potential states by 2-photon photoemission. *Appl. Phys. A*, **49**, 365–377.

11 Steinmann, W. and Fauster, Th. (1995) in *Laser Spectroscopy and Photochemistry on Metal Surfaces* (eds H.L. Dai and W., Ho), World Scientific, Singapore, Chapter 5.

12 Bokor, J. (1989) Ultrafast dynamics at semiconductor and metal-surfaces. *Science*, **246**, 1130–1134.

13 Haight, R. (1995) Electron dynamics at surfaces. *Surf. Sci. Rep.*, **21**, 275–325.

14 Petek, H. and Ogawa, S. (1997) Femtosecond time-resolved two-photon photoemission studies of electron dynamics in metals. *Prog. Surf. Sci.*, **56**, 239–310.

15 Ino, D., Watanabe, K., Takagi, N. and Matsumoto, Y. (2005) Electron transfer dynamics from organic adsorbate to a semiconductor surface: Zinc phthalocyanine on $TiO_2(110)$. *J. Phys. Chem. B*, **109**, 18018–18024.

16 Shen, Y.R. (1997) Wave mixing spectroscopy for surface studies. *Solid State Commun.*, **102**, 221–229.

17 Heintz, T.F., Chen, C.K., Richard, D. and Shen, Y.R. (1982) Spectroscopy of molecular monolayers by resonant 2nd-harmonic generation. *Phys. Rev. Lett.*, **48**, 478–481.

18 Yamaguchi, S. and Tahara, T. (2004) Precise electronic $\chi^{(2)}$ spectra of molecules adsorbed at an interface measured by multiplex sum frequency generation. *J. Phys. Chem. B*, **108**, 19079–19082.

19 Höfer, U. (1996) Nonlinear optical investigations of the dynamics of hydrogen interaction with silicon surfaces. *Appl. Phys. A*, **63**, 533–547.

20 Ishida, H., Mizoguchi, R., Onda, K., Hirose, C., Kano, S.S. and Wada, A. (2003) Second harmonic observation of Cu(111) surface: *in situ* measurements during molecular adsorption. *Surf. Sci.*, **526**, 201–207.

21 Hohlfeld, J., Wellershoff, S.-S., Güdde, J., Conrad, U., Jähnke, V. and Matthias, E. (2000) Electron and lattice dynamics following optical excitation of metals. *Chem. Phys.*, **251**, 237–258.

22 Shi, X., Borguet, E., Tarnovsky, A.N. and Eisenthal, K.B. (1996) Ultrafast dynamics and structure at aqueous interfaces by second harmonic generation. *Chem. Phys.*, **205**, 167–178.

23 Chang, Y.M., Xu, L. and Tom, H.W.K. (1997) Observation of coherent surface optical phonon oscillations by time-resolved surface second-harmonic generation. *Phys. Rev. Lett.*, **78**, 4649–4652.

24 Watanabe, K., Takagi, N. and Matsumoto, Y. (2002) Impulsive excitation of a vibrational mode of Cs on Pt(111). *Chem. Phys. Lett.*, **366**, 606–610.

25 Watanabe, K., Takagi, N. and Matsumoto, Y. (2005) Femtosecond wavepacket dynamics of Cs adsorbates on Pt(111): Coverage and temperature dependences. *Phys. Rev. B*, **71**, 085414-1–085414-9.

26 Petek, H., Nagano, H., Weida, M.J. and Ogawa, S. (2000) Quantum control of nuclear motion at a metal surface. *J. Phys. Chem. B*, **104**, 10234–10239.

27 Petek, H., Weida, M.J., Nagano, H. and Ogawa, S. (2000) Real-time observation of adsorbate atom motion above a metal surface. *Science*, **288**, 1402–1404.

28 Miller, A.D., Benzel, I., Gaffney, K.J., Garret-Roe, S., Liu, S.H., Szymanski, P. and Harris, C.B. (2002) Electron solvation in two dimensions. *Science*, **297**, 1163–1166.

29 Gahl, C., Bovensiepen, U., Frischkorn, C. and Wolf, M. (2002) Ultrafast dynamics of electron localization and solvation in ice layers on Cu(111). *Phys. Rev. Lett.*, **89**, 107402-1–107402-4.

30 Ino, D., Watanabe, K., Takagi, N. and Matsumoto, Y. (2005) Electronic structure and femtosecond electron transfer dynamics at noble metal/tris-(8-hydroxyquinoline) aluminum interfaces. *Phys. Rev. B*, **71**, 115427-1–115427-10.

31 Li, B., Zhao, J., Onda, K., Jordan, K.D., Yang, J. and Petek, H. (2006) Ultrafast interfacial proton-coupled electron transfer. *Science*, **311**, 1436–1440.

32 Matsumoto, Y., Watanabe, K. and Takagi, N. (2005) Excitation mechanism and ultrafast vibrational wavepacket dynamics of alkali-metal atoms on Pt(111). *Surf. Sci.*, **593**, 110–115.

33 Fuyuki, M., Watanabe, K. and Matsumoto, Y. (2006) Coherent surface phonon dynamics at K-covered Pt(111) surfaces investigated by time-resolved second harmonic generation. *Phys. Rev. B*, **74**, 195412-1–195412-6.

34 Fuyuki, M., Watanabe, K., Ino, D., Petek, H. and Matsumoto, Y. (2007) Electron-phonon coupling at an atomically defined interface: Na quantum well on Cu(111). *Phys. Rev. B*, **76**, 115427-1–115427-5.

35 Watanabe, K., Takagi, N. and Matsumoto, Y. (2004) Direct time-domain observation of ultrafast dephasing in adsorbate-substrate vibration under the influence of a hot electron bath: Cs adatoms on Pt(111). *Phys. Rev. Lett.*, **92**, 057401-1–057401-4.

36 Merlin, R. (1997) Generating coherent THz phonons with light pulses. *Solid State Commun.*, **102**, 207–220.

37 Matsumoto, Y. and Watanabe, K. (2006) Coherent vibrations of adsorbates induced by femtosecond laser excitation. *Chem. Rev.*, **106**, 4234–4260.

38 Watanabe, K., Takagi, N. and Matsumoto, Y. (2005) Mode-selective excitation of coherent surface phonons on alkali-covered metal surfaces. *Phys. Chem. Chem. Phys.*, **7**, 2697–2700.

20
Theoretical Aspects of Charge Transfer/Transport at Interfaces and Reaction Dynamics

Hisao Nakamura and Koichi Yamashita

20.1
Introduction and Theoretical Concepts

20.1.1
Introduction

Studies of molecular level charge transfer and bulk level charge transport processes have a long history [1–5]. Their fundamental mechanisms are fairly well described based on widely accepted theoretical models such as band theory [1], (classical) Boltzmann kinetics [5], Marcus theory [6] and so on. Recent development of computer technologies and theoretical methods allows us to perform large scale simulations combined with *ab initio* calculations [7] which can provide detailed analyses and theoretical data at the quantitative level for the above homogeneous charge transfer/transport (CTs) phenomena. Therefore interest in CTs has now shifted to processes on the nanoscale, and several state-of-the-art experimental techniques, for example, scanning tunneling spectroscopy (STS) [8, 9], time-resolved laser spectroscopy [10–12], and two-photon photoemission (2PPE) [10–14], shed light on the characterization of nanoscale CTs. As an example, one can see recent progress in molecular conducting junctions, which opens the door to the next generation electronic devices, "moletronics" [15–18]. In other words, it is only one application in the area of "nanoscale *heterogeneous* CTs", which is also the key to the elementary step in many other important fields of surface chemistry such as photocatalysis [19, 20], electrochemistry [21–23], solar photoconversion [24], and STM-chemistry [25–30].

Atomic level simulations and electronic structure calculations are necessary to understand the mechanisms and physical properties for these molecule/bulk interfacial CTs. However, unfortunately, a simple extension of standard theoretical models for homogeneous CTs is not always useful. While there are several difficulties in developing theoretical models (ideally possible to combine *ab initio* techniques) for interfacial CTs, the fundamental difficulties result from (i) the total system size often being (semi-) infinite (ii) the coexistence of locality and nonlocality in excited electron (charge) dynamics, and (iii) non-adiabaticity between electrons and nuclear motions during the CTs processes. In this chapter we introduce our recent theoretical studies for developing *ab initio* methods and a few practical applications, which are mainly focused on heterogeneous charge transport processes, as well as a brief outline of several fundamental theories.

Before proceeding to details of the theory and practical applications, we will try to give the basic concepts required to understand heterogeneousness in interfacial CTs with a few typical examples. To do so, we use the two terminologies "charge transfer" and "charge transport" distinctly throughout this review. Then we show that interfacial charge transport is as an important key step in surface chemistry. Detailed definitions are discussed in Section 20.1.3. In addition, we emphasize that these two terminologies are reserved to indicate dynamical processes of charge (electron or hole) in this chapter. Charge transfer, such as (charge) donation or back donation in the ground (thermal equilibrium) state, is termed *static* charge transfer.

20.1.2
Molecular Orbital Theory and Band Theory

First, we introduce the two basic frameworks of electronic structure theory, molecular orbital (MO) theory and band theory. Electronic structure theory can provide calculation of the total energy of a system. In addition, MO and band theories give one-electron states, which are often used to represent electron (hole) dynamics.

In MO theory, there are several methods to calculate the total energy, for example, Hartree–Fock (HF), Møller–Plesset perturbation (MP), configurational interaction (CI), and multi-reference self-consistent field (MCSF) [7]. The latter two methods are multi-configurational theory, which allows us to calculate excited states of the total electronic states with high accuracy. Comparing the total energy and total electronic wavefunction, the one-electron state is not always defined strictly, in particular, for the multi-configurational methods. However canonical MOs or natural MOs, (they are equivalent in the HF method), are often used to represent one-electron states due to their convenience. In this case an excited electron (hole) can be modeled by unoccupied (occupied) MOs. Density functional theory (DFT) is also a popular tool within the Kohn–Sham (KS) framework due to recent improvement in exchange-correlation (XC) functionals [31]. The HF and KS-DFT belong to mean-field level theory, and the KS orbitals are used as one-electron states in a similar way to MO theory. Usually the MOs are expanded in atomic orbitals (AOs), which are localized on each atom: thus an application of MO theory assumes the finiteness of the system implicitly, and the total number of electrons in the system is fixed with an integer value. The use of MOs under the above finite size condition is suitable if the charge dynamics maintains locality such as hopping from site to site. (As an example one can consider homogeneous charge transfer reactions in molecules or in solution.)

Although a nanoscale interface cannot be treated as a (small) finite size essentially, the total energy calculation by the MO theory is often adopted as a "cluster model" because of the locality of the chemical bond. For the same reason, the resulting density matrix in the focused region (interface) may be a good approximation for the real density matrix of the (infinite/semi-infinite) system if one treats the ground (or thermal equilibrium) state. The cluster model combined with highly accurate methods such as CI, MCSCF, and so on, is often applied to calculate the potential energy slope, not only for the ground state but also for excited states. However, the implicit assumption of locality sometimes leads to serious errors, in particular, for excited states. The model is not always valid to describe interfacial CTs.

Band theory is adopted to calculate bulk (periodic) systems [1]. Instead of AOs, the Bloch basis is used to expand the one-electron state (band state). The simplest Bloch basis is a set of plane waves, but other kinds of Bloch basis can also be adopted, for instance, a linear combination of AOs with the use of a phase factor relating to the wave vectors. While band theory can be extended to any multi-configuration method (CI, MCSCF, etc.), practical tools (program packages) are, at present, almost restricted to the DFT or HF levels. Band theory assumes a finite (integer) number of electrons only in the unit cell, and its main advantage is that the band state can express a quasi-free-electron state, that is, charge dynamics like a free-electron (hole) is naturally modeled. To perform a practical calculation for an interface (surface) system, the "slab model", which keeps the periodic boundary condition only in the directions parallel to the surface, is adopted. If the slab is sufficiently thick, both the potential energy obtained by calculation of the total energy and the density matrix at the interface are good approximations for a realistic semi-infinite system in the interface region over two dimensions. However, the slab model also assumes finiteness for the direction normal to the surface, hence similar limitations with the cluster model exist when treating charge dynamics across the interface.

As an extension of MO or band theories, several models have been proposed. For instance, to overcome problems relating to a fixed number of electrons and size truncation in the cluster model, chemical potential is introduced to electronic structure theory with a grand canonical ensemble, which is called the dipped adcluster model (DAM) [32, 33]. The DAM approach has been applied to the calculation of the potential energy surfaces (PESs) for several surface catalysis reactions. However this scheme is applicable only to static charge transfer and local electronic excitations. As an alternative approach, the density matrix formalism is a good choice to model charge dynamics, and Green's function theory is a promising route [34–36]. In this chapter, we will show the usefulness of Green's function theory and give a practical formulation, which can be combined with *ab initio* calculations, for interfacial CT processes with our recent applications [37–40].

20.1.3
Charge Transfer vs. Charge Transport

Until now, we have used two terminologies "transfer" and "transport" without giving distinct definitions. Both are based on the kinetic theory of electrons. Traditionally, the word "transfer" is used in chemistry and biology while "transport" is used to represent electron (hole) dynamics in the bulk; thus it is often seen in physics. At the nansocale interface, the two processes could be competing. Therefore it is convenient to define the two terminologies in terms of the theoretical models. The key to distinguishing between the two kinds of "charge movement" is the strength of the couplings between motions of electrons and nuclei, which are sometimes named as electron–nuclear (eN), electron–phonon (eph), or nonadiabatic couplings [14]. When the couplings are strong, movement of an electron (hole) is decoherent: thus an electron moves on each site maintaining its locality (hopping). This is defined as a "charge transfer" process. In the charge transfer, one can introduce an order parameter and define separate initial (donor) and final (acceptor) states by using it. The order parameter is taken as, for instance, the reaction coordinate or (environment) collective coordinate. Once the order parameter and resulting donor/acceptor states are set, one can apply Marcus theory by incorporating a few simplifications such as quadratic approximation for free energy profiles [6]. As an example, the charge transfer rate k_{ET} at the molecule/bulk interface can be expressed as

$$k_{ET} = \frac{2\pi}{\hbar} \int dE |V_{DA}(E)|^2 f(E) \rho(E) \frac{1}{\sqrt{4\pi k_B T}} \exp\left[-\frac{(\lambda-\Delta G(E))^2}{4\pi k_B T}\right] \quad (20.1)$$

where λ and ΔG are the reorganization energy and the difference in the free energy between the donor and the acceptor, k_B is the Boltzmann constant, T the temperature, $f(E)$ is an electron distribution function, $\rho(E)$ is the density of states of the bulk side, V_{DA} is the electronic coupling of the donor and acceptor which is a function of the electron energy E due to heterogeneousness. As stated in Section 20.1.2, the locality of charge movement enables application of MO theory. However, practical applications to heterogeneous systems are more complicated than homogeneous cases because of continuous E-dependence in the terms contained in Eq. (20.1). We note that the use of PESs obtained by the total electron wavefunctions are more useful than one-electron kinetics when both donor and acceptor states can be approximated as discrete states, and charge transfer occurs within quite a local region. However, if these conditions are fully satisfied, a system may be essentially treated as a (large) molecule rather than a nanoscale heterogeneous system.

Next we consider the "charge transport" process, which is opposite to charge transfer, that is, the nonadiabatic couplings are weak. Therefore the electron dynamics is coherent and maintains its nonlocality. As a typical example for nanoscale heterogeneous transport, we take an electron (hole) transport in a metal–molecule–metal junction, which is a fundamental unit of moletronics. To deal with metal–molecule–metal junctions, the Landauer formula is often adopted [41]. Since coherent dynamics ensures that the dynamics is quasi-free-electron-like, the mean free path length is a useful measurement for coherence. When the length of the bridge molecule is much less than the mean free path length, the contact (i.e., the bridge molecule and a few metal layers as the interface) can be a good conductor, keeping quantum confinement, and metal electrodes can be treated as ideal electron reservoirs. Recall that the quantum confinement caused by the existence of a bridge molecule provides a large difference from transport in the bulk. The Landauer formula gives the conductance G as:

$$G = G_0 T_0 = G_0 \sum_i \tau_i \qquad (20.2)$$

where G_0 is the unit of conductance, unit $e^2/\pi h$, T_0 is the transmission coefficient, and τ_i is the transmission probability for each channel. A method to estimate the transmission coefficient quantitatively at the atomic level is now desired, and Green's function theory can provide a practical scheme to carry out *ab initio* calculation for the conductance. This will be discussed in Section 20.2.

If one tries to find an analogue to Marcus theory, the order parameter of charge transport should be changed to the electron-coordinate itself. Then donor and acceptor can be set separately by introducing mediated bridge states. Assuming the super-exchange mechanism [4] (i.e., coherent tunneling through a mediated bridge) to estimate V_{DA} in Eq. (20.1), one can obtain a rough correspondence between the Marcus and Landauer expressions [42, 43]

$$G \sim (\text{const.}) k_{ET} \frac{1}{(FC)} \qquad (20.3)$$

where (FC) is the integral of the heterogeneous Franck–Condon factor,

$$\frac{1}{\sqrt{4\pi k_B T}} \int dE \exp\left[-\frac{(\lambda - \Delta G(E))^2}{4\pi k_B T}\right] \qquad (20.4)$$

Recall that Eq. (20.3) represents a quite rough correspondence, hence it will be useful only for a theoretical concept, which connects two limiting theoretical models. The two approaches should be understood as distinct models.

In real nanoscale heterogeneous systems, two processes are competing, just written as "CTs". So, which approach we select as a *starting point* to construct a model for the focused system is very important. It will depend on several factors relating to the physical properties but, ultimately, they depend on the strength of nonadiabatic couplings. As a few important factors, one can list the tunneling time at the interface or the bridge molecule, the residence time in the resonance state, and the locality of excited one-electron states [44]. The last factor relates to what type of electronic excitation triggers the CTs. Note that CTs cannot occur without electronic excitation or a change of order parameter. In the next section, we survey some typical categories of electronic excitations as well as concrete phenomena.

20.1.4
Electronic Excitation

To link the classification of electronic excitation to CTs, the important concept is locality of the one-electron (hole) state created by the initial excitation step. First let us consider electrode–molecule–electrode (E–M–E) conductance, although we already know that starting from the Landauer approach is better. To trigger the electron transport, the bias should be applied between the electrodes. Applying constant bias is equivalent to giving different chemical potentials (Fermi levels) to the two electrodes. Suppose that the applied bias is V_b, and the chemical potential in the left (right) electrode shifts to $E_F + V_b/2$ ($E_F - V_b/2$), where E_F is the Fermi level in the equilibrium (i.e., zero bias case). Then one can expect that excited electrons (holes) will be generated in the left (right) electrode by the applied bias: thus one can consider that electronic excitations trigger the charge transport through the bridge molecule. In this case the excited electrons and created holes are in bulk electrodes first (more strictly speaking, semi-infinite electrodes), hence the initially excited one-electron states should be nonlocal states. As a result, an advantage in starting from the Landauer approach is also expected from the point of view of created excited states.

As another example, we consider photochemistry on surfaces. When the interest is in a reaction (reconstruction) of adsorbed molecules (surface layers) triggered by an electronic excitation, the central problem is how an excited electron is localized into the adsorbate. For clarity, we restrict our consideration to one-electron excitation processes. In the following classification, proposed by Zhu [45], we consider four prototypes of an initial excitation step, as shown in Figure 20.1. The first three types, (a)–(c) are classified as *direct* excitation, which indicates that the excitation is a direct transition between initial and final states. In type (a), the initial state is the bulk (band) state, and the created excited state relates to an image surface state or molecular state coupled strongly to the substrate. In type (b), the initial state is similar to case (a), but the final state is a localized state on molecules or a state weakly coupled with the substrate. Type (c) is inter- (intra-) molecular excitation. Apparently it is not necessary to consider CTs in the first electronic excitation step for direct excitations. However, once the excited state is formed, it can be regarded as the initial state of a CTs process from adsorbate (surface layers) to the bulk, for example, solar cell energy conversion systems. Since the excited states in types (b) and (c) are localized and weakly coupled with the substrate, charge injection to the substrate should be close to a charge transfer process, that is, photoreaction (nuclear motion) is strongly correlated with charge injection. Therefore the Marcus approach will be a better start than the Landauer approach for construction of a theoretical model.

On the other hand, in type (d), the optical transition occurs between the bulk states. The reactive state is formed by the attachment (resonant) of the tunneling electron

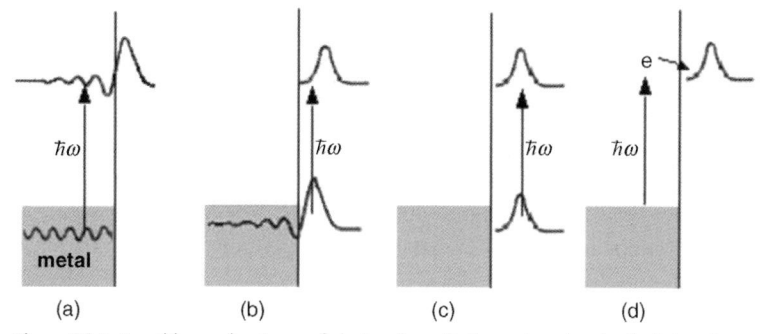

Figure 20.1 Possible mechanisms of electronic excitations at molecular/bulk interfaces.

created by excitation in the bulk, which is called a "hot electron": thus type (d) is categorized as an *indirect* excitation. Since the initial state is the bulk state, the process is charge transport. In the indirect excitation, reactive species are formed by charge transport from substrate to adsorbate, not by the electronic transition itself. If a photoreaction on a surface is triggered by the indirect excitation mechanism, one needs to introduce a model including charge transport processes, which should include the initial excitation step. Recent development of 2PPE techniques can distinguish between the mechanisms (a)–(d) for surface reactions. The major mechanism for photochemistry on metal surfaces is the indirect excitation (d). This leads to the importance of developing a theoretical model based on the Landauer approach for studying surface photochemistry.

20.1.5
Reaction Dynamics

Until now, we have focused on electron (hole) dynamics although the importance of nuclear motion was pointed out to distinguish transfer and transport. In Section 20.1.4, we took an example of correlations between photochemical reactions and CTs. Similar relations are also found in the other fields, for example, STM-chemistry. Therefore we now outline representative models focused on reaction dynamics (i.e., nuclear motions) at the surface (interface) *after* the CTs processes. If the amount of energy transfer by CTs is small, one can adopt only adsorbed molecular vibrations (phonons) as internal degree of freedoms (DoFs). However, if the energy transfer rate is sufficiently large, anharmonicity of nuclear motions should be considered. Hence, the global PESs for the internal coordinates of the reaction-center (adsorbed molecule) are required in order to deal with the reaction dynamics. Practically, reduced dimensional (typically one-dimensional) DoFs are set as the reaction coordinate to describe bond-breaking and so on, and the nuclear dynamics on the PESs is traced. Although there are many photo-induced and electron-stimulated (e.g., STM-chemistry) reactions on surfaces, the essential physics underlying the reaction dynamics is similar and modeled as an analogue to a unimolecular reaction.

When we concentrate on nuclear dynamics, it is sufficient to consider electronic structure within the reaction center, which consists of an adsorbate and a few surface atoms, that is, an "adcluster". Just as in unimolecular photoreaction dynamics, we can introduce the two-state-model for an adcluster. Before direct excitation or charge injection by indirect excitation, the nuclear configuration is governed by the PES of the ground state, which relates to the total energy of the lowest state of the adcluster. After direct excitation or charge injection, the nuclei move on the PES of the excited state and gain excess kinetic energy, which triggers a reaction. The main differences from unimolecular reaction are: (i) the PES of the excited state corresponds to the electronic state of the transient anion (or cation) while the ground state is an approximately neutral state. (ii) The transition mechanism does not necessarily correspond to the real transition between the two electronic states in the adcluster. (iii) There are, implicitly, many continuous manifolds of PESs between the above two symbolic states in the model. These manifolds represent electron–hole pair excitation states in the substrate, and fast radiationless relaxation to the PES of the ground state should be imposed. The above three differences are caused by renormalizing the existence of the substrate to the adcluster and replacing the direct excitation or charge injection in the indirect excitation mechanism with a simple transition between the two symbolic states. Point (iii) relates to electronic decay of a temporarily formed anion (cation) that is, back CTs to the substrate. In this sense, the two-state-model dynamics may be quite close to resonant Raman scattering [46–48].

Although quantitative calculation of the accurate PESs remains a difficult task (see Section 20.1.2), the two-state-model describes the essential reaction dynamic process and is useful for a qualitative understanding. When the reaction coordinate is set to the adsorbate–surface distance (one-dimension), the two-state-model is called the Menzel–Gomer–Redhead [49] and/or Antoniewicz [50] model. We refer to them as the MGR models. The MGR models are often used successfully to analyze photodesorption on metal surfaces by assuming a short residential time on the excited PES. There are several methods to simulate the quantum dynamics of the MGR models, for example, stochastic wavepacket [51], open density matrix methods [52], and so on.

We return to the relation between electronic excitation and CTs. To trigger a reaction, the amount of energy transfer must overcome a threshold of the bottleneck of the reaction. When the energy transfer rate to nuclear DoFs is large during the formation of a temporary anion (cation), the single excitation is sufficient to give the energy to overcome the threshold from the viewpoint of the two-state-model. On the other hand, multiple excitation is required if the rate per single excitation is smaller than the threshold. Since the single excitation dynamics includes one set of transitions (excitation/deexcitation) just as the resonant Raman scattering process, the dynamics is described by second-order perturbation with Frank–Condon approximation [53]. On the other hand, the multiple excitation dynamics includes plural sets of transitions: thus a higher order perturbation process is required. When the reaction is desorption, the former is called desorption induced by an electronic transition (DIET), and the latter is called desorption induced by multiple electronic transitions (DIMET) [54–58]. In order to avoid confusion, we use the terminologies DIET and DIMET even if a reaction is not desorption. Again, let us consider photoreaction by indirect excitation. In this case, the excitation and de-excitation in the two-state model relate to charge injection (to) and ejection (from) the adcluster, respectively. Therefore the charger transport and tunneling through the molecule (adcluster) correspond to a single excitation step in the model. The above consideration tells us that DIMET requires *sequential* charge transport, and the lifetime of excited nuclear motion should be sufficiently long compared to the tunneling time of an electron. Therefore the order of the perturbation to represent the DIMET dynamics is approximately the number of tunneling events, that is, the average number of the transport electrons needed to trigger the one event of the dynamics [58–60].

20.2
Electrode–Molecule –Electrode Junctions

20.2.1
Nonequilibrium Green's Function Formalism

The electrode–molecule–electrode (E–M–E) system is the fundamental unit of moletronics. While we denote the bridge part as a molecule, it can also be an atomic or molecular wire. Experimental studies have succeeded in measuring the I–V characteristics of small groups of molecules [16, 17, 61, 62]. The molecular conductance has functional properties as a device, for instance, nonlinear I–V curves from quantized conductance, negative differential resistance [63, 64], conductance increasing/decreasing by heating [65, 66]. Since it is difficult to manipulate, or specify, atomic structures for molecule–metal contacts experimentally, quantitative theoretical calculations on the charge transport of a realistic E–M–E system are highly desirable. One promising scheme is nonequilibrium Green's function (NEGF) theory [35, 67–69] combined with DFT (NEGF-DFT) [70–75]. Charge transport for an E–M–E system can be represented as the following Hamiltonian in the second quantization representation,

$$\begin{aligned} H &= H_C + H_{L_B} + H_{R_B} + H_T \\ &= \sum_{\mu\mu'}(H_C)_{\mu\mu'}d_\mu^\dagger d_{\mu'} + \sum_{\nu\nu'}(H_{L_B})_{\nu\nu'}c_\nu^{L\dagger}c_{\nu'}^L + \sum_{\nu\nu'}(H_{R_B})_{\nu\nu'}c_\nu^{R\dagger}c_{\nu'}^R \\ &+ \sum_{\mu\nu\nu'}(V_{\mu\nu}d_\mu^\dagger c_{\nu'}^L + V_{\mu\nu'}d_\mu^\dagger c_{\nu''}^R) + (H.C) \end{aligned} \quad (20.5)$$

where $d(d^\dagger)$, $c(c^\dagger)$ are one-electron annihilation (creation) operators corresponding to each site or atom. The first three terms are Hamiltonians of the central (C) region, left and right bulk electrodes, respectively. The last term, H_T, is the coupling between the C region and the electrodes. Interfacial charge transport requires an explicit treatment in the C region only. However, contributions by the connected deep bulk parts must be included implicitly as electron reservoirs. One of the great advantages in using Green's function theory is that the connected bulk parts can be normalized into the C region on equal footings by using the self-energy formalism.

Use of the time-independent Keldysh form [76] in the NEGF theory enables us to calculate one-electron (hole) density matrix of any nonequilibrium steady state. In the NEGF theory, the lesser and greater Green's functions, $G_{\mu\nu}^<(E)$ and $G_{\mu\nu}^>(E)$ are defined as follows:

$$G_{\mu\nu}^<(E) = i\int d(t-t') < d_\nu^\dagger(t')d_\mu(t) > e^{iE(t-t')}$$

$$G_{\mu\nu}^>(E) = -i\int d(t-t') < d_\mu(t)d_\nu^\dagger(t') > e^{iE(t-t')} \quad (20.6)$$

Hence, the integral of the lesser/greater Green's function is equal to the density matrix of the electron/hole. The density matrix of an electron is represented by

$$D_{\mu\nu} = \frac{-i}{\pi} \int dE G^{<}_{\mu\nu}(E). \tag{20.7}$$

Note that we incorporate the factor 2 for spin degeneracy, and we always include it without reference in this chapter. The Keldysh formalism leads to the following Keldysh–Ladanoff–Baym (KKB) equation:

$$G^{<(>)}(E) = G(E)\Sigma^{<(>)}(E)G^{\dagger}(E) \tag{20.8}$$

where $G(E)$ is the retarded Green's function, and is given as the formal solution of the Dyson equation,

$$G(E) = [E - H - \Sigma(E)]^{-1}. \tag{20.9}$$

The terms $\Sigma^{<}(E), \Sigma^{>}(E)$ and $\Sigma(E)$ are lesser, greater, and retarded self-energy terms, respectively. The word "retarded" for Green's function and self-energy is often omitted, and we adopt this omitted notation. In the present Hamiltonian system, the self-energy terms consist of the renormalized electrodes, which are denoted as L and R, and electron correlation denoted as ee

$$\begin{aligned}\Sigma &= \Sigma_L + \Sigma_R + \Sigma_{ee} \\ \Sigma &= \Sigma_L^{<(>)} + \Sigma_R^{<(>)} + \Sigma_{ee}^{<(>)}\end{aligned} \tag{20.10}$$

If electron correlation is incorporated within the DFT level, $\Sigma_{ee}^{<(>)}$ is equal to 0, and Σ_{ee} is the sum of the Hartree potential V_H and the exchange-correlation (XC) potential V_{XC}. Thus, Σ_{ee} can be implicitly included if the KS Hamiltonian is adopted for H_C, where the KS Hamiltonian is determined self-consistently by the density matrix of the nonequilibrium state.

The lead self-energy terms $\Sigma_{L/R}$ are the renormalized parts of the left and right electrodes. To represent Greens's function on the C region in the matrix form, the C region is divided into three blocks, L, R and central contact, c regions. The L and R regions should have bulk properties, and the c region is the contact. The electronic state in c can be changed by applied bias, due to induced polarization and net charge. The Green's function matrix (GFM) is expressed in AO basis as

$$\mathbf{G}_{CC}(E) = \begin{pmatrix} E\mathbf{S}_{LL} - \mathbf{H}_{LL} - \Sigma_L(E) & E\mathbf{S}_{Lc} - \mathbf{H}_{Lc} & 0 \\ E\mathbf{S}_{Lc}^{\dagger} - \mathbf{H}_{Lc}^{\dagger} & E\mathbf{S}_{cc} - \mathbf{H}_{cc} & E\mathbf{S}_{cR} - \mathbf{H}_{cR} \\ 0 & E\mathbf{S}_{cR}^{\dagger} - \mathbf{H}_{cR}^{\dagger} & E\mathbf{S}_{RR} - \mathbf{H}_{RR} - \Sigma_R(E) \end{pmatrix}^{-1} \tag{20.11}$$

where S is an overlap matrix, and H is now the KS-Hamiltonian. From now on, we adopt the terminology AO basis for both standard AOs or two-dimensional Bloch basis in terms of a linear combination of AOs. The left lead self-energy is written by using the clean surface GFM $\mathbf{G}_{L_B L_B}^{sur}$, which is constructed by the complete bulk Hamiltonian with a Direchlet boundary condition,

$$\Sigma_L(E) = (E\mathbf{S}_{L_B L} - \mathbf{H}_{L_B L})^{\dagger} \mathbf{G}_{L_B L_B}^{sur}(E)(E\mathbf{S}_{L_B L} - \mathbf{H}_{L_B L}). \tag{20.12}$$

The self-energy of the right electrode is similarly defined. How to define the lesser (greater) self-energy, $\Sigma_{L/R}^{<(>)}$, which represents the scatter-in (out) function of electrons provided by electrodes, is the central issue. Practical applications of the NEGF will be possible if the (generalized) Kadanoff–Baym ansatz is applicable [67–69, 77]. In the E–M–E system under constant bias V_b, electrodes are the electron reservoirs by means of the Landauer picture: thus they can be approximated as a non-interacting quasi-equilibrium system,

$$\Sigma_{L/R}^{<}(E) = -f_{L/R}(E)\left[\Sigma_{L/R}\left(E\pm\frac{V_b}{2}\right) - \Sigma_{L/R}^{\dagger}\left(E\pm\frac{V_b}{2}\right)\right] = if_{L/R}(E)\Gamma_{L/R}\left(E\pm\frac{V_b}{2}\right) \tag{20.13}$$

$$\Sigma_{L/R}^{>}(E) = (1-f_{L/R}(E))\left[\Sigma_{L/R}\left(E\pm\frac{V_b}{2}\right) - \Sigma_{L/R}^{\dagger}\left(E\pm\frac{V_b}{2}\right)\right]$$
$$= i(f_{L/R}(E)-1)\Gamma_{L/R}\left(E\pm\frac{V_b}{2}\right) \tag{20.14}$$

where $f_{L/R}$ are the Fermi functions for the left and right electrodes. Their Fermi levels are shifted to $\pm V_b/2$. The value E in the self-energy terms is also shifted by $\pm V_b/2$ because the bulk Hamiltonian should be shift up (down) due to the bias. The current I is formally expressed as a function of V_b:

$$I(V_b) = \frac{1}{\pi}\int dE \mathrm{Tr}[\Sigma_L^{<}(E)G_{CC}^{>}(E) - \Sigma_L^{>}G_{CC}^{<}(E)], \tag{20.15}$$

and the Landauer–Buttliker formula, which corresponds to Eq. (20.2), can be derived from Eqs. (20.8), (20.13) and (20.14),

$$\begin{aligned}I(V_b) &= \frac{1}{\pi}\int dE \mathrm{Tr}[\Gamma_L(E+V_b/2)G_{CC}(E)\Gamma_R(E-V_b/2)G_{CC}^{\dagger}(E)](f_L(E)-f_R(E))\\ &= \frac{1}{\pi}\int dE T_0(E,V_b)(f_L(E)-f_R(E))\\ &\sim G_0 T_0(E_F,0)V_b\end{aligned} \tag{20.16}$$

where the last equation is derived by the wide-band-limit (WBL) approximation. The eigenchannel τ_i is obtained by diagonalizing the transmission matrix $\mathbf{t}^{\dagger}\mathbf{t}$, where $\mathbf{t}$ is the transmission amplitude matrix and written as $\Gamma_R^{1/2}\mathbf{G}_{CC}\Gamma_L^{1/2}$. As stated above, the KS-Hamiltonian should be determined self-consistently through updating the change of V_H and V_{XC} in the NEGF-DFT. Recall that the updated part is only the c region, and the L and R blocks are fixed for the whole NEGF-SCF.

20.2.2
Efficient MO Approach

The NEGF-DFT methods to calculate charge transport in the E–M–E systems have recently been incorporated into several *ab initio* program codes with various practical models of systems [70–72, 73–78, 79]. There are two routes to construct the Hamiltonian in the C region that is, starting from the cluster model or the slab model. In addition, correct estimation of semi-infiniteness for electrodes is important for quantitative (*ab initio*) NEGF-DFT calculations, and it depends on the method of calculation of surface GFMs. The cluster model is flexible for incorporation into various well-established quantum chemistry programs, hence it is convenient to combine other methods in MO theory.

However, there are two disadvantages of the cluster model. The first is that modeling of the surface is somewhat artificial, and it is difficult to reproduce a suitable external potential that eliminates artificial surface effects. This could also lead to unphysical waveguide effects [40, 80], which is particularly important to predict energy-dependent terminal current or sensitive quantities against bias, for example, "inelastic electron tunneling spectroscopy" (IETS) signal, as shown in the section 20.2.4. The second is the difficulty in reproducing bulk properties by applying the same level cluster approximation to the electrodes. The alternative route is to adopt the slab model. So long as the slab is set sufficiently thick, it is free from the above two problems. However, the thick slab requires an estimation of large GFMs and, hence, the computational cost is very high. Furthermore, the numerical singularity of self-energy matrices and the instability for integration of the lesser Green's function lead to an increase in the computational cost due to the difficulty of convergence in the NEGF-SCF. More recently, we have proposed the "efficient MO approach" [38], which is intermediate between the cluster and the slab. We give a brief review of our scheme and show an application to the benzene-dithiol (BDT) molecule attached to Au(111) electrodes and an Au atomic wire on Au(001) electrodes.

Our efficient MO approach consists of new features, an embedding potential scheme, an O(N) method to calculate lead self-energy matrices, and the use of perturbation Green's functions (PT-GFs) expanded by the "restricted MO space" in the NEGF-SCF step. When the SCF cycle of NEGF-SCF is performed, the PT-GFMs in the MO basis, which are the eigenstates of the matrix $\mathbf{H}_{CC}$, are adopted. The PT-GFMs are diagonal and represented as follows:

$$\mathbf{G}_{CC}^{PT}(E) = \mathrm{diag}\left[\left\{E - \varepsilon_I^0 - \frac{i}{2}(\Gamma_{L,I}(E + V_b/2) + \Gamma_{R,I}(E - V_b/2))\right\}^{-1}\right] \quad (20.17)$$

where the label I is the index of the MO, and ε_I^0 is the MO energy. In each NEGF-SCF step, the density matrix is evaluated as an electron occupation number of MOs,

$$d_I = \frac{-i}{\pi}\int dE \mathbf{G}_{CC}^{<}(E) \quad (20.18)$$

then it is transformed to the AO (Bloch) basis to update the Hartree and XC potentials. The use of PT-GFs in MO basis allows much easier and faster evaluation of the density matrix than calculating Eq. (20.7) in the AO basis directly. In addition, one can introduce the WBL approximation by replacing E with the equilibrium Fermi level E_F or the golden-rule type approximation by replacing with ε_I^0 in the term $\Gamma_{L/R,I}$. These simplifications lead to further computational efficiency in the NEGF-SCF procedure.

To make the NEGF-SCF step even more efficient, the "restricted MO space" idea is proposed. The idea is similar to the scheme of the complete active space (CAS)-SCF method in quantum chemistry [81, 82]. The MOs, whose occupation number should be determined by NEGF-SCF, are the only "active" MOs, and their energies cover the region close to $E_F \pm V_b/2$. The "inactive" MOs, which are core orbitals, are always fully occupied. The MOs of much higher energy than E_F are "virtual" MOs, and their electron occupations are always equal to zero. In typical cases, the applied bias is within a few volts, and the active MOs in the restricted MO space are only about 10% of the MOs in the whole MO space. Note that orbital relaxation is allowed for all MOs because the Hamiltonian is updated. The fixed values in the inactive and virtual MOs are only occupation numbers.

Although the sub-block parts, $\mathbf{H}_{LL/RR}$, are incorporated in the bulk Hamiltonian, which can be obtained by separate band calculations, the long-range potential created by the deep bulk parts is not accounted for when constructing the c region. Our starting point is the assumption that the Hamiltonian matrix $\mathbf{H}_{CC}$ obtained by clipping the KS-Hamiltonian and the density matrix resulting from a standard KS-DFT for the $\tilde{W}$ region is a good approximation of the exact subpart C of the real semi-infinite system labeled as W, so long as the $\tilde{W}$ region is set sufficiently large. Usually the $\tilde{W}$ can be defined by adding two or three outer layers to the focused C region. Then the embedding potential $\mathbf{V}_{CC}^{emb}$ can be obtained as the potential, which enforces the agreement in the C region between the KS-DFT results of the $\tilde{W}$ and the NEGF results of the zero-bias, which is sometimes referred to as EGF. Once the embedding potential is determined, it is fixed in the whole NEGF-SCF for any bias: thus the calculation for the large W system is required only once. The embedding potential includes (at least a part coming from the outer layers) a long-range potential. We proposed a concrete procedure to construct $\mathbf{V}_{CC}^{emb}$ [38], and details can be found in the literature. In Figure 20.2, a brief schematic picture and flow chart of the procedure to construct the embedding potential are given.

The evaluation of self-energy matrices is also a computationally expensive task. To incorporate both correct bulk properties and semi-infiniteness (Direchlet boundary condition), we adopted the tight-binding-layer (TBL) method proposed by Sanvito et al. [72, 83] The TBL scheme requires a converged Hamiltonian matrix for the bulk, which can be obtained easily by *ab initio* band calculations, and solutions of generalized (complex) eigenvalues for each E value for the TBL block diagonal matrix derived from the bulk Hamiltonian. Therefore, the simple use of the TBL scheme is $O(N^3)$ for each E. The basic strategy is the reverse procedure of the k-sampling in band theory, as follows. The electrodes usually have smaller unit cells (highly two-dimensional periodicity) than the whole contact. We describe the method when the electrodes have $N_{cell} \times N_{cell}$ structure with Γ point approximation as an example. Instead of calculating the self-energies of the $N_{cell} \times N_{cell}$ structure directly, one can use a Fourier expansion of the self-energies of the 1×1 structure, $\Sigma^0(\vec{k}_{//})$, as follows:

$$\Sigma_{\mu'\nu'} = \sum_{\vec{k}_{//}} \sum \Sigma^0_{\mu\nu}(\vec{k}_{//}) e^{-i\vec{k}_{//} \cdot (\vec{R}_{\mu'} - \vec{R}_{\nu'})} \tag{20.19}$$

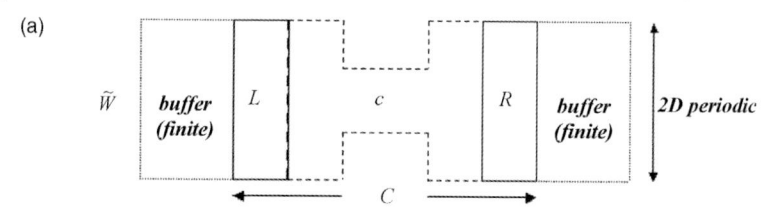

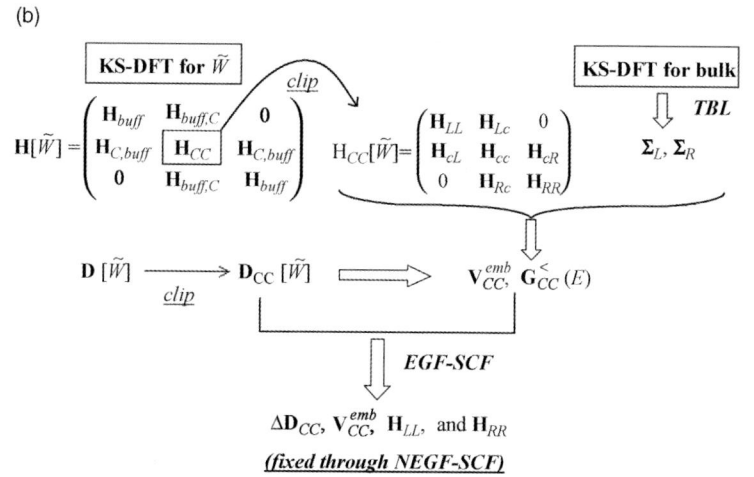

Figure 20.2 (a) Schematic picture of generic system to model the device part. (b) Flow chart of the preliminaries in the efficient MO approach. Recall that we need only the matrices in the C region after the preliminary procedure, and the elements on the buffer part are never used.

where μ' is the set of AOs of atoms in the $N_{cell} \times N_{cell}$ cell, but μ is in the unit 1×1 cell. The $\vec{k}_{//}$ vector is a wave vector parallel to the surface, and sufficient numbers of $\vec{k}_{//}$ should be sampled if the cell for Σ consists of many numbers of cells for Σ^0. Because Eq. (20.19) allows one to calculate two-dimensionally wide electrodes by using only the self-energy matrices of the smaller unit cell, it provides the $O(N)$ algorithm.

20.2.3
Ab Initio Calculations: Single Molecular Conductance and Waveguide Effects

As examples, we apply the presented efficient MO approach to the Au(111)/BDT/Au (111) system and atomic Au wire with Au(001) electrodes. The former is the first observed molecular conduction system [17] and one of the most popular benchmark systems for theoretical studies. First, we perform *ab initio* calculations for it. From now on, we set E_F to 0.

We adopted PBE XC functionals [84] for the DFT. The details of the parameters to carry out the *ab initio* (DFT) calculations, for example, basis set and pseudopotentials,

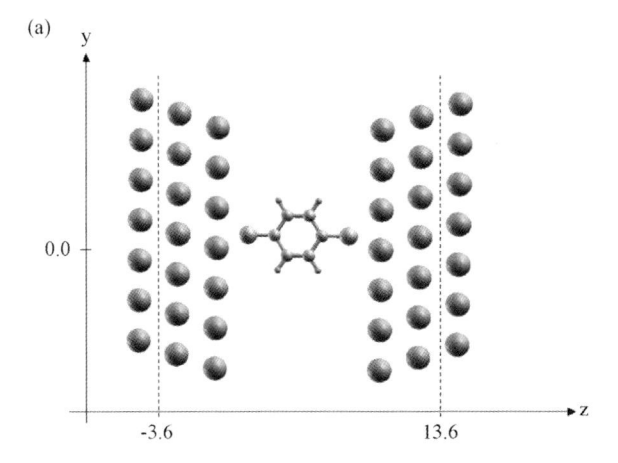

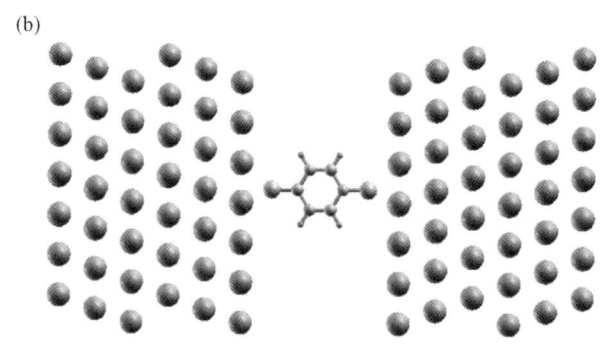

Figure 20.3 Modeled system for BDT on Au(111) electrodes. (a) The C region and (b) the $\widetilde{W}$ used to define the embedding potential. The largest spheres denote the Au atoms, and the BDT molecule is at the center. The y and z axes are inserted in (a). The dashed lines represent the boundaries of the c and L/R regions.

are given in Ref. [38]. The C region was modeled using three Au layers for each side and a single BDT molecule, as shown in Figure 20.3a. The c region consists of the top two layers for left and right and the BDT, and the $\widetilde{W}$ region is set by adding outer three layers on each side, as given in Figure 20.3b. The BDT is on the hollow sites, and the distance between the S atom and the surface is 1.97 Å. The (111) surface plane was constructed as a 4×4 (i.e., N_{cell} is set to 4), and the z axis is set as the direction of transport. The y axis is the $[1\bar{1}0]$ direction.

The contour plot of the mean-field potential in the C region for the EGF is shown in Figure 20.4, and the calculated induced potential by the applied bias, where V_b is 1.6 V, is given in Figure 20.5. We found that the embedding potential (and the resulting mean-field KS potential under the zero bias) converged rapidly for more than three additional outer layers of the W system. The I–V curve is given in Figure 20.6 and shows a rapid increase (decrease) at $V_b = \pm 0.75 \sim \pm 1.00$ V. Our I–V characteristic is in good agreement with other theoretical data, and gives

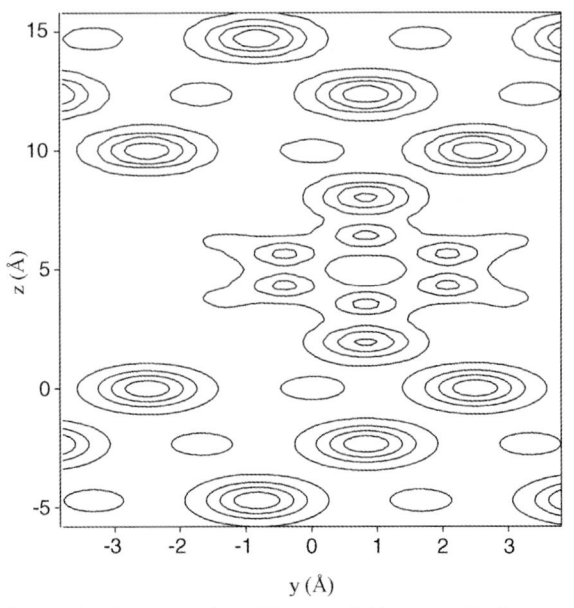

Figure 20.4 A contour plot of the mean-field potential in the C region obtained by the standard KS-DFT for the W region. The contours are separated by 20 eV. The x coordinate is fixed at the center of the system, and the (y,z) positions of the S atoms are set to (0.833, 1.97) and (0.833, 8.098).

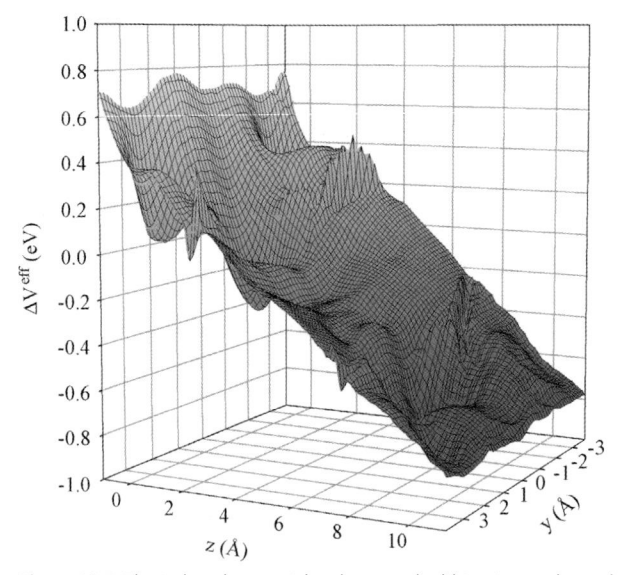

Figure 20.5 The induced potential with an applied bias (i.e., voltage drop) of 1.6 V.

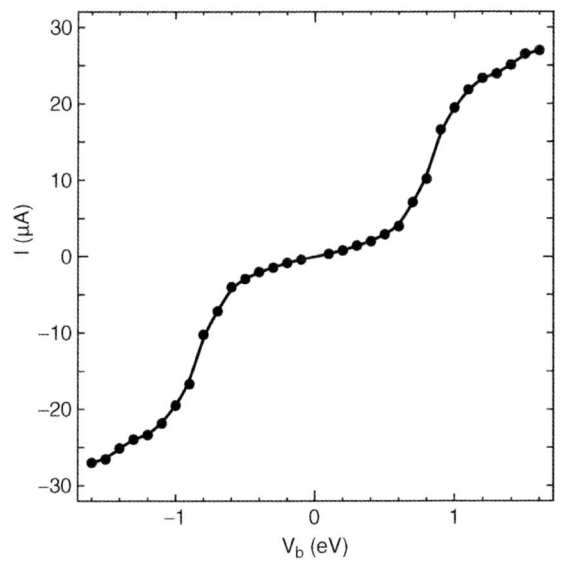

Figure 20.6 The I–V characteristic calculated by the present NEGF-DFT method.

qualitative agreement with experiments. In the present range of the applied bias, polarization or net charge in the contact is not important. However, the absolute value of the conductance is overestimated by more than one or two orders compared with the experiments of Reed et al. [16, 17] The large quantitative disagreement of the absolute value of conductance in organic molecular conductance between theory and experiment remains an open question and is debated [85].

In Figure 20.7, the induced Mulliken charges are shown as functions of V_b, for the average of the first Au layers, the average of the nearest three Au atoms of the hollow site, and the S atoms for the left and right sides, that are denoted as Au(L1), Au(R1), Au(LH), Au(RH), and S(L), S(R), respectively. Here we set the neutral charge to zero for each case. In the case of zero bias, BDT was negatively charged due to electron donation (i.e., static charge transfer) from the Au atoms nearest to the hollow site. Excess charge from the Au atoms in the left and right layers shows a clear dependence on V_b and, hence, the induced potential is dominated by the net charge on these Au atoms. Table 20.1 shows the orbital energies for the renormalized MOs (RMOs), which are defined as the eigenstates of the sub-block part of the Hamiltonian projected onto the BDT molecule, for each V_b. The labels HOMO and LUMO are those defined by a free BDT molecule. The LUMO level of the RMO is lower than the Fermi level, just as shown in Table 20.1. The occupation numbers of electrons in the HOMO, LUMO, and LUMO + 1 for the RMOs are 1.87, 1.53, and 0.31, respectively. Thus, LUNO + 1 provides a small contribution to the conduction by coupling with the broad self-energies, although the RMO energy is much higher than the bias. Compared to the free BDT, the LUMO is almost occupied, which relates to the rise in the I–V curve described above. It can be seen that the fluctuation in the RMO energies is not so strongly a function of bias in the present range.

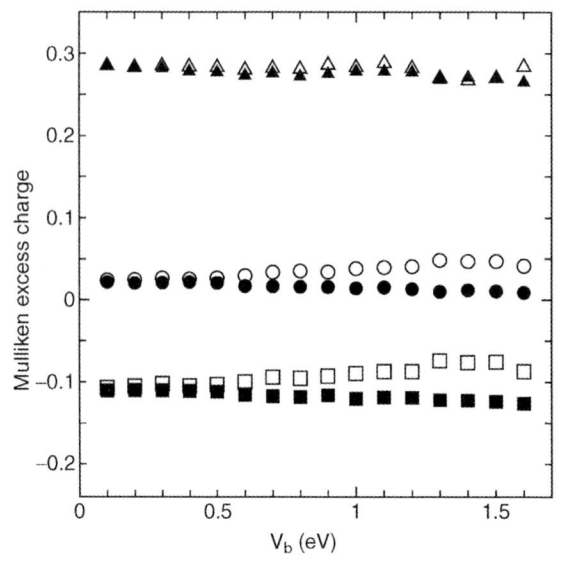

Figure 20.7 The Mulliken excess charge (in units of electron charge) on each atom (or atom average) as a function of V_b. The filled and open triangles denote S(L) and S(R), respectively. The filled and open circles denote the average of the Au atoms nearest to BDT for the left and right leads, that is, Au(L1) and Au(R1), respectively. The data for the Au(LH) and Au(RH) denoting the Au atoms of the left and right electrodes on the first layer are plotted as filled and open squares, respectively.

We give a brief analysis of the computational efficiency in our approach. To expand the $\tilde{W}$ region, the required number of AOs is 2026 but the C region can be expanded with 1546 AOs. Since we focused on the applied bias range of $V_b \in [-1.6, 1.6]$ V, the active MOs should cover the energy ± 0.8 V. We found that the sufficient size of restricted MO space to get converged results was $[-2.35, 2.65]$ V. The total number of active MOs was 185. Therefore, the time-consuming step becomes more than 50 times faster because each matrix inversion requires $O(N^3)$.

Table 20.1 Energy of the restricted MO as a function of the bias.

V_b	HOMO − 1	HOMO	LUMO	LUMO + 1
0.1	−2.82	−2.81	−1.24	1.79
0.3	−2.81	−2.74	−1.22	1.80
0.5	−2.84	−2.73	−1.26	1.77
0.7	−2.82	−2.73	−1.25	1.76
0.9	−2.75	−2.64	−1.19	1.83
1.1	−2.76	−2.57	−1.18	1.18
1.3	−2.80	−2.61	−1.23	1.79
1.5	−2.77	−2.54	−1.21	1.82

The units are volts for the bias and eV for the energies. The equilibrium Fermi level is set to 0.

Next we calculated the Au atomic wire system, which consists of six atoms in the wire and 3 × 3(001) electrodes [40]. Only the top apex layer on each side is taken as 2 × 2. To analyze the contributions of the periodic boundary condition for the direction normal to the transport direction, we introduced two models, model (I) and (II) for the Au(001)/Au$_6$/Au(001) system. Model (I) is entirely the 2D periodic system, that is, both the C region and the self-energy matrices are calculated by a Bloch-type AO basis maintaining the 2D periodic boundary condition. To incorporate the periodic boundary condition, we adopted Γ point sampling. In model (II), the lead self-energies are the same as those for model (I), but the C region is 1D nonperiodic. Since direct interactions between the wires in the neighboring cells are negligible, model (I) corresponds to a single wire and (semi) infinite bulk as the physical situation. The relating physical situation is that the wire attached on the finite 3 × 3 cross-section is then connected to the (semi-) infinite bulk. The schematic diagrams of the above physical situations for models (I) and (II) are given in Figure 20.8.

The calculated $T_0(E)$ in the model (I) is given in Figure 20.9 a. It is almost constant in the energy range [−0.2,0.2] eV and has the value 1.0 due to conduction of the 6s electron of Au. Our result of "single open eigenchannel" transmission agrees with the experimental results as well as the previous theoretical studies [66, 86]. $T_0(E)$ for model (II) is presented in Figure 20.9b. One can find the oscillation structure of $T_0(E)$ in the given energy window; the minimum value is about 0.7. The position of the Fermi level gives a value close to the bottom of $T_0(E)$. Recall that the differences between models (I) and (II) are periodicity for only the C region, and the self-energies are common. Since the unit cell size 3 × 3 is sufficient to avoid interaction between chains (and apex parts) in the neighboring cells, the local electronic structure in the contact is similar for each model.

The large fluctuation of $T_0(E)$ is caused by the "waveguide effect", which is a result of quantum confinement, that is, interference of transverse modes of one-electron wavefunctions. Another example of waveguide effects has been studied by Ke et al. with detailed analyses [80]. They concluded that the properties obtained by the integral of E, such as I–V curves are insensitive to the waveguide effects due to washing out of fluctuations during integration. However, their analysis was restricted to tunneling current (i.e., small $T_0(E)$) cases. When the system is conductive (e.g., a metallic wire), or the focus is on more sensitive quantities for the bias (e.g., IETS signals), unphysical fluctuation can lead to more serious errors. The waveguide effect is a clear examples of the importance of the correct models for both the C region and the electrodes. In particular, one needs careful modeling and checking of the results when the cluster model is adopted.

20.2.4
Inelastic Transport and Inelastic Electron Tunneling Spectroscopy

The interaction between electrons and the bridge molecular vibrations plays an important role in E–M–E junctions, and its effect on the change of current caused by energy dissipation from eN interactions is relevant experimentally [8, 65, 66]. Inelastic transport affects device characteristics and Joule heating, which ultimately

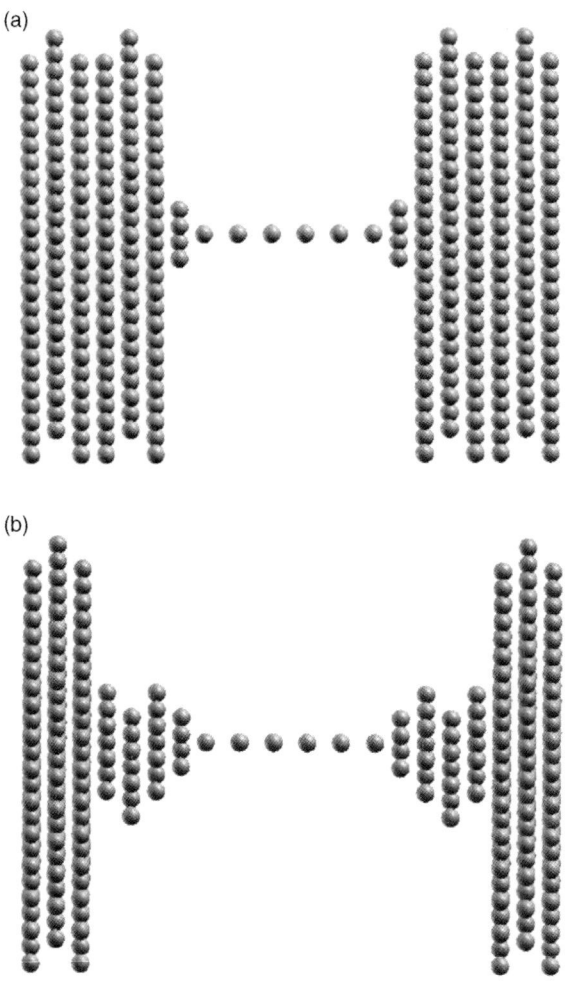

Figure 20.8 The schematic figures relating to periodic and non-periodic boundary conditions for the Au wire with Au(001) electrodes. (a) and (b) relate to model (I) and (II), respectively, see the text.

controls the mechanical stability of created devices. Another interesting function of inelastic transport is an electron-stimulated reaction, which is the fundamental process of STM-chemistry, and it is of promising use in nanotechnology [8, 27–30, 87]. The inelastic currents also provide an experimental technique of vibrational spectroscopy, IETS [88, 89]. The IETS signal is defined as the second derivative of the current (first derivative of the conductance) to the bias, and its usefulness has been successfully demonstrated.

When the molecule is rigid for the applied bias, approximations of harmonic motions for nuclei and linear couplings with electrons will be sufficient. Then we can adopt the (nonlocal) Holstein model [2], and the total Hamiltonian is expressed by

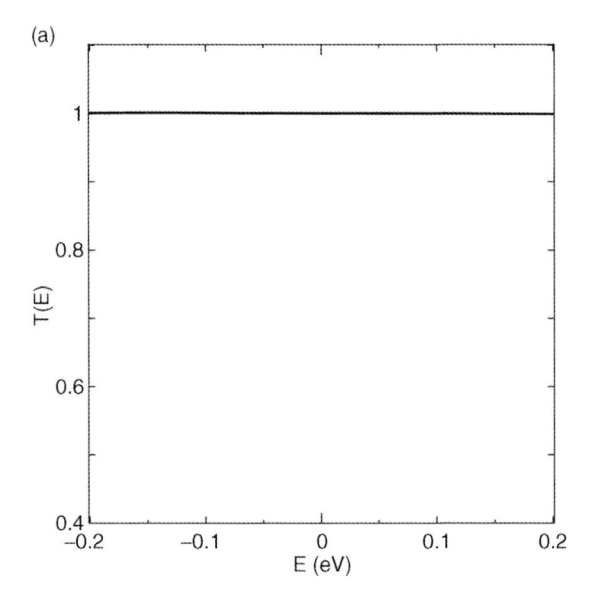

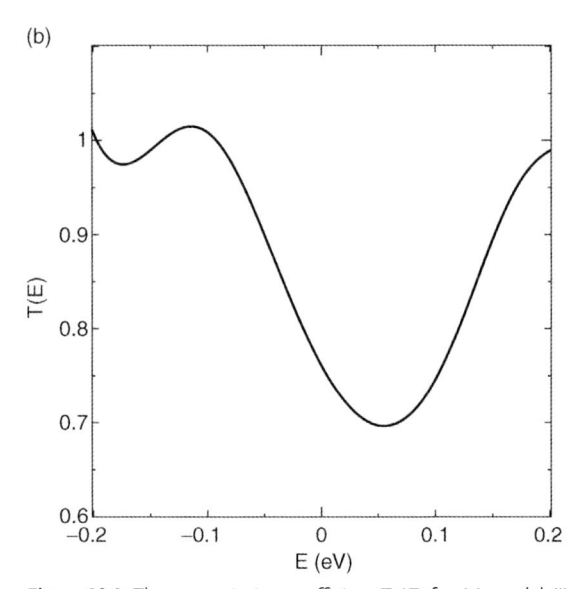

Figure 20.9 The transmission coefficient $T_0(E)$ for (a) model (I) and (b) model (II).

adding the two terms H_{eph} and H_{ph} to Eq. (20.15) as follows:

$$H = H_C + H_{L_B} + H_{R_B} + H_T + H_{\text{eph}} + H_{\text{ph}} \tag{20.20}$$

where H_{eph} and H_{ph} are the eph interaction term and the phonon Hamiltonian, respectively, and are written as

$$H_{\text{eph}} = \sum_{\alpha} M^{\alpha}_{\mu\mu'} d^{\dagger}_{\mu} d_{\mu'} (b^{\dagger}_{\alpha} + b_{\alpha}) \tag{20.21}$$

$$H_{\text{ph}} = \Omega_\alpha b_\alpha^\dagger b_\alpha \qquad (20.22)$$

The term $M_{\mu\mu'}^\alpha$ is the eph coupling constant, and b_α is the annihilation operator of the mode α, whose frequency and normal mode coordinate are represented by Ω_α and Q_α, respectively. The sites for electrons $\mu(\mu')$ coupled with phonons are restricted to the C region or a subpart of C. The focused modes should be sufficiently localized on the molecule in term of their definition. Practically, these internal modes can be calculated by means of a frozen-phonon approximation, where displaced atoms are atoms in the c region (or its subpart) denoted as a "vibrational box" though a check for convergence to the size of the vibrational box is necessary [90].

In general, the time to pass through the molecule is fast in the E–M–E system (in particular, when the electrodes are metal) so long as the bridge molecule is not too large; thus eph couplings are sufficiently weak (thus inelastic "transport", not "transfer"!). In other words, the transient anion is short-lived. In this case, perturbation expansion in the term of the HF diagram can be adopted to incorporate eph interactions into the NEGF formalism [91, 92]. If the coupling is strong and the transient anion is long-lived, the perturbation theory breaks down. In this case the present HF diagram expansion is also not suitable though the NEGF formalism is formally applicable.

The eph couplings with the mode α can be renormalized by eph self-energies from an HF diagram as follows:

$$\Sigma_{\alpha:\text{eph}}^<(E) = \frac{i}{2\pi}\int d\omega D_\alpha^<(\omega) \mathbf{M}^\alpha \mathbf{G}_{CC}^<(E-\omega)\mathbf{M}^\alpha \qquad (20.23)$$

$$\Sigma_{\alpha:\text{eph}}(E) = \frac{i}{2\pi}\int d\omega \{D_\alpha^<(\omega)[\mathbf{M}^\alpha \mathbf{G}_{CC}(E-\omega)\mathbf{M}^\alpha] + D_\alpha(\omega)[\mathbf{M}^\alpha \mathbf{G}_{CC}^<(E-\omega)\mathbf{M}^\alpha]$$
$$- D_\alpha(0)\mathbf{M}^\alpha \text{Tr}[\mathbf{M}^\alpha \mathbf{G}_{CC}^<(\omega)\mathbf{M}^\alpha]\} \qquad (20.24)$$

where $D_\alpha^<$ and D_α are lesser and retarded Green's functions of the phonon, determined by the KKB and Dyson equations, respectively. Since the equations of the electron and phonon Green's functions are coupled, one can solve them self-consistently combined with the Born approximation (self-consistent Born approximation: SCBA). The details of the SCBA and expressions for lesser/retarded Green's functions and self-energies for phonons are given in the literature [91, 93, 94]. The SCBA includes high order perturbation terms; thus multiple electron–phonon scatterings are partially incorporated. However a single electron–phonon scattering is most important when the SCBA is a good approximation. In this case, the lowest order expansion (LOE) by M^α (i.e., second order in M^α) is often valid, and it is much more convenient and practical than the SCBA for *ab initio* calculations [86, 94, 95]. We only show the resulting LOE expressions for a system constructed from symmetric electrodes for simplicity but it is extendable to a non-symmetric system [40].

In the LOE, the current can be divided into the three terms, I^0, δI^{el}, and I^{inel}, which are ballistic, elastic correction and inelastic terms, respectively. The last two terms are effects of inelastic transport. The elastic correction term relates to virtual phonon excitation and background scatterings caused by eph interactions. The corresponding terminal currents are given as follows:

$$i^0(E) = \frac{1}{\pi} T_0(E)(f_L(E) - f_R(E)) \tag{20.25}$$

$$\delta i^{el}(E) = \sum_\alpha \delta i^{el}_\alpha(E) = \sum_\alpha \frac{1}{\pi}\{T^{ec}_\alpha(2N_\alpha + 1)(f_L - f_R)$$
$$+ T^{ecLR}_\alpha(f_L - f_R)(f_L + -f_{L-} + f_{R+} - f_{R-})\} \tag{20.26}$$

$$i^{inel}(E) = \sum_\alpha i^{inel}_\alpha(E) = \sum_\alpha \frac{1}{\pi} T^{in}_\alpha\{2N_\alpha(f_L - f_R) + f_R + (1-f_L) + f_L(1-f_{R-})\} \tag{20.27}$$

where $f_\pm$ represents $f(E \pm \Omega_\alpha)$ for the Fermi functions. N_α is the nonequilibrium phonon distribution, which is a function of V_b, temperature T, and broadening parameter η. N_α is also determined by the LOE framework. The parameter η represents the dissipation to bath phonon in the electrodes. When dissipation to bath modes is zero (i.e., the electron energy is completely transferred to molecular motions), η should be set to 0 and called the "externally undamped limit". In contrast, if the phonon modes in the molecule are in completely thermal equilibrium by coupling with the bath, η is infinite, and the resulting N_α should be set to the Bose–Einstein distribution, N_{BE}. This is called the "externally damped limit". The functions in Eqs. (20.26) and (20.27) are then written as follows:

$$T^{ec}_\alpha(E) = 2\text{Re}\text{Tr}[\mathbf{M}^\alpha \mathbf{G}^0_{CC}(E) \mathbf{M}^\alpha \mathbf{G}^0_{CC}(E) \Gamma_R(E - V_b/2) \mathbf{G}^{0\dagger}_{CC}(E)$$
$$\times \Gamma_L(E + V_b/2) \mathbf{G}^0_{CC}(E)] \tag{20.28}$$

$$T^{ecLR}_\alpha = 2\text{Im}\,\text{Tr}[\mathbf{M}^\alpha \text{Im}\mathbf{G}^0_{CC}(E) \mathbf{M}^\alpha \mathbf{G}^0_{CC}(E) \Gamma_R(E - V_b/2) \mathbf{G}^{0\dagger}_{CC}(E)$$
$$\times \Gamma_L(E + V_b/2) \mathbf{G}^0_{CC}(E)] \tag{20.29}$$

$$T^{in}_\alpha(E) = \text{Tr}[\mathbf{M}^\alpha \mathbf{G}^0_{CC}(E) \Gamma_R(E - V_b/2) \mathbf{G}^{0\dagger}_{CC}(E) \mathbf{M}^\alpha \mathbf{G}^{0\dagger}_{CC}(E) \Gamma_L(E + V_b/2) \mathbf{G}^0_{CC}(E)] \tag{20.30}$$

where $\mathbf{G}^0_{CC}$ is the retarded GFM without eph interactions.

The ab initio calculation of inelastic transport in the LOE was also applied to models (I) and (II) of Au(001)/Au$_6$/Au(001). In Figure 20.10a and b, the change in conductance, δG, caused by inelastic transport is shown for model (I) and (II), respectively. Here we only included the highest alternative bond length (ABL) mode in the wire, and the undamped limit is assumed. The calculated value of Ω_α for the ABL mode was 128.3 cm^{-1}. The term δG is the sum of the elastic correction δG^{el}, which is the derivative of the elastic correction current to V_b, and the inelastic term δG^{inel}, which is a derivative of the inelastic current term. The term δG^{inel} is always positive mathematically from the definition, but the sign of δG^{el} depends on the system. In both models, δG^{el} is negative and dominates δG, that is, eph interactions provide the conductance-drop for the Au(001)/Au$_6$/Au(001) system.

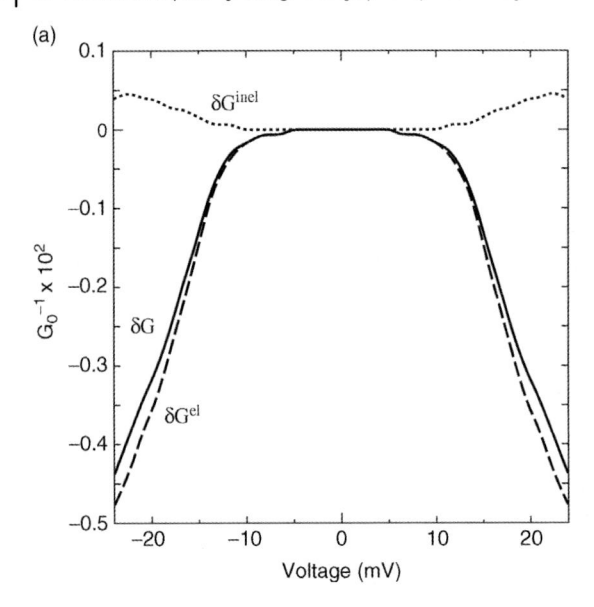

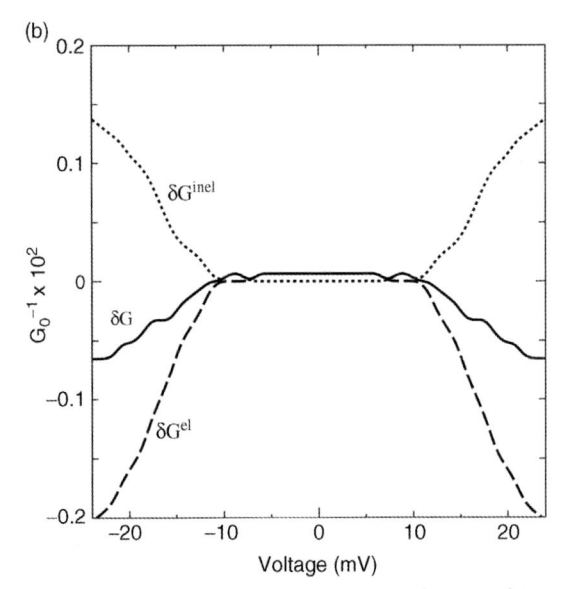

Figure 20.10 The change in conductance as a function of the voltage for (a) model (I) and (b) model (II), in (a) and (b). The total change $\frac{dI}{dV_b}$ is labeled as δG and shown by the solid line. The elastic term $\frac{d(\delta I^{el})}{dV_b}$ and the inelastic term $\frac{d(I^{inel})}{dV_b}$ are denoted as (δG^{el}) and (δG^{inel}) inset and given by the dashed and dotted lines, respectively.

The decrease in the conductance is close to linear when the bias is larger than Ω_α. Comparing with model (I), the magnitude of δG^{el} becomes small (about 1/3 of the value in model (I)) due to enhancement of δG^{inel}. The eph couplings in the wire are dominated by the local electronic structure. As a result, the difference in M^α between models (I) and (II) is small: thus the main reason for the difference is waveguide effects, that is, properties relating to inelastic transport are sensitive not only to the local geometry (e.g., adsorbed site) but also to the shape of the contact. To create molecular devices, increases or decreases in conductance by eph interactions are quite important. The LOE form tells us that the sign of $T_\alpha^{ec,ecLR}$ and their magnitude relative to T_α^{in} are key factors. Although the simple rule for a rough estimation of increase/decrease in conductance is proposed, it is not certain that the rule is general for any kind of phonon modes and applicable to any systems. Hence *ab initio* calculation in the LOE is a powerful tool to understand inelastic transport processes.

Although we introduced the Holstein model, it is not always suitable. When one focuses on vibrational excitation dynamics, the use of the Holstein model and perturbation expansion is not applicable to DIET type dynamics in the two-state mode because the linear eph coupling in the Holstein model cannot treat multiphonon excitation by single electron excitation. Readers should distinguish two perturbation concepts, where the first is a high order perturbation scheme to represent electron transitions between two (symbolic) states in the two-state model, as stated in Section 20.1.5, and the other is, as given in the present section, perturbation expansion for the self-energies in the Green's function theory. If DIET is dominant to the dynamics in the focused inelastic transport process, one needs to modify the model by including higher order coupling terms than linear couplings. To control chemical reactions on surfaces by injected electrons, a DIET type mechanism should be considered.

20.3
Photochemistry on Surfaces

20.3.1
Theoretical Model of Hot Electron Transport and Reaction Probability

Photo-induced reaction on a metal surface usually consists of several elementary reactions and it is difficult to model the whole reaction process. However, any reactions need to be triggered by electronic excitation. As stated in Section 20.1.4, the major mechanism is indirect excitation; thus we focus on modeling the indirect excitation reaction. Since desorption from the surface is one of the simplest processes and can be a prototype for other complex surface reactions, DIET or DIME are clearly the best to study [10, 48, 53, 57, 96]. In photochemistry, continuous wave or nanosecond lasers lead to DIET, where desorption increase linearly with fluence. In contrast, the DIMET process is caused by intense and short laser pulses on the picosecond or femtosecond time scale, with nonlinear dependence on fluence. Since the fluence is proportional to the number of created hot electrons in the bulk, linear and nonlinear dependences reflect whether desorption is triggered by tunneling of a single electron or sequential tunneling of multiple electrons. From the discussion in Section 20.1.5, this is consistent with the definitions of DIET and DIMET. There have been many studies of DIET/DIMET dynamics within the two-state model, where electronic excitations between the two PESs are often reduced to a two-temperature model, that is, the concept of thermal electron bath and electronic temperature is introduced phenomenologically [48].

While the electron excitation is a quite important step for the surface chemistry, the two-state model usually omits a detailed excitation (and de-excitation) mechanism in the reaction process. One of the most important issues in reaction dynamics is to estimate reaction probability, and the main features of indirect excitation are that the resulting reaction probability is a continuous function of photon energy. It is insensitive to incident direction and polarization of light [97]. In the reaction caused by the indirect excitation, the electronic transition in the two-state model should be described by electron–hole pair creation and charge transport, that is, hot electron dynamics. To account for these features, a more rigorous theoretical treatment of hot electron dynamics, in particular, quantitative estimation of heterogeneous charge transport, is necessary. Hence, we focus on the DIET under continuous wave to develop the theory [37].

To treat charge transport, the Landauer approach should be a good start, and we have seen the usefulness of the NEGF framework to perform practical (*ab initio*) calculations in the previous section. First, we set the Hamiltonian to represent an indirect DIET system. Referring to Eq. (20.5), we introduce the Hamiltonian:

$$
\begin{aligned}
H &= H_C + H_{L_B} + H_T + H_{eN} \\
&= \sum_{\mu\mu'} (H_C)_{\mu\mu'} d_\mu^\dagger d_{\mu'} + \sum_{vv'} (H_{L_B})_{vv'} c_v^{L\dagger} c_{v'}^L + \sum_{\mu vv'} V_{\mu v} d_\mu^\dagger c_v^L + (H.C) + H_{eN}
\end{aligned}
$$
(20.31)

The terms H_C, H_{L_B}, and H_T are the same as those in Eq. (20.20), but there is no right bulk part because the system is a molecule adsorbed on a surface. The C region is also divided into the two regions, L and c.

The term H_{eN} is an electron–nuclear interaction, where we use eN, instead of eph, because nuclear motion is now an anharmonic large amplitude motion. The first task is to model H_{eN}. In the direct excitation mechanism, the energy transfer between an electron and nuclear DoFs is caused by the attachment of a hot electron and formation of a transient anion: thus H_{eN} can be defined as follows:

$$H_{eN} = \eta_{eN} \left\{ \sum_{\mu\mu'} \langle \mu | \phi^{res} \rangle \langle \phi^{res} | \mu' \rangle (d_\mu^\dagger d_\mu - \langle n \rangle_\phi) \right\}$$
(20.32)

The orbital ϕ^{res} is a resonant orbital, in which a hot electron is attached to form a temporary anion. We assume that desorption is triggered by single orbital resonance to ϕ^{res} through an extension to multi-resonant levels is straightforward. The term $\langle n \rangle_\phi$ is the occupation of the electron in the ground state, and is required to keep the Pauli principle. The coefficient η_{eN} is the reactive eN coupling factor and assumed to be constant. By using the KKB and Dyson equations, we can write down lesser, greater, and retarded Green's functions formally. For simplicity, we omit the contribution of the retarded self-energy of the eN term in the retarded Green's function. To combine the DFT, we adopt the KS-Hamiltonian for H_C. In the present case, it is practically quite difficult to include electron-correlation in the nonequilibrium case; thus we always adopt the Hartree and XC potentials of the ground state. The self-energy Σ_L is the same as that in Section 20.2.1. From Eq. (20.32), the eN self-energy is easily obtained as follows:

$$\Sigma_{eN}(E) = i\mathrm{Im}\left\{\eta_{eN}\sum_{\mu'\nu'}\langle\mu|\phi^{res}\rangle\langle\phi^{res}|\mu'\rangle G_C(E)\langle\nu'|\phi^{res}\rangle\langle\phi^{res}|\nu\rangle\right\}\theta_h(E-E_F)$$

$$= \frac{i}{2}\Gamma_{eN}(E)\theta_h(E-E_F)$$

(20.33)

where θ_h is the Heaviside function, and we set the real part to 0, that is, the level shift of a resonant orbital by attachment of a hot electron is omitted.

To determine lesser (greater) eN self-energy, another condition is required. We introduce an assumption that is called electronic transition state theory (eTST) by analogy with transition state theory. The eTST assumes that a reaction is promoted by attachment of a transient electron to ϕ^{res} *without* any energetic threshold, and there is no recrossing of the transient electron. This leads to the following lesser and greater self-energy terms:

$$\Sigma_{eN}^<(E) = 0 \tag{20.34}$$

$$\Sigma_{eN}^>(E) = \frac{i}{2}\Gamma_{eN}(E) \tag{20.35}$$

Then we can obtain the cumulative reaction probability, $P_{cum}(E)$, as an in-flux current to the resonant orbital, which is represented by the NEGF framework as follows:

$$P_{cum}(E) = -\frac{1}{2\pi}\mathrm{Tr}\left[\Sigma_{eN}^< G_{CC}^> - \Sigma_{eN}^> G_{CC}^<\right] \tag{20.36}$$

Eq. (20.36) is a relation of (cumulative) reaction probability, that is, reaction dynamics, and hot electron dynamics. Because the rate of energy transfer is much higher in DIET than DIMET per single hot electron attachment, the eTST is suitable for a DIET process as a first approximation.

Although we have written the lesser (greater) Green's functions as functions of E, they should be also functions of the photon energy ω. The photon-energy dependence is caused by generation of photo-excited hot electrons in the bulk, and transport to the surface region. From the viewpoint of theoretical formulation, this gives a way to determine the lesser (greater) self-energy of the substrate $\Sigma_L^{<(>)}(E,\omega)$, which now depends on the photon energy and should be defined explicitly because we want to apply the NEGF-DFT calculation. To define $\Sigma_L^{<(>)}(E,\omega)$ we introduce an *ad hoc* extension of the Kadanoff–Baym ansatz,

$$\Sigma_L^<(E,\omega) = f(E,\omega)\left(\Sigma_L^\dagger(E) - \Sigma_L(E)\right) \tag{20.37}$$

The function $f(E,\omega)$ is the pseudoequilibrium distribution of photo-excited hot electrons, and we assume its existence. Introducing Eq. (20.37) means fictitious partitioning of the scenario, that is, (i) electron excitation in bulk, these electrons are called primary electrons, (ii) transport of excited electrons to the surface, including generation of secondary electrons by electron–electron scattering during the transport; and (iii) tunneling of hot electrons to the resonant orbital, that is already formulated. The pseudoequlibrium distribution is far from the Fermi function (this is trivial because we treat photo-excitation!), and written as

$$f(E,\omega) = (\text{const}) \frac{A(\omega)[N_1(E) + N_2(E)]}{\text{Tr}[\text{Im}G^{\text{sur}}_{L_B L_B}(E)]} \tag{20.38}$$

where $A(\omega)$ is the absorbance of the substrate, which can be calculated by the Fresnel formula using a complex refractive index. The terms N_1 and N_2 are the numbers of primary and secondary electrons reaching the surface region. To calculate N_1 and N_2, we employed semi-classical transport theory developed by Berglund and Spicer [98]. The details can be found in Ref. [37]. Now we arrived at the final formula for the reaction probability $P(\omega)$, which forms a connection between two processes, i.e., between the charge transport and reaction dynamics described by the two-state-model,

$$\begin{aligned}P(\omega) &= \int_{E_F}^{E_F+\omega} dE P_{\text{cum}}(E,\omega) \\ &= \int_{E_F}^{E_F+\omega} dE \left\{ \frac{1}{\pi} \text{Tr}\left[\Gamma_{\text{eN}}(E) G_C(E) \text{Im}\Sigma_L(E) G_C^\dagger(E)\right] f(E,\omega) \theta_h(E-E_F) \right\}\end{aligned} \tag{20.39}$$

Recall that Eq. (20.39) can be estimated by *ab initio* NEGF-DFT calculations.

20.3.2
Photodesorption Mechanism of Nitric Oxide on an Ag(111) Surface

Adsorption of NO on a metal surface is important in heterogeneous catalysis [99], and the photodesorption of NO on an Ag(111) surface is an intensely studied DIET process. The action spectra (reaction probability) have been measured by several groups, and the data analyzed [100–102]. Recently, Carlisle and King have observed four different ordered dimer phases by STM, named as the α, β, γ and δ phases [103]. Furthermore theoretical calculations also support a stable dimer as an adsorbate, the photoactive species can be considered as a dimer. Recently Kidd *et al.* carried out a comparative study of the action spectra of NO and OCS photodesorption on Ag(111), and analyzed it using a phenomenological model [100]. Their analysis can be summarized as follows:

(1) Two resonance levels are located at 1.2 and 3.9 eV above E_F. The lower level is needed to reproduce a long tail in the action spectra at the long wavelength (∼600 nm). The presence of the higher level results in the rapid rise in the action spectra at 350 nm (∼3.6 eV).

(2) Because of the existence of the lower resonance level, the contribution of secondary electrons must be considered.
(3) The resonance widths of the lower and higher levels are 1.0 and 0.5 eV, respectively.

However, a resonance level, whose position is close to 3.9 eV, has not been observed experimentally, and the predicted parameters such as resonance width are results of fitting. Therefore, a more detailed analysis based on first principles is desired, and we applied our *ab initio* scheme to NO photodesorption [39].

Since, experimentally, the most stable phase is the δ phase, we performed DFT calculations to find the details of the adsorbed structure of dimer in the δ phase, which was modeled by a 2 × 4 supercell with four Ag layers. For these calculations, we adopted the PBE XC function and the polarized double zeta (DZP) level basis set. According to the STM image, the adsorption sites of the dimers can be selected as the bridge and the threefold hollow sites, that are shown in Figure 6 of Ref. [102]. We adopted two possible models for the ordered phase. One consists of bridge and fcc sites, which we called the $δ_1$ model, and the other has bridge and hcp sites and is called the $δ_2$ model. The geometry optimization was carried out for the above two models. The resulting dimer structures and related adsorption energies are given in Table 20.2. The bond axis of N−N is parallel to the surface in both cases. The height from the surface to the N atoms, which is denoted as d_h is 2.27 Å for the bridge and 2.11 Å for the dimers on the threefold site. The differences between the two N−O distances, and the angles made by the N−N and N−O axes of each dimer are negligible: thus only one set of N−O distances and N−N−O angles is given for each dimer in Table 20.2.

From now on, we focus only on the $δ_1$ model when refering to the δ phase, because the difference between the two models is equivalent to the difference between a dimer on the fcc or hcp site and is not important in the analysis. In addition, we estimated the contribution of the dimmer–dimer interaction on the stability of the δ phase. The stabilization energy due to the interaction is only 0.03 eV, that is, the interaction is so weak that formation of bonding between the dimers is negligible. Hence, the dimers on the fcc and bridge site in the δ phase can be treated separately in the hot electron transport process. To apply our NEGF-DFT approach, we set the C

Table 20.2 The calculated results of geometric structures and adsorption energies E_{ad} of dimers for the $δ_1$ and $δ_2$ models.

Site	N−N	N−O	N−N−O	d_h	E_{ad}
fcc ($δ_1$)	1.62	1.22	116.5	2.11	1.47
bridge ($δ_1$)	1.64	1.21	115.9	2.27	1.48
hcp ($δ_2$)	1.63	1.21	116.4	2.11	1.44
bridge ($δ_1$)	1.64	1.21	116.4	2.27	1.48

The heights of the dimers from the surface are also given as d_h. All distances, angles, and energies are in Å, degrees, and eV, respectively.

(a)

(b)

Figure 20.11 The atomic structures for (a) the fcc model and (b) the bridge model. The largest balls are the Ag atoms. The Ag atoms, which give the threefold or bridge sites, are denoted as bright balls. The N and O atoms are small bright and light gray balls, respectively.

region for the dimer on the fcc and bridge, which are denoted as the fcc model and the bridge model, respectively. The structure of the C region is outlined in Figure 20.11.

In Figure 20.12, the numbers of electrons reaching the surface, $N(E,\omega)$, for $\omega = 3.0$ (a) and 3.9 eV (b) are shown. Since the inter-band transition occurs in the case of $\omega = 3.9$ eV, the ratio of electrons close to E_F (recall we set it to 0) becomes large; thus $N(E,\omega = 3.9)$ decreases much faster than $N(E,\omega = 3.0)$. In Figure 20.13(a) and (b), we plot the absorption coefficient $\alpha(\omega)$ and the absorbance $A(\omega)$ as functions of ω, respectively. Since $N(E,\omega)$ is roughly proportional to $A(\omega)$, there is a possibility that $N(E,\omega)$ will also have a sharp peak, which causes an unphysical peak in the action

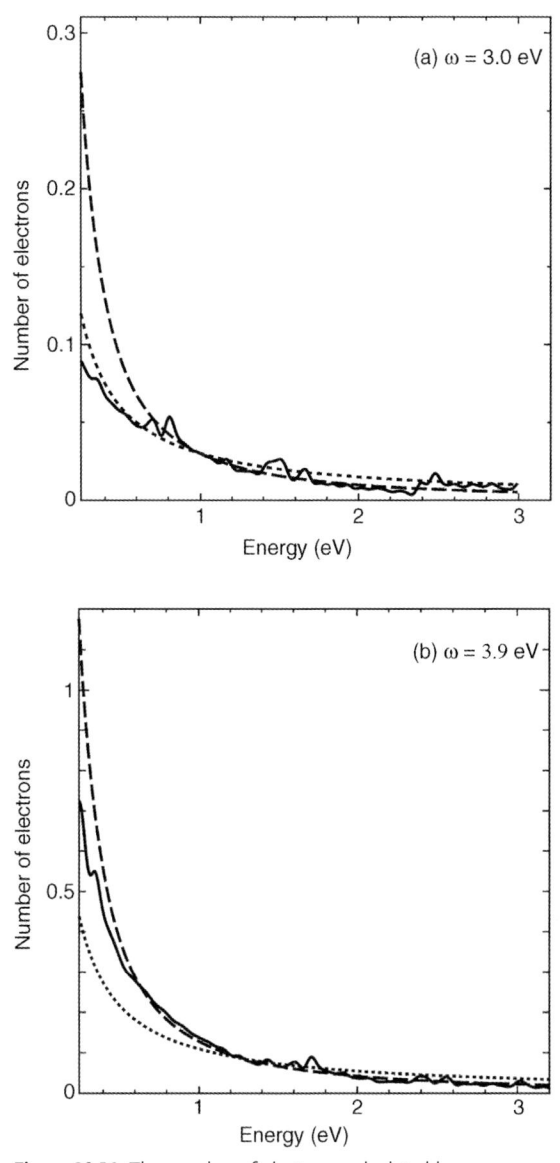

Figure 20.12 The number of electrons calculated by our theoretical model, $N(E,\omega)$ and by the empirical form, $N^{\text{eff}}(E,\omega)$, with $\omega =$ (a) 3.0 eV and (b) 3.9 eV. In each panel, the solid line is N, the dashed line is N^{eff} with $\beta = 1.6$. For comparison, the value of N^{eff} with $\beta = 1.0$ is plotted as the dotted line.

spectra. To avoid such an unphysical peak, previous theoretical works introduced a modified absorbance, $A^{\text{eff}}(\omega)$, which is termed an effective absorbance [100, 102]. In our approach, the term $N(E,\omega)$ is determined by microscopic view without any phenomenological modifications. The E-dependence in $N(E,\omega)$ includes both

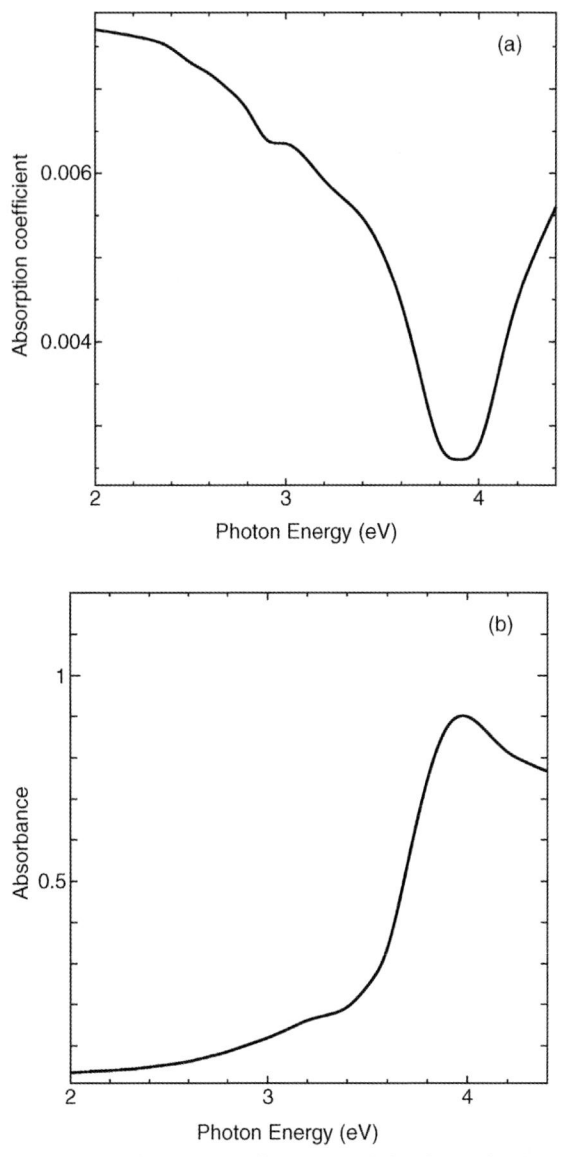

Figure 20.13 Absorption coefficient (a) and absorbance (b) plotted as a function of photon energy.

inter/intra-band transitions, and we found that it eliminated a possible unphysical peak in the action spectra.

The values of $N(E,\omega)$ used for the analysis by Kidd et al. [100] are also shown for ω 3.0 and 3.9 eV in Figure 20.11a and b, respectively. To distinguish $N(E,\omega)$ determined by Eq. (20.38), we denote it as N^{eff}, which is a phenomenological empirical form [104]

$$N^{\text{eff}}(E, \omega) \propto \frac{A^{\text{eff}}(\omega)}{\omega} \left(\frac{\omega}{E}\right)^{\beta} \qquad (20.40)$$

where a concrete form of N^{eff} is given in Ref. [102]. The value of β, which is a fitting parameter of the action spectra, can be interpreted as the contribution of secondary electrons. With increasing β, the contribution of secondary electrons becomes important. When β is set to 1, no secondary electrons contribute. In the NO/Ag (111), β was set to 1.6 by Kidd et al. [100]. For a comparison we give N^{eff} obtained by setting β to 1 and 1.6. When E is larger than 1.0 eV, our $N(E,\omega)$ and N^{eff} obtained using $\beta = 1.6$ agree well. In other words, our formulation by Eq. (20.38) can give two empirical parameters $N^{\text{eff}}(\omega)$ and β by the microscopic viewpoint without any arbitrariness. When E is close to 0, the difference between N and N^{eff} increases rapidly. If the resonant level is close to 0, this difference leads to serious disagreement for the action spectra obtained by the phenomenological model and our model. However, as shown later, the resonance level is sufficiently higher than 1.0 eV above the E_F; thus the disagreement of N and N^{eff} in the low energy region does not play any important role in the action spectra.

Now we need to introduce a resonant orbital. By calculating the free NO dimers relating to the fcc and bridge models, we found that only the LUMO + 1 level was a suitable resonant orbital for both cases. We denote these resonant orbitals as $\phi_{\text{fcc}}^{\text{res}}$ and $\phi_{\text{brd}}^{\text{res}}$, respectively. The selected two orbitals form clear π-type bonding for N–N and N–O axes, and one can consider that the resonance for NO photodesorption first triggers decomposition of a dimer to two monomers, rather than directly initiating cleavage of a bond between NO and Ag atoms. In our calculation, each absorbed dimer has only one resonant orbital as a candidate. The projected density of states (PDOS) for $\phi_{\text{fcc}}^{\text{res}}$ and $\phi_{\text{brd}}^{\text{res}}$ are given in Figure 20.14a and b), respectively. In Figure 20.14a, there is one large peak at $E = 1.98$ eV. Another small peak close to $E = 2.5$ eV can be regarded as part of the tail of the large peak, that is, not another resonant level. This structure of the PDOS is similar to Figure 20.14b. Therefore we conclude that the resonant level is single for both fcc and bridge models, as well as the resonant orbital. The resonant energy $E_{\text{brd}}^{\text{res}}$ and width $\gamma_{\text{brd}}^{\text{res}}$ are 1.97 and 0.85 eV, respectively. The difference between $E_{\text{fcc}}^{\text{res}}$ and $E_{\text{brd}}^{\text{res}}$ or $\gamma_{\text{fcc}}^{\text{res}}$ and $\gamma_{\text{brd}}^{\text{res}}$ is negligible, and they can be treated as almost the same narrow resonance positions for the photon-energy dependence, even if the photoactive species is assumed to be a dimer.

The action spectra (reaction probability), $P_{\text{fcc}}(\omega)$ and $P_{\text{brd}}(\omega)$, were calculated for 12 photon energies, in the photon energy range of 2.1–4.1 eV, for the fcc and bridge models, respectively. We estimated the rate $P_{\text{brd}}(\omega)/P_{\text{fcc}}(\omega)$ and found that it is almost 50% for all the given photon energies. Therefore the dimer on the fcc site is more reactive than that on the bridge site for photodesorption in the present model. The calculated and experimental action spectra are plotted in Figure 20.15. In addition we plotted $P_\delta(\omega)$, which is the sum of the fcc and bridge models, as the reaction probability of the δ phase. Note that in Figure 20.15, they are scaled by setting the values to unity at $\omega = 3.6$ eV for comparison. In the complete range of ω, the theoretical results are in reasonably good agreement with experiment as a function

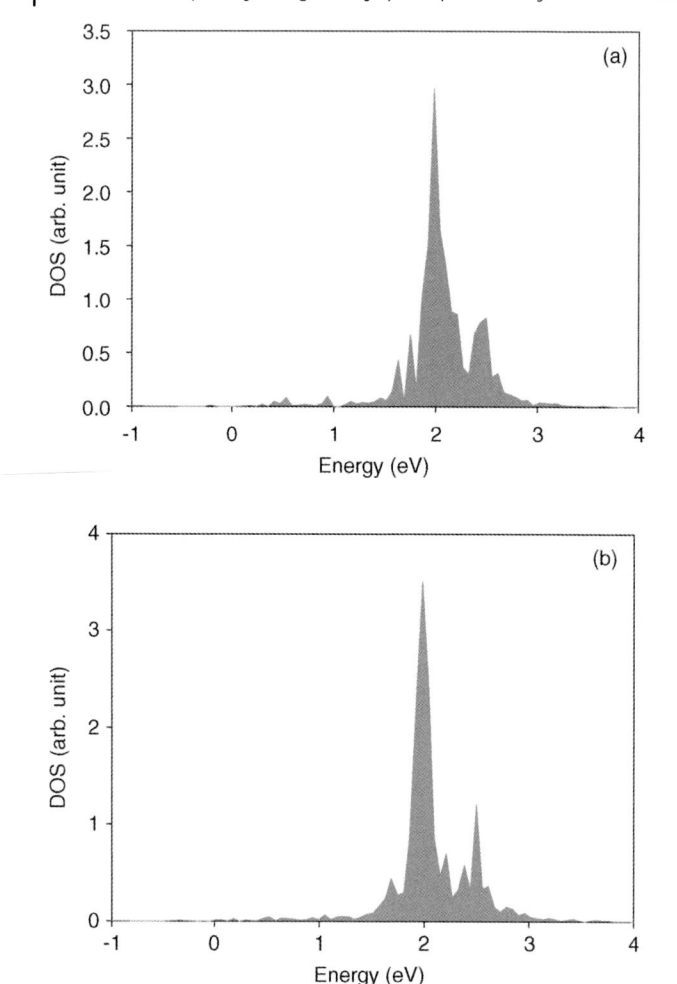

Figure 20.14 PDOS of the resonant orbital for the fcc model (a) and the bridge model (b).

of ω. In particular, the experimental features: (i) a long tail in the region $\omega \sim 2.0$ eV and (ii) a rapid increase at $\omega \sim 3.6$ eV, are perfectly reproduced.

We applied our approach under the assumption that the monomer is the photoactive species, and compared the results for the case of the dimer to check whether or not another photoactive species exists. The procedure for the calculation is the same with the case of dimer, although we considered only the fcc site, which is more stable than the bridge in monomer adsorption. The resonant orbital is π^* for the free NO molecule, and the resulting action spectra is shown in Figure 20.16 with the experimental and theoretical dimer action spectra. At low photon energy, they agree well, but the increase at $\omega \sim 3.6$ eV seems to be too rapid in the monomer model.

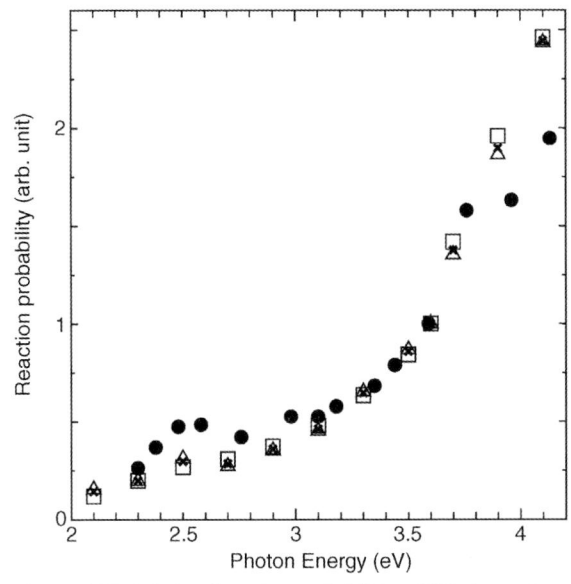

Figure 20.15 Calculated reaction probabilities of NO desorption for the fcc model (open triangles), the bridge model (open squares), and the total δ phase (crosses). The experimental action spectrum is also plotted as filled circles. All the data are scaled by setting the reaction probabilities to 1 at $\omega = 3.6$ eV.

From the *ab initio* calculations and analysis based on our NEGF theory, we can suggest a mechanism of the interfacial charge transport for NO photodesorption

(1) The photoactive species is the NO dimer
(2) Although the dimer phase consists of two kinds of structures, the resonant level and width are similar, hence, the dimer phase can be treated as a single resonant level system.
(3) The long tail of the action spectrum is caused by low resonant energy, which is about 2 eV above the Fermi level.
(4) The effect of secondary electrons gradually increases, reaching ~40% at high photon energy. In addition, the interference for tunneling of hot electrons contributes to the increase in the action spectrum.

Furthermore, the results support that NO photodesorption and decay of a dimer itself should be competing channels because the introduced resonant orbital is the only one for a dimer which has anti-bonding character for both N–O and NO–NO. This is consistent with the experimental results that N_2O desorption is also observed by substrate-mediated excitation, and its action spectrum exactly mirrors that for NO photodesorption. Again we emphasize our results and analyses are obtained by *ab iniio* calculations within the model of NEGF-DFT focusing on hot electron transport and a charge injection process. This is an example to show the importance and

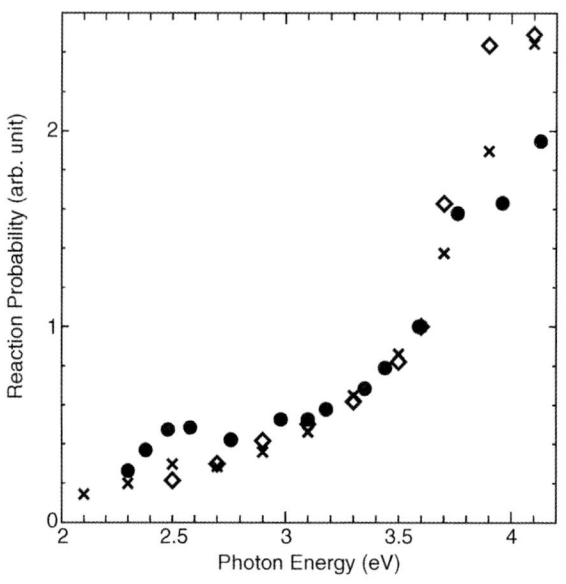

Figure 20.16 Calculated reaction probabilities for an assumed photoactive monomer (open diamonds). The reaction probability of NO for the total δ phase (crosses) and the experimental action spectrum (filled circles), which are as in Figure 20.15, are plotted for comparison. All the data are scaled by normalizing to a value of 1 at $\omega = 3.6$ eV.

necessity of a theoretical model which connects reaction dynamics and excited electron dynamics followed by heterogeneous transport.

20.4 Summary and Outlook

Throughout this chapter, we have tried to set the problems in a unified theoretical viewpoint. To do so, we rearranged the terminologies used in each research field, particularly transfer and transport, which are often treated in terms of Marcus and Landauer theories. We showed which approach is the better description as a starting point for several systems by focusing on the classification of electronic excitation mechanisms. In addition, we gave a conceptual insight of the connection between traditional models of reaction dynamics, which is close to molecular science, and CTs in heterogeneous (surface) systems.

We have focused on the charge transport at interfaces and nonadiabatic interactions between injected electrons and nuclear motions. Our purpose is to establish practical models, which enable us to perform *ab initio* calculations. We adopted the NEGF formalism, and developed theoretical models combined with a practical *ab initio* scheme by means of DFT. We chose two systems as examples, the E–M–E junction and photoreaction on metal surfaces.

One of the advantages in adopting the NEGF for charge transport is that semi-infiniteness, which is a most important feature of nanoscale heterogeneous systems, can be renormalized by self-energies, and the non-equilibrium one-particle state of continuous electronic excitations/de-excitations can be expressed formally. In addition, within DFT (or, more widely, the mean-field approximation) level, the surface Green's functions are obtained exactly, that is, the retarded Green's function can be estimated by *ab initio* calculations as a function of energy. However, although we have shown successful models by NEGF formalism, an extension to general interfacial CTs is *not* straightforward. To apply the NEGF, lesser (greater) self-energies are required, and determination of them usually requires the (generalized) Kadanoff–Baym ansatz; thus the possibility of application of the NEGF depends on the detail of electronic excitation processes, and how to introduce the ansatz as suitable physical models. Even if the NEGF is applicable, the combination with DFT, that is, NEGF-DFT (or other mean-field theory such as Hartree–Fock) is not often sufficient for the requirements of highly accurate quantitative estimations. This is because the strict applicability of DFT is limited to the ground state; thus the extension to excited states gives a crude approximation although it is often useful and qualitatively sufficient.

As stated in Sections 20.1 and 20.2, the separation of "transfer" and "transport" is the classification relating to coherent and decoherent limiting cases, respectively. Each case can be described by a model expressed in terms of Landauer (coherent limit) and Marcus (decoherent limit) approaches. However, many heterogeneous CTs processes are intermediate between transfer and transport, or consist of both transfer and transport steps in the whole process while the process is ultra-fast. In the present study we tried to start from the Landauer approach, then extended it to incorporate decoherence for the complex heterogeneous CTs. One of the applications was a calculation of the IETS, where the LOE formulation was adopted, and the other was photodesorption on metal surfaces. The former is formally rigorous, but application is limited to the weak eph coupling, that is, weak decoherence only. Furthermore, it is difficult to develop the scheme for the case where nuclear motion is (anharmonic) large amplitude motion during transport. To model the latter case, we proposed a fictitious separation of the scenario for the entire photodesoprtion process. Though the fictitious partitioning combined with NEGF-DFT is a promising model, the procedure required some premises or *ad hoc* (crude) assumptions, for example, indirect excitation is assumed, eTST, omitting details of reaction dynamics, extension of the Kadanoff–Baym ansatz. Generally, it is not always possible to introduce such a tricky fictitious separation, and the required theory should predict the details of a process without any pre-knowledge of the mechanism. One of the other frameworks, time-dependent-DFT may be listed because the formulation is very simple and independent of the strength of nonadiabatic couplings. Several applications to heterogeneous systems have been carried out, for instance, charge injection of a dye-sensitized TiO_2 solar cell [105–107] or nanoscale conductance [61, 108]. However, there are a few serious difficulties in performing practical (*ab initio*) calculations, for example, incorporation of semi-infiniteness of the substrate into the time propagation of electronic wavefunctions and representing the decoherence of nuclear motions by their stochastic property [109, 110]. As another promising formalism, time-dependent NEGF is also extensively studied, but practical applications are limited to a few cases [111].

Comparing homogeneous charge transfer and reactions, the theoretical understanding of heterogeneous CTs is very insufficient although it is a common problem underlying many important chemical and physical processes. The difficulty arises from the coexistence of locality and nonlocality in the electronic state to be considered and competition between coherence and decoherence in the particle dynamics as well as the resulting nonadiabatic reaction dynamics. In this sense, the heterogeneous CTs just border on molecular science (chemistry) and condensed physics (physics), and it is challenging to develop a theoretical model that is comparable with recent advances in experiments, and the model should be unified with traditional theoretical models in the fields of both chemistry and physics.

Acknowledgments

This research was supported by a Grant-in-aid for Scientific Research in Priority Area, "Molecular Nano Dynamics" (Grant No. 16072206) from the Ministry of Education, Culture, Sports, Science and Technology of Japan (MEXT). One of the authors (HN) would like to give thanks for financial support by a Grant-in-aid for Scientific Research in Priority Area, "Electron transport through a linked molecule in nanoscale" (Grant No. 20027002) from MEXT, and a Grant-in-aid for Scientific Research (C) (Grant No. 20613002) from the Japan Society for the Promotion of Science.

References

1 Harrison, W.A. (1980) *Solid State Thoery*, Dover, New York.
2 Holstein, T. (1959) Studies of polaron motion: Part II. The "small" polaron. *Ann. Phys.*, **8**, 343–389.
3 Jortner, J. and Bixon, M. (1999) *Electron-Transfer (Part 2)* (eds I. Prigogine and S.A. Rice), John Wiley & Sons Inc.
4 May, V. and Kuhn, O. (2000) *Charge Energy Transfer Dynamics in Molecular Systems*, Wiley-VHC, Weinheim.
5 Ziman, J.N. (1960) *Electrons and Phonons*, Oxford University Press, Oxford.
6 Marcus, R.A. and Sutin, N. (1985) Electron transfers in chemistry and biology. *Biochim. Biophys. Acta*, **811**, 265–322.
7 Cramer, C.J. (2004) *Essentials of Computational Chemistry*, John Wiley & Sons Ltd.
8 Mayne, A.J., Dujardin, G., Comtet, G. and Riedel, D. (2006) Electronic control of single-molecule dynamics. *Chem. Rev.*, **106**, 4355–4378.
9 Stipe, B.C., Rezaei, M.A. and Ho, W. (1999) A variable-temperature scanning tunneling microscope capable of single-molecule vibrational spectroscopy. *Rev. Sci. Instrum.*, **70**, 137–143.
10 Arnolds, H. (2004) Femtosecond laser-induced reactions with O-2 on Pt{111}. *Surf. Sci.*, **548**, 151–156.
11 Bauer, C., Abid, J.P., Fermin, D. and Girault, H.H. (2004) Ultrafast chemical interface scattering as an additional decay channel for nascent nonthermal electrons in small metal nanoparticles. *J. Chem. Phys.*, **120**, 9302–9315.
12 Her, T.H., Finlay, R.J., Wu, C. and Mazur, E. (1998) Surface femtochemistry

of CO/O-2/Pt(111): The importance of nonthermalized substrate electrons. *J. Chem. Phys.*, **108**, 8595–8598.

13 Zhao, J., Li, B., Onda, K., Feng, M. and Petek, H. (2006) Solvated electrons on metal oxide surfaces. *Chem. Rev.*, **106**, 4402–4427.

14 Zhu, X.Y. (2004) Electronic structure and electron dynamics at molecule-metal interfaces: implications for molecule-based electronics. *Surf. Sci. Rep.*, **56**, 1–83.

15 Adams, D.M., Brus, L., Chidsey, C.E.D., Creager, S., Creutz, C., Kagan, C.R., Kamat, P.V., Lieberman, M., Lindsay, S., Marcus, R.A., Metzger, R.M., Michel-Beyerle, M.E., Miller, J.R., Newton, M.D., Rolison, D.R., Sankey, O., Schanze, K.S., Yardley, J. and Zhu, X.Y. (2003) Charge transfer on the nanoscale: current status. *J. Phys. Chem. B*, **107**, 6668–6697.

16 Reed, M.A., Zhou, C., Deshpande, M.R., Muller, C.J., Burgin, T.P., Jones, L. and Tour, J.M. (1998) The electrical measurement of molecular junctions. *Mol. Electron.: Sci. Technol.*, **852**, 133–144.

17 Reed, M.A., Zhou, C., Muller, C.J., Burgin, T.P. and Tour, J.M. (1997) Conductance of a molecular junction. *Science*, **278**, 252–254.

18 Rocha, A.R., Garcia-Suarez, V.M., Bailey, S.W., Lambert, C.J., Ferrer, J. and Sanvito, S. (2005) Towards molecular spintronics. *Nature Mater.*, **4**, 335–339.

19 Subramanian, V., Wolf, E. and Kamat, P.V. (2001) Semiconductor-metal composite nanostructures. To what extent do metal nanoparticles improve the photocatalytic activity of TiO_2 films? *J. Phys. Chem. B*, **105**, 11439–11446.

20 Subramanian, V., Wolf, E.E. and Kamat, P.V. (2004) Catalysis with TiO_2/gold nanocomposites. Effect of metal particle size on the Fermi level equilibration. *J. Am. Chem. Soc.*, **126**, 4943–4950.

21 Janik, M.J., Taylor, C.D. and Neurock, M. (2007) First principles analysis of the electrocatalytic oxidation of methanol and carbon monoxide. *Top. Catal.*, **46**, 306–319.

22 Owens, B.B., Passerini, S. and Smyrl, W.H. (1999) Lithium ion insertion in porous metal oxides. *Electrochim. Acta*, **45**, 215–224.

23 Taylor, C., Kelly, R.G. and Neurock, M. (2006) First-principles calculations of the electrochemical reactions of water at an immersed Ni(111)/H_2O interface. *J. Electrochem. Soc.*, **153**, E207–E214.

24 Wurfel, P. (2005) *Physics of Solar Cells*, Wiley-VCH, Weinheim.

25 Cranney, M., Mayne, A.J., Laikhtman, A., Comtet, G. and Dujardin, G. (2005) STM excitation of individual biphenyl molecules on Si(100) surface: DIET or DIEF? *Surf. Sci.*, **593**, 139–146.

26 Galperin, M. and Beratan, D.N. (2005) Simulation of scanning tunneling microscope images of 1,3-cyclohexadiene bound to a silicon surface. *J. Phys. Chem. B*, **109**, 1473–1480.

27 Kawai, M., Komeda, T., Kim, Y., Sainoo, Y. and Katano, S. (2004) Single-molecule reactions and spectroscopy via vibrational excitation. *Philos. Trans. R. Soc. London, Ser. A*, **362**, 1163–1171.

28 Komeda, T., Kim, Y., Fujita, Y., Sainoo, Y. and Kawai, M. (2004) Local chemical reaction of benzene on Cu(110) via STM-induced excitation. *J. Chem. Phys.*, **120**, 5347–5352.

29 Sainoo, Y., Kim, Y., Okawa, T., Komeda, T., Shigekawa, H. and Kawai, M. (2005) Excitation of molecular vibrational modes with inelastic scanning tunneling microscopy processes: Examination through action spectra of cis-2-butene on Pd(110). *Phys. Rev. Lett.*, **95**, 246102-1–246102-4.

30 Stipe, B.C., Rezaei, M.A. and Ho, W. (1998) Single-molecule vibrational spectroscopy and microscopy. *Science*, **280**, 1732–1735.

31 Koch, E. and Holthausen, M.C. (2001) *A Chemist's Guide to Density Functional Theory*, Wiley-VCH, Weinheim.

32 Nakatsuji, H. and Nakai, H. (1993) Dipped adcluster model study for molecular and dissociative

chemisorptions of O_2 on Ag surface. *J. Chem. Phys.*, **98**, 2423–2436.

33 Nakatsuji, H., Nakai, H. and Fukunishi, Y. (1991) Dipped adcluster model for chemisorptions and catalytic reactions on a metal-surface - image force correction and applications to Pd-O_2 adclusters. *J. Chem. Phys.*, **95**, 640–647.

34 Bruus, H. and Flensberg, K. (2004) *Many-Body Quantum Theory in Condensed Matter Physics*, Oxford University Press, Oxford.

35 Datta, S. (1995) *Electronic Transport in Mesoscopic Systems*, Cambridge University Press, Cambridge.

36 Kadanoff, L.P. and Baym, G. (1962) *Quantum Statistical Mechanics*, Benjamin, New York.

37 Nakamura, H. and Yamashita, K. (2005) Electron tunneling of photochemical reactions on metal surfaces: nonequilibrium Green's function-density functional theory approach to photon energy dependence of reaction probability. *J. Chem. Phys.*, **122**, 194706-1–195706-13.

38 Nakamura, H. and Yamashita, K. (2006) An efficient molecular orbital approach for self-consistent calculations of molecular junctions. *J. Chem. Phys.*, **125**, 194106-1–194106-12.

39 Nakamura, H. and Yamashita, K. (2006) Theoretical study of the photodesorption mechanism of nitric oxide on a Ag(111) surface: A nonequilibrium Green's function approach to hot-electron tunneling. *J. Chem. Phys.*, **125**, 084708-1–084708-12.

40 Nakamura, H. and Yamashita, K. (2008) Systematic study on quantum confinement and waveguide effects for elastic and inelastic currents in atomic gold wire. *Nano Lett.*, **8**, 6–12.

41 Landauer, R. (1970) Electrical resistance of disordered one-dimensional lattices. *Philos. Mag.*, **21**, 863–867.

42 Nitzan, A. (2001) A relationship between electron-transfer rates and molecular conduction. *J. Phys. Chem. A*, **105**, 2677–2679.

43 Traub, M.C., Brunschwig, B.S. and Lewis, N.S. (2007) Relationships between nonadiabatic bridged intramolecular, electrochemical, and electrical electron-transfer processes. *J. Phys. Chem. B*, **111**, 6676–6683.

44 Lindstrom, C.D. and Zhu, X.Y. (2006) Photoinduced electron transfer at molecule-metal interfaces. *Chem. Rev.*, **106**, 4281–4300.

45 Zhu, X.Y. (2004) Charge transport at metal-molecule interfaces: A spectroscopic view. *J. Phys. Chem. B*, **108**, 8778–8793.

46 Seideman, T. (2003) Current-driven dynamics in molecular-scale devices. *J. Phys.: Condens. Matter*, **15**, R521–R549.

47 Seideman, T. and Guo, H. (2003) Quantum transport and current-triggered dynamics in molecular tunnel junctions. *J. Theor. Comput. Chem.*, **2**, 439–458.

48 Zimmermann, F.M. and Ho, W. (1995) State-resolved studies of photochemical dynamics at surfaces. *Surf. Sci. Rep.*, **22**, 127–247.

49 Menzel, D. and Gomer, R. (1964) Electron-impact desorption of carbon monoxide from tungsten. *J. Chem. Phys.*, **41**, 3329–3351.

50 Antoniewicz, P.R. (1980) Model for electron-stimulated and photon-stimulated desorption. *Phys. Rev. B*, **21**, 3811–3815.

51 Saalfrank, P. (1996) Stochastic wave packet vs. direct density matrix solution of Liouville-von Neumann equations for photodesorption problems. *Chem. Phys.*, **211**, 265–276.

52 Saalfrank, P., Nest, M., Andrianov, I., Klamroth, T., Kroner, D. and Beyvers, S. (2006) Quantum dynamics of laser-induced desorption from metal and semiconductor surfaces, and related phenomena. *J. Phys.: Condens. Matter*, **18**, S1425–S1459.

53 Guo, H., Saalfrank, P. and Seideman, T. (1999) Theory of photoinduced surface reactions of admolecules. *Prog. Surf. Sci.*, **62**, 239–303.

54 Misewich, J.A., Heinz, T.F. and Newns, D.M. (1992) Desorption induced by multiple electronic-transitions. *Phys. Rev. Lett.*, **68**, 3737–3740.

55 Saalfrank, P. (1997) Open-system quantum dynamics for laser-induced DIET and DIMET. *Surf. Sci.*, **390**, 1–10.

56 Saalfrank, P. (2004) Theory of photon- and STM-induced bond cleavage at surfaces. *Curr. Opin. Solid. St. M.*, **8**, 334–342.

57 Saalfrank, P., Boendgen, G., Corriol, C. and Nakajima, T. (2000) Direct and indirect, DIET and DIMET from semiconductor and metal surfaces: What can we learn from 'toy models'? *Faraday Discuss.*, **117**, 65–83.

58 Salam, G.P., Persson, M. and Palmer, R.E. (1994) Possibility of coherent multiple excitation in atom-transfer with a scanning tunneling microscope. *Phys. Rev. B*, **49**, 10655–10662.

59 Gao, S.W., Lundqvist, B.I. and Ho, W. (1995) Hot-electron-induced vibrational heating at surface – importance of a quantum-mechanical description. *Surf. Sci.*, **341**, L1031–L1036.

60 Gao, S.W., Persson, M. and Lundqvist, B.I. (1997) Theory of atom transfer with a scanning tunneling microscope. *Phys. Rev. B*, **55**, 4825–4836.

61 Gebauer, R. and Car, R. (2004) Kinetic theory of quantum transport at the nanoscale. *Phys. Rev. B*, **70**, 125324-1–125324-5.

62 Joachim, C., Gimzewski, J.K. and Aviram, A. (2000) Electronics using hybrid-molecular and mono-molecular devices. *Nature*, **408**, 541–548.

63 Chen, J., Reed, M.A., Rawlett, A.M. and Tour, J.M. (1999) Large on-off ratios and negative differential resistance in a molecular electronic device. *Science*, **286**, 1550–1552.

64 Chen, J., Wang, W., Reed, M.A., Rawlett, A.M., Price, D.W. and Tour, J.M. (2000) Room-temperature negative differential resistance in nanoscale molecular junctions. *Appl. Phys. Lett.*, **77**, 1224–1226.

65 Agrait, N., Untiedt, C., Rubio-Bollinger, G. and Vieira, S. (2002) Onset of energy dissipation in ballistic atomic wires. *Phys. Rev. Lett.*, **88**, 216803-1–216803-4.

66 Agrait, N., Untiedt, C., Rubio-Bollinger, G. and Vieira, S. (2002) Electron transport and phonons in atomic wires. *Chem. Phys.*, **281**, 231–234.

67 Jauho, A.P., Wingreen, N.S. and Meir, Y. (1994) Time-dependent transport in interacting and noninteracting resonant-tunneling systems. *Phys. Rev. B*, **50**, 5528–5544.

68 Lake, R. and Datta, S. (1992) Nonequilibrium greens-function method applied to double-barrier resonant-tunneling diodes. *Phys. Rev. B*, **45**, 6670–6685.

69 Wingreen, N.S., Jauho, A.P. and Meir, Y. (1993) Time-dependent transport through a mesoscopic structure. *Phys. Rev. B*, **48**, 8487–8490.

70 Brandbyge, M., Mozos, J.L., Ordejon, P., Taylor, J. and Stokbro, K. (2002) Density-functional method for nonequilibrium electron transport. *Phys. Rev. B*, **65**, 165401-1–165401-17.

71 Ke, S.H., Baranger, H.U. and Yang, W.T. (2004) Electron transport through molecules: Self-consistent and non-self-consistent approaches. *Phys. Rev. B*, **70**, 085410-1–085410-12.

72 Rocha, A.R., Garcia-Suarez, V.M., Bailey, S., Lambert, C., Ferrer, J. and Sanvito, S. (2006) Spin and molecular electronics in atomically generated orbital landscapes. *Phys. Rev. B*, **73**, 085414-1–085414-22.

73 Stokbro, K., Taylor, J., Brandbyge, M. and Ordejon, P. (2003) TranSIESTA - A spice for molecular electronics. *Annals of the New York Academy of Sciences III*, **1006**, 212–226.

74 Taylor, J., Guo, H. and Wang, J. (2001) *Ab initio* modeling of quantum transport properties of molecular electronic devices. *Phys. Rev. B*, **63**, 245407-1–245407-13.

75 Xue, Y.Q., Datta, S. and Ratner, M.A. (2002) First-principles based matrix Green's function approach to molecular

electronic devices: general formalism. *Chem. Phys.*, **281**, 151–170.

76 Keldysh, L.V. (1965) Diagram Technique for Nonequilibrium Processes. *Sov. Phys. Jetp-USSR*, **20**, 1018–1026.

77 Lipavsky, P., Spicka, V. and Velicky, B. (1986) Generalized Kadanoff-Baym Ansatz for Deriving Quantum Transport-Equations. *Phys. Rev. B*, **34**, 6933–6942.

78 Damle, P., Ghosh, A.W. and Datta, S. (2002) First-principles analysis of molecular conduction using quantum chemistry software. *Chem. Phys.*, **281**, 171–187.

79 Faleev, S.V., Leonard, F., Stewart, D.A. and van Schilfgaarde, M. (2005) *Ab initio* tight-binding LMTO method for nonequilibrium electron transport in nanosystems. *Phys. Rev. B*, **71**, 195422-1–195422-18.

80 Ke, S.H., Baranger, H.U. and Yang, W.T. (2005) Models of electrodes and contacts in molecular electronics. *J. Chem. Phys.*, **123**, 114701–1114701–1114708.

81 Roos, B.O. and Taylor, P.R. (1980) A complete active space scf method (Casscf) using a density-matrix formulated super-Ci approach. *Chem. Phys.*, **48**, 157–173.

82 Ruedenberg, K., Schmidt, M.W., Gilbert, M.M. and Elbert, S.T. (1982) Are atoms intrinsic to molecular electronic wavefunctions. 1. The fors model. *Chem Phys*, **71**, 41–49.

83 Sanvito, S., Lambert, C.J., Jefferson, J.H. and Bratkovsky, A.M. (1999) General Green's-function formalism for transport calculations with spd Hamiltonians and giant magnetoresistance in Co- and Ni-based magnetic multilayers. *Phys. Rev. B*, **59**, 11936–11948.

84 Perdew, J.P., Burke, K. and Ernzerhof, M. (1996) Generalized gradient approximation made simple. *Phys. Rev. Lett.*, **77**, 3865–3868.

85 Koentopp, M., Chang, C., Burke, K. and Car, R. (2008) Density functional calculations of nanoscale conductance. *J. Phys.: Condens. Matter*, **20**, 083203-1–083203-21.

86 Frederiksen, T., Brandbyge, M., Lorente, N. and Jauho, A.P. (2004) Inelastic scattering and local heating in atomic gold wires. *Phys. Rev. Lett.*, **93**, 256601-1–256601-4.

87 Kim, Y., Komeda, T. and Kawai, M. (2002) Single-molecule reaction and characterization by vibrational excitation. *Phys. Rev. Lett.*, **89**, 126104–126104-4.

88 Ho, W. (2002) Single-molecule chemistry. *J. Chem. Phys.*, **117**, 11033–11061.

89 Ho, W. (2004) Single molecule vibrational spectroscopy with the STM. *Abs. Papers Am. Chem. Soc.*, **227**, U825.

90 Sergueev, N., Roubtsov, D. and Guo, H. (2005) Ab initio analysis of electron-phonon coupling in molecular devices. *Phys. Rev. Lett.*, **95**, 146803-1–146803-4.

91 Frederiksen, T., Paulsson, M., Brandbyge, M. and Jauho, A.P. (2007) Inelastic transport theory from first principles: Methodology and application to nanoscale devices. *Phys. Rev. B*, **75**, 235441-1–235441-8.

92 Galperin, M., Ratner, M.A. and Nitzan, A. (2007) Molecular transport junctions: vibrational effects. *J. Phys.: Condens. Matter*, **19**, 103201-1–103201-81.

93 Galperin, M., Ratner, M.A. and Nitzan, A. (2004) Inelastic electron tunneling spectroscopy in molecular junctions: Peaks and dips. *J. Chem. Phys.*, **121**, 11965–11979.

94 Viljas, J.K., Cuevas, J.C., Pauly, F. and Hafner, M. (2005) Electron-vibration interaction in transport through atomic gold wires. *Phys. Rev. B*, **72**, 245415-1–245415-18.

95 Paulsson, M., Frederiksen, T. and Brandbyge, M. (2005) Modeling inelastic phonon scattering in atomic- and molecular-wire junctions. *Phys. Rev. B*, **72**, 201101-1–201101-4.

96 Hellsing, B. and Zhdanov, V.P. (1994) Substrate-mediated mechanism for photoinduced chemical-reactions on metal-surfaces. *J. Photochem. Photobiol. A*, **79**, 221–232.

97 Zhu, X.Y., White, J.M., Wolf, M., Hasselbrink, E. and Ertl, G. (1991)

Polarization probe of excitation mechanisms in surface photochemistry. *Chem. Phys. Lett.*, **176**, 459–466.

98 Berglund, C.N. and Spicer, W.E. (1964) Photoemission studies of copper and silver – theory. *Phys. Rev.*, **136**, A1030–A1034.

99 Brown, W.A. and King, D.A. (2000) NO chemisorption and reactions on metal surfaces: A new perspective. *J. Phys. Chem. B*, **104**, 2578–2595.

100 Kidd, R.T., Lennon, D. and Meech, S.R. (1999) Comparative study of the primary photochemical mechanisms of nitric oxide and carbonyl sulfide on Ag(111). *J. Phys. Chem. B*, **103**, 7480–7488.

101 Kidd, R.T., Lennon, D. and Meech, S.R. (2000) Surface plasmon enhanced substrate mediated photochemistry on roughened silver. *J. Chem. Phys.*, **113**, 8276–8282.

102 So, S.K., Franchy, R. and Ho, W. (1991) Photodesorption of No from Ag(111) and Cu(111). *J. Chem. Phys.*, **95**, 1385–1399.

103 Carlisle, C.I. and King, D.A. (2001) Direct molecular imaging of NO monomers and dimers and a surface reaction on Ag{111}. *J. Phys. Chem. B*, **105**, 3886–3893.

104 Weik, F., Demeijere, A. and Hasselbrink, E. (1993) Wavelength dependence of the photochemistry of O_2 on Pd(111) and the role of hot-electron cascades. *J. Chem. Phys.*, **99**, 682–694.

105 Craig, C.F., Duncan, W.R. and Prezhdo, O.V. (2005) Trajectory surface hopping in the time-dependent Kohn-Sham approach for electron-nuclear dynamics. *Phys. Rev. Lett.*, **95**, 163001-1–163001-4.

106 Duncan, W.R. and Prezhdo, O.V. (2005) Electronic structure and spectra of catechol and alizarin in the gas phase and attached to titanium. *J. Phys. Chem. B*, **109**, 365–373.

107 Duncan, W.R., Stier, W.M. and Prezhdo, O.V. (2005) Ab initio nonadiabatic molecular dynamics of the ultrafast electron injection across the alizarin-TiO_2 interface. *J. Am. Chem. Soc*, **127**, 7941–7951.

108 Gebauer, R. and Car, R. (2004) Current in open quantum systems. *Phys. Rev. Lett.*, **93**, 160404-1–160404-4.

109 Miyamoto, Y. and Sugino, O. (2000) First-principles electron-ion dynamics of excited systems: H-terminated Si(111) surfaces. *Phys. Rev. B*, **62**, 2039–2044.

110 Sugino, O. and Miyamoto, Y. (1999) Density-functional approach to electron dynamics: Stable simulation under a self-consistent field. *Phys. Rev. B*, **59**, 2579–2586.

111 Kurth, S., Stefanucci, G., Almbladh, C.O., Rubio, A. and Gross, E.K.U. (2005) Time-dependent quantum transport: A practical scheme using density functional theory. *Phys. Rev. B*, **72**, 035308-1–035308-13.

21
Dynamic Behavior of Active Ag Species in NOx Reduction on Ag/Al$_2$O$_3$

Atsushi Satsuma and Ken-ichi Shimizu

21.1
Introduction

21.1.1
NOx Reduction Technologies for Diesel and Lean-Burn Gasoline Engines

In view of the reduction of total CO$_2$ emissions from automobiles, the production of diesel and lean-burn gasoline engine cars is expected to increase because of their greater fuel economy than conventional gasoline engine cars. Especially, the total CO$_2$ emission (well-to-tank and tank-to-wheel) from diesel hybrid cars is almost equivalent to fuel-cell engine cars using H$_2$ produced from fossil fuels. Further reduction of CO$_2$ emission can be expected by the use of biodiesel fuels. However, serious drawbacks of diesel engines are emissions of particulate materials (PM) and nitrogen oxides (NOx: NO and NO$_2$). Removal of PM is well established by the use of ceramic filters (DPF: diesel particulate filter) [1]. For NOx removal, the following three technologies are being developed and coming onto the market: NSR (NOx storage-reduction) system, urea-SCR (selective catalytic reduction by urea) and HC-SCR (selective catalytic reduction by hydrocarbons). The NSR system developed by Toyota [2–4] is characterized as an evolved three-way catalyst combined with NOx storage. The catalyst contains NOx storage materials, such as BaO, which store NOx as nitrates during lean conditions. When turning the engine to rich conditions for a short period, the stored NOx is released from the storage materials and reduced over noble metals by hydrocarbons and CO. The improvement in SOx tolerance of NSR catalysts is a further problem to be solved. Urea-SCR is thought to be effective for large-scale diesel engine cars such as tracks and buses [5]. The injection of urea into a catalytic converter results in the generation of NH$_3$ via hydrolysis, and then NOx in exhausts is effectively reduced to N$_2$ by generated NH$_3$ over zeolite-based catalysts. The effective reduction of NO at low temperatures is the merit of the urea-SCR technology, however, this technology still has some problems to be solved, such as refilling with urea solution, appropriate injection of urea solution, and so on.

21.1.2
Selective Catalytic Reduction of NOx by Hydrocarbons Over Ag/Al$_2$O$_3$

Selective catalytic reduction of NOx by hydrocarbons (HC-SCR) is also thought to be a potential method for removing NOx from lean-burn and diesel exhausts. In 1990, Iwamoto et al. [6] and Held et al. [7] discovered that NOx reduction under excess oxygen could be accomplished using hydrocarbons as reducing agents over Cu-MFI. Unburned hydrocarbons in exhausts or fuels can be used as reductants, which is the main advantage of the HC-SCR system. Since the air/fuel ratio in exhausts of lean burn and diesel engines is very high, the catalysts for the HC-SCR system should work in the presence of excess oxygen, around 10%, and the following functions are required: (i) promotion of the reduction of NO (less than 1000 ppm) by HC and (ii) the suppression of the undesired combustion of HC with excess O$_2$ (around 10%). Through numerous screening runs after the discovery of the HC-SCR technology, various types of catalysts, which can be classified into three categories: ion-exchanged zeolites, supported precious metals, and metal oxide-based catalysts, have been found [1, 8, 9].

Ion-exchanged zeolites, especially Cu-MFI, show very high performance in HC-SCR. However, deactivation due to instability of the zeolite framework under hydrothermal conditions is a serious problem. Supported precious metal catalysts, such as Pt/Al$_2$O$_3$, show high stability and high tolerance to sulfur oxides (SOx) and water vapor. They show very high activity at lower temperatures, however, the formation of N$_2$O and a narrow temperature range for NOx reduction are the problems to be solved. Metal oxide-based catalysts show high stability and moderate tolerance to SOx and water vapor. Among a number of catalysts reported, Ag/Al$_2$O$_3$ is thought to be the most promising candidate for practical applications. Several reports on engine bench tests have demonstrated a high possibility of Ag/Al$_2$O$_3$ for practical applications [10–14], but low HC-SCR activity at lower temperatures should be improved.

Miyadera first reported the high efficiency of Ag/Al$_2$O$_3$ catalysts in NO reduction to N$_2$ by alcohols [15, 16]. By the use of ethanol as a reductant, Ag/Al$_2$O$_3$ catalysts showed 80–100% conversion of NO in the range 300–400 °C, as shown in Figure 21.1. He et al. demonstrated a diesel engine bench test by using ethanol-SCR over Ag/Al$_2$O$_3$ [17]. Ag/Al$_2$O$_3$ showed average NO$_x$ conversion of greater than 80% in the temperature range 300–400 °C. For a space-velocity of 30 000 h^{-1}, the light-off temperature was 270 °C and the highest NOx conversion was 92.3% at 400 °C. Thomas et al. tested SCR by various alcohols for the exhaust from a 5.9 L diesel engine using a 7.0 L Ag/Al$_2$O$_3$ catalyst system with a reductant injector [14]. The low price of Ag is also attractive for practical use. The Ag/Al$_2$O$_3$ catalyst shows very high SCR activity when alcohols are used as reductants, however, it is not effective for SCR by light hydrocarbons such as methane and propane.

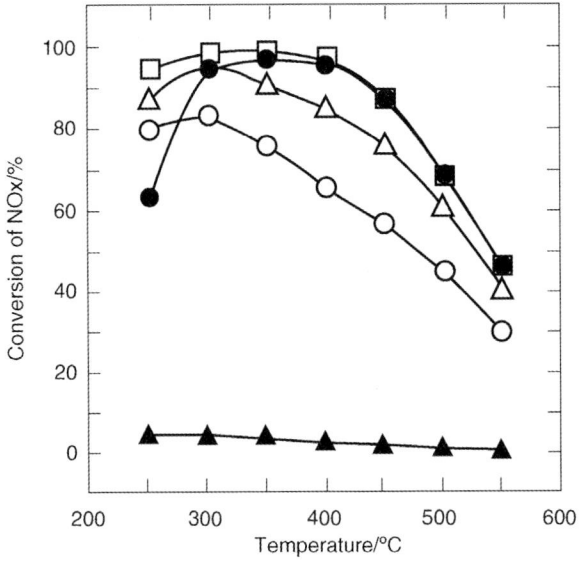

Figure 21.1 Conversion of NO by oxygen-containing organic compounds over 2 wt.%Ag/Al$_2$O$_3$ catalyst. Reaction conditions: 500 ppm NO, 1000 ppm (as CHx) reductant (333 ppm), 10% O$_2$, 10% CO$_2$, N$_2$ balance, 10% H$_2$O, SV = 6400 h^{-1}. Symbols: (○) 1-propanol, (□) 2-propanol, (△) ethanol, (●) acetone, and (▲) methane [16].

21.2
Hydrogen Effect of HC-SCR Over Ag/Al$_2$O$_3$

21.2.1
Boosting of HC-SCR Activity of Ag/Al$_2$O$_3$ by Addition of H$_2$

The most significant and surprising promotion effect of Ag/Al$_2$O$_3$ can be obtained by the addition of a small amount of H$_2$ into the reaction atmosphere. Satokawa first reported the promotion effect of H$_2$ on the HC-SCR activity of Ag/Al$_2$O$_3$ at lower temperatures [18, 19]. As shown in Figure 21.2, the addition of H$_2$ into propane-SCR atmosphere boosted the NO conversion to N$_2$. The effect is greater at lower temperatures, below 773 K. For example, the conversion of NO at 673 K was nearly zero, but it increased to about 50% by co-feeding 909 ppm of H$_2$. The promotion effect of H$_2$ on NO and propane conversions was reversible, depending on the presence of H$_2$ in the gas phase (Figure 21.3). Interestingly, NO conversion is nearly zero when only hydrogen is used as the reductant. Therefore, hydrogen does not act as a reducing agent, but plays the role of "promoter" for the NO reduction by hydrocarbons [19]. Actually, the steep increase in propane conversion with the introduction of H$_2$ suggests that the activation of hydrocarbon reductants is a key step.

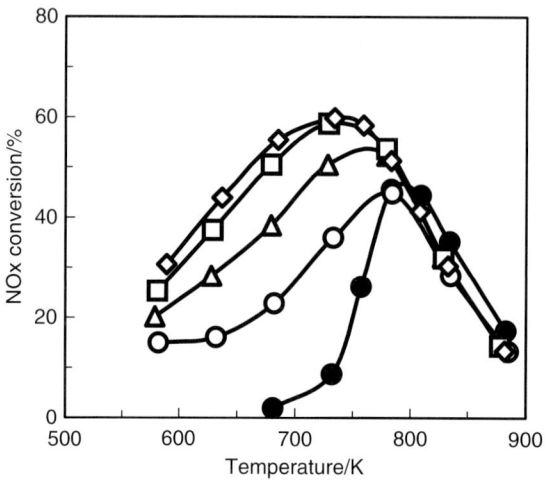

Figure 21.2 NO$_x$ conversion over Ag/Al$_2$O$_3$ at various H$_2$ concentrations. Feed: 91 ppm NO, 91 ppm C$_3$H$_8$, 9.1% O$_2$, 9.1% H$_2$O, and 0 ppm (●), 227 ppm (○), 451 ppm (△), 909 ppm (□), or 1818 ppm (◇) H$_2$ in He. GHSV = 44 000 h^{-1} [19].

This remarkable role of hydrogen is called the "hydrogen effect" [1] and has been re-examined by various research groups, such as Burch et al. [1, 20], Sazama et al. [21], Klingstedt et al. [22], Richter et al. [23] and Zhang et al. [24]. Burch et al. reported superior H$_2$-HC-SCR performance of Ag/Al$_2$O$_3$ by using octane as a reductant [1, 20]. As shown in Figure 21.4, the conversion of NO increased from 450 °C in the absence

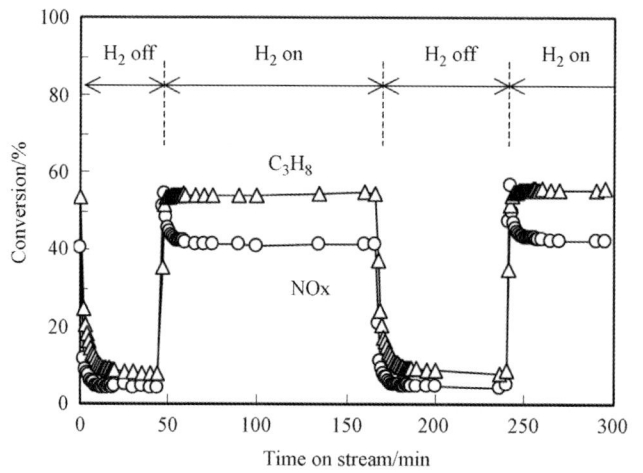

Figure 21.3 Effect of H$_2$ on the NO$_x$ and C$_3$H$_8$ conversions at 673 K as a function of time over Ag/Al$_2$O$_3$. Feed: 91 ppm NO, 91 ppm C$_3$H$_8$, 9.1% O$_2$, 0 or 455 ppm H$_2$, and He as balance. GHSV: 44 000 h^{-1} [19].

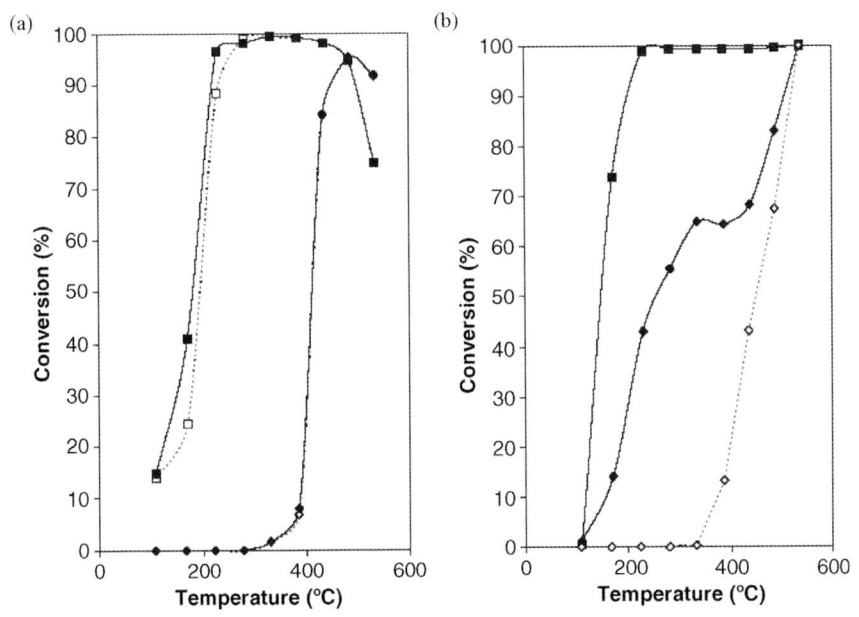

Figure 21.4 The effect of H_2 on NOx reduction by octane on Ag/Al_2O_3 catalyst. (a) (●) NO conversion octane alone; (◇) NOx conversion octane alone; (■) NO conversion with 0.72% H_2; (□) NOx conversion with 0.72% H_2. (b) (■) H_2 conversion; (◆) octane conversion with 0.72% H_2; (◇) octane conversion in the absence of H_2 [20].

of H_2, while sufficient NO conversion was observed from 200 °C. Surprisingly, the NO conversion started from a temperature 250 °C lower than in the absence of H_2, and nearly 100% conversion of NO was achieved over the wide temperature range of 200–500 °C. They also confirmed that the conversion of octane increased with the addition of H_2. The "hydrogen effect" is also achieved by using a wide range of reductants, C2–C4 hydrocarbons [18], ethanol [24], and NH_3 [25, 26]. The "hydrogen effect" is of practical interest because (i) the temperature of diesel exhaust ranges from 200 to 400 °C, (ii) the light olefins, the main components of unburned hydrocarbons, or higher paraffins, such as cetane, the main component of diesel fuel, can be used as reductants. The "hydrogen effect" was also observed over Ag-zeolite catalysts, such as Ag/MFI and Ag/BEA [27]. Furthermore, the co-feeding of hydrogen is also preferable for SO_x tolerance [28, 29].

21.2.2
Surface Dynamics of Ag Species Analyzed by *in situ* UV–Vis

For a better understanding of the "hydrogen effect", *in situ* analysis is applied to the Ag/Al_2O_3 under the working state of HC-SCR. In the UV–Vis region, LMCT bands and d–d transfer bands of various elements can be observed, and thus UV–Vis spectra may give directly the states of active elements on solid surfaces. Figure 21.5 shows the

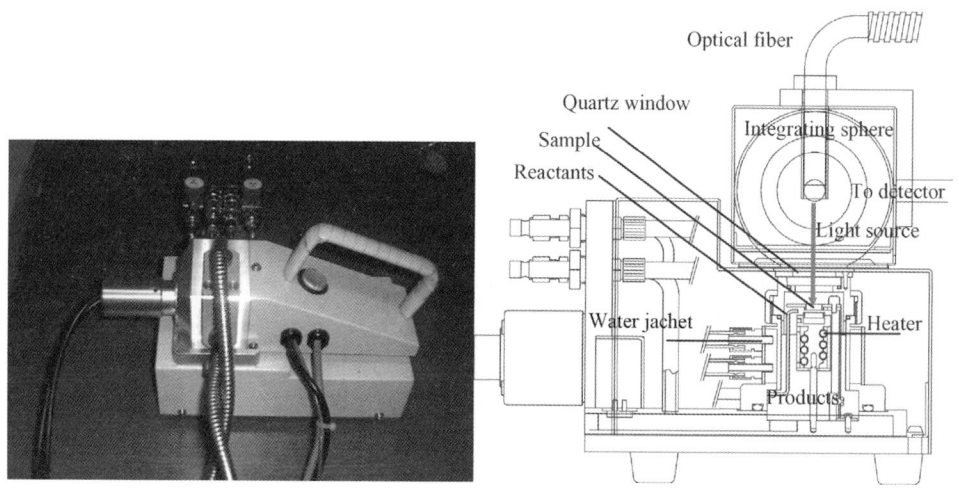

Figure 21.5 The exterior and cross-section of the *in situ* UV–Vis cell [30].

exterior and the cross-section of an *in situ* UV–Vis cell (JASCO VHR-630) developed by our group and the JASCO corporation [30]. The apparatus can be used outside the main optical system of the UV–Vis spectrophotometer (JASCO V-570). Inside the lower unit is a diffuse reflectance sample cell with a heating system, which is connected to a gas flow system. In the upper unit, the light source is led to the center of an integrating sphere by an optical fiber.

The transient reaction experiment was carried out under the same conditions as the flow reaction of the HC-SCR. Figure 21.6 shows the *in situ* UV–Vis spectra thus measured [31]. It is accepted that the bands above 40 000 cm^{-1} correspond to the 4d^{10}

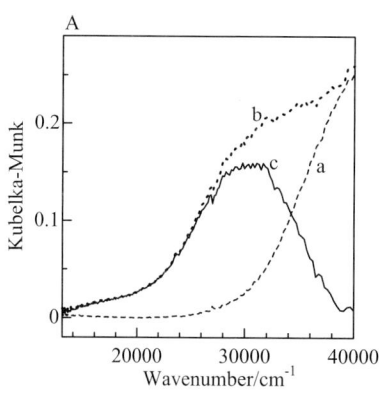

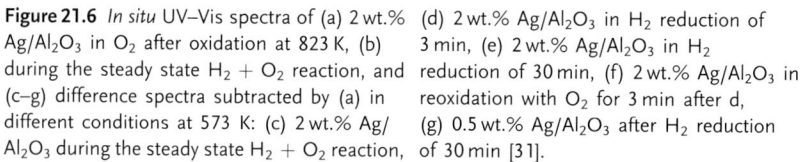

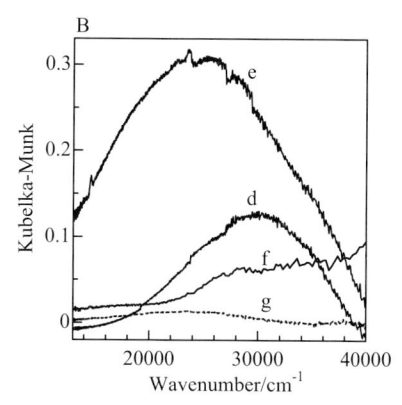

Figure 21.6 *In situ* UV–Vis spectra of (a) 2 wt.% Ag/Al$_2$O$_3$ in O$_2$ after oxidation at 823 K, (b) during the steady state H$_2$ + O$_2$ reaction, and (c–g) difference spectra subtracted by (a) in different conditions at 573 K: (c) 2 wt.% Ag/Al$_2$O$_3$ during the steady state H$_2$ + O$_2$ reaction, (d) 2 wt.% Ag/Al$_2$O$_3$ in H$_2$ reduction of 3 min, (e) 2 wt.% Ag/Al$_2$O$_3$ in H$_2$ reduction of 30 min, (f) 2 wt.% Ag/Al$_2$O$_3$ in reoxidation with O$_2$ for 3 min after d, (g) 0.5 wt.% Ag/Al$_2$O$_3$ after H$_2$ reduction of 30 min [31].

to $4d^9 5s^1$ transition of Ag^+ ions, and the bands between 25 000 and 40 000 cm^{-1} are due to silver clusters with different size and oxidation states [21, 27, 31]. In a flow of 10% O_2 (spectrum a), a band centered above 40 000 cm^{-1} due to Ag^+ ions was observed. During the steady-state of $H_2 + O_2$ reaction, a broad band centered around 30 000 cm^{-1} assignable to $Ag_n^{\delta+}$ ($n \leq 8$) clusters was observed in a difference spectrum (spectrum c). When 2 wt.% Ag/Al_2O_3 was exposed to a flow of H_2 (0.5%) for 3 min (spectrum d), a broad band in a similar position was monitored. We have reported that bands at 32 800 and 38 000 cm^{-1} were observed in the UV–Vis spectrum of Ag-MFI zeolite reduced by 0.5%H_2 at 573 K and assigned them to the Ag_4^{2+} cluster. In the EXAFS spectra during $H_2 + O_2$ reaction and during H_2 reduction for 3 min described later, the coexistence of the Ag_4^{2+} cluster and highly dispersed Ag^+ ions is suggested by the low Ag–Ag coordination number (0.6) and Ag–Ag distance, indicative of the Ag_4^{2+} cluster as well as a large Ag–O contribution. With an increase in H_2 reduction time ($t = 30$ min, spectrum e), a broad band centered around 25 000 cm^{-1} assignable to large Ag_n ($n > 4$) clusters and a shoulder around 28 600 cm^{-1} due to the Ag_4^{2+} cluster were observed. When the flowing gas was switched from H_2 to O_2 (10%), the intensity of the band due to the Ag clusters, including $Ag_n^{\delta+}$ and Ag_n ($n > 4$), decreased (spectrum e to f). This result is consistent with the EXAFS, which also indicated the re-dispersion of Ag clusters to Ag^+ ion under O_2. The states of surface Ag were dependent on the Ag content. The UV–Vis spectrum of 0.5 wt.% Ag/Al_2O_3 (spectrum g) was almost unchanged after 30 min of H_2 reduction. This indicates that reductive agglomeration of Ag^+ to Ag cluster does not occur on this sample. In contrast, Ag/Al_2O_3 containing higher Ag (5 and 10 wt.%) showed that the bands are observed at lower wavenumbers at 25 000 cm^{-1} assignable to the large Ag clusters and at 17 500 cm^{-1} assignable to Ag particles (figures are not shown). At higher Ag loadings, reduced Ag species of larger size are dominant.

The structure of surface Ag species was also identified by *in situ* Ag K-edge EXAFS measurements. Figure 21.7 shows Fourier transforms of k^3-weighted EXAFS of Ag/Al_2O_3 at 573 K [31]. The EXAFS of 2 wt.% Ag/Al_2O_3 after oxidation at 773 K (spectrum b) shows a large Ag–O contribution. From curve-fitting analysis, the Ag–Ag coordination number was negligibly small (0.1), indicating that Ag^+ ions dispersed on the alumina surface are dominant. Under the steady-state $H_2(0.5\%)$ $O_2(10\%)$ reaction (spectrum a), the Ag–Ag shell with coordination number 0.6 and bond distance 2.74 Å was observed. In a flow of 0.5% H_2 for 3 min (spectrum c), the bond distance of the Ag–Ag contribution was the same (2.74 Å). According to the Ag K-edge EXAFS measurements of Ag-MFI zeolite, the Ag–Ag shell with a bond distance of 2.74 Å is tentatively assigned to a small $Ag_n^{\delta+}$ cluster such as Ag_4^{2+}. This assignment is based on the following results of Ag-MFI zeolite reduced by 0.5% H_2 at 573 K [32]: (i) the coordination number of the Ag–Ag shell with bond distance 2.73 Å was around 3 (3.3), (ii) the average valence of Ag is +0.5 from H_2-TPR. The average valence of Ag cluster on alumina was also around +0.6 from a microcalorimetric study of H_2 adsorption on Ag/Al_2O_3: The H_2/Ag ratio was around 0.19 for the H_2 adsorption with heats above 30 kJ mol^{-1} [31]. As shown in Figure 21.7, the intensity of the Ag–Ag peak increased while that of the Ag–O shell decreased with increase in H_2 reduction time. After the steady-state (30 min, spectrum d), the Ag–Ag contribution

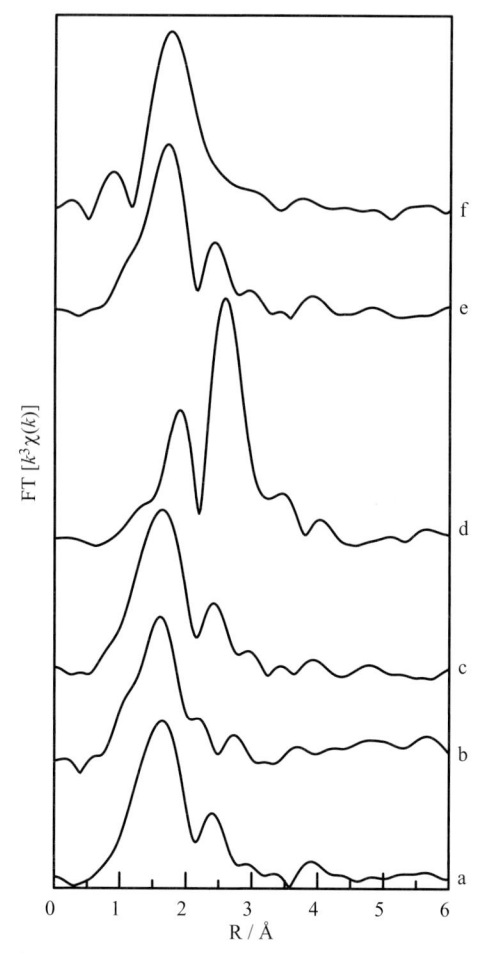

Figure 21.7 Fourier transforms of Ag K-edge *in situ* EXAFS measured at 573 K: (a) 2 wt.% Ag/Al$_2$O$_3$ during the steady state H$_2$ + O$_2$ reaction, (b) 2 wt.% Ag/Al$_2$O$_3$ in He after oxidation at 773 K, (c) 2 wt.% Ag/Al$_2$O$_3$ in H$_2$ reduction of 3 min, (d) 2 wt.% Ag/Al$_2$O$_3$ in H$_2$ reduction of 30 min, (e) 2 wt.%Ag/Al$_2$O$_3$ in re-oxidation with O$_2$ for 3 min after d, and (f) 0.5 wt.% Ag/Al$_2$O$_3$ after H$_2$ reduction of 30 min [31].

with coordination number of 3.2 and bond distance of 2.83 Å was observed. The Ag—Ag shell with a bond distance of 2.83 Å is assigned to large Ag cluster because the distance is shorter than that of bulk Ag and is the same as that reported for silver clusters in Ag-X zeolite formed by H$_2$ reduction at 573 K [33]. When the flowing gas was switched from 0.5% H$_2$ to 10% O$_2$, the features of spectrum d changed quickly to spectrum e. The Ag—Ag coordination number decreased from 3.2 to 0.7 and the Ag—O coordination number increased from 1.6 to 4.4. This indicates that oxidation with O$_2$ results in the re-dispersion of Ag cluster to Ag$^+$ ion. In the case of 0.5 wt.%

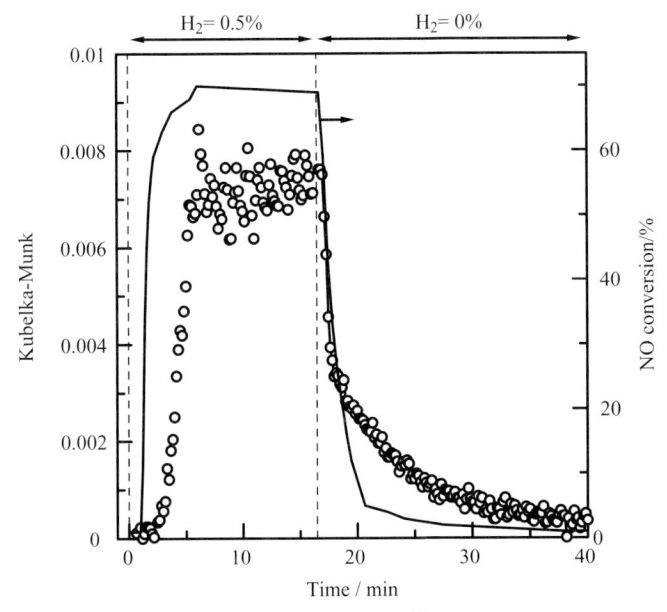

Figure 21.8 Effect of hydrogen switching on/off on NO conversion and the UV–Vis band height at 28 500 cm^{-1} during propane–SCR over at 573 K. Conditions: 0.1% NO, 0.1% C_3H_8, 0% or 0.5% H_2, 10% O_2, catalyst weight = 50 mg [31].

Ag/Al$_2$O$_3$, on the other hand, the Ag–Ag contribution was negligibly small, even after 30 min of H$_2$ reduction treatment (spectrum f), indicating that reductive agglomeration of Ag$^+$ to silver cluster does not occur. These results are consistent with the *in situ* UV–Vis spectra shown in Figure 21.6: The small Ag$_n^{\delta+}$ clusters are formed by slight reduction of 2 wt.% Ag/Al$_2$O$_3$, further reduction or higher Ag loading results in the formation of large Ag clusters, Ag clusters are reversibly re-dispersed in the oxidative atmospheres, and there is no change of Ag$^+$ ion 0.5 wt.% Ag/Al$_2$O$_3$ by reduction with H$_2$.

Figure 21.8 shows the time dependence of NO conversion for the C$_3$H$_8$-SCR at 573 K over 2 wt.% Ag/Al$_2$O$_3$. The NO conversion was only 1% in the C$_3$H$_8$-SCR condition ($t = 0$ min), but significantly increased to 68% upon addition of H$_2$. Upon removal of H$_2$ from the reaction mixture at $t = 16.5$ min, the NO conversion decreased and at $t = 43$ min reached nearly the same conversion level as before the addition of H$_2$. In the same figure, the time course of the band height at 28 500 cm^{-1} is also plotted as an indicator of the relative amount of Ag$_n^{\delta+}$ clusters. The reversible change in the NO conversion is in harmony with the time-dependence of the band assignable to the Ag$_n^{\delta+}$ cluster. The band of the Ag$_n^{\delta+}$ cluster was not observed in the propane-SCR condition, but increased after the addition of H$_2$, steeply decreased on removal of H$_2$ in the reaction atmosphere, and finally reached nearly the original level. The results suggest that the formation of Ag$_n^{\delta+}$ cluster is essential for the "hydrogen effect" on HC-SCR.

21.3
The Role of Surface Adsorbed Species Analyzed by *in situ* FTIR

21.3.1
Reaction Scheme of HC-SCR Over Ag/Al$_2$O$_3$

Dynamic analysis of Ag species by *in situ* UV–Vis and *in situ* EXAFS indicated the contribution of Ag cluster to the "hydrogen effect". But we still have the question as to why the addition of hydrogen promotes the activity of HC-SCR. In this section, studies on the dynamic analysis of surface adsorbed species by *in situ* FTIR are summarized.

The measurement of *in situ* FTIR spectra during H$_2$-HC-SCR over Ag/Al$_2$O$_3$ was carried out using a quartz *in situ* IR cell connected to a gas flow apparatus, as shown in Figure 21.9 [24]. The self-supporting catalyst disk (20 mm in diameter, about 0.08 g) is mounted at the center of the cell, and the thermocouple can be positioned very near to the catalyst disk. CaF$_2$ or KBr windows are fixed on both sides of the cell, which is surrounded by water jackets to cool the windows. A reference spectrum of the catalyst wafer, which was taken in flowing He, was subtracted from each spectrum. The compositions of the feeds are the same as in the flow reaction apparatus. The kinetics of the surface adsorbed species can be determined through measuring transient phenomena of spectra by a stopped-flow technique.

Figure 21.10 shows the IR spectrum of adsorbed species on Ag/Al$_2$O$_3$ after exposing the catalyst to a flow of NO + O$_2$, followed by a flow of n-hexane + O$_2$ on Ag/Al$_2$O$_3$ at 623 K [35]. Just after the pretreatment in NO + O$_2$ (0 min), the bands assignable to unidentate nitrate (1554, 1292 cm^{-1}) and bidentate nitrate (1580, 1246 cm^{-1}) and a small band due to weakly adsorbed NO$_2$ (1624 cm^{-1}) were

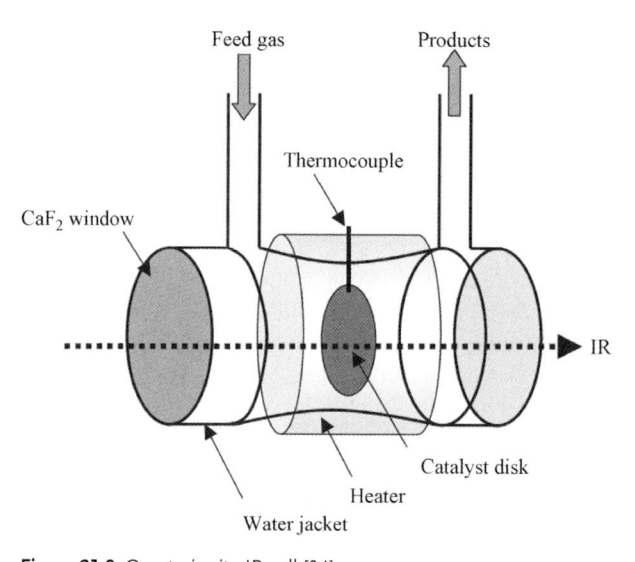

Figure 21.9 Quartz *in situ* IR cell [34].

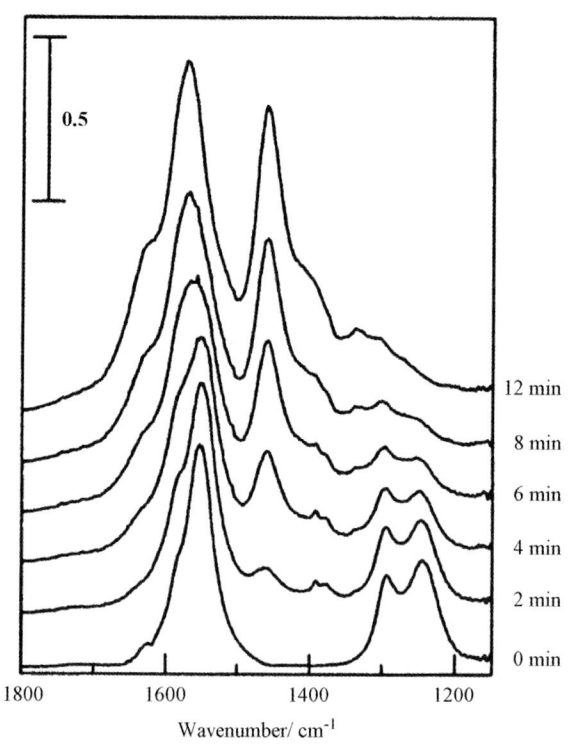

Figure 21.10 Dynamic changes in the IR spectra as a function of time in flowing n-hexane + O_2 on Ag/Al_2O_3 at 623 K. Before the measurements, the catalyst was pre-exposed to a flow of NO + O_2 for 120 min at 623 K [35].

observed. The formation of nitrates proceeds by NO oxidation to NO_2 followed by adsorption of NO_2 on basic oxygen sites. The surface concentration of these species can be estimated from the band intensities and the extinction coefficients of these bands determined by separate experiments [34]. After switching the flow gas to n-hexane + O_2, the nitrate bands gradually decreased and new bands appeared at around 1570–1580 and 1460 cm^{-1}, assigned to v_{as}(COO) and v_s(COO) of adsorbed acetate. Shoulder bands assignable to formate (1378 and 1390 cm^{-1}) and carbonate (1630, 1410, 1300–1336 cm^{-1}) were also observed. From the time courses of the surface concentration of these species, the surface reaction rates of the adsorbed species in the stopped-flow experiments were estimated and compared with those of gas phase products in n-hexane-SCR over Ag/Al_2O_3. Figure 21.11 shows Arrhenius plots of the reaction rate of nitrate consumption measured by *in situ* FTIR, and the steady state reaction rates of NO reduction and n-hexane conversion. In the temperature range 573–648 K, the initial rates of nitrates consumption were close to the steady-state rates of NO reduction. The apparent activation energy of nitrates consumption was 61 kJ mol^{-1}, which is comparable to that for NO reduction in the gas phase (67 kJ mol^{-1}). The good agreement of these rates clearly indicates that

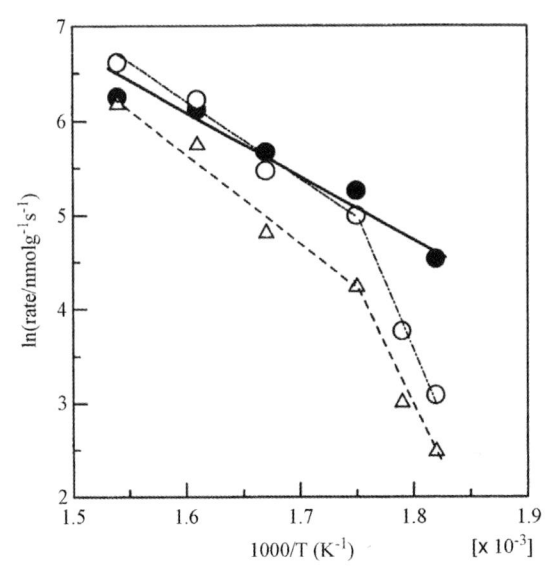

Figure 21.11 Arrhenius plots of (●) the initial rate of NO_3^- consumption in n-hexane + O_2, and the steady-state rates of (○) NO reduction to N_2 and (△) n-hexane conversion to CO, for NO + n-hexane + O_2 reaction over Ag/Al_2O_3. Condition: NO = 1000 ppm, HC = 6000 ppm C, O_2 = 10% [35].

N_2 is formed via surface nitrates as intermediates. It should be noted that the apparent activation energy for NO reduction increased to 156 kJ mol^{-1} below 573 K. The apparent activation energy for n-hexane conversion to COx also increased below 573 K. The higher activation energy at lower temperature suggests self-poisoning of nitrates on Ag/Al_2O_3, which can be correlated to the low HC-SCR activity of Ag/Al_2O_3 below 573 K.

As well as the surface NOx species, the reaction rates of hydrocarbon-derived species were also estimated by *in situ* FTIR. Figure 21.12 shows the transient behavior of the IR spectra of surface adsorbed species after NO + n-hexane + O_2 reaction followed by treatment in a flow of NO + O_2 [35]. During the NO + n-hexane + O_2 reaction, at 0 min in the figure, bands due to acetate (1460 cm^{-1}), formate (1378 and 1390 cm^{-1}) and unidentate and bidentate nitrates (1292 and 1246 cm^{-1}, respectively) were observed. Minor bands are assigned to carbonyl (1720 cm^{-1}), carbonate (1630, 1410, 1300–1336 cm^{-1}), -NCO species bound to Ag^+ ions (2232 cm^{-1}), and -CN species (2162 and 2130 cm^{-1}). After the flowing gas was switched to NO + O_2, the bands due to acetate (1460 cm^{-1}) decreased, while nitrates bands progressively appeared. The result indicate that acetate is a reactive species toward nitrates. The surface oxygenated hydrocarbons, including acetate [35–41], acrylate [42], and enolate species [43], are generally accepted as possible intermediates for HC-SCR. The presence of acetate species is also supported by DFT calculation reported by Gao et al. [41]. From a comparison of the theoretical IR spectra of 10 possible model oxygenated compounds with the experimental spectra in a flow of ethanol + O_2, they

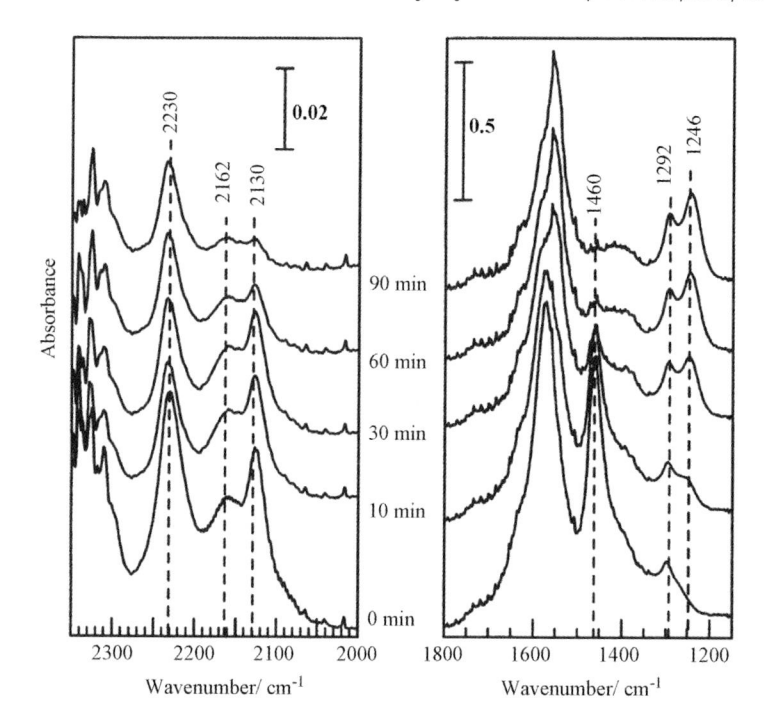

Figure 21.12 Dynamic changes in the IR spectra as a function of time in a flow of NO + O_2 on Ag/Al_2O_3 at 623 K. Before the measurement, the catalyst was pre-exposed to a flow of NO + n-hexane + O_2 for 120 min at 623 K [35].

concluded that the acetate species is the most likely structure on the Ag/Al_2O_3 surface. The reaction of nitrates with oxygenated hydrocarbon species is proposed to be a crucial step in the initial stage of the HC-SCR reaction. The bands due to Ag^+–NCO and –CN decreased in a flow of both NO + O_2 and O_2, indicating the reaction between these species.

From the kinetic analysis of these species, the main reaction pathways on Ag/Al_2O_3 catalyst can be depicted as shown in Scheme 21.1. These reaction pathways are common to the other alumina-based HC-SCR catalysts [36–40]. The selective reduction of NO to N_2 initially proceeds by the parallel reactions: (i) the oxidation of NO to surface nitrates, and (ii) the oxidation of hydrocarbons to surface oxygenates, followed by selective reaction between these species. The surface reaction between

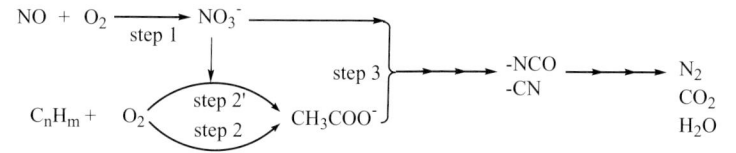

Scheme 21.1 Proposed reaction mechanism of HC-SCR over alumina-based catalysts.

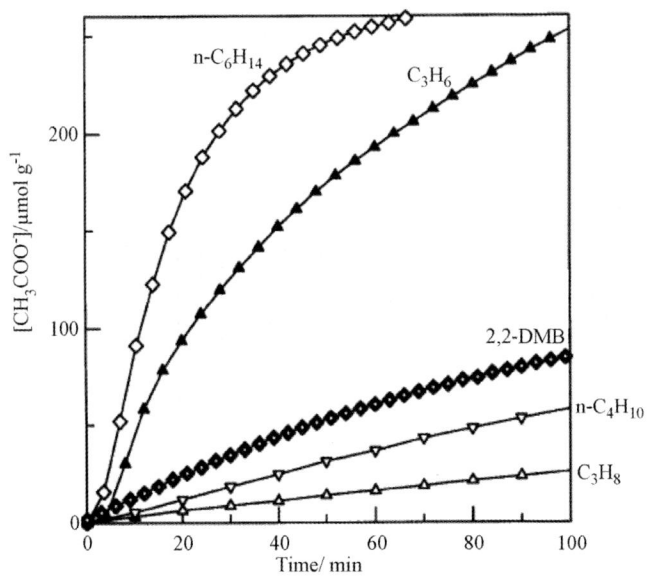

Figure 21.13 Dynamic changes in the IR spectra as a function of time in flowing HC + O_2 on Ag/Al$_2$O$_3$ at 623 K. Conditions: HC = 6000 ppm C, O_2 = 10% [40].

nitrates and oxygenates leads to the formation of NCO and CN species, and finally leads to the formation of N_2 through oxidation or hydrolysis of these nitrogen-containing species.

A sufficient reaction rate of NO to N_2 can be achieved in n-hexane-SCR, however, the reaction rates are too low at lower temperatures when light hydrocarbons are used as reductants. Figure 21.13 shows the dynamic changes in the acetate formation in flowing hydrocarbons + O_2 on Ag/Al$_2$O$_3$ at 623 K after adsorption of nitrates in a flow of NO + O_2 [40]. The rate of acetate formation is very low when propane and n-butane are used as reductants. It is clear that the low activity for partial oxidation of light hydrocarbons is the main problem with the use of Ag/Al$_2$O$_3$ as an HC-SCR catalyst.

21.3.2
Effect of H$_2$ Addition on Reaction Pathways of HC-SCR Over Ag/Al$_2$O$_3$

Shibata et al. clearly demonstrated that the boosting of the HC-SCR activity in the presence of H_2 is mainly attributable to the promotion of partial oxidation of hydrocarbons to surface oxygenates [39], and this conclusion is now generally accepted [44]. Figure 21.14 shows the dynamic changes of surface adsorbed species on Ag/Al$_2$O$_3$. Even in a flow of propane + O_2, the adsorbed nitrates (1251 and 1296 cm^{-1}) cannot be removed in the absence of H_2. In the presence of H_2, the nitrates gradually decreased and then the bands assignable to acetate (1457 cm^{-1}) appeared, indicating formation of oxygenated species significantly boosted in the

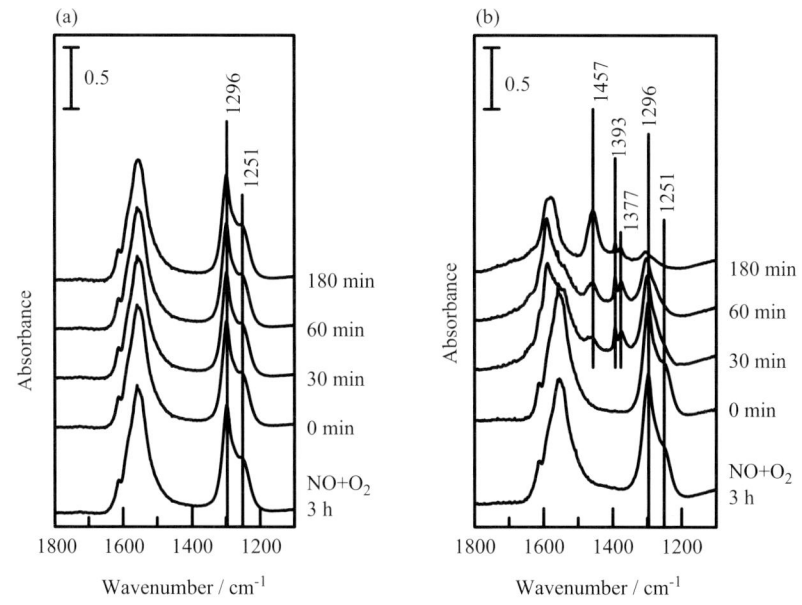

Figure 21.14 Dynamic changes in the IR spectra as a function of time in a flow of propane + O_2 in the (a) absence and (b) presence of H_2 on Ag/Al_2O_3 at 473 K. Before the measurement, the catalyst was pre-exposed to a flow of NO + O_2 for 3 h at 473 K. [40].

presence of H_2. Figure 21.15 shows the time course of the acetate and nitrates concentrations in the presence and absence of H_2 [40]. Clearly, the formation of acetate by the oxidation of propane and the consumption of nitrates were much higher in the presence of H_2 than in the absence of H_2. From the slope of the transient responses, the initial rate of nitrate consumption was below 0.1 nmol g^{-1} s^{-1} in the absence of H_2, but increased to 75 nmol g^{-1} s^{-1} in the presence of H_2 at a surface concentration of nitrates of 280 µmol g^{-1}. This rate corresponds well to the consumption rate of gaseous NO (68 nmol g^{-1} s^{-1}) in the steady state. The comparison of the surface reaction rates should reflect the "hydrogen effects" observed in the gas phase NO conversion. The formation of acetate by propane oxidation (step 2 and 2′ in Scheme 21.1) and nitrates by NO oxidation (step 1) were remarkably promoted by the addition of H_2. The acceleration of step 3 by the addition of H_2 is caused by the increases in concentration terms of adsorbed species, which is due to the promotion of steps 1 and 2′. Comparing the promotion effect on NO oxidation (step 1) and propane oxidation (step 2 and 2′), the former rate is already sufficient (66 nmol g^{-1} s^{-1}), even in the absence of H_2, but the latter became two orders of magnitude faster on addition of H_2 (from less than 0.1 to 19 nmol g^{-1} s^{-1} for step 2, and to 40 nmol g^{-1} s^{-1} for step 2′). From the kinetic analysis of the adsorbed species, the promotion effect of the addition of H_2 on the propane-SCR is attributed to the remarkable promotion of partial oxidation of propane to mainly surface acetate.

Figure 21.15 Time dependence of (○, ●) nitrate and (□, ■) acetate concentration in (○, □) propane + O_2 and (●, ■) propane + O_2 + H_2 on Ag/Al$_2$O$_3$ at 473 K. Before the measurement, the catalyst was pre-exposed to a flow of NO + O_2 for 3 h at 473 K [40].

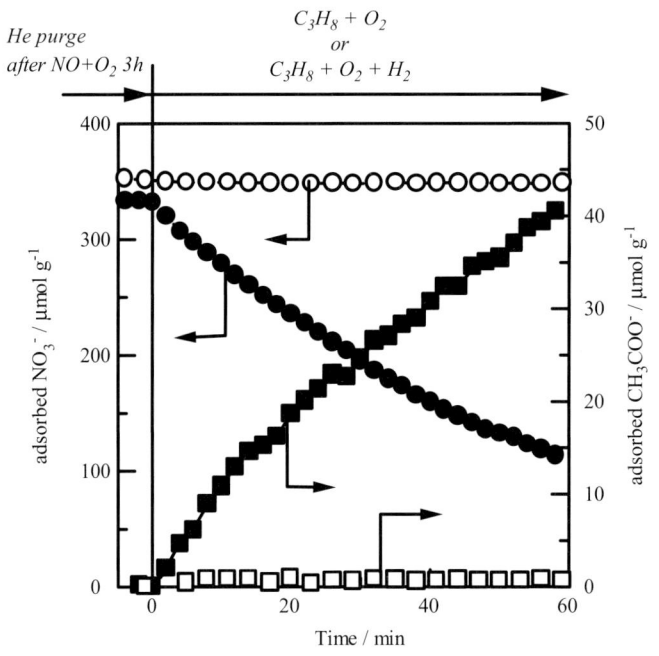

21.4
Relation Between Ag Cluster and Oxidative Activation of Hydrocarbons

21.4.1
Debates on Role of Ag Clusters

Although *in situ* analysis by UV–Vis and EXAFS indicated that the Ag cluster is the active species for H_2-HC-SCR, there are negative results reported about the role of Ag cluster as active species [45]. Sazama *et al.* investigated the dynamic analysis of the formation and re-dispersion of Ag clusters on alumina under realistic reaction conditions using a homemade *in situ* UV–Vis spectrometer [21]. They also confirmed the presence of an additional band at around 31 000 cm^{-1}, assignable to Ag$_n^{\delta+}$ clusters having cubic symmetry, [46] under the decane-SCR in the presence of H_2, while only bands at 41 600 and 46 600 cm^{-1}, attributable to isolated Ag$^+$ ions, were observed without H_2. Figure 21.16 shows the response of NO conversion and the intensity of the UV–Vis band assignable to the Ag cluster (31 000 cm^{-1}) during hydrogen switching on/off. The NO$_x$ conversion and the band of the Ag$_n^{\delta+}$ cluster increased sharply after H_2 addition. After the hydrogen was switched off, NO$_x$

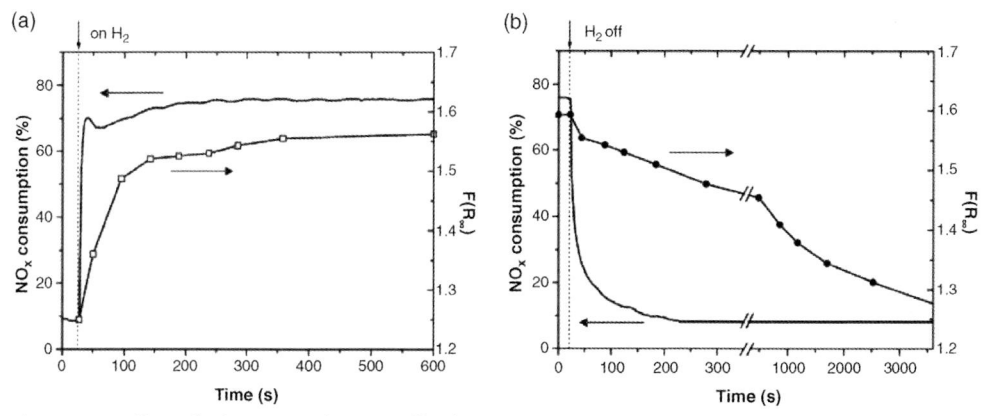

Figure 21.16 Effect of hydrogen switching on/off in decane-SCR-NO over Ag/Al$_2$O$_3$ at 523 K, 0.1% NO, 0.06% decane, 6% O$_2$ and 0 or 0.2% H$_2$. GHSV = 60 000 h^{-1}. NOx consumption and intensity of the UV–Vis band at 31 000 cm^{-1} (323 nm) as a function of time and hydrogen switch (a) on and (b) off [21].

conversion decreased with longer response within 4 min. Comparing the responses of the Ag$_n^{\delta+}$ cluster and NO conversion, the dispersion of Ag$_n^{\delta+}$ clusters to Ag$^+$ ions was significantly slower than the decrease in NO conversion. From this result, they claimed that the formation of Ag cluster is not essential for the enhancement of HC-SCR activity in the presence of H$_2$.

By using *in situ* EXAFS, Been *at al.* analyzed Ag species of Ag/Al$_2$O$_3$ under the H$_2$-octane-SCR conditions [47]. In their EXAFS spectra, shown in Figure 21.17, the change of Ag structure in the presence of H$_2$ was not significant. The Ag$_n^{\delta+}$ (n ca. 3 from the coordination number) cluster was already present on the alumina support under octane-SCR conditions without H$_2$ (shoulder in spectrum b). Furthermore, the Ag$_n^{\delta+}$ cluster was also observed in the co-presence of CO. Although the promoting effect of octane-SCR was only observed in the presence of H$_2$, the Ag$_n^{\delta+}$ cluster is present under the octane-SCR without H$_2$ and also under the octane-SCR with CO. They concluded that the "hydrogen effect" could not be attributed to changes in Ag particle size.

Table 21.1 summarizes the relationships between reaction conditions, presence of Ag$_n^{\delta+}$ cluster, and hydrogen effect on HC-SCR. Sazama *et al.* and Burch *et al.* have pointed out that the "hydrogen effect" cannot be attributed only to the formation of Ag$_n^{\delta+}$ cluster because of (i) the delay in the disappearance of Ag cluster after H$_2$ cut-off [21], and (ii) the formation of Ag$_n^{\delta+}$ cluster by other non-effective reductants such as octane and CO [47]. However, most importantly, the "hydrogen effect" was not observed in the absence of Ag$_n^{\delta+}$ clusters. It is clearly shown that neither Ag cluster nor "hydrogen effect" were observed in H$_2$-propane-SCR over 0.5 wt.%Ag/Al$_2$O$_3$. Therefore, the formation of Ag$_n^{\delta+}$ cluster is not a sufficient condition, but a necessary condition. There must be another factor for the "hydrogen effect" than the formation of Ag$_n^{\delta+}$ cluster, which cannot be achieved by the co-presence of CO or hydrocarbons.

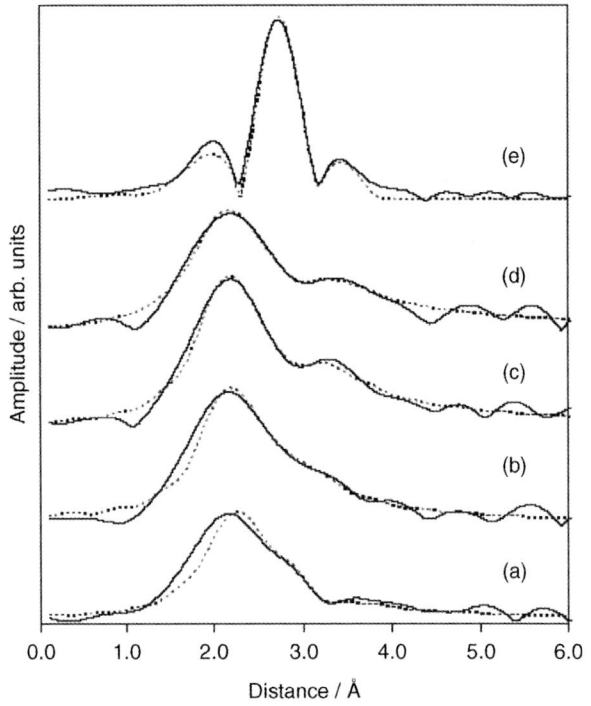

Figure 21.17 Comparison of the experimental (solid line) and fitted (dashed line) pseudo-radial distribution (lower) functions from 2%Ag/alumina (a) as received, under SCR reaction conditions at 225 °C with (b) no co-reductant (c) 0.72% hydrogen and (d) 0.72% carbon monoxide and (e) at 224 °C in 8% hydrogen/helium [47].

What is another factor for the "hydrogen effect"? In relation to this question, Sazama et al. published a very interesting communication [48]. Table 21.2 shows the effect of hydrogen and hydrogen peroxide on the decane-SCR over Ag/Al_2O_3 at 473 K. The conversion of NO is negligible without co-feeding of hydrogen or hydrogen

Table 21.1 Relationships between reaction conditions, presence of Ag cluster, and hydrogen effect on HC-SCR.

Reaction conditions	Presence of $Ag_n^{\delta+}$ cluster	Hydrogen effect on HC-SCR	Reference
H_2-HC-SCR in steady state	Yes	Yes	[31]
0–20 min after H_2 cut-off	Yes	No	[21, 31]
CO-HC-SCR	Yes	No	[47]
Octane-SCR	Yes	No	[47]
H_2-HC-SCR over 0.5 wt.%Ag/Al_2O_3	No	No	[31]

Table 21.2 Comparison of the effect of hydrogen and hydrogen peroxide on the decane-SCR reaction over Ag/Al_2O_3 at 473 K [48].

	Reducing agent		
	Decane	Decane + H_2O_2	Decane + H_2
X NO[a]	2.5	60.0	49.5
Yield of N_2	0	11.8	21.0
Yield of NO_2	2.5	48.2	28.5

[a]Conversion of NO to nitrogen and nitrogen dioxide.

peroxide, but the NO conversion increased significantly to 60% in the presence of 0.2% H_2. The promotion effect was also observed in decane-SCR with 0.2% of hydrogen peroxide with an increase in decane conversion from nearly zero to 12%. The results suggest a role for highly reactive hydroxy and hydroperoxy radicals formed from hydrogen peroxide on HC-SCR. In the co-presence of hydrogen peroxide, the NO_2 yield was higher than that with H_2, which may be due to the reaction between hydroxy radicals as follows: $HO_2 + NO \rightarrow NO_2 + OH$. The similarity of the effect of H_2 and hydrogen peroxide on the decane-SCR performance strongly suggests the role of hydroperoxy-like species during HC-SCR in the presence of H_2.

Since the addition of hydrogen promotes partial oxidation of hydrocarbons, the hydrogen effect can be rationalized by the formation of reactive oxygen species. It is well known that the addition of hydrogen promotes the partial oxidation of hydrocarbons such as CH_4 [49], ethane [50], and so on. For example, the formation of peroxide (O_2^{2-}) on iron phosphate catalyst is the cause of the promotion of the selective oxidation of methane or ethane by oxygen with hydrogen [49, 50]. A similar mechanism is proposed for H_2-assisted allyl alcohol epoxidation over titanosilicate catalyst; the formation of O_2^- intermediates via $H_2 + O_2$ reaction is proposed as a key step [51]. In these cases, hydrogen reductively activates molecular oxygen as the electron donor as well as the proton donor. In general, the term "reductive oxidation" stands for the following process: $O_2 + 2H^+ + 2e^- \rightarrow [O] + H_2O$, where [O] represents reactive oxygen species. This idea can be extended to the H_2-HC-SCR over Ag/Al_2O_3.

Reductive activation of molecular oxygen on an Ag cluster was confirmed by EPR spectroscopy. Figure 21.18 shows EPR spectra recorded at 77 K for Ag/Al_2O_3 after various pre-treatments [31]. After the propane-SCR reaction at 573 K (spectrum a), the sample showed no EPR signal. After the $H_2 + O_2$ reaction at 573 K (spectrum b), 2 wt.% Ag/Al_2O_3 showed an EPR spectrum with anisotropic g values $g_{xx} = 2.001$, $g_{yy} = 2.009$, and $g_{zz} = 2.030$. This signal can be identified as O_2^- (super oxide) ion on a silver site [52]. The EPR signals due to O_2^- ion were observed even after the H_2-propane-SCR reaction at 573 K (spectrum c). These results clearly indicate the reductive activation of molecular oxygen to the reactive oxygen species, O_2^-. It is important to note that no signal was observed after CO (0.5%) + O_2(10%) reaction on 2 wt.% Ag/Al_2O_3 at 573 K (spectrum d). In the case of 0.5 wt.% Ag/Al_2O_3, no EPR

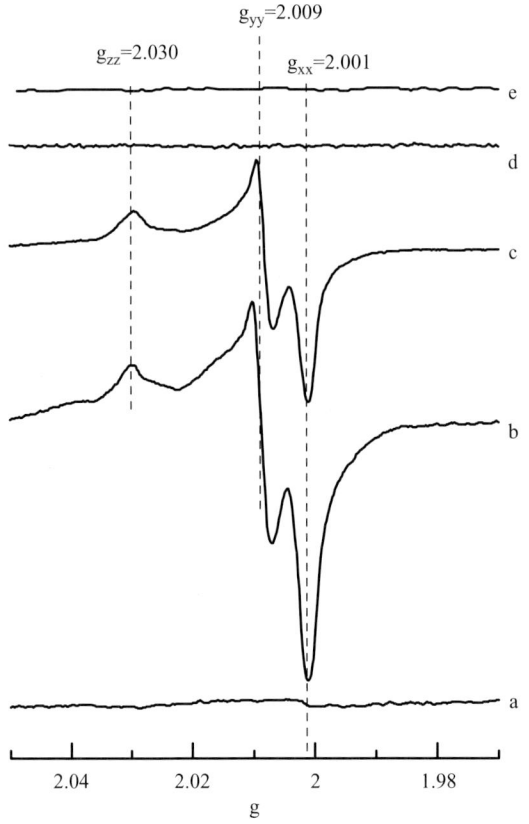

Figure 21.18 EPR spectra recorded at 77 K for Ag/Al$_2$O$_3$ after various pre-treatments at 573 K: (a) 2 wt.% Ag/Al$_2$O$_3$ after propane-SCR (b) 2 wt.% Ag/Al$_2$O$_3$ after H$_2$ + O$_2$ reaction, (c) 2 wt.% Ag/Al$_2$O$_3$ after H$_2$-propane-SCR, (d) 2 wt.% Ag/Al$_2$O$_3$ after CO + O$_2$ reaction, (e) 0.5 wt.% Ag/Al$_2$O$_3$ after H$_2$-C$_3$H$_8$-SCR [31].

signals due to O$_2^-$ ion were observed after the H$_2$-propane-SCR reaction (spectrum e) nor after H$_2$ + O$_2$ reaction (not shown). The EPR results provide direct evidence for the reductive activation of molecular oxygen into reactive oxygen species, superoxide anion (O$_2^-$), which is known to be reactive toward C–H bonds of hydrocarbons.

21.4.2
Reductive Activation of O$_2$ and Promoted HC-SCR on Ag Cluster

In this final section, the role of H$_2$ in the reductive activation of O$_2$ on Ag cluster is discussed. Baba et al. reported that silver cations in Ag-exchanged zeolites [53] and silver salts of heteropoly acids [54] are reduced by hydrogen to generate protons and silver clusters according to the following scheme.

$$Ag^+ + 1/2 H_2 \rightarrow Ag^0 + H^+$$

$$(n-1)Ag^0 + Ag^+ \rightarrow Ag_n^+$$

$$Ag_n^+ + 1/2 H_2 \rightarrow Ag_n^0 + H^+$$

On the Ag/Ag$_2$O$_3$ catalyst, the production of OH groups as a result of generation of protons on an alumina surface was also observed in the IR spectrum [31]. After dehydration at 823 K followed by exposure to a flow of H$_2$ (0.5%) at 423 K, a broad band characteristic of a new O−H stretching vibration was observed in the 3700–3000 cm^{-1} region, while the intensity of the band due to the deformation vibration of water at 1617 cm^{-1} was relatively small. This indicates the formation of OH groups as well as water as minor species. Due to the low signal/noise ratio in the O−H stretching vibration region, the formation of OH species was also confirmed by reduction with D$_2$. The resulting IR spectrum shows broad bands due to a new O−D stretching vibration at 2642, 2672, 2704, 2744, and 2772 cm^{-1}, which correspond to the O−H bands with positions at 3494, 3552, 3588, 3672, and 3720 cm^{-1}. The bands at lower stretching frequency assignable to the O−H (O−D) group having higher Brønsted acidity are more intense than those at higher frequencies. Furthermore, the intensity of the band due to the DOD deformation vibration of water is absent, indicating the formation of H$^+$ (D$^+$). These IR results indicate that Ag$^+$ species are reduced by hydrogen to generate acidic protons together with the formation of Ag$_n^{\delta+}$ clusters.

As described in Section 21.3.2, the H$_2$ addition promotes the oxidation of hydrocarbon to surface oxygenates and the oxidation of NO to nitrates. It is clear, that the co-presence of protons and reduced Ag species is indispensable for reductive oxidation of O$_2$ to yield reactive oxygen species (O$_2^-$). This clearly explains why CO is not effective as a co-reductant and why the formation of Ag clusters is not sufficient to improve the catalytic activity. Over 0.5 wt.% Ag/Al$_2$O$_3$, showing no activity for NO reduction, monomeric Ag$^+$ species are not reduced to clusters, and O$_2^-$ is not produced because of the absence of reduced Ag species. The co-presence of CO leads to the formation of Ag clusters, though the condition is not sufficient because of the absence of protons. The co-presence of both the dissociated hydrogen as acidic proton and Ag cluster are indispensable for the O$_2$ activation.

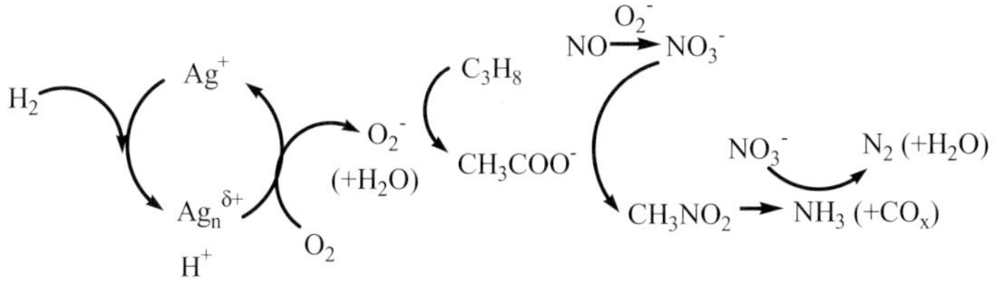

Scheme 21.2 Proposed reaction scheme of H$_2$-HC-SCR over Ag/Al$_2$O$_3$ catalyst.

As discussed above, the reaction mechanism of H_2-HC-SCR is proposed as shown in Scheme 21.2. The initial steps of the reaction are: (i) H_2 dissociation on a Ag site, (ii) spillover of H atom to form proton, (iii) aggregation of isolated Ag to form the reduced $Ag_n^{\delta+}$ cluster, (iv) O_2 reduction with $Ag_n^{\delta+}$ and H^+ to yield O_2^- and H_2O. Then (v) O_2^- activates the C—H bond of hydrocarbons to yield hydrocarbon radicals, which will be finally converted to acetate ions, and (vi) NO oxidation to NO_2. These intermediates react to produce N_2: (vii) acetate reaction with NO_2 to yield nitromethane, which is then converted to NH_3 via NCO species, followed by (viii) reduction of NO_2 with adsorbed NH_3 (NH_4^+) to produce N_2. The reaction scheme thus obtained can rationalize the role of $Ag_n^{\delta+}$ cluster as a necessary condition and the role of protons as another condition for the "hydrogen effect".

References

1 Burch, R. (2004) Knowledge and know-how in emission control for mobile applications. *Catal. Rev.*, **46**, 271–333.

2 Takahashi, N., Shinjoh, H., Iijima, T., Suzuki, T., Yamazaki, K., Yokota, K., Suzuki, H., Miyoshi, N., Matsumoto, S., Tanizawa, T., Tanaka, T., Tateishi, S. and Kasahara, K. (1996) The new concept 3-way catalyst for automotive lean-burn engine: NO_x storage and reduction catalyst. *Catal. Today*, **27**, 63–69.

3 Matsumoto, S. (1996) $DeNO_x$ catalyst for automotive lean-burn engine. *Catal. Today*, **29**, 43–45.

4 Matsumoto, S. (2004) Recent advances in automobile exhaust catalysts. *Catal. Today*, **90**, 183–190.

5 Koebel, M., Elsener, M. and Kleemann, M. (2000) Urea-SCR: a promising technique to reduce NO_x emissions from automotive diesel engines. *Catal. Today*, **59**, 335–345.

6 Iwamoto, M., Yahiro, H., Yu-u, Y., Shundo, S. and Mizuno, N. (1990) Selective reduction of NO by lower hydrocarbons in the presence of O_2 and SO_2 over copper ion-exchanged zeolites. *Shokubai (Catalyst)*, **32**, 430–433.

7 Held, W., König, A., Richter, T. and Pupper, L. (1990) Catalytic NOx reduction in net oxidizing exhaust gas. SAE Paper, 900496.

8 Iwamoto, M. and Hamada, H. (1991) Removal of nitrogen monoxide from exhaust gases through novel catalytic processes. *Catal. Today*, **10**, 57–71.

9 Iwamoto, M. (2000) Air pollution abatement through heterogeneous catalysis. *Stud. Surf. Sci. Catal.*, **130**, 23–47.

10 Kikuchi, T. and Kumagai, M. (1998) Selective reduction of NO_x in diesel engine exhaust over supported Co, Cu catalysts. *Sekiyu Gakkaishi*, **41**, 173–174.

11 Nakatsuji, T., Yasukawa, R., Tabata, K., Ueda, K. and Niwa, M. (1998) Catalytic reduction system of NO_x in exhaust gases from diesel engines with secondary fuel injection. *Appl. Catal. B: Environ.*, **17**, 333–345.

12 Eränen, K., Lindfors, L.-E., Niemi, A., Elfving, P. and Cider, L. (2000) Influence of hydrocarbons on the selective catalytic reduction of NOx over Ag/Al_2O_3 - laboratory and engine tests. SAE paper, 012813.

13 Lindfors, L.-E., Eranen, K., Klingstedt, F. and Murzin, D.Yu. (2004) Silver/alumina catalyst for selective catalytic reduction of NO_x to N_2 by hydrocarbons in diesel powered vehicles. *Top. Catal.*, **28**, 185–189.

14 Thomas, J.F., Lewis, S.A., Bunting, G.B., Storey, J.M., Graves, R.L. and Park, P.W. (2005) Hydrocarbon selective catalytic reduction using a silver-alumina catalyst

with light alcohols and other reductants. SAE paper, 011082.

15 Miyadera, T. and Yoshida, K. (1993) Alumina-supported catalysts for the selective reduction of nitric oxide by propene. *Chem. Lett.*, 1483–1484.

16 Miyadera, T. (1993) Alumina-supported silver catalysts for the selective reduction of nitric oxide with propene and oxygen-containing organic compounds. *Appl. Catal. B: Environ.*, **2**, 199–205.

17 He, H. and Yu, Y. (2005) Selective catalytic reduction of NO_x over Ag/Al_2O_3 catalyst: from reaction mechanism to diesel engine test. *Catal. Today*, **100**, 37–47.

18 Satokawa, S. (2000) Enhancing the $NO/C_3H_8/O_2$ reaction by using H_2 over Ag/Al_2O_3 catalysts under lean-exhaust conditions. *Chem. Lett.*, 294–295.

19 Satokawa, S., Shibata, J., Shimizu, K., Satsuma, A. and Hattori, T. (2003) Enhancing the low temperature activity by adding H_2 over Ag/Al_2O_3 catalyst for the selective reduction of NO by light hydrocarbons under lean-exhaust conditions. *Appl. Catal. B: Environ.*, **42**, 179–186.

20 Burch, R., Breen, J.P., Hill, C.J., Krutzsch, B., Konrad, B., Jobson, E., Cider, L., Eranen, K., Klingstedt, F. and Lindfors, L.-E. (2004) Exceptional activity for NO_x reduction at low temperatures using combinations of hydrogen and higher hydrocarbons on Ag/Al_2O_3 catalysts. *Top. Catal*, **30–31**, 19–25.

21 Sazama, P., Cǎpek, L., Drobná, H., Sobalík, Z., Dědeček, J., Arve, K. and Wichterlová, B. (2005) Enhancement of decane-SCR-NO_x over Ag/alumina by hydrogen. Reaction kinetics and *in situ* FTIR and UV–vis study. *J. Catal.*, **232**, 302–317.

22 Klingstedt, F., Arve, K., Eränen, K. and Murzin, D.Y. (2006) Toward improved catalytic low-temperature NO_x removal in diesel-powered vehicles. *Acc. Chem. Res.*, **39**, 273–282.

23 Richter, M., Bentrup, U., Eckelt, R., Schneider, M., Pohl, M.-M. and Fricke, R. (2004) The effect of hydrogen on the selective catalytic reduction of NO in excess oxygen over Ag/Al_2O_3. *Appl. Catal. B: Environ.*, **51**, 261–274.

24 Zhang, X., He, H. and Ma, Z. (2007) Hydrogen promotes the selective catalytic reduction of NO_x by ethanol over Ag/Al_2O_3. *Catal. Commun.*, **8**, 187–192.

25 Richter, M., Fricke, R. and Ecklt, R. (2004) Unusual activity enhancement of NO conversion over Ag/Al_2O_3 by using a mixed NH_3/H_2 reductant under lean conditions. *Catal. Lett.*, **94**, 115–118.

26 Shimizu, K. and Satsuma, A. (2007) Reaction mechanism of H_2-promoted selective catalytic reduction of NO with NH_3 over Ag/Al_2O_3. *J. Phys. Chem. C*, **111**, 2259–2264.

27 Satsuma, A., Shibata, J., Shimizu, K. and Hattori, T. (2005) Ag cluster as active species for HC-SCR over Ag-zeolites. *Catal. Surv. Asia*, **9**, 75–85.

28 Shimizu, K., Higashimata, T., Tsuzuki, M. and Satsuma, A. (2006) Effect of hydrogen addition on SO_2-tolerance of silver-alumina for SCR of NO with propane. *J. Catal.*, **239**, 117–124.

29 Breen, J.P., Burch, R., Hardacre, C., Hill, C.J., Krutzsch, B., Bandl-Konrad, B., Jobson, E., Cider, L., Blakeman, P.G., Peace, L.J., Twigg, M.V., Preis, M. and Gottschling, M. (2007) An investigation of the thermal stability and sulphur tolerance of $Ag/\gamma\text{-}Al_2O_3$ catalysts for the SCR of NO_x with hydrocarbons and hydrogen. *Appl. Catal. B: Environ.*, **70**, 36–44.

30 Satsuma, A., Shibata, J., Wada, A., Shinozaki, Y. and Hattori, T. (2003) In-situ UV-visible spectroscopic study for dynamic analysis of silver catalyst. *Stud. Surf. Sci. Catal.*, **145**, 235–238.

31 Shimizu, K., Tsuzuki, M., Kato, K., Yokota, S., Okumura, K. and Satsuma, A. (2007) Reactive activation of O_2 with H_2 reduced silver clusters a key step in the H_2-promoted selective catalytic reduction of

NO with C_3H_8 over Ag/Al_2O_3. *J. Phys. Chem. C*, **111**, 950–959.

32 Shibata, J., Shimizu, K., Takada, Y., Shichi, A., Yoshida, H., Satokawa, S., Satsuma, A. and Hattori, T. (2004) Structure of active Ag clusters in Ag-zeolites for SCR of NO by propane in the presence of hydrogen. *J. Catal.*, **227**, 367–374.

33 Suzuki, Y., Matsumoto, N., Aitani, T., Miyanaga, T. and Hoshino, H. (2005) *In situ* infrared and EXAFS studies of an Ag cluster in zeolite X. *Polyhedron*, **24**, 685–691.

34 Satsuma, A. and Shimizu, K. (2003) *In-situ* FT/IR study of selective catalytic reduction of NO over alumina-based catalysts. *Prog. Energy Combust. Sci.*, **29**, 71–84.

35 Shimizu, K., Shibata, J., Yoshida, H., Satsuma, A. and Hattori, T. (2001) Silver-alumina catalysts for selective reduction of NO by higher hydrocarbons: structure of active sites and reaction mechanism. *Appl. Catal. B: Environ.*, **30**, 151–162.

36 Shimizu, K., Kawabata, H., Satsuma, A. and Hattori, T. (1998) Formation and reaction of surface acetate on Al_2O_3 during NO reduction by C_3H_6. *Appl. Catal. B: Environ.*, **19**, L87–L92.

37 Shimizu, K., Kawabata, H., Satsuma, A. and Hattori, T. (1999) Role of acetate and nitrates in the selective catalytic reduction of NO by propene over alumina catalyst as investigated by FTIR. *J. Phys. Chem. B*, **103**, 5240–5245.

38 Shimizu, K., Kawabata, H., Maeshima, H., Satsuma, A. and Hattori, T. (2000) Intermediates in the selective reduction of NO by propene over $Cu-Al_2O_3$ catalysts: transient *in situ* FTIR study. *J. Phys. Chem. B*, **104**, 2885–2893.

39 Shimizu, K., Shibata, J., Satsuma, A. and Hattori, T. (2001) Mechanistic causes of hydrocarbon effect on the activity of $Ag-Al_2O_3$ catalyst for selective reduction NO. *Phys. Chem. Chem. Phys.*, **3**, 880–884.

40 Shibata, J., Shimizu, K., Satokawa, S., Satsuma, A. and Hattori, T. (2003) Promotion effect of hydrogen on surface steps in SCR of NO by propane over alumina-based silver catalyst as examined by transient FT-IR. *Phys. Chem. Chem. Phys.*, **5**, 2154–2160.

41 Gao, H., Yan, T., Yu, Y. and He, H. (2008) DFT and DRIFTS studies on the adsorption of acetate on the Ag/Al_2O_3 catalyst. *J. Phys. Chem. C*, **112**, 6933–6938.

42 Iglesias-Juez, A., Hungría, A.B., Martínez-Arias, A., Fuerte, A., Fernández-Garcia, M., Anderson, J.A., Conesa, J.C. and Soria, J. (2003) Nature and catalytic role of active silver species in the lean NO_x reduction with C_3H_6 in the presence of water. *J. Catal.*, **217**, 310–323.

43 Yu, Y., He, H. and Feng, Q. (2003) Novel enolic surface species formed during partial oxidation of CH_3CHO, C_2H_5OH, and C_3H_6 on Ag/Al_2O_3: An *in situ* DRIFTS study. *J. Phys. Chem. B*, **107**, 13090–13092.

44 Breen, J.P. and Burch, R. (2006) A review of the effect of the addition of hydrogen in the selective catalytic reduction of NO_x with hydrocarbons on silver catalysts. *Top. Catal.*, **39**, 53–58.

45 Shimizu, K. and Satsuma, A. (2006) Selective catalytic reduction of NO over supported silver catalysts - practical and mechanistic aspects. *Phys. Chem. Chem. Phys.*, **8**, 2677–2695.

46 Texter, J., Hastreiter, J.J. and Hall, J.L. (1983) Spectroscopic confirmation of the tetrahedral geometry of tetraaquasilver $(+)$ ion $(Ag(H_2O)_4^+)$. *J. Phys. Chem.*, **87**, 4690–4693.

47 Breen, J.P., Burch, R., Hardacre, C. and Hill, C.J. (2005) Structural investigation of the promotional effect of hydrogen during the selective catalytic reduction of NOx with hydrocarbons over Ag/Al_2O_3 catalysts. *J. Phys. Chem. B*, **109**, 4805–4807.

48 Sazama, P. and Wichterlova, B. (2005) Selective catalytic reduction of NOx by hydrocarbons enhanced by hydrogen peroxide over silver/alumina catalysts. *Chem. Commun*, **38**, 4810–4811.

49 Wang, Y., Otsuka, K. and Ebitani, K. (1995) *In situ* FTIR study on the active oxygen species for the conversion of methane to methanol. *Catal. Lett.*, **35**, 259–263.

50 Wang, Y. and Otsuka, K. (1995) Catalytic oxidation of methane to methanol with H_2-O_2 Gas mixture at atmospheric pressure. *J. Catal.*, **155**, 256–267.

51 Setti, V.N., Manikandan, P., Srinivas, D. and Ratonasamy, P. (2003) Reactive oxygen species in epoxidation reactions over titanosilicate molecular sieves. *J. Catal.*, **216**, 461–467.

52 Clarkson, R.B. and McCellan, S. (1978) The character of adsorption of molecular oxygen(1-) on supported silver surfaces. *J. Phys. Chem.*, **82**, 294–297.

53 Baba, T., Komatsu, N., Sawada, H., Yamaguchi, Y., Takahashi, T., Sugisawa, H. and Ono, Y. (1999) ^{1}H magic angle spinning NMR evidence for dissociative adsorption of hydrogen on Ag^+-exchanged A- and Y-zeolites. *Langmuir*, **15**, 7894–7896.

54 Baba, T., Nimura, M., Ono, Y. and Ohno, Y. (1993) Solid-state proton MAS NMR study on the highly active protons in partially reduced trisilver dodecatungstophosphate ($Ag_3PW_{12}O_{40}$). *J. Phys. Chem.*, **97**, 12888–12893.

22
Dynamic Structural Change of Pd Induced by Interaction with Zeolites Studied by Means of Dispersive and Quick XAFS

Kazu Okumura

22.1
Introduction

X-ray absorption fine structure (XAFS) is a useful technique for the analysis of the local structure of heterogeneous catalysts, whose structural information is otherwise difficult to obtain. In the conventional XAFS technique, the data collection is carried out under static conditions using double monochromators that are moved stepwise. Along with the recent development of various techniques for high-speed measurement, that is, QXAFS (quick XAFS) and DXAFS (dispersive XAFS), *in situ* measurements were realized. In QXAFS, the double monochromator is moved continuously to obtain a monochromatic X-ray beam. On the other hand, in DXAFS, the whole energy region is collected at the same time using a bent polychromator, as shown in Figure 22.1. Bragg and Laue configurations are chosen, depending on the X-ray energy region for XAFS measurements. Using these techniques combined with an appropriate cell and a gas-flow line, *in situ* measurements of the various chemical processes, such as formation of the active sites in heterogeneous catalysts and clustering of metal atoms, have been realized [1, 2].

In recent years, much effort has been devoted to the design of various *in situ* cells and studies combined with other techniques, that is, XRD and a mass spectrometer. Hannemann *et al.* proposed a versatile *in situ* cell for fluorescence/transmission EXAFS and XRD of heterogeneous catalysts in the gas and liquid phases [3]. Clausen *et al.* and Sankar *et al.* used a capillary cell for combined XRD-EXAFS studies [4, 5]. A capillary-type high pressure cell was also designed by Bare *et al.* [6]. Grunwaldt *et al.* designed various types of *in situ* cell and applied them to kinetic studies of the reduction of CuO/ZnO and PdO/ZrO_2 catalysts [7]. They reported the two-dimensional distribution of Rh with different valence states in the course of the partial oxidation of methane, which was realized through the combination of XAFS, mass spectroscopy and a CCD camera [8]. The local structure of Al in zeolites was measured by *in situ* Al K-edge XAFS investigation [9–13]. In this chapter, we will focus on our studies concerning the structural change of Pd induced by the metal–support interaction with zeolite supports. Small metal clusters occluded in a zeolite pore

Molecular Nano Dynamics, Volume II: Active Surfaces, Single Crystals and Single Biocells
Edited by H. Fukumura, M. Irie, Y. Iwasawa, H. Masuhara, and K. Uosaki
Copyright © 2009 WILEY-VCH Verlag GmbH & Co. KGaA, Weinheim
ISBN: 978-3-527-32017-2

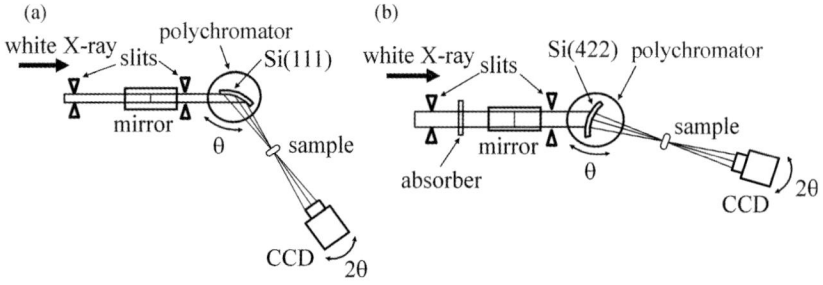

Figure 22.1 An energy-dispersive XAFS instrument installed in SPring-8 BL28B2. Bragg ($E_{x\text{-}ray}$ < 12 keV) (a), and Laue configurations ($E_{x\text{-}ray}$ > 12 keV) (b).

have been studied primarily from the viewpoint of the formation of uniform active sites for catalytic reactions. The introduction of metals into zeolites has been achieved through various techniques including chemical vapor deposition (CVD), the ship-in-bottle method and decarbonylation of carbonyl clusters in zeolite pores [14–18]. Palladium clusters in zeolites have been paid special attention since Pd exhibits high catalytic activity in many valuable reactions, such as the selective reduction of NO, the total combustion of hydrocarbons and organic reactions. Recently, small metal Pd clusters were obtained through the introduction of Pd into the pores of Na-Y zeolite, followed by a successive reduction at low temperatures. Furthermore, it was found that the location of the metal Pd cluster in X and Y-zeolites was significantly affected by the calcination temperature [19, 20]. We have focused on the influence of the Brønsted acid site as well as the structure of zeolite on the generation of metal Pd clusters and highly dispersed PdO. This is because it has been revealed that the structure and the acid sites of zeolites remarkably affect the generation of active sites for catalytic reactions. The catalytic performance of Pd was found to be highlyly dependent on the structure and composition of the zeolite support [21, 22]. This suggested that the strong metal–support interaction between PdO and Brønsted acid sites is an important factor not only in the generation of the active Pd center but also in its catalytic performance. The interaction was directly evidenced from the structural change in Pd induced by the Brønsted acid sites of zeolites in an oxidative or reductive atmosphere. However, the formation and structure of the active Pd species or its precursor and the role of the Brønsted acid sites associated with Pd are rather ambiguous. In this study, DXAFS and QXAFS techniques were utilized to follow the dynamic structural change of Pd induced by interaction with the Brønsted acid sites of zeolites.

22.2
Formation and Structure of Highly Dispersed PdO Interacted with Brønsted Acid Sites [23–25]

First, H-ZSM-5 zeolites with different Al content were employed as supports for palladium and the structure of Pd was measured by Pd K-edge EXAFS in the static

22.2 Formation and Structure of Highly Dispersed PdO Interacted with Brønsted Acid Sites

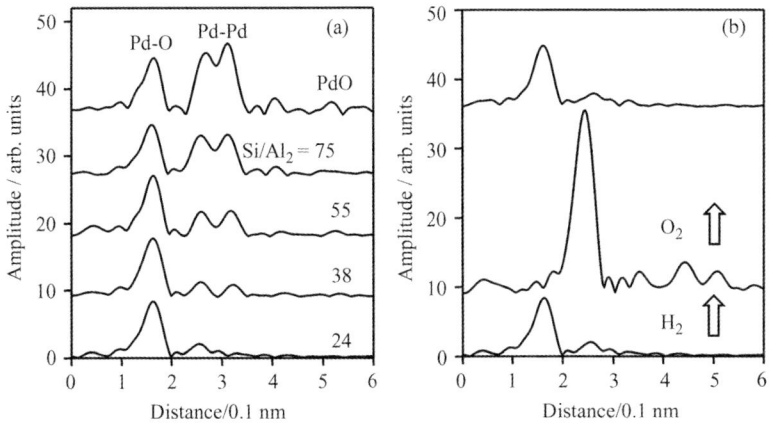

Figure 22.2 k^3-weighted Pd K-edge EXAFS Fourier transforms of 0.4 wt% Pd loaded on H-ZSM-5 with different Al concentrations (a) and H-ZSM-5 (Si/Al$_2$ = 24) treated with H$_2$, and O$_2$ at 773 K (b).

mode. The catalyst exhibited high activity for the selective reduction of NO with methane in an atmosphere of excess O$_2$ [26]. Figure 22.2a shows the Fourier transforms of the $k^3\chi(k)$ EXAFS for Pd/H-ZSM-5 with different Al content.

All samples were oxidized under an oxygen flow at 773 K for 3 h before the measurement. With increasing Al content of the H-ZSM-5, the intensity of the Pd–Pd peaks appearing at 0.26 and 0.31 nm gradually decreased, and finally disappeared for the Pd/H-ZSM-5 with the highest Al content (Si/Al$_2$ = 24), where high activity toward the NO–CH$_4$–O$_2$ reaction was observed. The results indicated that the size of PdO is a function of the amount of Brønsted acid in H-ZSM-5 and decreases with an increase in the acid amount, since the intensity of the Pd–Pd shell seems to reflect the size of PdO. On the other hand, for the Pd–O bond observed at 0.16 nm (phase shift uncorrected) in Figure 22.2a, the spectra for bulk PdO and highly dispersed PdO on H-ZSM-5 are quite similar. In addition, the coordination number and bond distance of Pd–O determined by curve fitting analysis on highly dispersed PdO agreed well with those for bulk PdO, implying that the local structure of highly dispersed PdO is closely similar to that of bulk PdO. Therefore, it can be noted that the role of the Brønsted acid sites of H-ZSM-5 is not to provide ion-exchange sites for Pd^{2+} but to stabilize the dispersed state of PdO. Based on the analysis, the local structure of Pd in the oxidized Pd/H-ZSM-5 was proposed by Liu et al. as Pd surrounded by four oxygen atoms in a square planar arrangement, a part of which came from the zeolite structure [27]. In order to confirm the ability of zeolite Brønsted acid sites to anchor PdO, the regeneration of dispersed PdO with repeated reduction and oxidation treatments was followed by EXAFS. The experiment was conducted on the Pd/H-ZSM-5 (Si/Al$_2$ = 24) where highly dispersed PdO was observed by the oxidation treatment, as described above. Figure 22.2b shows the EXAFS FT spectrum measured after the reduction of previously oxidized Pd/H-ZSM-5. The formation of metal Pd was confirmed by the appearance of an intense peak at 0.24 nm (phase shift uncorrected). The particle size of the metal Pd calculated from the Pd–Pd coordination number (CN = 10.6) was estimated to be >3 nm, which was far larger than the zeolite pore diameter. Thus, with treatment under H_2, the highly dispersed PdO was reduced and migrated to form aggregated Pd metal particles on the external surface of H-ZSM-5. The reduced sample was subsequently oxidized again under oxygen flow at 773 K for 3 h. The spectrum measured after oxidation was identical with that measured after initial oxidation treatment, as shown in Figure 22.2b. Therefore, the aggregated Pd de-aggregated and returned to the H-ZSM-5 pores as the highly dispersed PdO. This behavior of Pd demonstrates the high mobility of PdO and the presence of strong interaction between the Brønsted acid sites of H-ZSM-5 and PdO. Probably, the acid–base interaction between highly dispersed PdO and Brønsted acid sites of zeolite promoted the de-aggregation and fixation of highly dispersed PdO.

22.3
Energy-Dispersive XAFS Studies on the Spontaneous Dispersion of PdO and Reversible Formation of Stable Pd Clusters in H-ZSM-5 and H-Mordenite [28, 29]

As described in the previous section, we found that the agglomerated metal Pd spontaneously migrated into zeolite pores to form the molecular-like dispersed PdO on acid sites of H-ZSM-5 under an O_2 atmosphere at elevated temperatures. From these findings, it was postulated that the Brønsted acid sites played a key role in the generation of metal Pd clusters. In order to further reveal the dynamic behavior of Pd with zeolite supports, we have tried to follow the local structure of Pd in the course of the clustering process of metal Pd supported on H-form zeolites in an atmosphere of H_2. For these purposes, an *in situ* DXAFS experiment was applied to the determination of the Pd structure during temperature programmed reduction. Figure 22.3a and c show the coordination number (CN) of the nearest neighboring Pd–Pd (metal Pd) peak determined based on the curve-fitting analysis of the EXAFS spectra for H-ZSM-5 and H-Mordenite, respectively.

A slight increase in the Pd–Pd was observed from the beginning of the reduction. At the same time, the CN of the Pd–O bond decreased, suggesting that reduction of PdO to metal Pd took place up to 440 K. After completion of the reduction, the CN of the metal Pd–Pd remained constant at 4 between 440 K and 620 K on both H-ZSM-5 and H-Mordenite. The appearance of this plateau indicated the generation of a stable Pd cluster in this temperature region. From the CN value, the Pd cluster was estimated to consist of six atoms. On further heating of the samples under flowing 8% H_2, the CN increased steeply from 620 to 770 K. Probably, the Pd_6 cluster migrated into the external surface of the zeolites to form agglomerated metal Pd particles. The change in CN on Pd/H-Mordenite was similar to that on Pd/H-ZSM-5, except that the growth of metal Pd observed above 620 K was steeper on the H-Mordenite. After the measurements given in Figure 22.3a and c, the samples were oxidized at 773 K for 4 h in an O_2 flow. Then the *in situ* cell was cooled to room temperature and temperature programmed reduction was carried out in an 8% H_2

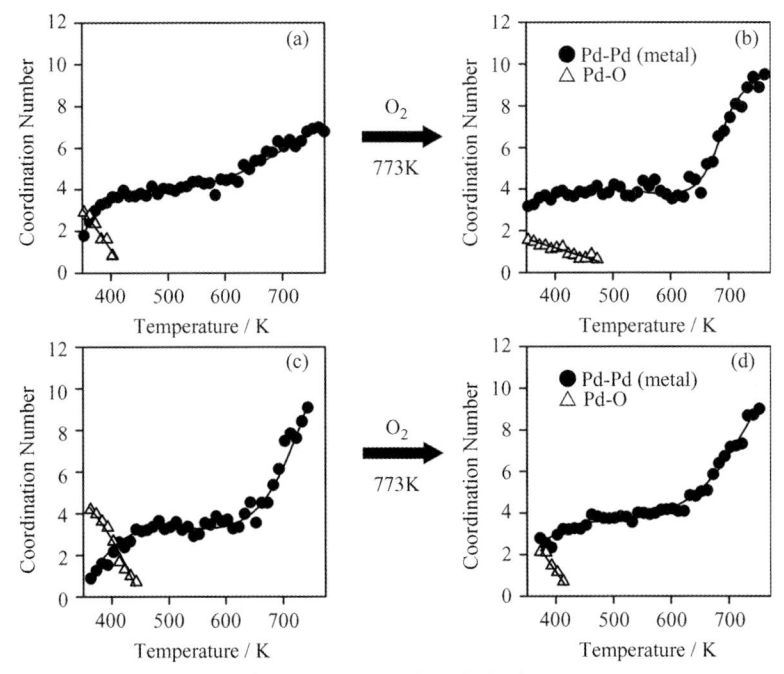

Figure 22.3 Dependence of coordination number of Pd–Pd (●) and Pd–O (△) on the temperature measured in an 8% H_2 flow; Pd/H-ZSM-5 (a,b); Pd/H-Mordenite (c,d).

flow again. Figure 22.3b and d shows the CN of H-ZSM-5 and H-Mordenite measured during the second run, respectively. A similar pattern for the change of CN (Pd–Pd) to that in the first runs was observed on both H-ZSM-5 and H-Mordenite. That is to say, a plateau of CN (Pd–Pd) was observed from 440 to 620 K, similar to the first run of the experiment. Therefore, it was confirmed that the generation of a stable metal Pd cluster was reversible upon oxidation with O_2 at 773 K and successive reduction with H_2, as illustrated in Figure 22.4. The phenomenon

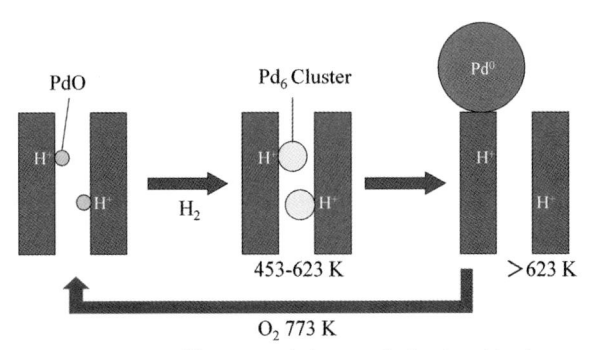

Figure 22.4 Reversible structural change of Pd induced by the interaction with acid sites of H-ZSM-5 and H-Mordenite.

could be explained by the re-dispersion of the agglomerated metal Pd onto the acid sites of zeolites through oxidation at 773 K in an O_2 flow, as evidenced in the previous section.

22.4
In Situ QXAFS Studies on the Dynamic Coalescence and Dispersion Processes of Pd in USY Zeolite [30]

As described in the previous section, we found that Pd_6 clusters were obtained through the introduction of Pd into the pores of H-ZSM-5 and H-Mordenite and a successive reduction in a H_2 stream. The formation of the Pd_6 clusters was observed to be reversible upon repeated treatments with O_2 and H_2. Such regeneration of clusters was promoted through the spontaneous migration of molecular PdO onto the acid sites of H-type zeolites under an O_2 atmosphere, which was evident from the EXAFS analysis. Resasco et al. also reported that the morphology of the oxidized Pd species strongly depended on the acidity of the support [17]. From these findings, it was postulated that the acid sites played a key role in the generation of metal Pd clusters inside zeolites. In fact, in contrast to the H-type zeolites, the severe aggregation of Pd progressed over the Na-type zeolite without the formation of stable metal clusters. Here, the dynamic behavior of Pd in the pore of a USY zeolite was measured in atmospheres of H_2 (TPR) and O_2 (TPO). The Pd clusters generated in the supercage of an FAU-type zeolite were active and reusable in Heck reactions [31]. The Pd or bimetallic Pd–Pt supported on the USY zeolite has been found to exhibit high sulfur tolerance in the hydrogenation of aromatics and hydrodesulfurization activity [32]. For this purpose, the QXAFS technique was first applied to detect the detailed structural change of Pd in the USY zeolite.

TPR and TPO measurements were repeated one after another to follow the structural change of Pd induced by reduction and oxidation; first, the changes in the structure of Pd loaded on the USY zeolite were measured by using the QXAFS technique in an atmosphere of hydrogen (first TPR). The Fourier transforms of the $k^3\chi(k)$ EXAFS collected after every 10 K are given in Figure 22.5.

Initially, a single Pd–O peak could be seen at 0.16 nm, which was directly assigned to the Pd–O bond characteristic of PdO. However, the Pd–Pd bonds that are characteristic of bulk PdO were not seen in the spectrum; this implies the formation of highly dispersed PdO in the initial stage. The intensity of the Pd–O peak gradually decreased. This was accompanied by an increase in the temperature, and a new peak attributable to the Pd–Pd bond of Pd metal appeared at 0.24 nm as a result of the reduction of the dispersed PdO. On further increasing the temperature above 673 K, a new peak appeared at 0.18 nm (phase shift uncorrected), which corresponded to a longer bond in comparison to the covalent Pd–O bond of PdO. The peak could probably be assigned to oxygen in the framework of the USY zeolite (denoted as Pd–$O_{surface1}$) by considering that Pd was already reduced to Pd^0. The CNs of these bonds were determined by curve-fitting analysis and the data are summarized in Figure 22.6.

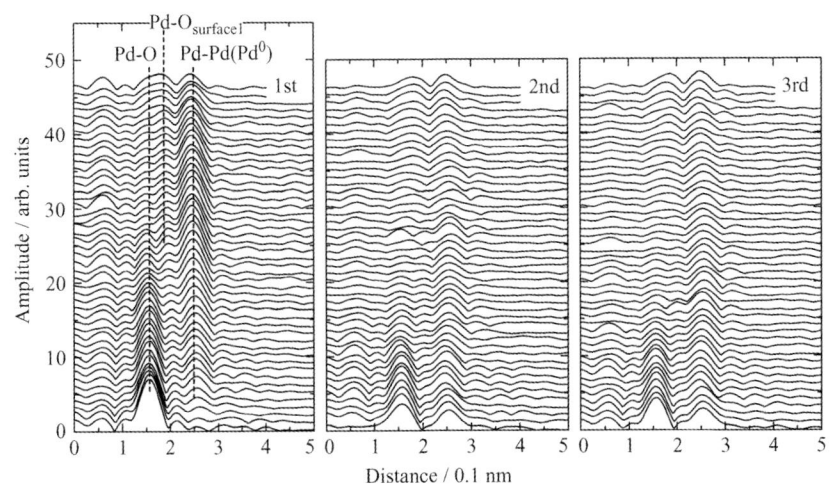

22.4 In Situ QXAFS Studies on the Dynamic Coalescence

Figure 22.5 Pd K-edge EXAFS Fourier transforms for Pd/USY measured in the TPR (r.t.–773 K). Fourier transforms range, 30–120 nm^{-1}.

In the initial step of the first TPR, the CN of the Pd–O bond (△) decreased up to 523 K, which alternated with an increase in the CN of the Pd–Pd bond (●) up to 7.5 at 673 K. The CN of the Pd–Pd bond decreased with further increase in the temperature, indicating the dispersion of previously agglomerated Pd metal at elevated temperatures. At the same time, the CN of the Pd–O$_{surface1}$ bond (□) increased, accompanied by an increase in the temperature. This fact suggested that a strong interaction between the framework of the USY zeolite and the Pd led to the dispersion of the Pd metal. This phenomenon appears to be interesting, taking into account that heating at a high temperature usually results in severe sintering of the metal.

The sample was cooled to room temperature and the first TPO experiment was subsequently carried out after switching the flowing gas to an 8% O$_2$ flow. The CNs determined from the spectra are given in Figure 22.7 (1st).

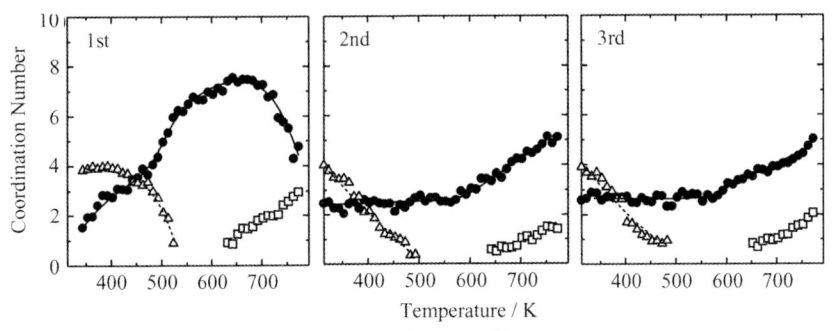

Figure 22.6 Coordination numbers determined by curve fitting analysis plotted as a function of temperature measured in TPRs. (●) Pd–Pd (metal), (△) Pd–O, (□) Pd–O$_{surface1}$.

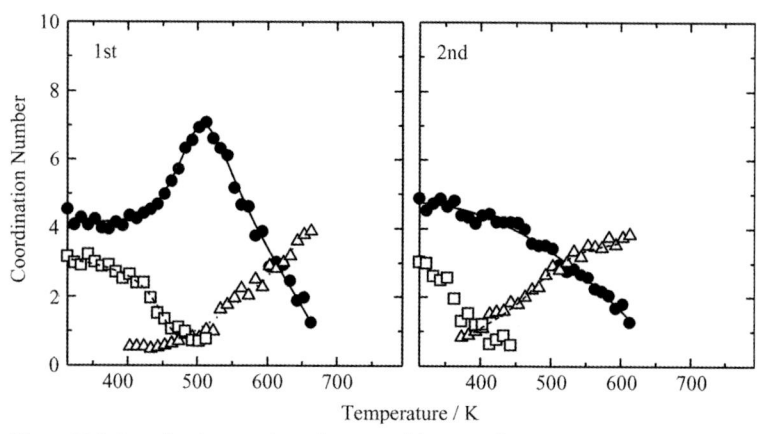

Figure 22.7 Coordination numbers determined by curve fitting analysis plotted as a function of temperature measured in TPOs. (●) Pd−Pd (metal), (△) Pd−O, (□) Pd−O$_{surface1}$.

The CN of Pd−O$_{surface1}$ decreased up to 500 K. On the other hand, the CN of the Pd metal reached 7.1 at 513 K. This change implied that removal of the Pd metal clusters from the framework of the zeolite and agglomeration took place in the initial step. On further increasing the temperature, the CNs of Pd−Pd decreased and, in turn, the CN of covalent Pd−O increased. Despite the disappearance of the Pd−Pd bond attributable to the Pd metal at 773 K, the Pd−Pd bond of PdO did not appear, suggesting the formation of highly dispersed PdO.

After the first TPO experiment, the sample was cooled to room temperature and the second TPR experiment was carried out after switching the flowing gas to an 8% H$_2$ flow. The CNs determined by the curve-fitting analysis are shown in Figure 22.6 (2nd). As can be seen in the Fourier transforms, the Pd−Pd bond of Pd metal was already observed at 313 K. With increasing temperature, the covalent Pd−O decreased and finally disappeared at 500 K to yield Pd^0. The CN of Pd metal obtained at 500–573 K was as low as 2.7, indicating the formation of extremely small metal Pd clusters having about 4 atoms. Subsequently, the CN(Pd−Pd) of the Pd clusters started to increase to a temperature above 600 K. At the same time, the Pd−O$_{surface1}$ bond appeared. The appearance of the Pd−O$_{surface1}$ bond suggested a strong interaction between the Pd clusters and the zeolite wall, as observed in the first TPR. This change implied that the Pd$_4$ clusters agglomerated and migrated onto the wall of the USY zeolite.

The changes in the CNs in the second TPO experiment are summarized in Figure 22.7 (2nd). The profile of the second TPO was different from that of the first TPO in that the oxidation of Pd clusters progressed without an increase in the CN (Pd−Pd). As a result, the removal of the Pd clusters from the zeolite framework and their oxidation took place simultaneously in the second TPO. The curve-fitting analysis of the EXAFS data obtained in the third TPR run is included in Figure 22.6 (3rd). The changes in the CNs were very similar to those of the second run. In other words, the formation of Pd$_4$ clusters was observed after the disappearance of the

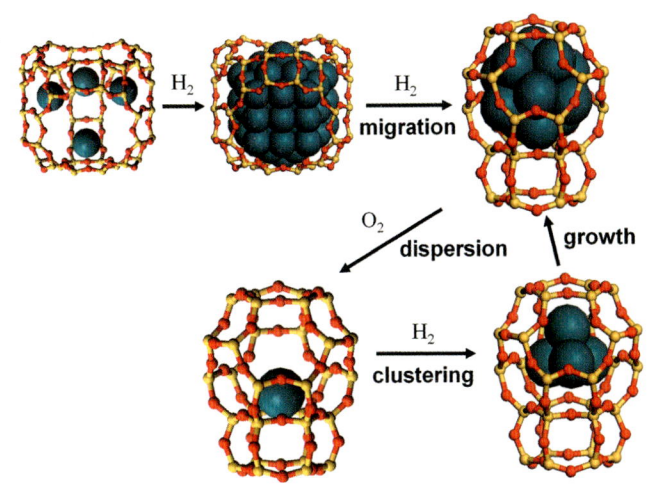

Figure 22.8 A proposed structural change of Pd in the course of temperature programmed reduction and oxidation.

Pd–O bond; this was followed by an increase in the CN(Pd–Pd) and the appearance of the Pd–O$_{surface1}$ bond at an elevated temperature. This implied that the formation of the Pd$_4$ clusters was reversible upon repeated oxidation and reduction treatments.

The proposed structural change of Pd in the USY zeolite is summarized in Figure 22.8. In the first TPR process, dispersed PdO was reduced to Pd^0, followed by agglomeration to give Pd clusters inside the supercage of USY. In the subsequent first TPR, the Pd^0 clusters migrated and were dispersed on the acid sites of the sodalite cage as the PdO form. In the second TPR process, Pd was reduced to give Pd$_4$ clusters inside the sodalite cage, followed by agglomeration to larger clusters. The Pd$_4$ clusters were generated repeatedly in further TPO and TPR processes.

22.5
Time-Resolved EXAFS Measurement of the Stepwise Clustering Process of Pd Clusters at Room Temperature [33]

Finally, the time-resolved QXAFS technique was applied to follow the clustering process of Pd in the cages of H-USY zeolite. The measurements carried out in atmospheres of H$_2$ and O$_2$ were repeated successively at room temperature; first, the changes in the structure of the Pd loaded on the H-USY zeolite were measured using the QXAFS technique in an H$_2$ atmosphere. The Fourier transforms of the $k^3\chi(k)$ data of 0.4 wt%-Pd(NH$_3$)$_4^{2+}$/H-USY collected after every 0.6 min are given in Figure 22.9.

Although it was difficult to distinguish between Pd–O and Pd–N due to the similarity of the backscatters, the peak that appeared at 0.16 nm in the initial stage may be assigned either to the Pd–N bond of Pd(NH$_3$)$_4^{2+}$ or the Pd–O bond of H$_2$O

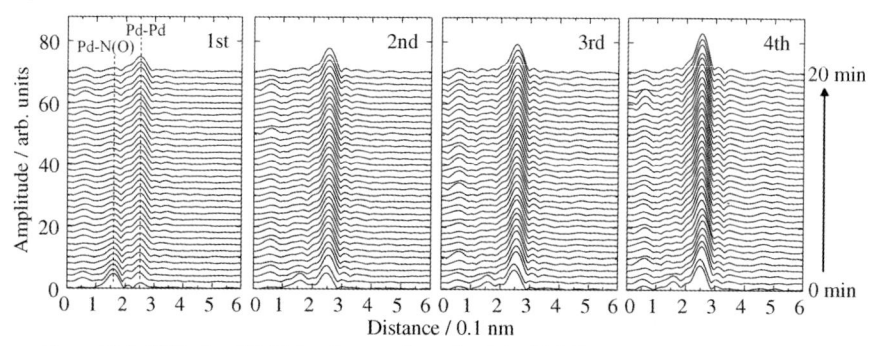

Figure 22.9 Pd K-edge EXAFS Fourier transforms for 0.4 wt%-Pd/H-USY measured in the atmosphere of 8% H_2. Fourier transforms range, 30–130 nm^{-1}.

coordinated to Pd^{2+}. The intensity of the Pd–O(N) peak gradually decreased with time. This was accompanied by an increase in a new peak attributable to the Pd–Pd bond of Pd metal at 0.24 nm as a result of the reduction of Pd^{2+}. In this process, the color of the sample gradually changed from white to the gray that is characteristic of Pd^0. Then the flowing gas was switched to 8% O_2 for 20 min, followed again by 8% H_2; at this point, the second QXAFS measurement was carried out. In the second step, a small Pd–O bond could be seen in the initial stage, indicating that the metal Pd clusters generated in the first step were partially oxidized by exposure to 8% O_2. The intensity of the Pd–O bond decreased while that of the Pd–Pd bond increased quickly within 3 min after switching to 8% H_2. It was clear that the intensity of the metal Pd–Pd bond became larger than in the first step. A similar change was observed on further switching the flowing gas to O_2 and then to H_2 (third and fourth steps). A comparison of the spectra at 20 min. revealed that the intensity of the Pd–Pd bond increased in a stepwise fashion, suggesting that the sizes of the Pd clusters increased with the repetition of the O_2 and H_2 exposures.

The CNs of the Pd–O(N) and Pd–Pd bonds were determined by curve-fitting analysis; the data are summarized in Figure 22.10.

In the first H_2 exposure, the CN of the Pd–O(N) bond decreased up to about 10 min, accompanied by an increase in the CN of the Pd–Pd bond. The growth of the Pd–Pd bond stopped when the CN of Pd–Pd reached 5.1. A small Pd–O bond appeared on exposure to O_2, as can be seen at the beginning of the second exposure to H_2. In the second exposure, the partially oxidized Pd clusters were quickly reduced on exposure to H_2, in less than 2 min; the CN reached 6.7, which was larger than after the first process. In the third and fourth runs, a similar change in the CN was observed, where the ultimate CN of Pd–Pd reached 7.8 and 8.7 after the third and fourth runs, respectively. The progressive increase in the CN with the number of H_2 exposure cycles suggests that stepwise growth of Pd clusters occurred in the H-USY support.

Information on the valence state of Pd can be obtained from the analysis of the X-ray absorption near-edge structure (XANES) region. The calculated ratio of Pd^0 and the total amount of Pd in the 0.4 wt%-Pd/H-USY are summarized in Figure 22.11.

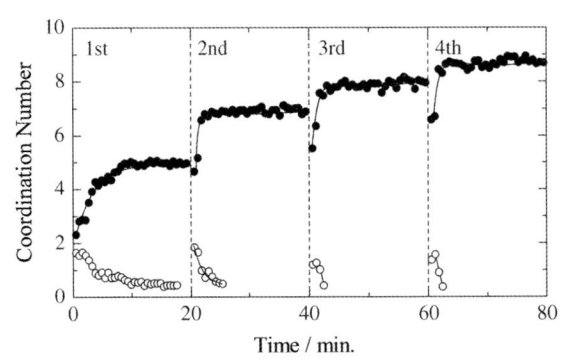

Figure 22.10 CNs of Pd–O(N) and nearest-neighboring Pd–Pd bonds determined by curve-fitting analysis of 0.4 wt%-Pd/H-USY plotted as a function of time measured in an 8% H_2 flow: (●) Pd–Pd (metal); (○) Pd–O or Pd–N.

It can be seen that the reduction of Pd^{2+} proceeded slowly and was complete in 20 min. from the start of the introduction of H_2 in the first exposure. Unlike in the first H_2 exposure, the reduction of Pd in the second, third, and fourth exposures was quickly completed, in less than 3 min. In the latter cases, 10%–20% of the Pd content was oxidized after the introduction of O_2 and before the admission of H_2. These facts are consistent with the change in the Pd–O and Pd–Pd bonds, as observed in the Fourier transforms of Figure 22.9. The oxidation state of Pd in the first H_2 exposure was kinetically analyzed using the data of Figure 22.11: the first-order rate constant k was determined to be 0.28 min^{-1}. In addition, the first-order rate constant of the CN of the Pd–Pd bond was determined independently, based on the data of Figure 22.10. The obtained k value was 0.35 min^{-1}, which is close to that of the kinetic constant k for the reduction of Pd^{2+}, suggesting the reduction of Pd^{2+} and the coalescence of Pd clusters progressed simultaneously in H-USY.

As demonstrated here, *in situ* QXAFS was effectively applied to follow precisely the clustering process of Pd in zeolites to show that bare Pd clusters were

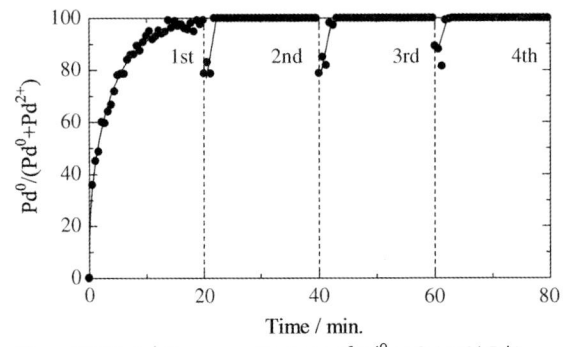

Figure 22.11 Relative concentrations of Pd^0 in 0.4 wt%-Pd/H-USY plotted as a function of time, measured in an 8% H_2 flow.

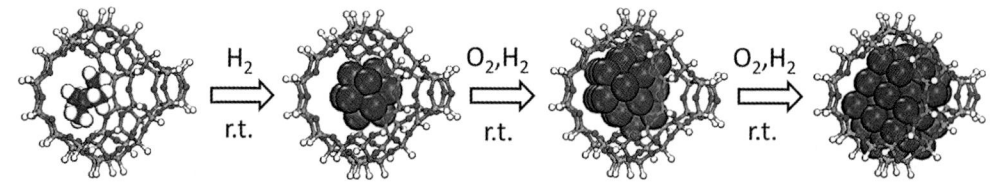

Figure 22.12 A proposed stepwise growth of Pd clusters in H-USY at room temperature.

easily formed at room temperature on exposure to H_2. The maximum CN was close to that of Pd_{13} clusters (CN = 5.5), corresponding to the one-shell structure of a cuboctahedron. The formed clusters were stable up to 443 K, as evidenced by temperature-programmed measurements (not shown). Another finding was that the Pd clusters grew in a stepwise manner on repeated exposure to O_2 and H_2 to form larger clusters after the second and third stages of H_2 flow, as illustrated in Figure 22.12.

22.6
Summary

We successfully applied the DXAFS and QXAFS techniques to follow the versatile structural change of Pd induced by interaction with zeolite supports. That is to say, from the measurements using different kinds of zeolites, the dynamic behavior as well as the formation of various Pd clusters was revealed. In an oxygen atmosphere, the spontaneous dispersion of PdO was observed on H-ZSM-5 and H-Mordenite. Furthermore, it was revealed that the crystal structure and acid sites of zeolites had a profound effect on the generation of stable metal Pd clusters in an atmosphere of H_2. In the case of H-ZSM-5 and H-Mordenite, the formation of Pd_6 cluster was observed. The formation was repeatedly observed upon oxidation and successive reduction with H_2. In the case of Pd loaded on USY zeolite, complex behavior of Pd was observed through interaction with the framework structure in the temperature programmed measurement under an atmosphere of H_2 and O_2. Furthermore, time-resolved measurement revealed the clustering and stepwise growth process of Pd over Pd^{2+}/USY at room temperature. I believe that the finding obtained here sheds light on the importance of the metal–support interaction in Pd/zeolite catalysts.

Acknowledgments

The author is very grateful to SPring-8 staff: Dr T. Uruga, Dr H. Tanida, Dr T. Honma, Mr K. Kato, and Ms S. Hirayama for technical support. The present work is supported by the Grant-in-Aid for Scientific Research (KAKENHI) in Priority Area "Molecular Nano Dynamics" from the Ministry of Education, Culture, Sports, Science and Technology.

References

1 Neylon, M.K., Marshall, C.L. and Kropf, A.J. (2002) *In situ* EXAFS analysis of the temperature-programmed reduction of Cu-ZSM-5. *J. Am. Chem. Soc.*, **124**, 5457–5476.

2 Asakura, K. (2003) Recent development in the *in-situ* XAFS and related work for the characterization of catalysts in Japan. *Catal. Surv. Asia*, **7**, 177–182.

3 Hannemann, S., Casapu, M., Grunwaldt, J.-D., Haider, P., Trüssel, P., Baiker, A. and Welter, E. (2007) A versatile *in situ* spectroscopic cell for fluorescence/transmission EXAFS and X-ray diffraction of heterogeneous catalysts in gas and liquid phase. *J. Synchrotron Rad.*, **14**, 345–354.

4 Sankar, G., Thomas, J.M., Rey, F. and Greaves, G.N. (1995) Probing the onset of crystallization of a microporous catalyst by combined X-ray absorption spectroscopy and X-ray diffraction. *Chem. Commun.*, 2549–2550.

5 Clausen, B.S. and Topsoe, H. (1991) *In situ* high pressure, high temperature XAFS studies of Cu-based catalysts during methanol synthesis. *Catal. Today*, **9**, 189–196.

6 Bare, S.R., Yang, N., Kelly, S.D. and Mickelson, G.E. (2007) Design and operation of a high pressure reaction cell for *in situ* X-ray absorption spectroscopy. *Catal. Today*, **126**, 18–26.

7 Grunwaldt, J.-D., Caravati, M., Hannemann, S. and Baiker, A. (2004) X-ray absorption spectroscopy under reaction conditions: suitability of different reaction cells for combined catalyst characterization and time-resolved studies. *Phys. Chem. Chem. Phys.*, **6**, 3037–3047.

8 Grunwaldt, J.-D., Hannemann, S., Schroer, C.G. and Baiker, A. (2006) 2D-mapping of the catalyst structure inside a catalytic microreactor at work: partial oxidation of methane over Rh/Al_2O_3. *J. Phys. Chem. B*, **110**, 8674–8680.

9 van Bokhoven, J.A., Koningsberger, D.C., Kunkeler, P. and van Bekkum, H. (2002) Influence of steam activation on pore structure and acidity of zeolite beta: an Al K edge XANES study of aluminum coordination. *J. Catal.*, **211**, 540–547.

10 van Bokhoven, J.A., van der Eerden, A.M.J. and Prins, R. (2004) Local Structure of the Zeolitic Catalytically Active Site During Reaction. *J. Am. Chem. Soc.*, **126**, 4506–4507.

11 Omegna, A., Prins, R. and van Bokhoven, J.A. (2005) Effect of temperature on aluminum coordination in zeolites H-Y, H-USY and amorphous silica-alumina: an *in-situ* Al K edge XANES study. *J. Phys. Chem. B*, **109**, 9280–9283.

12 Bugaev, L.A., van Bokhoven, J.A., Sokolenko, A.P., Latokha, Y.V. and Avakyan, L.A. (2005) Local structure of aluminum in zeolite mordenite as affected by temperature. *Phys. Chem. B*, **109**, 10771–10778.

13 van Bokhoven, J.A., Lee, T.-L., Drakopoulos, M., Lamberti, C., Thieβ F S. and Zegenhagen, J. (2008) Determining the aluminum occupancy on the active T-sites in zeolites using X-ray standing waves. *Nature. Mater.*, **7**, 551–555.

14 Weber, W.A. and Gates, B.C. (1998) Rhodium supported on faujasites: effects of cluster size and CO ligands on catalytic activity for toluene hydrogenation. *J. Catal.*, **180**, 207–217.

15 Brabec, L. and Nováková, J. (2001) Ship-in-bottle synthesis of anionic Rh carbonyls in faujasites. *J. Mol. Catal. A*, **166**, 283–292.

16 Gurin, V.S., Petranovskii, N.P. and Bogdanchikova, N.E. (2002) Metal clusters and nanoparticles assembled in zeolites: an example of stable materials with controllable particle size. *Mater. Sci. Eng. C*, **C19**, 327–331.

17 Jacobs, G., Ghadiali, F., Posanu, A., Borgna, A., Alvarez, W.E. and Resasco, D.E. (1999) Characterization of the morphology of Pt

clusters incorporated in a KL zeolite by vapor phase and incipient wetness impregnation. Influence of Pt particle morphology on aromatization activity and deactivation. *Appl. Catal. A*, **188**, 79–98.

18 Wen, B., Sun, Q. and Sachtler, W.M.H. (2001) Function of Pd_n^0 clusters, Pd^{2+} (oxo-) ions, and PdO clusters in the catalytic reduction of NO with methane over Pd/MFI catalysts. *J. Catal.*, **204**, 314–323.

19 Moller, K., Koningsberger, D.C. and Bein, T. (1989) Stabilization of metal ensembles at room temperature: palladium clusters in zeolites. *J. Phys. Chem.*, **93**, 6116–6120.

20 Bergeret, G., Gallezot, P. and Imelik, B. (1981) X-ray study of the activation, reduction, and reoxidation of palladium in Y-type zeolites. *J. Phys. Chem.*, **85**, 411–416.

21 Okumura, K. and Niwa, M. (2002) Support effect of zeolite on the methane combustion activity of palladium. *Catal. Surv. Japan*, **5**, 121–126.

22 Okumura, K., Matsumoto, S., Nishiaki, N. and Niwa, M. (2003) Support effect of zeolite on the methane combustion activity of palladium. *Appl. Catal. B*, **40**, 151–159.

23 Okumura, K., Amano, J., Yasunobu, N. and Niwa, M. (2000) X-ray absorption fine structure study of the formation of the highly dispersed PdO over ZSM-5 and the structural change of Pd induced by adsorption of NO. *J. Phys. Chem. B*, **104**, 1050–1057.

24 Okumura, K. and Niwa, M. (2000) Regulation of the dispersion of PdO through the interaction with acid sites of zeolite studied by extended X-ray absorption fine structure. *J. Phys. Chem. B*, **104**, 9670–9675.

25 Okumura, K., Amano, J. and Niwa, M. (1999) Studies on the formation and structure of highly dispersed PdO interacted with brønsted acid sites on zeolite by EXAFS. *Chem. Lett.*, 997–998.

26 Nishizaka, Y. and Misono, M. (1994) Essential role of acidity in the catalytic reduction of nitrogen monoxide by methane in the presence of oxygen over palladium-loaded zeolites. *Chem. Lett.*, 2237–2238.

27 Wang, J., Liu, C., Fang, Z., Liu, Y. and Han, Z. (2004) DFT study of structural and electronic properties of PdO/HZSM-5. *J. Phys. Chem. B*, **108**, 1653–1659.

28 Okumura, K., Yoshimoto, R., Uruga, T., Tanida, H., Kato, K., Yokota, S. and Niwa, M. (2004) Energy-dispersive XAFS studies on the spontaneous dispersion of PdO and the formation of stable Pd clusters in zeolites. *J. Phys. Chem. B*, **108**, 6250–6255.

29 Okumura, K., Yoshimoto, R., Yokota, S., Kato, K., Tanida, H., Uruga, T. and Niwa, M. (2005) Spontaneous dispersion of PdO and generation of metal Pd cluster in zeolites studied by means of *in situ* DXAFS. *Phys. Scr.*, **T115**, 816–818.

30 Okumura, K., Kato, K., Sanada, T. and Niwa, M. (2007) In-situ QXAFS studies on the dynamic coalescence and dispersion processes of Pd in the USY zeolite. *J. Phys. Chem. C*, **111**, 14426–14432.

31 Okumura, K., Nota, K., Yoshida, K. and Niwa, M. (2005) Catalytic performance and elution of Pd in the Heck reaction over zeolite supported Pd cluster catalyst. *J. Catal.*, **231**, 245–253.

32 Yasuda, H., Sato, T. and Yoshimura, Y. (1999) Influence of the acidity of USY zeolite on the sulfur tolerance of Pd-Pt catalysts for aromatic hydrogenation. *Catal. Today*, **50**, 63–71.

33 Okumura, K., Honma, T., Hirayama, S., Sanada, T. and Niwa, M. (2008) Stepwise growth of Pd clusters in USY zeolite at room temperature analyzed by QXAFS. *J. Phys. Chem. C*, **112**, 16740–16747.

**Part Four
Single Crystals**

23
Morphology Changes of Photochromic Single Crystals
Seiya Kobatake and Masahiro Irie

23.1
Introduction

Molecular materials that change shape and/or size reversibly in response to external stimuli such as light have attracted much attention as photomechanical actuators because the materials can allow remote operation without any direct contact. The photomechanical phenomena are potentially induced by the photoisomerization of constituent molecules. Reversible photo-transformation reactions of a chemical species between two isomers having different absorption spectra are called photochromism [1, 2]. The two isomers differ from one another not only in the absorption spectra but also in various physical and chemical properties, such as refractive indices, dielectric constants, and oxidation–reduction potentials. The instant property changes upon photoirradiation without processing lead to their use in various optoelectronic devices, such as optical memory [3–6], photoswitching [7, 8], display materials [9, 10], and nonlinear optics [11, 12]. In addition to the above electronic property changes, the photochromic compounds change their geometrical structures during photoisomerization. In this chapter, we focus on the geometrical structure changes and describe the photochromic reactions of diarylethene derivatives in the single-crystalline phase and their photomechanical phenomena.

Although many photochromic compounds have already been reported, compounds that show photochromic reactions in the crystalline phase are rare [13]. Typical photochromic compounds such as spiropyran and azobenzene do not show any photochromism in the crystalline phase because large geometrical structure changes are prohibited in the crystals. Typical examples of crystalline photochromic compounds are paracyclophanes [14], triarylimidazole dimer [15, 16], diphenylmaleronitrile [17], aziridines [18], 2-(2,4-dinitrobenzyl)pyridine [19–22], *N*-salicylideneanilines [23–25], and triazenes [26]. Figure 23.1 shows some examples of crystalline photochromic compounds. In many cases, their photogenerated isomers are thermally unstable.

Thermally irreversible and fatigue-resistant photochromic diarylethene crystals have been developed in the past decade. The colored isomers are stable in the crystals, even at $100\,°C$ and hardly return to the initial colorless isomers in the dark. The thermally

Molecular Nano Dynamics, Volume II: Active Surfaces, Single Crystals and Single Biocells
Edited by H. Fukumura, M. Irie, Y. Iwasawa, H. Masuhara, and K. Uosaki
Copyright © 2009 WILEY-VCH Verlag GmbH & Co. KGaA, Weinheim
ISBN: 978-3-527-32017-2

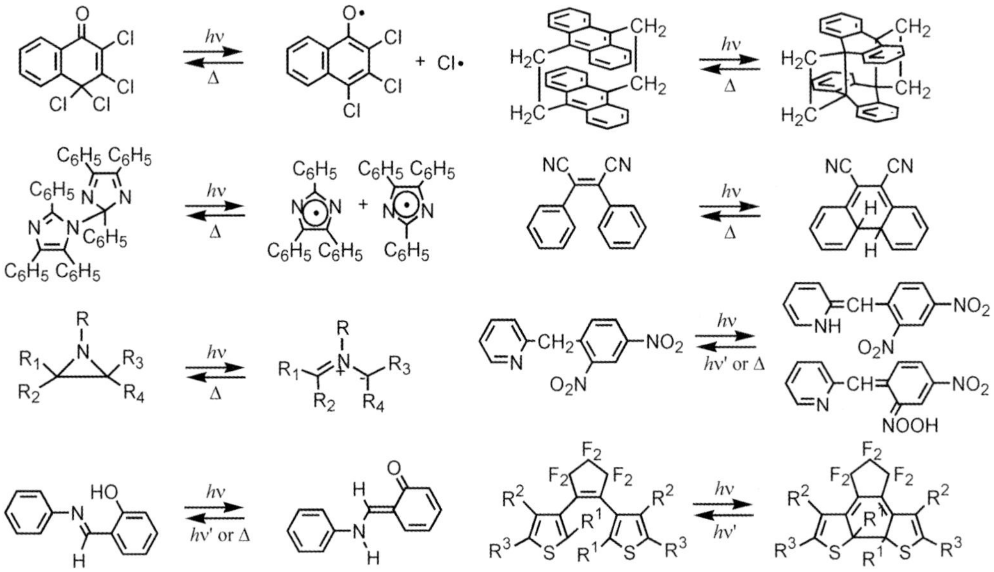

Figure 23.1 Typical examples of photochromic crystals.

irreversible crystalline photochromic materials are potentially applicable not only to optical memory, photoswitching, and display, but also to photomechanical actuators.

23.2
Photochromic Diarylethene Crystals

Some diarylethene derivatives were found to undergo reversible photochromic reactions in the single-crystalline phase as shown in Figure 23.2. Figure 23.3 shows the typical color changes of several diarylethene single crystals [27]. Upon irradiation with ultraviolet light, the colorless crystals change to yellow, red, blue, or green, depending on the molecular structure of the diarylethenes. The colors of the crystals are due to the formation of the closed-ring isomers. The colors remain stable in the dark, but they disappear on irradiation with visible light. The photoinduced coloration/decoloration cycles of the crystals can be repeated more than 10^4 times without any destruction of the crystals. Upon irradiation with ultraviolet light, the light penetrates the crystals in the bulk, and the photochromic reaction takes place not only on the crystal surface but also inside the crystal.

23.3
X-Ray Crystallographic Analysis

The molecular and geometrical structure changes in the diarylethene crystals can be directly observed by X-ray crystallographic analysis. When diarylethene crystal **3** was

Figure 23.2 Diarylethene derivatives that show photochromism in the crystalline phase.

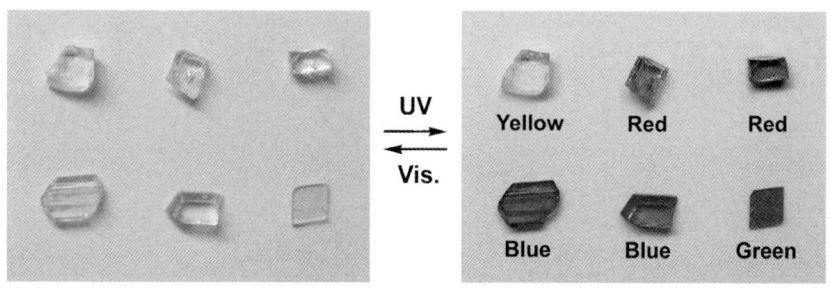

Figure 23.3 Photochromism of diarylethene derivatives in the single-crystalline phase.

irradiated with polarized 360 nm light, the unit cell dimension changed according to the photochromic reaction [28]. All unit cell lengths and unit cell volumes tended to decrease during the photocyclization. This corresponds to the decrease in the molecular volume by the transformation from the open-ring isomers to the closed-ring ones. The crystal irradiated for 24 h was analyzed by single-crystal X-ray diffraction [28]. The difference Fourier electron density map of the crystal indicates the existence of two quite high electron density peaks corresponding to the sulfur atoms of the photogenerated closed-ring isomer. The locations of these peaks are close to positions expected for the closed-ring isomer photogenerated in a conrotatory mode. Electron density peaks corresponding to two carbons at the reacting points also appeared. Figure 23.4 shows ORTEP drawings of the molecular structures that are a mixture of open- and photogenerated closed-ring isomers. The occupancy factor for the photogenerated closed-ring isomer was 0.084(2), which indicates that about 8% of the molecules in the crystal underwent the photocyclization reaction upon 360-nm light irradiation. The molecular structure of the photogenerated closed-ring isomer was compared with that of the isolated closed-ring

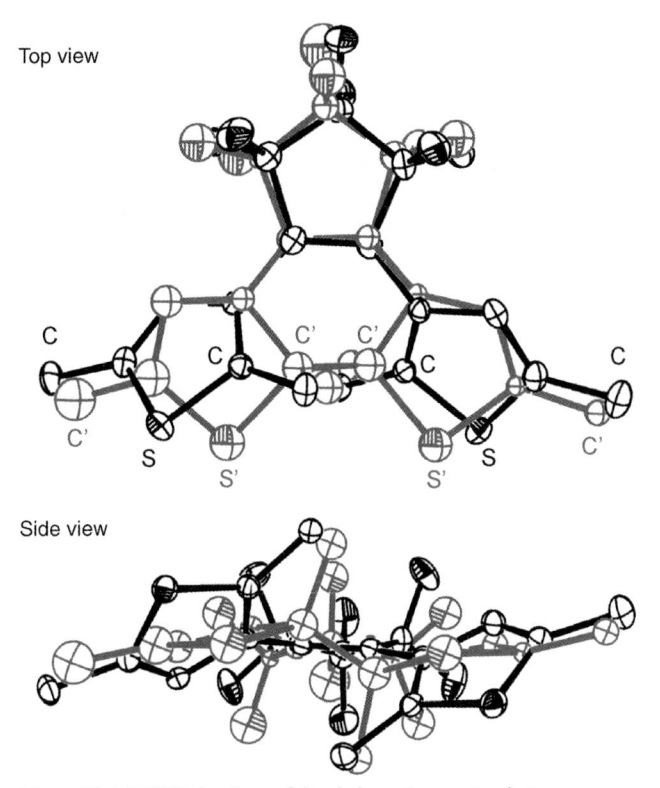

Figure 23.4 ORTEP drawings of diarylethene **3** upon irradiation with ultraviolet light. The open-ring isomer and the photogenerated closed-ring isomer are shown in black and gray, respectively.

isomer. The structural difference between the photogenerated closed-ring isomer in crystal **3** and the isolated closed-ring isomer in the crystal appeared as the difference of the distance between two sulfur atoms [29]. The structure of the closed-ring isomer produced in the open-ring form crystal is distorted. The structure difference was reflected in the absorption maximum of the closed-ring isomer. The closed-ring isomer in the closed-ring form crystal had an absorption maximum at 485 nm and an edge at 610 nm [29]. However, the photogenerated closed-ring isomer in the open-ring form crystal shifted to longer wavelength. The maximum was observed at 535 nm with an edge at 650 nm. The red shift of the absorption maximum of the closed-ring isomer is ascribed to the strained structure [29].

23.4
Reactivity in the Crystal

It is of interest to know the reactivity in the crystalline phase. In most cases, the cyclization quantum yields of diarylethenes in crystal were twice as large as those in solution. The low quantum yield in solution is due to the presence of molecules in photoinactive parallel conformation. On the other hand, the cyclization quantum yields in the crystal were around unity (100%). This means that photon energy absorbed in the crystal is used quantitatively for the cyclization reaction. The single crystal utilizes all absorbed photons for the coloration chemical reaction. X-ray crystallographic analysis of the crystals indicated that diarylethene molecules in the crystals were fixed in the antiparallel conformation. The distances between the reactive carbon atoms were estimated to be 3.48–3.96 Å, which are close enough for the conrotatory cyclization reactions. Figure 23.5 shows a correlation between the cyclization quantum yields of diarylethenes in crystals and the distances between the reactive carbon atoms of the diarylethenes [30]. When the distance is larger than 4.2 Å, the photocyclization reaction in the crystal is suppressed. The reaction process was analyzed based on *ab initio* and DFT calculation of the initial geometries, the

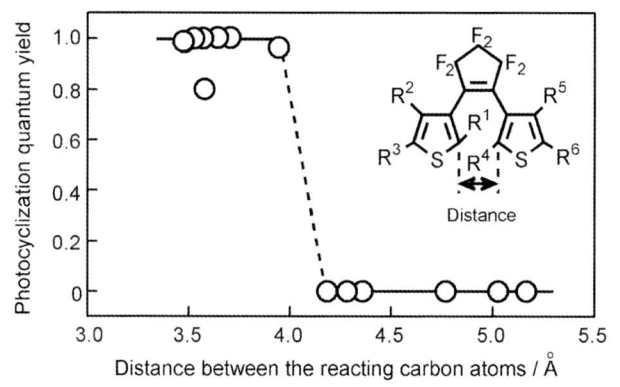

Figure 23.5 Relationship between photocyclization quantum yield and distance between the reacting carbon atoms.

relaxation from the Franck–Condon states, the shapes of the potential energy surface of the ground states, and the geometry change by the large amplitude motions [31]. The large cyclization quantum yield in the crystalline phase can be ascribed to three factors. One is a high population of the photoreactive antiparallel conformation in the crystalline phase, in which the distance between the reacting carbon atoms is less than 4 Å. All photoexcited molecules in the antiparallel conformation fixed in the crystal lattice readily undergo the photocyclization reactions. Other factors are the very low activation energy of the conrotatory cyclization reaction [32, 33] and the rapid cyclization rate, less than 10 ps [34]. The rapid reaction rate prevails over other relaxation processes, such as radiative and nonradiative transitions from the excited states to the ground state.

23.5
Photomechanical Effect

The photoinduced deformation phenomenon of materials is called a photomechanical effect, and it has been so far reported for photoresponsive polymer films and gels [35–43]. When azobenzene is isomerized from the trans form to the cis form, the length of the molecule is shortened from 0.90 to 0.55 nm. The size change of the molecule on photoirradiation is expected to alter the shape of the polymers which contain the azobenzene molecules. However, it is not the case in polymer systems. The transformation in polymer films does not change the polymer shape because of the large free volumes of the polymer bulk. Suitable organization or assembly of the molecules is required for the photoinduced deformation of materials.

The reversible shape change in molecular materials was found for the first time in 2001 by using azobenzene-containing liquid crystal elastomers [39]. Figure 23.6

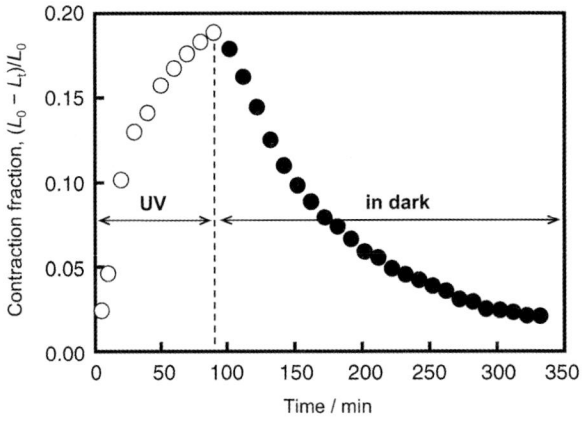

Figure 23.6 Contraction fraction of azobenzene-containing liquid crystal elastomers, $(L_0 - L_t)/L_0$, at 298 K against the time upon irradiation with ultraviolet light and in the dark. L_0 and L_t represent lengths of the elastomers in the initial state and after the time, respectively.

shows the fractional contraction of the elastomers, $(L_0 - L_t)/L_0$, at 298 K against the time of irradiation with ultraviolet light and in the dark for recovery [39]. L_0 and L_t represent the lengths of the elastomers in the initial state and after the time, respectively. The elastomers contract by almost 20% upon photoisomerization of azobenzene chromophores. After shutting off the illumination, the elastomers recovered to their initial length. The effect can give rise to directed bending of elastomer films when the chromophores are selectively excited with linearly polarized light [40] or unidirectionally aligned in the film by a rubbing procedure [41, 42]. In these cases, the light-induced trans–cis photoisomerization of the azobenzene chromophores reduces the ordering of the liquid-crystal material, which can give rise to macroscopic contraction or bending. However, the phenomenon occurs only around the phase transition temperature. The response time of these systems is rather slow, and the deformed states are unstable because the cis-azobenzene isomers relax back to trans-azobenzenes.

23.6
Crystal Surface Changes

The colorless single crystals of diarylethenes change color by the formation of closed-ring isomers upon irradiation with ultraviolet light. As described in Section 23.3, the component diarylethene molecules shrink during the photoisomerization. The molecular-scale shape change induces nano-scale morphological changes of the crystal surfaces. The morphology changes were detected by an atomic force microscope (AFM) [44]. Two crystal surfaces, (100) and (010), of diarylethene crystal 7 were used for observation of the surface morphological changes. The crystal surface before photoirradiation was flat (Figure 23.7a). Upon irradiation for more than 10 s with

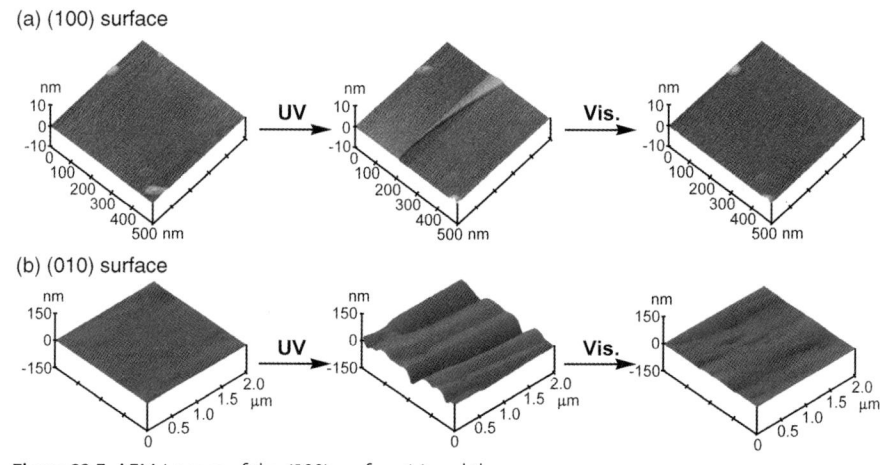

Figure 23.7 AFM images of the (100) surface (a) and the (010) surface (b) of crystal **7** upon alternate irradiation with ultraviolet and visible light.

ultraviolet light, steps appeared on the (100) surface. No step formation was discerned during the irradiation for the initial 10 s but appeared after the induction period. The step height was 1.0 nm. The step disappeared by bleaching upon irradiation with visible light ($\lambda > 500$ nm). When the irradiation time was prolonged, the number of steps increased and steps with heights of 2.0 and 3.0 nm appeared. The height was always a multiple of the minimum step height in 1.0 nm, and any steps with a height lower than the unit height of 1.0 nm were not observed. Each step of 1.0 nm requires reactions to depths of about 600 molecular layers. The morphological change was reversible and correlated with the color change of the crystal. The AFM images of the (010) surface before and after ultraviolet light irradiation are shown in Figure 23.7b. Upon irradiation for 15 s with ultraviolet light, the crystal turned blue and valleys appeared on the crystal surface. The depth of the valleys was estimated to be 10 to 50 nm. The valley almost disappeared by bleaching upon irradiation with visible light ($\lambda > 500$ nm). The morphological change was again reversible and correlated with the color change.

23.7
Photoreversible Crystal Shape Changes

The crystal surface morphology changes encouraged us to study the shape change of bulk crystals upon photoirradiation. Diarylethene molecular crystals with sizes ranging from 10 to 100 µm exhibit rapid and reversible macroscopic changes in shape and size induced by ultraviolet and visible light [45]. The changes occur about five orders of magnitude faster than the response time of the azobenzene-containing liquid crystal elastomers [39–43].

Single crystals of diarylethenes **4** and **8** with sizes on the 10–100 µm scale were prepared by sublimation of the compounds on glass plates. Figure 23.8 shows a

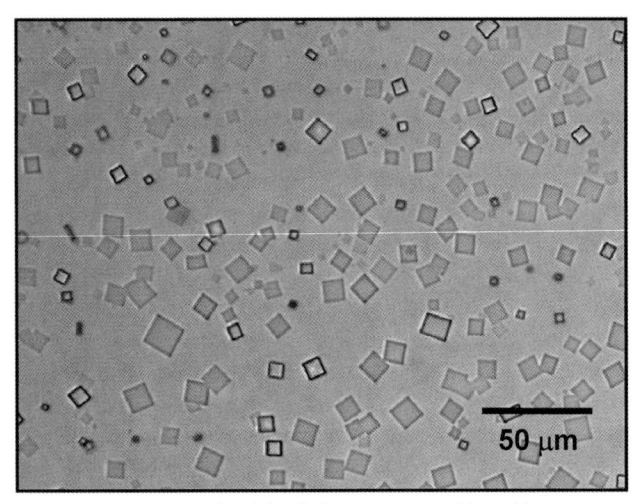

Figure 23.8 Microcrystals of diarylethene **8** prepared by sublimation.

microscopic picture of the microcrystals of diarylethene **8**. Upon irradiation with ultraviolet light, the molecules in the crystals underwent a cyclization reaction that transformed open-ring isomers into closed-ring isomers. The colors of crystals **4** and **8** turned to violet and blue, respectively. The colors were stable in the dark, but disappeared on irradiation with visible light [46, 47].

The crystal shape changes during the photochromic reactions were observed directly with a microscope. Figure 23.9 illustrates the deformations of diarylethene single crystals **4** and **8** on alternate irradiation with ultraviolet ($\lambda = 365$ nm) and visible ($\lambda > 500$ nm) light. As shown in Figure 23.9a, a rectangular single crystal of **4**

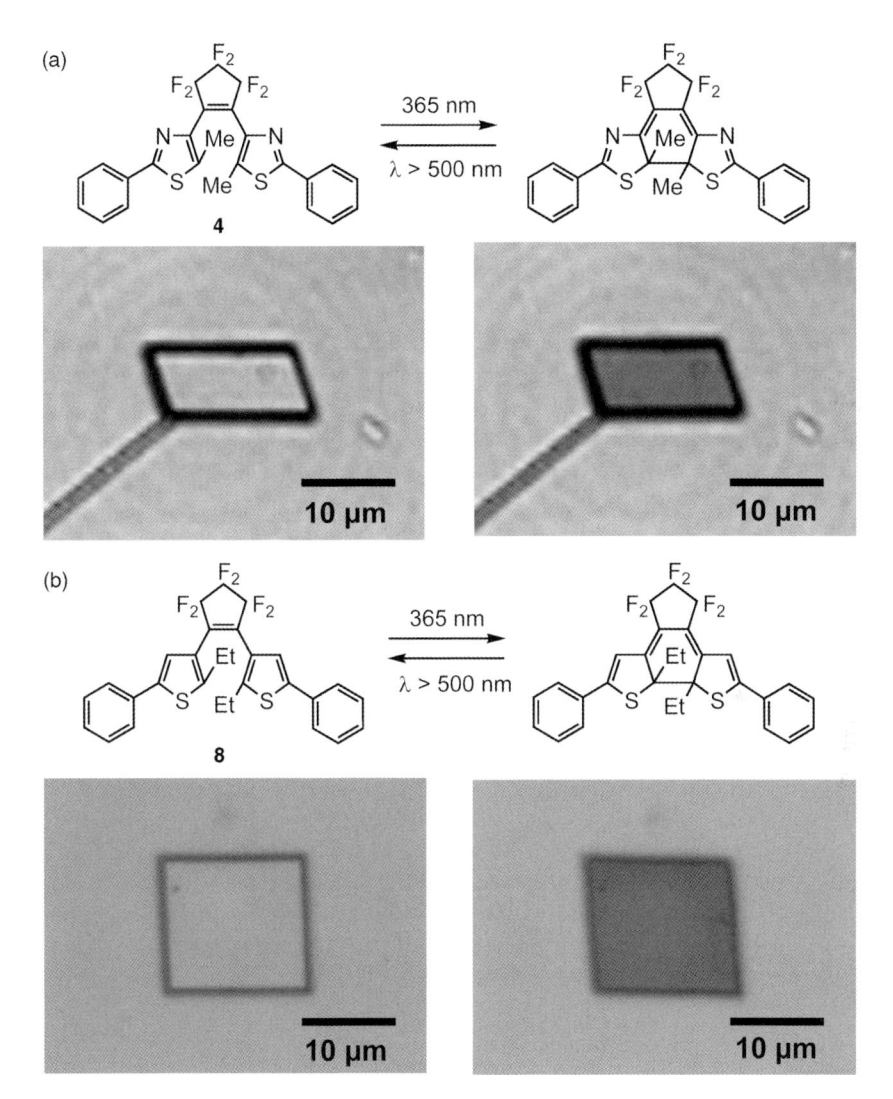

Figure 23.9 Photoresponsive deformation of diarylethene crystals **4** (a) and **8** (b).

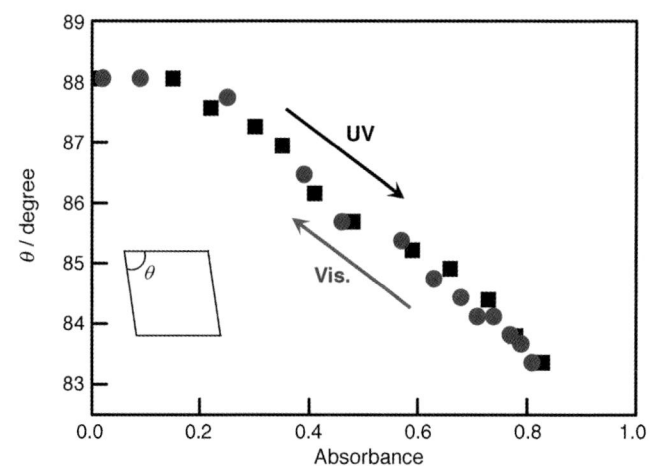

Figure 23.10 Relationship between the corner angle of the single crystal of **8** and the absorption intensity of the crystal measured at 600 nm upon alternate irradiation with ultraviolet light (■) and visible light (●).

induced contraction and expansion by as much as 7% on irradiation with ultraviolet and visible light, respectively. Ultraviolet irradiation of a single crystal of **8** changed its corner angles from 88° and 92° to 82° and 98°, respectively, changing its shape from squares to lozenges, as shown in Figure 23.9b. The angle changes can be repeated for more than 20 cycles of alternate irradiation with ultraviolet and visible light without any evidence of a change in the performance of the crystal.

Figure 23.10 shows the time dependence of the color and shape changes on alternate irradiation with ultraviolet and visible light by showing the relation between the absorption intensity of the crystal at 600 nm wavelength and its corner angle. The absorption intensity increases with the amount of photogenerated closed-ring isomers in the crystal, which reaches 70% of all molecules in the photostationary state. The angle initially remains unchanged and then decreases by as much as 5° to 6°. No hysteresis between the forward and reverse processes was observed. An interesting correlation between the absorption maximum and the shape change was observed. The absorption maximum initially remained constant at 625 nm, but then shifted to 585 nm as the crystal shape changed. The spectral shift is attributable to the interaction of adjacent closed-ring isomers [29]. An induction period for the changes in the crystal shape is necessary for the photomechanical phenomenon to take place.

The crystallinity of the small crystal was evaluated from its melting point and the order parameter of the visible absorption in the photostationary state. The melting point of crystal **8** before photoirradiation was 164 °C, and decreased to 45–55 °C upon irradiation with ultraviolet light. The decrease in the melting point upon ultraviolet irradiation is due to the coexistence of two isomers in the same crystal. The crystal became colorless again on irradiation with visible light. The recovery of the melting

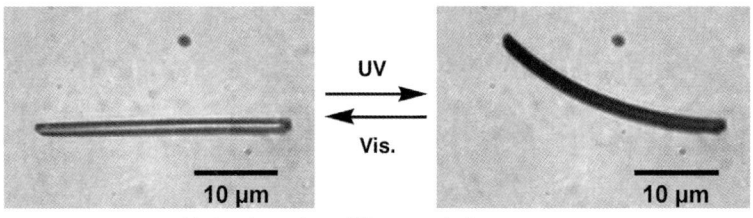

Figure 23.11 Reversible bending of a rod-like crystal of **4** upon alternate irradiation with ultraviolet and visible light.

point (164 °C) upon visible-light irradiation indicates that crystal **8** remains highly crystalline after a cycle of irradiation with ultraviolet and visible light. The order parameter $((A_{//} - A_\perp)/(A_{//} + 2A_\perp))$ at 600 nm in the photostationary state was 0.53, which is identical to the value measured with a large crystal [47]. The constant order parameter during photochromic reactions also indicates that the crystallinity is maintained even in the photostationary state.

Rod-like crystals of **4** were also prepared by sublimation. X-ray crystallographic analysis revealed that the thin plate-like crystal and the rod-like crystal have the same crystal structure. The rod-like crystal mounted at one end on a glass surface bent upon irradiation with ultraviolet light, as shown in Figure 23.11, with the bending moving towards the direction of the incident light [45, 48]. This effect is due to a gradient in the extent of photoisomerization caused by the high absorbance of the crystal, so that the shrinkage of the irradiated part of the crystal causes bending, just as in a bimetal. The bent rod-like crystal became straight again upon irradiation with visible light. Such reversible bending could be repeated over 80 cycles. The power produced during bending can move a gold

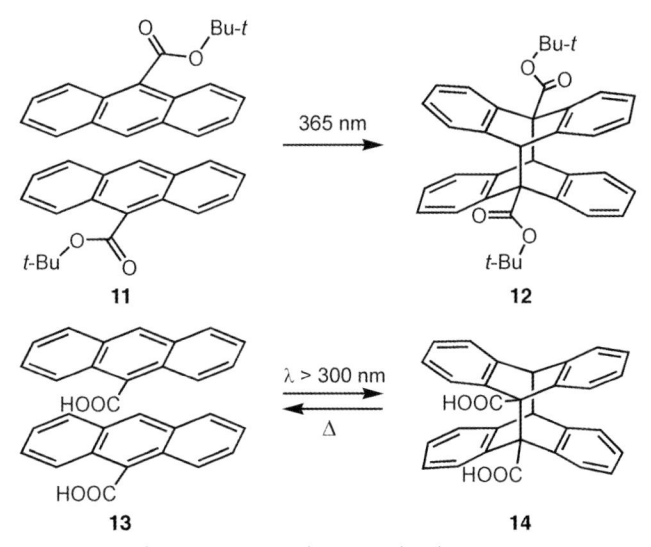

Figure 23.12 Photoresponsive anthracenecarboxylates.

microparticle with a weight 90 times greater than the single crystal over a distance of 30 μm. The bending of the crystal was found to be almost complete within 25 μs, whereas the photoreaction of diarylethenes in crystals takes place in less than 10 ps [34]. This very fast response time of the bending is about 10^5 times faster than those of the azobenzene-containing elastomer systems [39–43], and is comparable to the response time of piezoelectric devices.

The shape change of molecules upon photoirradiation is directly linked to the macroscale shape change of the crystals. The suitable arrangement of molecules in crystals induces a cooperative motion of the crystal lattice and results in the mechanical motion of the crystals. The specific molecular packing in the crystals is therefore considered to play an important role in macroscale motion.

Bardeen et al. have reported analogous effects with 200-nm-diameter nanorods of 9-*tert*-butylanthracenecarboxylate **11** [49] and 9-anthracenecarboxylic acid **13** (Figure 23.12) [50]. Both crystals were prepared by solvent annealing in Al_2O_3 templates [51]. Crystals of **11** were shown to expand irreversibly by as much as 15% along the long axis in a photochemical reaction to give dimer **12**. Exposure of **13** to ultraviolet light resulted in formation of the thermally unstable *syn* 4 + 4 photodimer **14**, which spontaneously returns to the starting anthracenecarboxylic acid [50]. These photomechanical materials can be useful for applications to photomechanical actuators in many fields of electronic, photonic, mechanical, medical, and functional materials [52].

References

1 Brown, G.H. (1971) *Photochromism*, Wiley-Interscience, New York.
2 Dürr, H. and Bouas-Laurent, H. (1990) *Photochromism: Molecules and Systems*, Elsevier, Amsterdam.
3 Irie, M. (1994) *Photo-reactive Materials for Ultrahigh Density Optical Memory*, Elsevier, Amsterdam.
4 Irie, M. (2000) Photochromism: memories and switches. *Chem. Rev.*, **100** (5).
5 Irie, M. and Matsuda, K. (2001) Memories, in *Electron Transfer in Chemistry*, Vol. 5 (ed. V. Balzani), Wiley-VCH, Weinheim, pp. 215–242.
6 Irie, M. (2002) High-density optical memory and ultrafine photofabrication, in *Nano-Optics* (eds S. Kawata, M. Ohtsu and M. Irie), Springer, Berlin, pp. 137–150.
7 Irie, M. (2001) Photoswitchable molecular systems based on diarylethenes, in *Molecular Switchings* (ed. B.L. Feringa), Wiley-VCH, Weinheim, pp. 37–62.
8 Matsuda, K. and Irie, M. (2002) Photoswitching of intermolecular magnetic interaction using photochromic compounds, in *Chemistry of Nano-molecular Systems-Toward the Realization of Molecular Devices* (eds T. Nakamura, T. Matsumoto, H. Tada and K.-I. Sugiura), Springer, Berlin, pp. 25–40.
9 Yao, J., Hashimoto, K. and Fujishima, A. (1992) Photochromism induced in an electrolytically pretreated MoO_3 thin-film by visible-light. *Nature*, **355**, 624–626.
10 Bechinger, C., Ferrer, S., Zaban, A., Sprague, J. and Gregg, B.A. (1996) Photoelectrochromic windows and displays. *Nature*, **383**, 608–610.
11 Nakatani, K. and Delaire, J.A. (1997) Reversible photoswitching of second-order nonlinear optical properties in an organic photochromic crystal. *Chem. Mater.*, **9**, 2682–2684.

12 Delaire, J.A. and Nakatani, K. (2000) Linear and nonlinear optical properties of photochromic molecules and materials. *Chem. Rev.*, **100**, 1817–1845.

13 Scheffer, J.R. and Pokkuluri, P.R. (1990) *Photochemistry in Organized & Constrained Media* (ed. V. Ramamurthy), VCH, New York, p. 185.

14 Golden, J.H. (1961) Bi(anthracene-9,10-dimethylene)(tetrabenzo-[2,2]-paracyclophane]. *J. Chem. Soc.*, 3741–3748.

15 Maeda, K. and Hayashi, T. (1970) The mechanism of photochromism, thermochromism and piezochromism of dimers of triarylimidazolyl. *Bull. Chem. Soc. Jpn.*, **43**, 429–438.

16 Kawano, M., Sano, T., Abe, J. and Ohashi, Y. (1999) The first *in situ* direct observation of the light-induced radical pair from a hexaarylbiimidazolyl derivative by X-ray crystallography. *J. Am. Chem. Soc.*, **121**, 8106–8107.

17 Ichimura, K. and Watanabe, S. (1976) Photocyclization of diphenylmaleonitrile in crystalline state. *Bull. Chem. Soc. Jpn.*, **49**, 2220–2223.

18 Trozzolo, A.M., Leslie, T.M., Sarpotdar, A.S., Small, R.D., Ferraudi, G.J., DoMinh, T. and Hartless, R.L. (1979) Photochemistry of some 3-membered heterocycles. *Pure Appl. Chem.*, **51**, 261–270.

19 Sixl, H. and Warta, R. (1985) Reaction-mechanism of photochromic 2-(2′,4′-dinitrobenzyl)pyridine (DNBP) single-crystals. *Chem. Phys.*, **94**, 147–155.

20 Eichen, Y., Lehn, J.-M., Scherl, M., Haarer, D., Fischer, J., DeCian, A., Corval, A. and Trommsdorff, H.P. (1995) Photochromism dependent on crystal packing - photoinduced and thermal proton-transfer processes in single-crystals of 6-(2,4-dinitrobenzyl)-2,2′-bipyridine. *Angew. Chem., Int. Ed. Engl.*, **34**, 2530–2533.

21 Schmidt, A., Kababya, S., Appel, M., Khatib, S., Botoshansky, M. and Eichen, Y. (1999) Measuring the temperature width of a first-order single crystal to single crystal phase transition using solid-state NMR: Application to the polymorphism of 2-(2,4-dinitrobenzyl)-3-methylpyridine. *J. Am. Chem. Soc.*, **121**, 11291–11299.

22 Naumov, P., Sekine, A., Uekusa, H. and Ohashi, Y. (2002) Structure of the photocolored 2-(2′,4′-dinitrobenzyl)-pyridine crystal: two-photon induced solid-state proton transfer with minor structural perturbation. *J. Am. Chem. Soc.*, **124**, 8540–8541.

23 Hadjoudis, E., Vittorakis, M. and Moustakali-Mavridis, I. (1987) Photochromism and thermochromism of schiff-bases in the solid-state and in rigid glasses. *Tetrahedron*, **43**, 1345–1360.

24 Harada, J., Uekusa, H. and Ohashi, Y. (1999) X-ray analysis of structural changes in photochromic salicylideneaniline crystals. Solid-state reaction induced by two-photon excitation. *J. Am. Chem. Soc.*, **121**, 5809–5810.

25 Amimoto, K., Kanatomi, H., Nagakari, A., Fukuda, H., Koyama, H. and Kawato, T. (2003) Deuterium isotope effect on the solid-state thermal isomerization of photo-coloured cis-keto species of N-salicylideneaniline. *Chem. Commun.*, 870–871.

26 Mori, Y., Ohashi, Y. and Maeda, K. (1989) Crystal structure and photochemical behavior of 2,2,4,6-tetraphenyldihydro-1,3,5-triazine and its inclusion compounds. *Bull. Chem. Soc. Jpn.*, **62**, 3171–3176.

27 Kobatake, S. and Irie, M. (2004) Single-crystalline photochromism of diarylethenes. *Bull. Chem. Soc. Jpn.*, **77**, 195–210.

28 Yamada, T., Kobatake, S. and Irie, M. (2000) X-ray crystallographic study on single-crystalline photochromism of 1,2-bis(2,5-dimethyl-3-thienyl)-perfluorocyclopentene. *Bull. Chem. Soc. Jpn.*, **73**, 2179–2184.

29 Kobatake, S., Morimoto, M., Asano, Y., Murakami, A., Nakamura, S. and Irie, M. (2002) Absorption spectra of colored

isomer of diarylethene in single crystals. *Chem. Lett.*, 1224–1225.

30 Kobatake, S., Uchida, K., Tsuchida, E. and Irie, M. (2002) Single-crystalline photochromism of diarylethenes: reactivity–structure relationship. *Chem. Commun.*, 2804–2805.

31 Asano, Y., Murakami, A., Kobayashi, T., Kobatake, S., Irie, M., Yabushita, S. and Nakamura, S. (2003) Theoretical study on novel quantum yields of dithienylethenes cyclization reactions in crystals. *J. Mol. Struct.: THEOCHEM*, **625**, 227–234.

32 Nakamura, S. and Irie, M. (1988) Thermally irreversible photochromic systems. A theoretical study. *J. Org. Chem.*, **53**, 6136–6138.

33 Uchida, K., Nakayama, Y. and Irie, M. (1990) Thermally irreversible photochromic systems. Reversible photocyclization of 1,2-bis(benzo[b]-thiophen-3-yl)ethane derivatives. *Bull. Chem. Soc. Jpn.*, **63**, 1311–1315.

34 Miyasaka, H., Nobuto, T., Itaya, A., Tamai, N. and Irie, M. (1997) Picosecond laser photolysis studies on a photochromic dithienylethene in solution and in crystalline phases. *Chem. Phys. Lett.*, **269**, 281–285.

35 Smets, G., Braeken, J. and Irie, M. (1978) Photomechanical effects in photochromic systems. *Pure Appl. Chem.*, **50**, 845–856.

36 Eisenbach, C.D. (1980) Isomerization of aromatic azo chromophores in poly(ethyl acrylate) networks and photomechanical effect. *Polymer*, **21**, 1175–1179.

37 Matějka, L., Ilavsk,ý M., Dušek, K. and Wichterle, O. (1981) Photomechanical effects in crosslinked photochromic polymers. *Polymer*, **22**, 1511–1515.

38 Irie, M. (1990) Photoresponsive polymers. *Adv. Polym. Sci.*, **94**, 27–67.

39 Finkelmann, H., Nishikawa, E., Pereira, G.G. and Warner, M. (2001) A new opto-mechanical effect in solids. *Phys. Rev. Lett.*, **87**, 015501.

40 Yu, Y., Nakano, M. and Ikeda, T. (2003) Directed bending of a polymer film by light - miniaturizing a simple photomechanical system could expand its range of applications. *Nature*, **425**, 145.

41 Ikeda, T., Nakano, M., Yu, Y., Tsutsumi, O. and Kanazawa, A. (2003) Anisotropic bending and unbending behavior of azobenzene liquid-crystalline gels by light exposure. *Adv. Mater.*, **15**, 201–205.

42 Yu, Y., Nakano, M., Shishido, A., Shiono, T. and Ikeda, T. (2004) Effect of cross-linking density on photoinduced bending behavior of oriented liquid-crystalline network films containing azobenzene. *Chem. Mater.*, **16**, 1637–1643.

43 Jiang, H., Kelch, S. and Lendlein, A. (2006) Polymers move in response to light. *Adv. Mater.*, **18**, 1471–1475.

44 Irie, M., Kobatake, S. and Horichi, M. (2001) Reversible surface morphology changes of a photochromic diarylethene single crystal by photoirradiation. *Science*, **291**, 1769–1772.

45 Kobatake, S., Takami, S., Muto, H., Ishikawa, T. and Irie, M. (2007) Rapid and reversible shape changes of molecular crystals on photoirradiation. *Nature*, **446**, 778–781.

46 Kuroki, L., Takami, S., Shibata, K. and Irie, M. (2005) Photochromism of single crystals composed of dioxazolylethene and dithiazolylethene. *Chem. Commun.*, 6005–6007.

47 Kobatake, S., Shibata, K., Uchida, K. and Irie, M. (2000) Photochromism of 1,2-bis(2-ethyl-5-phenyl-3-thienyl)-perfluorocyclopentene. Conrotatory thermal cycloreversion of the closed-ring isomer. *J. Am. Chem. Soc.*, **122**, 12135–12141.

48 McBride, J.M. (2007) Crystal tennis rackets. *Nature*, **446**, 736–737.

49 Al-Kaysi, R.O., Müller, A.M. and Bardeen, C.J. (2006) Photochemically driven shape changes of crystalline organic nanorods. *J. Am. Chem. Soc.*, **128**, 15938–15939.

50 Al-Kaysi, R.O. and Bardeen, C.J. (2007) Photoinduced shape changes of crystalline organic nanorods. *Adv. Mater.*, **19**, 1276–1280.

51 Al-Kaysi, R.O. and Bardeen, C.J. (2006) General method for the synthesis of crystalline organic nanorods using porous alumina templates. *Chem. Commun.*, 1224–1226.

52 Garcia-Garibay, M.A. (2007) Molecular crystals on the move: from single-crystal-to-single-crystal photoreactions to molecular machinery. *Angew. Chem. Int. Ed.*, **46**, 8945–8948.

24
Direct Observation of Change in Crystal Structures During Solid-State Reactions of 1,3-Diene Compounds

Akikazu Matsumoto

24.1
Introduction

24.1.1
Crystal Engineering Renaissance

Solid-state organic reactions often provide a high regio- or stereoselectivity because the structure of a product is strictly determined by the crystal structure of the reactant, that is, the reaction proceeds under crystalline-lattice control [1–10]. Especially, topochemical reactions promise the formation of products with a highly controlled structure, which can be predicted on the basis of the structure of a reactant. The crystallographic investigation of topologically controlled reactions was first carried out by Schmidt and coworkers during the early 1960s, and the principles of the photodimerization, that is, the Schmidt rule, as well as the concept of crystal engineering, were established [11, 12]. They examined a relationship between the crystal structure of the reactant and the chemical structure of the product to discuss topochemical principles for the reactions. Since the finding of the topochemical polymerization of diacetylenes [13] and diolefins [14] via a chain or stepwise reaction mechanism, many studies on the solid-state polymerization of various kinds of monomers have been carried out [15, 16]. For example, Tieke [17] reported the radiation polymerization of butadiene derivatives crystallized in perovskite-type layer structures, but the details of topochemical polymerization of 1,3-diene monomers have still been unknown until recent years.

Thirty years later, we discovered the topochemical polymerization of various 1,3-diene monomers giving a highly stereoregular polymer in the form of polymer crystals. When ethyl (Z,Z)-muconate was photoirradiated in the crystalline state, a tritactic polymer was produced [18, 19], in contrast to the formation of an atactic polymer by conventional radical polymerization in an isotropic state. Thereafter, comprehensive investigation was carried out, for example, the design of monomers, the crystal structure analysis of monomers and polymers, and polymerization reactivity control, in order to reveal the features of the polymerization of 1,3-diene monomers [20–23]. Eventually, it was revealed that the solid-state photoreaction

Molecular Nano Dynamics, Volume II: Active Surfaces, Single Crystals and Single Biocells
Edited by H. Fukumura, M. Irie, Y. Iwasawa, H. Masuhara, and K. Uosaki
Copyright © 2009 WILEY-VCH Verlag GmbH & Co. KGaA, Weinheim
ISBN: 978-3-527-32017-2

pathway depended importantly on the chemical structure of the reacting 1,3-diene compound. As a result of the solid-state photoreaction, one of the corresponding EE-isomers, [2 + 2] cyclodimer, and 1,4-polymer were obtained by unimolecular, bimolecular, and polymerization reactions, respectively. The pathway is selectively determined by the structure of the substituent, that is, the molecular packing fashion in the crystals. A crystal engineering approach has changed since the 1990s, and been renewed as a strategy for the rational design of organic solid architecture utilizing supramolecular chemistry [24–26]. Furthermore, studies on organic solid-state reactions and crystal structure analysis have been accelerated by the development of methods and apparatus for crystal structure analyses [27–29].

In this chapter, we describe the isomerization, dimerization, and topochemical polymerization of benzyl muconates with various kinds of substituents on the benzyl ester group (Scheme 24.1), in order to reveal the solid-state reaction mechanism by the direct observation of crystal structures which change during the reactions.

24.2
EZ-Photoisomerization

24.2.1
Model of Photoisomerization

Only a limited number of isomerizations of olefins between the E and Z forms in the crystalline state have been reported because of the difficulty in the inevitable

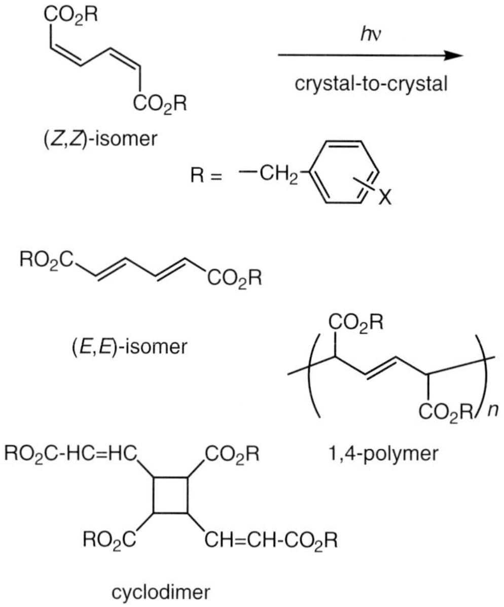

Scheme 24.1 Possible pathways for the solid-state photoreactions of benzyl (Z,Z)-muconates.

change in the size and shape of the space occupied by the substituents of a double bond [30–32]. The isomerization of polyenes is an important photochemical process in biological systems, and a bicycle-pedal model as the volume-conserving reaction mechanism was first pointed out by Warshel in 1976 [33]. Later, Liu et al. [34] proposed the hula-twist process to explain the results of the picosecond time-resolved kinetics of the reaction in Rhodopsin. The hula-twist model has been applied to various reactions in confined media such as a viscous fluid, a rigid matrix, an organic glass, and organic crystals [35, 36]. Reactions proceeding according to the bicycle-pedal [37] and hula-twist models, as well as reactions including crankshaft motion [38–40], require only a small change in the molecular shape, differing from the conventional one-bond-flip motion, which is usually observed during many reactions in solution (Figure 24.1).

Several years ago, we found that the isomerization of n-butylammonium (Z,Z)-muconate produces the corresponding EE-isomer in a high yield in the crystalline state under photoirradiation [41]. This solid-state photoisomerization was revealed to be a one-way reaction and no EZ-isomer was formed during the reaction, while unsaturated compounds such as olefins, polyenes, and azo compounds generally undergo reversible one-bonded photoisomerization to form a mixture of isomers. Previously, we pointed out the possibility that the isomerization of the muconic derivatives in the solid state follows the bicycle-pedal reaction mechanism, but the details of the molecular dynamics of the reaction had not been clarified. Saltiel et al. [42–44] and Liu et al. [45, 46] have independently discussed volume-conserving reaction mechanisms for the isomerization of 1,4-diphenyl-1,3-butadienes, and

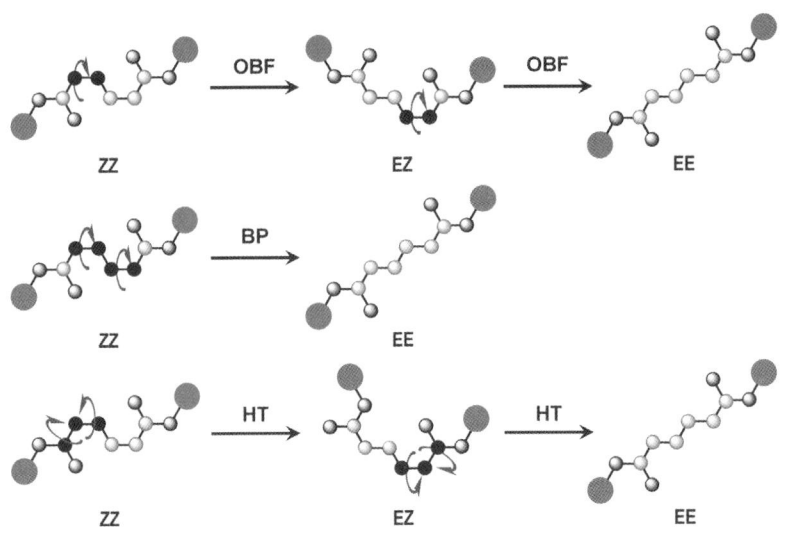

Figure 24.1 Photoisomerization mechanism of (Z,Z)-muconate to the (E,E)-muconate via one-bond-flip (OBF), bicycle-pedal (BP), and hula-twist (HT) models. The OBT and HT models include two-step reactions, while the BP reaction provides the product by a one-step reaction.

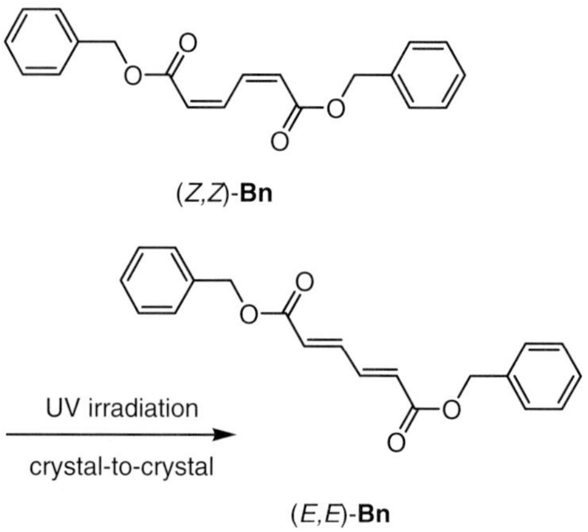

Scheme 24.2 One-way photoisomerization of benzyl muconate from the (Z,Z)- to (E,E)-isomer in the crystalline state.

1,6-disubstitued hexatrienes [47] in the crystalline state and other confined media. They also referred to the isomerization mechanism of the muconate derivatives. More recently, we succeeded in clarification of the reaction mechanism for the solid-state photoisomerization by the direct observation of a change in the single crystal structure during the photoisomerization of benzyl (Z,Z)-muconate [(Z,Z)-**Bn**] to the (E,E)-muconate [(E,E)-**Bn**] in the solid state (Scheme 24.2) [48].

24.2.2
Photoisomerization of Benzyl Muconate

The (Z,Z)-**Bn** provided three polymorphic forms of crystals [(Z,Z)-**Bn**-α, β, and γ] when it was recrystallized from *n*-hexane or chloroform the same at room temperature. Each polymorph is preferentially obtained, depending on the recrystallization conditions, such as the temperature and evaporation rate. For example, the (Z,Z)-**Bn**-α crystals most frequently appeared when the solution of (Z,Z)-**Bn** was slowly evaporated, while a fast evaporation sometimes results in the growth of the crystals as (Z,Z)-**Bn**-β. The structures of these polymorphs were determined by single crystal X-ray analyses. The packing structures of (Z,Z)-**Bn** in the crystals are shown in Figure 24.2. The crystal systems of all the polymorphs are monoclinic and the space group is $P2_1/c$ or $P2_1/n$. The lattice volumes (833.8–841.7 Å) and packing densities (1.272–1.284 g cm^{-3}) are very similar to each other.

The photoisomerization reactivity of (Z,Z)-**Bn** depended significantly on the polymorphic structures. After photoirradiation of the powdered (Z,Z)-**Bn**-α crystals for 5 h at room temperature, the conversion to (E,E)-**Bn** was 71%, but the crystals with the other forms had a lower reactivity (13 and 6% conversions for (Z,Z)-**Bn**-β and γ, respectively) under the same conditions. In general, isomerization requires a reaction space to change its molecular conformation. However, the crystallographic data for the three polymorphs indicated a similar average density for the molecular

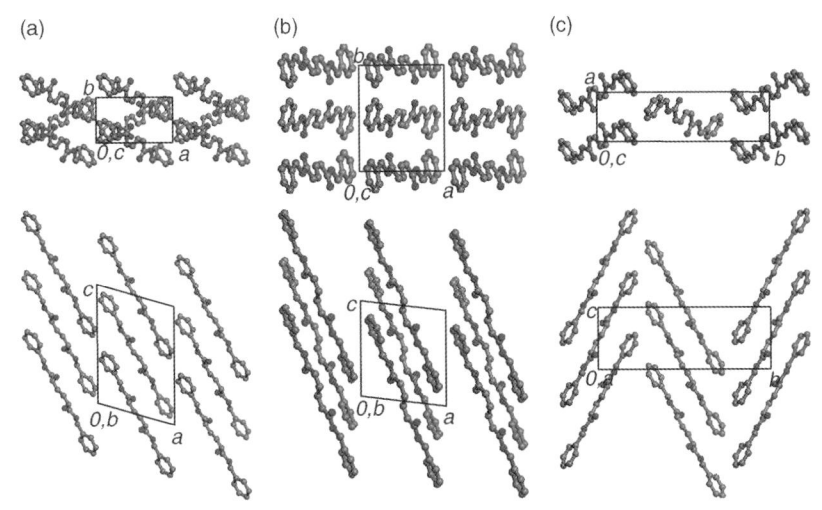

Figure 24.2 Top and side views of single crystal structures of polymorphic (Z,Z)-**Bn** crystals. (a) (Z,Z)-**Bn**-α, (b) (Z,Z)-**Bn**-β, and (c) (Z,Z)-**Bn**-γ.

packing in the crystals. The total volumes of the void spaces in each unit cell were 51.8, 61.2, and 83.6 Å^3 for (Z,Z)-**Bn**-α, β, and γ, respectively. These results disagree with the order of the isomerization reactivity.

24.2.3
Change in Crystal Structures During Photoisomerization

To reveal the reaction mechanism, we directly monitored the change in the single crystal structure of (Z,Z)-**Bn** during the photoisomerization using a band path filter to irradiate with light longer than 300 nm, corresponding to the absorption edge of the (Z,Z)-**Bn** crystals. As a result, the crystal structure of (Z,Z)-**Bn** after UV light irradiation was successfully solved. The space group of the product crystals was the same as the crystal before photoirradiation, and the b- and c-axis lengths as well as the cell volume increased slightly during the initial stage of the reaction. The crystal of (Z,Z)-**Bn** after photoirradiation has cell volume and density values ($V = 858.8$ Å^3, $\rho = 1.247$ g cm^{-3}) identical to those of the crystal of (E,E)-**Bn** prepared by recrystallization ($V = 859.5$ Å^3, $\rho = 1.246$ g cm^{-3}), but the packing mode of the molecules in the crystals is different. Namely, the crystal of (Z,Z)-**Bn** has the space group $P2_1/c$ while the crystal of (E,E)-**Bn** has $P2_1$ [48].

For the crystal structure after photoirradiation, a disordered structure was observed around the diene moiety. This disordered structure contains both (Z,Z)-**Bn** and (E,E)-**Bn** as the molecular structures, as shown by the ORTEP drawing in Figure 24.3. Both the ZZ- and EE-isomers included in the crystal after photoirradiation have similar molecular shapes and planar conformations. The direct observation of a crystal structure during the reaction has revealed that the diene moiety changes its geometric structure from the ZZ-isomer into the EE-isomer consistent with the

(a)

(b)

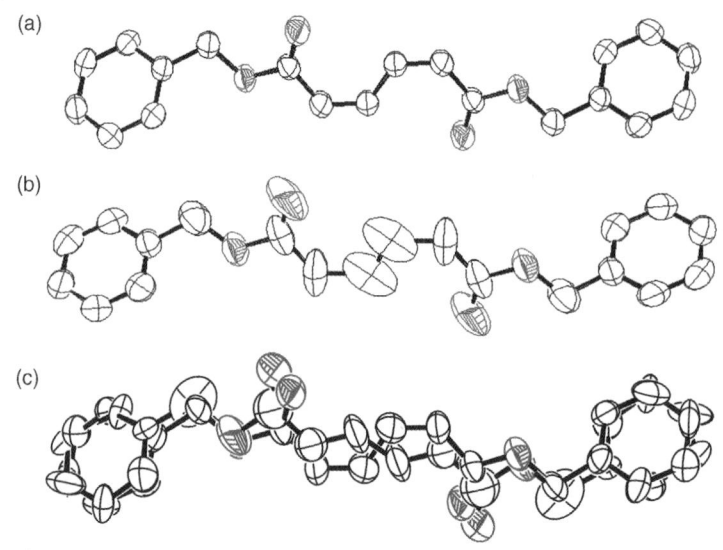

(c)

Figure 24.3 ORTEP drawings of (a) (Z,Z)-**Bn** crystal, (b) disordered structure observed after 27 h photoirradiation, and (c) separated into (Z,Z)-**Bn** and (E,E)-**Bn** with a site occupancy factor of 35% for (E,E)-**Bn**.

bicycle-pedal model. The change in the volume and shape of the reacting molecule [49–51] is small, as expected for a reaction proceeding via the bicycle-pedal mechanism.

The topochemical isomerization of (Z,Z)-**Bn** is accompanied by the expansion of the lattice lengths and the cell volume, but the space group of the crystals does not change. As a result, a structural strain may be produced by a mismatch in the structures of the ZZ- and EE-derivatives in the crystal lattice. When the (E,E)-**Bn** molecules are produced in the crystal lattice of the (Z,Z)-**Bn** molecules during the solid-state reaction, a structural strain is accumulated in the (E,E)-**Bn** molecules as a strained molecular conformation. By comparison of the structures of the recrystallized or calculated molecules, it was revealed a small change in the molecular structure of (Z,Z)-**Bn** in the product crystal, and the structural strain is found at the bond angles in the produced EE-molecules. Thus, the analysis of an intermediate single crystal structure during the reaction has revealed the evolution of a significant strain in the crystal lattice during the topochemical EZ-isomerization.

Furthermore, a structural change in the (Z,Z)-**Bn**-α crystals was observed at a higher conversion. The X-ray diffraction lines after photoirradiation for 120 h agreed well with those for the recrystallized (E,E)-**Bn**. The molecular conformation of (E,E)-**Bn** includes an asymmetric structure with a bent benzyl moiety, being different from the symmetric conformation of the (Z,Z)-**Bn** molecules, and also the (E,E)-**Bn** molecules as the cocrystals during the photoisomerization. These results indicate that a phase transition occurs in the final stage of the isomerization process. The (E,E)-**Bn** molecules can change their conformation in the solid state using the void space around the benzyl moiety during the phase transition. Namely, the primary EE-isomer produced by photoirradiation has a molecular length similar to that of the original (Z,Z)-**Bn** isomer in the initial stage of the reaction, but the cell length increases along the directions of the b- and c-axes during the reaction, leading to an increase in the total volume of the crystals. Drastic changes in molecular conformation and packing structure occur during the crystal-to-crystal transition to the stable crystal structure of (E,E)-**Bn**, accompanying a phase separation.

Thus, we clarified the isomerization process of the muconates according to a bicycle-pedal model. The isomerization occurs via a topochemical reaction process which does not require significant movement of the atoms. The void space included in the crystals plays an important role in the phase transition rather than in the isomerization. In future, the photochemical process of polyene systems performed in confined media will be further clarified using a simple model such as the isomerization of the muconates.

24.3 [2 + 2] Photodimerization

24.3.1 [2 + 2] Photodimerization of 1,3-Dienes

Since the works of Schmidt et al. [52, 53], the [2 + 2] photodimerization of various olefin and diene compounds has been investigated to control reactivity in the solid-state by advanced molecular design using crystal engineering and host–guest chemistry. For example, halogen–halogen, donor–acceptor, and phenyl–perfluorophenyl interactions, as well as hydrogen bonding, are used to control the organization of reacting olefins. The molecular design includes the introduction of appropriate functional groups or atoms, which interact with each other to make the desired molecular assembly in the crystals [54–57]. Another approach using inclusion crystals or templates has also been reported [58, 59]. MacGillivray et al. [58] succeeded in the supramolecular construction of molecular ladders in the solid state using a linear template approach. They designed cocrystals of resorcinol with butadiene and hexatriene derivatives, in which two resorcinol molecules preorganize two polyene molecules through two hydrogen bond interactions, appropriate for [2 + 2] cycloaddition. In this design, two polyenes are positioned at a distance less than 4.2 Å by the templates, leading to the production of the targeted ladderane with the fused cyclobutane framework.

24.3.2 [2 + 2] Photodimerization of Benzyl Muconates

We have examined the [2 + 2] photodimerization of the fluorine-substituted benzyl esters of (Z,Z)- and (E,E)-muconic acids in the solid state (Scheme 24.3) [60]. The muconates undergo [2 + 2] cyclodimerization, EZ-isomerization, and

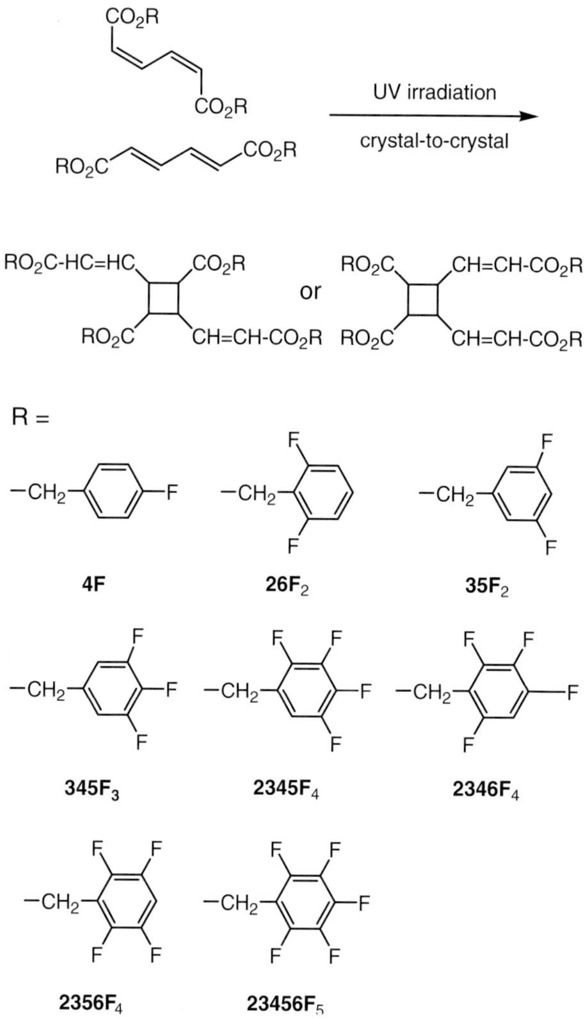

Scheme 24.3 [2 + 2] photodimerization of fluorine-substituted benzyl (Z,Z)- and (E,E)-muconates in the crystalline state.

polymerization, depending on the position and number of the fluorine substituents and the EZ-structure of the diene moiety, as well as the kind of irradiation. We clarified the crystal structure of these compounds in order to discuss the relationship between the molecular packing of the muconates and the photoreaction behavior.

The (Z,Z)-isomers provided either the corresponding dimers or (E,E)-isomers during UV irradiation. During the photoreaction of (E,E)-isomers, some crystals provided the dimer and others showed no reaction. Neither (Z,Z)- nor (E,Z)-isomers were formed because the photoisomerization of muconates occurs via a one-way reaction mechanism. On the basis of the results of single crystal structure analysis, various molecular packing structures were observed, depending on the number and position of the fluorine substituents. In the molecular packing structure of (E,E)-**3454F$_3$** and (E,E)-**2356F$_4$**, we can see one-dimensional zigzag hydrogen bonding with a ladder structure through mutual CH−F interactions as well as cyclic CH−O intermolecular interaction. In most crystals, the characteristic patterns consisting of cyclic CH−F and CH−O interactions between adjacent molecules are observed. In these crystals, close F−F contact (2.82–3.09 Å) was also observed, but such a close contact may be interpreted as the result of simple close packing, being different from clear interactions for Cl−Cl and Br−Br contacts.

There are two possible structures for the products obtained during the photodimerization of the muconates. One is a *syn*-product with mirror-symmetry, and another is an *anti*-product with centro-symmetry (Figure 24.4). The pathway for the reaction is determined by the molecular packing in the crystal. Two reacting double bonds are required to take an appropriate position and direction for the process of dimerization. The reaction needs a specific geometrical structure for the translational packing of reacting double bonds on exactly parallel locations at a smaller stacking distance [61, 62]. The results of the photodimerization and the distance between double bonds ($d_{CC\text{-dim}}$) for the muconates are summarized in Table 24.1.

The dimerization of the muconates occurred when the crystals included a $d_{CC\text{-dim}}$ value less than 4.2 Å, as expected. In the crystals of (Z,Z)-**4F**, the $d_{CC\text{-dim}}$ for *syn*-dimer formation is 4.00 Å, which is allowed to undergo [2 + 2] dimerization, but another distance (4.58 Å) is too long to form an *anti*-dimer. The values for (E,E)-**2356F$_4$** and (E,E)-**23456F$_5$** (4.20 and 4.28 Å) are at the boundary between reactive and non-reactive crystal structures, and the lack of dimer formation is due to the disadvantageous stacking angles instead of an ideal 90°. Some crystals have an acceptable $d_{CC\text{-dim}}$ value for both reactions providing *syn*- and *anti*-dimers, for example, $d_{CC\text{-dim}} = 3.8$–4.1 Å to produce both dimers of (Z,Z)-**26F$_2$**, (Z,Z)-**35F$_2$**, and (Z,Z)-**345F$_3$**. However, the NMR data of the isolated products clarified that (Z,Z)-**26F$_2$** and (Z,Z)-**345F$_3$** give only *anti*-dimers and (Z,Z)-**4F** and (Z,Z)-**35F$_2$** give only *syn*-dimers. For all the cases shown in this study, only one type of dimer was produced, but not a mixture of both dimers. A high selectivity during *syn*- and *anti*-dimer formation is quite consistent with the $d_{CC\text{-dim}}$ values. Namely, the reaction proceeds favorably according to the path accompanying a smaller $d_{CC\text{-dim}}$ value when both the $d_{CC\text{-dim}}$ values are less than 4.2 Å.

Furthermore, we revealed the photoreaction mechanism of the [2 + 2] photodimerization of (E,E)-**345F$_3$** on the basis of the direct observation of single-crystal-to-single-crystal transformation during the reaction [63]. The photodimerization occurred randomly in the columnar assembly of monomer molecules in the crystals of (E,E)-**345F$_3$**, followed by formation of a trimer. When we carefully carried out the photoreaction of (E,E)-**345F$_3$**, a disordered structure was observed. The structure was separated into the monomer and a product with an occupancy factor of 28% for the latter. The product appeared to have a ladder structure but, in an actual case, the dimer including a cyclobutane ring was randomly formed in the lattice. The molecule (E,E)-**345F$_3$** has double bonds at four crystallographically equivalent positions for cyclobutane ring formation. A cyclobutane ring can be randomly formed in the

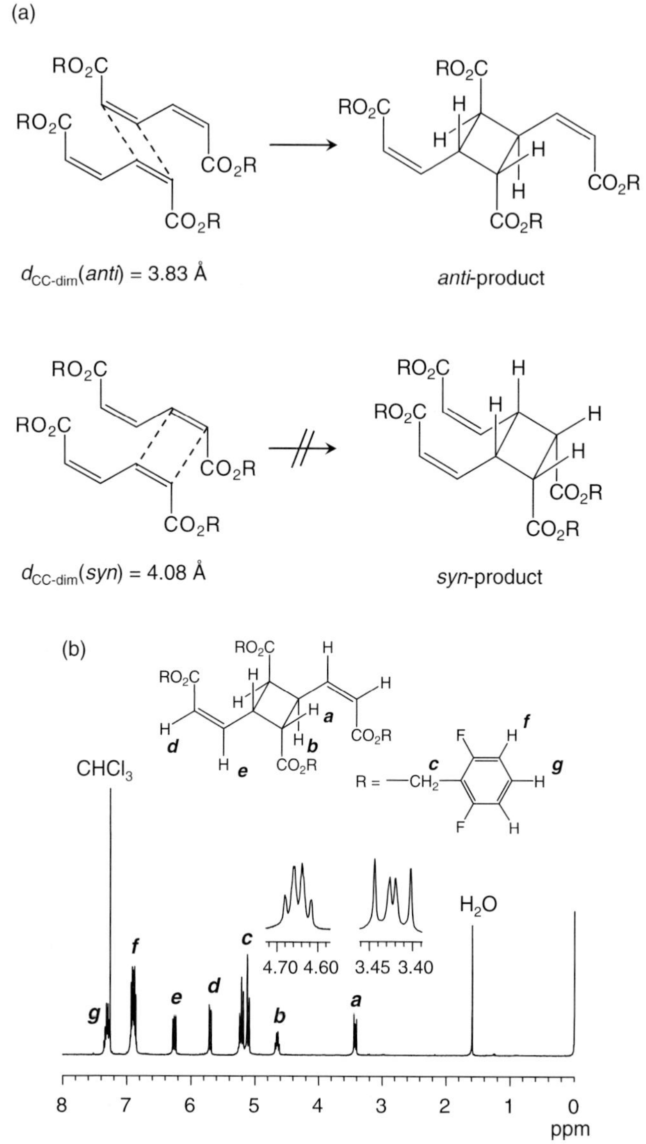

Figure 24.4 (a) Structure of syn- and anti-type [2 + 2] cyclodimers of (Z,Z)-**26F$_2$**, (b) ^{1}H NMR spectrum of the obtained anti-dimer. The dimer was isolated by column chromatography after UV irradiation.

columnar structure consisting of the translational arrangement of the (E,E)-**345F$_3$** molecules, which have a centro-symmetrical structure. When we investigated the photoreactions up to high conversion of (E,E)-**345F$_3$**, a trimer was also confirmed, in addition to the formation of a dimer, in the NMR spectrum of the photoproducts

Table 24.1 Structure of obtained dimers and stacking parameters for [2 + 2] photodimerization of fluorine-substituted muconates in the crystalline state.

Monomer	Ester	Dimer structure	$d_{CC\text{-dim}}$ (Å)	
			syn	anti
(Z,Z)-4F	4-fluorobenzyl	syn-type	4.00	4.58
(Z,Z)-26F$_2$	2,6-difluorobenzyl	anti-type	4.08	3.83
(Z,Z)-35F$_2$	3,5-difluorobenzyl	syn-type	3.83	3.91
(Z,Z)-345F$_3$	3,4,5-trifluorobenzyl	anti-type	3.93	3.90
(Z,Z)-2345F$_4$	2,3,4,5-tetrafluorobenzyl	no dimer	7.01	5.27
(Z,Z)-2346F$_4$	2,3,4,6-tetrafluorobenzyl	no dimer	5.20	5.74
(E,E)-4F	4-fluorobenzyl	syn-type	3.88	4.31
(E,E)-26F$_2$	2,6-difluorobenzyl	no dimer	5.43	4.76
(E,E)-35F$_2$	3,5-difluorobenzyl	syn-type	3.97	4.34
(E,E)-345F$_3$	3,4,5-trifluorobenzyl	syn-type	3.90	4.54
(E,E)-2356F$_4$	2,3,5,6-tetrafluorobenzyl	no dimer	5.13	4.20
(E,E)-23456F$_5$	2,3,4,5,6-pentafluorobenzyl	no dimer	5.20	4.28

(Figure 24.5). The trimer was isolated as a white powder. A trace of a higher oligomer was also detected in the NMR spectra at higher conversion.

24.4
Topochemical Polymerization

24.4.1
Features of Topochemical Polymerization

The topochemical polymerization of 1,3-diene monomers proceeds under photo-, X-, and γ-ray irradiation or upon heating [64–70], similar to the solid-state polymerization of diacetylene compounds [71–73]. The polymerization proceeds via a radical chain-reaction mechanism and the propagating radicals are readily detected by EPR spectroscopy during polymerization in the crystalline state, because termination between the propagating radicals occurs less frequently in the solid state. The stacking structure of the monomers used for the topochemical polymerization is evaluated using the following parameters: the intermolecular distance between carbons that react to form a new bond during the topochemical polymerization (d_{CC}), the stacking distance along the column (d_s), and the angles between the stacking direction and the molecular plane in orthogonally different directions (θ). The polymerization principle (the 5 Å rule) that the d_s values are in an exclusively limited region of 4.74–5.21 Å for the 1,3-diene monomers [74, 75] has features similar to the empirical rules for the polymerization of diacetylene compounds [72].

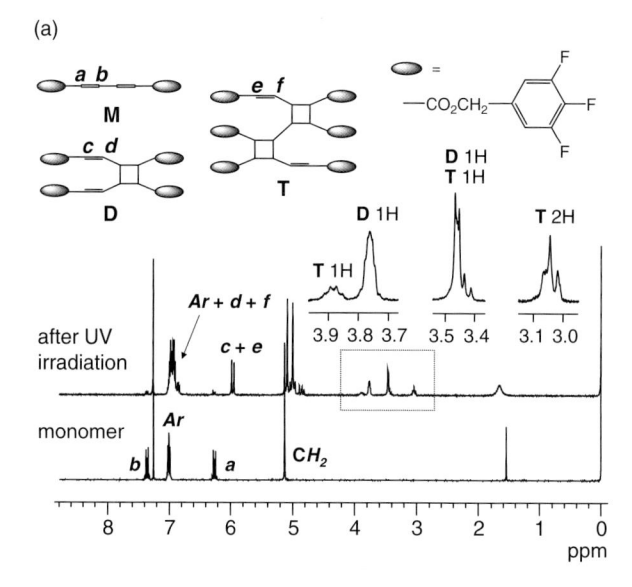

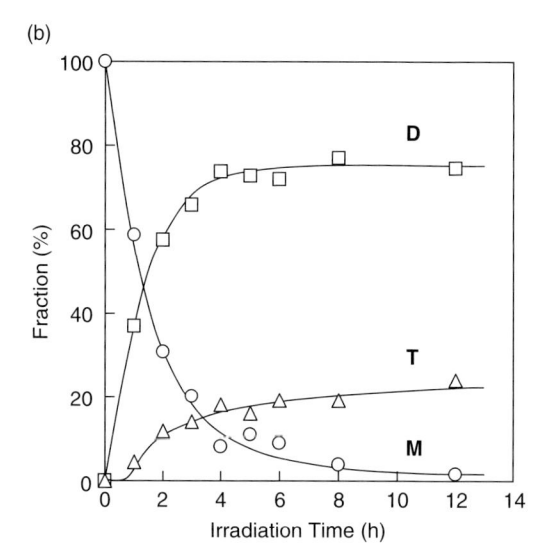

Figure 24.5 (a) ^1H NMR spectra of (E,E)-**345F$_3$** and photoproducts at 92% conversion after UV irradiation, (b) time-dependence of the fractions of monomer, dimer, and trimer during the photoreaction.

24.4.2
Monomer Stacking Structure and Polymerization Reactivity

We determined the single crystal structures for the monomer and polymer crystals of several kinds of benzyl muconates (Scheme 24.4) [76–79]. The polymer crystals were

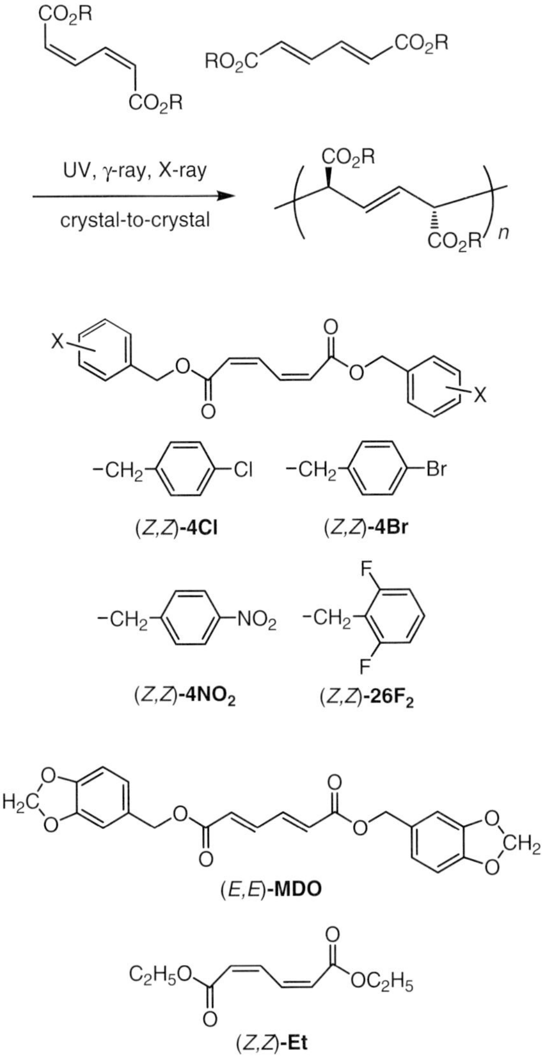

Scheme 24.4 Topochemical polymerization of (Z,Z)- and (E,E)-muconates in the solid state.

prepared by the γ-radiation polymerization of the monomer single crystals. The changes in the selected unit cell lengths and volume before and after the polymerization are summarized in Table 24.2. The d_s and fiber period (FP) values are lattice lengths parallel to a fiber axis for the monomer and polymer crystals, respectively. A model for the solid-state polymerization of 1,3-diene monomers is illustrated in Figure 24.6. When the polymerization proceeds in a domino-type reaction mechanism, all the monomer molecules include a conrotatory molecular motion for the formation of the polymer chain. Each molecule rotates and changes its conformation to make a new covalent bond between molecules with the least translational movement of the center of the molecular mass.

24 Direct Observation of Change in Crystal Structures

Table 24.2 Change in the crystal structures and polymerization rates for solid-state polymerization of muconates

Monomer	Lattice length (Å)		Cell volume (Å³)		Polymerization rate constant $k \times 10^3$ (s^{-1})
	Monomer(d_s)	Polymer(FP)	Monomer	Polymer	
(Z,Z)-Et	4.931	4.839 (−1.9%)	554.3	525.3 (−5.2%)	5.16
(Z,Z)-4Cl	5.122	4.863 (−5.1%)	923.0	886.5 (−4.0%)	1.15
(E,E)-MDO	4.432	4.702 (+6.1%)	941.9	920.0 (−2.3%)	0.75
(Z,Z)-4Br	5.21	4.856 (−6.8%)	953.0	927.8 (−2.6%)	0.25
(Z,Z)-4NO$_2$	5.239	4.855 (−7.3%)	947.6	927.8 (−2.1%)	0.10
(Z,Z)-26F$_2$	4.177	4.716 (+12.9%)	871.3	905.5 (+3.9%)	∼0

When monomer molecules stack translationally in a column with a d_s value larger than the FP of the corresponding polymer, shrinking is observed along the column direction. In fact, the change in the lattice lengths along the fiber axis was −1.9 to −7.3% for the crystals of (Z,Z)-Et, (Z,Z)-4Cl, (Z,Z)-4Br, and (Z,Z)-4NO$_2$. The other two axis directions involved expansion, except for shrinking in the perpendicular axis for (Z,Z)-Et. Conversely, the fiber axis length increases by +6.1 and +12.9% during the polymerization of (E,E)-MDO and (Z,Z)-26F$_2$, respectively. Simultaneously, a large shrinking occurs in the orthogonal direction. Overall, the unit cell volume of these crystals is reduced during the polymerization by 2–5%, except for the volume-expanding polymerization of (Z,Z)-26F$_2$. Volume shrinking is usually observed for the addition polymerization of unsaturated monomers because it is a covalent bond-forming reaction. In Table 24.2, we notice the exactly identical FP values for the polymer crystals, irrespective of the d_s values for the monomers; that is, FP 4.84–4.86 Å and 4.70–4.72 Å for the polymers obtained by the shrinking and expanding polymerizations, respectively. This indicates that the polymer chains have a specific conformation in the crystals, depending on the polymerization mode.

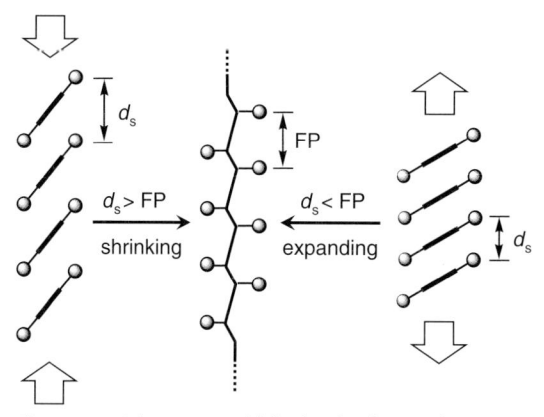

Figure 24.6 Schematic model for the shrinking and expanding of crystals during topochemical polymerization in the solid state.

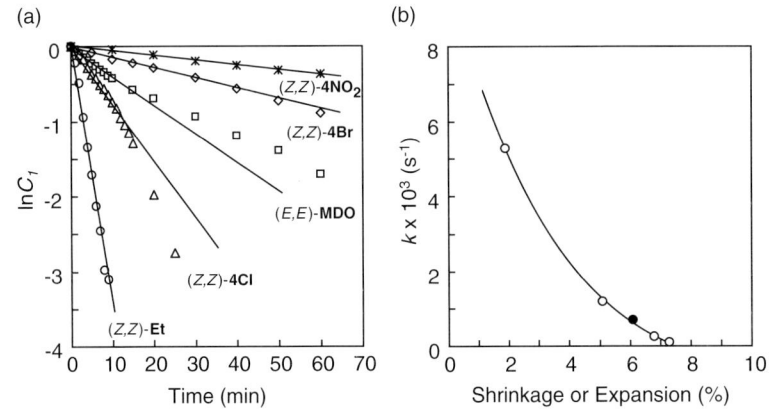

Figure 24.7 (a) Semilogarithmic plots of monomer fraction versus UV irradiation time for the topochemical polymerization of muconates, (b) relationship between relative polymerization rate and degree of shrinking (open circles) or expansion (closed circle) during topochemical polymerization.

Previously, a zigzag-type reaction mechanism was proposed for the diacetylene polymerization [55] and the step-wise [2 + 2] photopolymerization of diolefin compounds [80], which have a large stacking angle (i.e., a small tilt angle) in the columnar structure of the monomers. However, we can conclude that the polymerization of the muconates proceeds via a domino-type polymerization mechanism, irrespective of the shrinking and expanding polymerizations.

Next, we investigated the relationship between the stacking structure and the polymerization reactivity. The polymerization was carried out using powdered crystals embedded in a KBr pellet under UV irradiation. The conversion was determined from a change in the absorption intensity of the C=C stretching around 1600 cm^{-1} in the IR spectrum [66, 76]. The first-order plot of the conversion is shown in Figure 24.7, as a function of the UV irradiation time. Acceleration of polymerization was observed for the shrinking polymerization of (Z,Z)-**4Cl**, while the polymerization rate decreased during the expanding polymerization of (E,E)-**MDO**. This is due to a different mode for quenching the evolved strain through a crystal lattice change and a polymer conformational stress. The polymerization rate depended significantly on the magnitude of change in the lattice length during polymerization. The polymerization rate constant, k, becomes greater as the change in the lattice is smaller. Interestingly, this unique relationship is true for both shrinking and expanding polymerizations.

24.4.3
Shrinking and Expanding Crystals

We further examined a continuous change in the crystal structure and the strain accumulated in the crystals during the shrinking and expanding polymerizations [79].

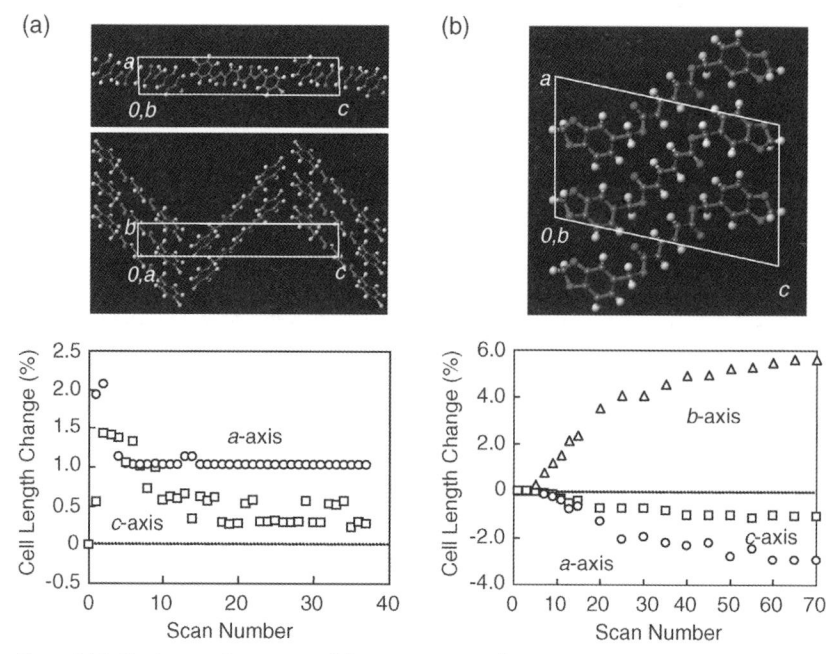

Figure 24.8 Single crystal structure of the monomers and a change in the lattice lengths during the continuous X-ray radiation of (a) (Z,Z)-**4Cl** and (b) (E,E)-**MDO**. Polymer chains are formed along the b-axis. Circle: a-axis, triangle: b-axis, square: c-axis.

The change in the X-ray diffraction profiles of the muconates was investigated by continuous X-ray radiation. Figure 24.8 shows the single crystal structure of (Z,Z)-**4Cl** and (E,E)-**MDO** and a change in the lattice lengths during the X-ray radiation polymerization. A temporary increase was observed up to the about 2% increment in the lattice length along the a- and c-axes during the initial stage of the polymerization of (Z,Z)-**4Cl**. The lattice lengths then gradually decreased and approached the values for the single crystals of poly(**4Cl**); that is, +0.9 and 0.1% increase. In the diffraction profile of (E,E)-**MDO**, the shift of each line was more clearly observed and all the axes change their length monotonically without exhibiting a temporary peak. An increase in the b-axis length as the fiber axis was observed due to the expanding polymerization, and a decrease in the other orthogonal axes. An induction period was detected in the structural change.

24.4.4
Accumulation and Release of Strain During Polymerization

The thermally-induced polymerization of the 1,3-diene monomers occurs at a much lower rate than the rates of the UV- and X-ray-induced polymerizations. As a result, we can readily observe an induction period during the initial stage of the polymerization and the subsequent acceleration of the rate by IR microscope

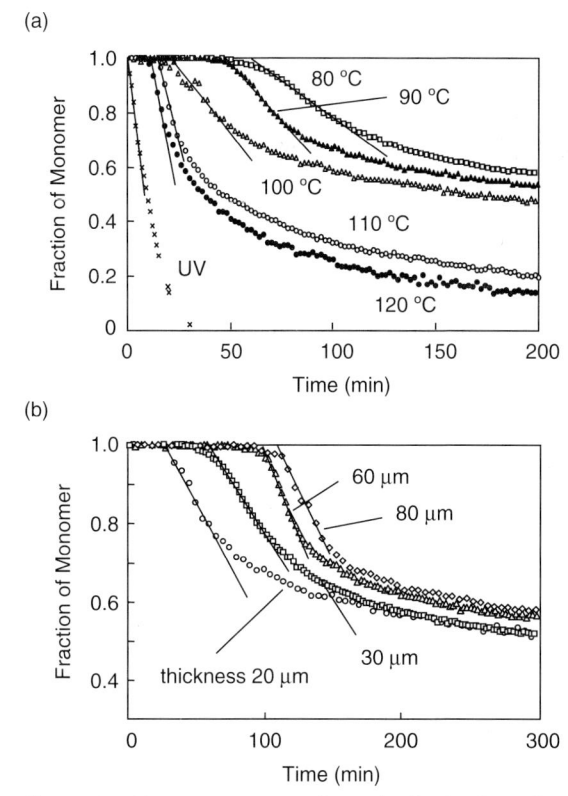

Figure 24.9 (a) Time-conversion relationship for the thermally-induced polymerization of (Z,Z)-**4Cl** in the dark at 80–120 °C, (b) photopolymerization under UV irradiation, and the effect of crystal size.

spectroscopy under temperature control [81]. The platelet crystals of (Z,Z)-**4Cl** were set on a thin KBr plate for the IR measurement, and the light was induced from the 100 face of the crystals. In the time dependence of the absorbance of some selected bands in the IR spectrum observed during the polymerization, we detected an induction period. The length of the induction period depended on the temperature, the kind of monomer, and also the size of the crystals. The results are represented as a change in the monomer fraction as a function of the polymerization time (Figure 24.9). The length of the induction period was dependent on the temperature; for example, about 10 min at 120 °C and 1 h at 80 °C. The polymerization of (Z,Z)-**4Br** occurred similarly with a longer induction period for the polymerizations at 110 and 100 °C. In contrast to the observation of the induction period during the polymerization of (Z,Z)-**4Cl** and (Z,Z)-**4Br**, no induction period was detected during the polymerization of (E,E)-**MDO** at each temperature in the range 110–130 °C. The polymerization rate decreased with the increase in the polymerization time.

In contrast to the results for the polymerization using the larger crystals (about 40–100 μm in length), the induction period was drastically reduced at each temperature when microcrystals prepared by a precipitation method were used. In order to further examine the effect of the crystal size on the length of the induction period, we prepared various platelet or block-like crystals with different thicknesses and similar sizes. We found that an increase in the thickness of the crystals lengthened the induction period. The size effect on the induction period is related to the difference in the occurrence of the phase transition of the crystals. Eventually, it was concluded that the formation of polymer chains in the monomer crystals leads to the evolution of strain in the crystals during the initial stage of the polymerization. The accumulated strain induces the phase transition from the monomer crystal phase to the polymer crystal phase just after the induction period. In other words, the strain accumulated during the induction period is a trigger for the phase transition and the acceleration of the polymerization.

24.4.5
Homogeneous and Heterogeneous Polymerizations

Previously, we found that the di(4-alkoxybenzyl) esters of (E,E)- and (Z,Z)-muconic acids have different geometric structures, but the obtained polymers have completely identical chemical structures with the same tacticity (Scheme 24.5) [82–84]. We investigated the molecular dynamics during the polymerization of (E,E)-4MeO, (Z,Z)-4MeO, monitored by *in situ* single and powder X-ray diffraction experiments in order to discuss a reaction mechanism for the solid-state polymerization

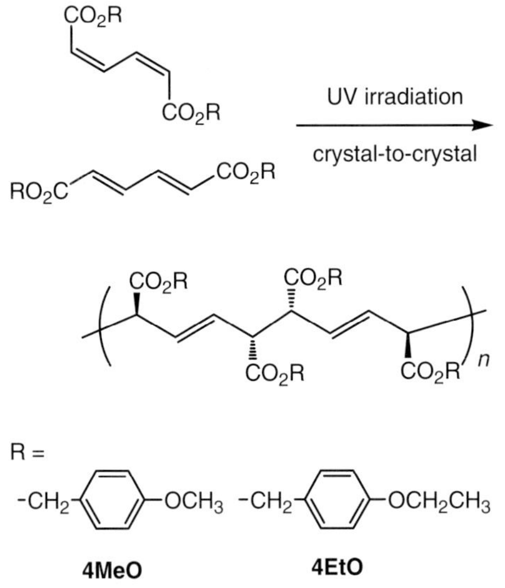

Scheme 24.5 Formation of syndiotactic polymers during the solid-state polymerization of alkoxybenzyl (Z,Z)- and (E,E)-muconates.

that proceeds via a chain reaction mechanism with or without crystal phase separation [85].

The change in the lattice parameters before and after the polymerization was 0.2–3.0% during the polymerization of (E,E)-4MeO, much smaller than that of (Z,Z)-4MeO (2.7–6.8%). Interestingly, both polymer crystals have completely identical structures regarding not only the stereochemical structure (tacticity) but also the crystal structure (molecular conformation). In other words, both polymer crystals have the same crystallographic parameters. This indicates that a greater movement of atoms is required in the crystals of (Z,Z)-4MeO than in (E,E)-4MeO during the polymerization. We examined the change in the crystal lattice for (E,E)-4MeO and (Z,Z)-4MeO by continuous X-ray radiation at room temperature [85]. In the powder X-ray diffraction profiles of the (E,E)-4MeO crystals observed during the polymerization, all diffraction lines continuously shifted from a position for the monomer to that for the polymer without broadening of the line. On the other hand, the diffraction lines of (Z,Z)-4MeO showed a discontinuous change under the same radiation conditions. The profiles of the diffraction from the (Z,Z)-4MeO crystals consisted of the lines due to the monomer and the polymer accompanied by a crystal phase separation, despite highly penetrating X-ray radiation conditions. From the diffraction profile of the (E,E)-4MeO crystals, the c-axis length along the fiber axis of the resulting polymer crystals decreased by about 6%. The a-axis length first increased by about 1% and then decreased, finally leading to shrinking along that axis direction. The b-axis length gradually increased. The expanding and shrinking of the lattice length agree well with the results of single crystal structure analysis of the monomer and polymer. Such a continuous change in the crystal lattice had already been observed for the polymerization of other muconate monomers, for example (Z,Z)-4Cl and (E,E)-MDO [79]. On the other hand, the polymerization of (Z,Z)-4MeO induced a crystal phase separation from the initial stage of the reaction up to at least about 20% conversion. At higher conversion, no diffraction line due to the monomer phase was observed. The polymer phase included a small change in all the axis lengths, while the lattice parameters of the monomer phase changed drastically.

We also tried to determine the structural change in the crystals during the polymerization more precisely by *in situ* X-ray single crystal structure analysis and eventually observed a disordered structure around the diene moiety of (E,E)-4MeO, as shown in the ORTEP drawings in Figure 24.10. The disordered structure was divided into the monomer and polymer structures. Furthermore, we noticed a maximum value for the lattice parameters of (E,E)-4MeO (a- and c-axis lengths, and cell volume) after starting the polymerization within 1 h of radiation, that is, below 30% conversion. The lattice lengths initially include an expansion and consequently the cell volume also temporarily increases. Such a temporary increase in the lattice constants is due to the coexistence of the monomer and polymer molecules in the common crystal lattice. The coexistence of monomer and polymer structures in a solid solution should cause a strain in the molecular conformation of the monomer or polymer as the intermediate structure of the cocrystals. The initial change in the lattice parameters is ascribed to the polymer chain produced in the monomer crystal.

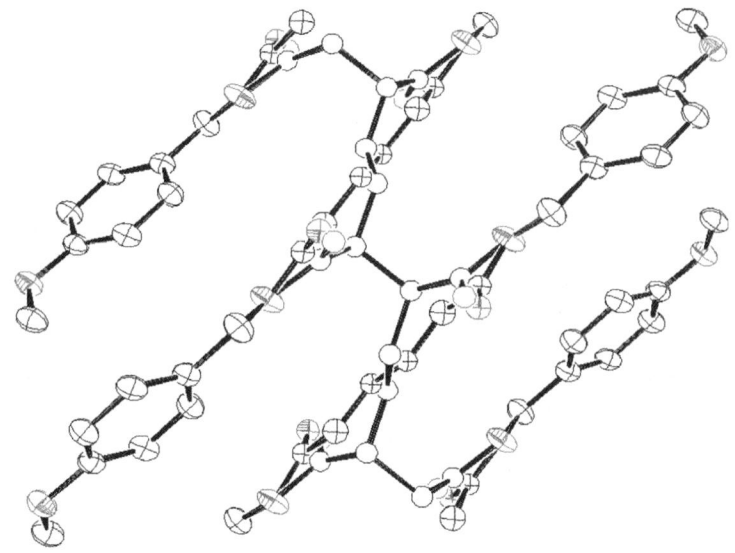

Figure 24.10 ORTEP drawing of the (E,E)-4MeO crystal as the intermediate structure during the polymerization with a site occupancy factor of 53% for poly[(E,E)-4MeO].

Topochemical polymerization includes two kinds of polymerization mechanisms as the extreme cases, as shown in Figure 24.11. One is the homogeneous reaction mechanism, as was seen in the polymerization of (E,E)-4MeO. In this model, polymerization occurs at random positions in the crystals, and both the monomer and polymer molecules include structural strain during the reaction. This forms a solid solution and no phase separation is observed at the intermediate stage of the reaction. Therefore, we can directly observe a change in the single crystal structure from 0 to 100% conversion throughout the reaction. Another is the heterogeneous reaction mechanism, as observed for (Z,Z)-4MeO. The reaction starts preferentially

Figure 24.11 Schematic model for homogeneous and heterogeneous polymerization mechanism.

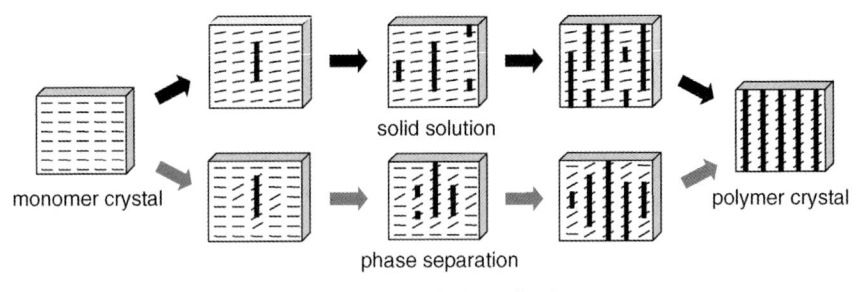

near specific defect sites and accompanies the nucleation of a polymer phase. The produced polymer forms a new domain in the monomer crystals in the process of the polymerization, and consequently, phase separation is observed at the intermediate stage. In the latter model, the strain is possibly concentrated around the interface of the monomer and polymer phases, and the polymerization is accelerated near the phase boundary. In this model, a conformational change in the unreacted monomer domains is emphasized for the homogeneous polymerization model. Such a change often results in the drastic phase transition of the whole crystals from the starting structure into the structure of the product. Furthermore, the proposed model is specific to polymerization that proceeds via a chain reaction mechanism in the solid state. This model includes the strain accumulated in all the crystals during the reaction because of the presence of long-chain polymer molecules as the products. For the reactions of low-molecular-weight compounds yielding the corresponding low-molecular-weight products, such as the isomer or dimer products, the reactivity in the solid state has exclusively been explained by the mobility of the atom in the solid state. A free volume surrounding a reacting center is also important because the intermolecular space can act as a buffer to reduce the strain evolved during the reactions.

Thus, we have clarified the polymerization mechanism of the muconates during solid-state polymerization via a crystal-to-crystal transformation by the direct observation of a change in the crystal structures during the polymerization. We revealed a change in the structure of the muconate crystals accompanying the shrinking and expanding of the lattice lengths, on the basis of the X-ray single crystal structure analysis of the monomers and polymers as well as a change in the transient structure during the continuous X-ray radiation. During the polymerization, an initially evolved strain induces the temporary expansion of the crystal. A strain is accumulated in both the formed polymer chain and the remaining monomer crystal during the initial stage of the reaction and is finally released after the polymerization. Furthermore, we observed a definite induction period during the initial stage of the thermally-induced polymerization. Such polymerization behavior was observed for the shrinking-type polymerization, but not for the expanding polymerization. Furthermore, we have concluded that the solid-state polymerization of 1,3-diene monomers is divided into a homogeneous polymerization which proceeds in the solid solution of the remaining monomer and the resulting polymer and a heterogeneous polymerization in which a crystal phase separation occurs between the monomer and polymer crystal domains during the reaction. According to the former model, the monomer crystal phase changes continuously into the polymer crystal phase without any phase separation when the least movement of atoms induces less strain in the crystals. We succeeded in the direct observation of the expanding crystal lattices and cell volume in the initial stage of polymerization, followed by the subsequent shrinking during the progress of the homogeneous reaction. Whereas, a phase separation was observed during the reaction according to the heterogeneous polymerization model, in which a considerable movement of atoms is required. The monomer stacking structure determines the reaction path during the solid-state polymerization.

24.5
Conclusion

For a long time, it has been believed that molecules can react only in fluid media such as solution, liquid and gaseous states. However, recent progress in crystal engineering and crystallography has revealed that various reactions can be performed successfully in the solid state. During investigation of solid-state organic chemistry and materials science, including polymer chemistry, the direct observation of a continuous change in the conformational and chemical structures of reactant molecules during reactions in the crystalline state can give us important information. In the project of "Molecular Nano Dynamics", we have investigated a change in the crystal structures and the accumulation of a strain in the crystals during the reaction to reveal the mechanism of the topochemical polymerization, EZ-photoisomerization, and [2 + 2] photodimerization of muconic esters as the 1,3-diene monomers via a crystal-to-crystal transformation. We proposed a solid-state reaction mechanism for the crystal-lattice controlled reactions of various unsaturated organic compounds based on the results of in situ X-ray crystal structure analysis during the polymerization.

In addition, we revealed the structures and the intercalation behavior of several stereoregular polymers [86, 87] obtained by the topochemical polymerization during the course of the project. Furthermore, we developed a double intercalation method, using an alkylamine and pyrene as the guests, to control fluorescence properties [88]. Some collaboration with the other research groups in the project was also successfully carried out, regarding the fabrication of the nanocomposite consisting of a crystalline organic polymer and fine metal particles [89–91], and crystal engineering using naphthylmethylammonium supramolecular synthons [92–95]. Studies on the two-dimensional polymerization of the muconates to fabricate polymer thin films are also continuing. We also investigated diacetylene polymerization [96–99], which is useful for the fabrication of conjugating polymers, on the basis of the results obtained from the studies on the diene polymerization. The direct observation of molecular motion in the crystals will bring new insight to the mechanism of solid-state reactions and also the design of new functional solid materials.

Acknowledgments

The author acknowledges Daisuke Furukawa, Dr Yutaka Mori, Takako Ueno, Katsuya Onodera, Natsuko Nishizawa, and the other members of the research group of Osaka City University. The author also acknowledges Prof. Seiya Kobatake, Osaka City University, Prof. Mikiji Miyata, Prof. Norimitsu Tohnai, and Dr Ichiro Hisaki, Osaka University, Dr Toru Odani, Prof. Hidetoshi Oikawa, and Prof. Hachiro Nakanishi, Tohoku University, Prof. Shuji Okada, Yamagata University, Prof. Masato Suzuki, Nagoya Institute of Technology, and Dr Ken Nakajima, Tokyo Institute of Technology for their fruitful collaborations in the project "Molecular Nano Dynamics". The author gratefully thanks Prof. Kunio

Oka, Osaka Prefecture University, for his kind assistance with the γ-radiation experiment. This work was supported by Grants-in-Aid for Scientific Research on Priority Areas (Area No. 432, No. 16072215) and for Scientific Research (No. 16350067) from the Ministry of Education, Culture, Sports, Science and Technology (MEXT) of Japan.

References

1 Thomas, J.M. (1974) Topography and topology in solid-state chemistry. *Phil. Trans. R. Soc. London Series A*, **277**, 251–286.
2 Green, B.S., Arad-Yellin, R. and Cohen, M.D. (1986) Stereochemistry and organic solid-state reactions. *Top. Stereochem.*, **16**, 131–218.
3 Ramamurthy, V. and Venkatesan, K. (1987) Photochemical reactions of organic crystals. *Chem. Rev.*, **87**, 433–481.
4 Toda, F. (1995) Solid state organic chemistry: efficient reactions, remarkable yields, and stereoselectivity. *Acc. Chem. Res.*, **28**, 480–486.
5 Ohashi, Y. (1998) Real-time in-situ observation of chemical reactions. *Acta Crystallogr. Sec. A.*, **54**, 842–849.
6 Tanaka, K. and Toda, F. (2000) Solvent-free organic synthesis. *Chem. Rev.*, **100**, 1025–1074.
7 Toda, F. (ed.) (2002) *Organic Solid-State Reactions*, Kluwer, London.
8 Kobatake, S. and Irie, M. (2004) Single-crystalline photochromism of diarylethenes. *Bull. Chem. Soc. Jpn.*, **77**, 195–210.
9 Braga, D. and Grepioni, F. (2005) Making crystals from crystals: a green route to crystal engineering and polymorphism. *Chem. Commun.*, 3635–3645.
10 Toda, F.(ed.) (2005) Organic solid state reactions, *Top. Curr. Chem.*, **254**.
11 Cohen, M.D. and Schmidt, G.M.J. (1964) Topochemistry part I. A survey. *J. Chem. Soc.*, 1996–2000.
12 Schmidt, G.M.J. (1971) Photodimerization in the solid state. *Pure Appl. Chem.*, **27**, 647–678.
13 Wegner, G. (1977) Solid-state polymerization mechanisms. *Pure Appl. Chem.*, **49**, 443–454.
14 Hasegawa, M. (1995) Photodimerization and photopolymerization of diolefin crystals. *Adv. Phys. Org. Chem.*, **30**, 117–171.
15 Paleos, M. (ed.) (1992) *Polymerization in Organized Media*, Gordon and Breach Science, Philadelphia.
16 Matsumoto, A. *Handbook of Radical Polymerization*, (2002) (eds K. Matyjaszewski and T.P. Davis), Wiley, New York, pp. 691–773.
17 Tieke, B. (1985) Polymerization of butadiene and butadiyne (diacetylene) derivatives in layer structures. *Adv. Polym. Sci.*, **71**, 79–151.
18 Matsumoto, A., Matsumura, T. and Aoki, S. (1994) Stereospecific polymerization of diethyl (Z,Z)-hexa-2,4-dienedioate in the crystalline state. *J. Chem. Soc. Chem. Commun.*, 1389–1390.
19 Matsumoto, A., Matsumura, T. and Aoki, S. (1996) Stereospecific polymerization of dialkyl muconates through free radical polymerization: isotropic polymerization and topochemical polymerization. *Macromolecules*, **29**, 423–432.
20 Matsumoto, A. and Odani, T. (2001) Topochemical polymerization of 1,3-diene monomers and features of polymer crystals as organic intercalation materials. *Macromol. Rapid Commun.*, **22**, 1195–1215.
21 Matsumoto, A. (2001) Stereospecific polymerization of 1,3-diene monomers in the crystalline state. *Prog. React. Kinet. Mech.*, **26**, 59–110.

22. Matsumoto, A. (2003) Polymer structure control based on crystal engineering for materials design. *Polym. J.*, **35**, 93–121.
23. Matsumoto, A. (2005) Reactions of 1,3-diene compounds in the crystalline state. *Top. Curr. Chem.*, **254**, 263–305.
24. Desiraju, G.R. (1989) *Crystal Engieering: The Design of Organic Solids*, Elsevier, Amsterdam.
25. MacNicol, D.D., Toda, F. and Bishop, R. (eds) (1996) Comprehensive supramolecular chemistry, *Solid State Supramolecular Chemistry: Crystal Engineering, vol.* **6**, Pergamon, Oxford.
26. Steed, J.W. and Atwood, J.L. (2000) *Supramolecular Chemistry*, Wiley, Chichester, p. 389.
27. Ohashi, Y. (ed.) (1993) *Reactivity in Molecular Crystals*, Kodansha-VCH, Tokyo.
28. Braga F D. and Grepioni F F. (eds) (2007) *Making Crystals by Design: Methods, Techniques and Applications*, Wiley-VCH, Weinheim.
29. Garcia-Garibay, M.A. (2007) Molecular crystals on the move: from single-crystal-to-single- crystal photoreactions to molecular machinery. *Angew. Chem. Int. Ed.*, **46**, 8945–8947.
30. Alfimov, S.M. and Razumov, V.F. (1978) Amorphism of polycrystal phase-transition in course of cis-trans photoisomerization. *Mol. Cryst. Liq. Cryst.*, **49**, 95–97.
31. Kinbara, K., Kai, A., Maekawa, Y., Hashimoto, Y., Naruse, S., Hasegawa, M. and Saigo, K. (1996) Photoisomerization of ammonium a,b-unsaturated carboxylates in the solid state: effect of the hydrogen-bond network on the reactivity. *J. Chem. Soc. Perkin. Trans. 2*, 247–253.
32. Moorthy, J.N., Venkatakrishnan, P., Savitha, G. and Weiss, R.G. (2006) Cis-to-trans and trans-to-cis isomerizations of styrylcoumarins in the solid state. Importance of the location of free volume in crystal lattices. *Photochem. Photobio. Sci.*, **5**, 903–913.
33. Warshel, A. (1976) Bicycle-pedal model for 1st step in vision process. *Nature*, **260**, 679–683.
34. Liu, R.S.H. and Asato, A.F. (1985) Photochemistry of polyenes. 22. The primary process of vision and the structure of bathorhodopsin: a mechanism for photoisomerization of polyenes. *Proc. Natl. Acad. Sci. U S A.*, **82**, 259–263.
35. Liu, R.S.H. (2002) Introduction to the symposium-in-print: photoisomerization pathways, torsional relaxation and the hula twist. *Photochem. Photobiol.*, **76**, 580–583.
36. Liu, R.S.H. and Hammond, G.S. (2005) Reflection on medium effects on photochemical reactivity. *Acc. Chem. Res.*, **38**, 396–403.
37. Saltiel, J., Krishna, T.S.R., Turek, A.M. and Clark, R.J. (2006) Photoisomerization of *cis,cis*-1,4- diphenyl-1,3-butadiene in glassy media at 77 K: the bicycle-pedal mechanism. *Chem. Commun.*, 1506–1508.
38. Harada, J. and Ogawa, K. (2006) Torsional motion of stilbene-type molecules in crystals. *Top. Stereochem.*, **25**, 31–47.
39. Khuong, T.A.V., Zepeda, G., Sanrame, C.N., Dang, H., Bartberger, M.D., Houk, K.N. and Garcia-Garibay, M.A. (2004) Crankshaft motion in a highly congested bis(triarylmethyl)peroxide. *J. Am. Chem. Soc.*, **126**, 14778–14786.
40. Khuong, T.A.V., Nunez, J.E., Godinez, C.E. and Garcia-Garibay, M.A. (2006) Crystalline molecular machines: a quest toward solid-state dynamics and function. *Acc. Chem. Res.*, **39**, 413–422.
41. Odani, T., Matsumoto, A., Sada, K. and Miyata, M. (2001) One-way EZ-Isomerization of Bis(*n*-butyl- ammonium) (Z,Z)-Muconate under Photoirradiation in the Crystalline State. *Chem. Commun.*, (2001) 2004–2005.
42. Saltiel, J., Krishna, T.S.R. and Clark, R.J. (2006) Photoisomerization of *cis,cis*-1,4-Diphenyl-1,3- butadiene in the Solid State: the bicycle-pedal mechanism. *J. Phys. Chem. A.*, **110**, 1694–1697.
43. Saltiel, J., Krishna, T.S.R., Laohhasurayotin, S., Fort, K. and Clark, R.J. (2008) Photoisomerization of *cis,cis*- to *trans,trans*-1,4-Diaryl-1,3-butadienes in the

solid state: the bicycle-pedal mechanism. *J. Phys. Chem. A*, **112**, 199–209.

44 Saltiel, J., Bremer, M.A., Laohhasurayotin, S. and Krishna, T.S.R. (2008) Photoisomerization of *cis,cis*- and *cis,trans*-1,4-di-*o*-tolyl-1,3-butadiene in glassy media at 77 K: one-bond-twist and bicycle-pedal mechanisms. *Angew. Chem. Int. Ed.*, **47**, 1237–1240.

45 Yang, L.-Y., Liu, R.S.H., Wendt, N.L. and Liu, J. (2005) Steric effects in hula-twist photoisomerization. 1,4-dimethyl- and 2,3-dimethyl-1,4-diphenylbutadienes. *J. Am. Chem. Soc.*, **127**, 9378–9379.

46 Liu, R.S.H., Yang, l.-Y. and Liu, J. (2007) Mechanisms of photoisomerization of polyenes in confined media: from organic glasses to protein binding cavities. *Photochem. Photobiol.*, **83**, 2–10.

47 Sonoda, Y., Kawanishi, Y., Tsuzuki, S. and Goto, M. (2005) Crystalline-state *Z,E*-photoisomerization of a series of (*Z,E,Z*)-1,6-diphenylhexa-1,3,5-triene 4,4′-dicarboxylic acid dialkyl esters. Chain length effects on the crystal structure and photoreactivity. *J. Org. Chem.*, **70**, 9755–9763.

48 Furukawa, D., Kobatake, S. and Matsumoto, A. (2008) Direct observation of change in molecular structure of benzyl (*Z,Z*)-muconate during photoisomerization in the solid state. *Chem. Commun.*, 55–57.

49 Cohen, M.D. (1975) Photochemistry of organic solids. *Angew. Chem. Int. Ed. Eng.*, **14**, 386–393.

50 Gavezzotti, A. (1983) The calculation of molecular volumes and the use of volume analysis in the investigation of structured media and of solid-state organic reactivity. *J. Am. Chem. Soc.*, **105**, 5220–5255.

51 Ohashi, Y. (1988) Dynamical structure-analysis of crystalline-state racemization. *Acc. Chem. Res.*, **21**, 268–274.

52 Lahav, M. and Schmidt, G.M.J. (1967) Topochemistry. Part XXIII. The solid-state photochemistry at 25° of some muconic acid derivatives. *J. Chem. Soc. B*, 312–317.

53 Green, B.S., Lahav, M. and Schmidt, G.M.J. (1971) Topochemistry. Part XXXI. Formation of cycloocta-1,5-cis,cis-dienes from 1,4-disubstituted s-trans-butadienes in the solid state. A contribution to the problem of C_4-versus C_8-cyclodimerisation. *J. Chem. Soc. B.*, 1552–1564.

54 Enkelmann, V., Wegner, G., Novak, K. and Wagener, K.B. (1993) Single-crystal-to-single-crystal photodimerization of cinnamic acid. *J. Am. Chem. Soc.*, **115**, 10390–10391.

55 Coates, G.W., Dunn, A.R., Henling, L.M., Dougherty, D.A. and Grubbs, R.H. (1997) Phenyl-perfluorophenyl stacking interactions: a new strategy for supermolecule construction. *Angew. Chem. Int. Ed. Engl.*, **36**, 248–251.

56 Honda, K., Nakanishi, F. and Feeder, N. (1999) Kinetic and mechanistic study on single-crystal-to- single-crystal photodimerization of 2-benzyl-5-benzylidenecyclopentanone utilizing X-ray diffraction. *J. Am. Chem. Soc.*, **121**, 8246–8250.

57 Ohba, S., Hosomi, H. and Ito, Y. (2001) *In situ* X-ray observation of pedal-like conformational change and dimerization of *trans*-cinnamamide in cocrystals with phthalic acid. *J. Am. Chem. Soc.*, **123**, 6349–6352.

58 Gao, X., Friščić, T. and MacGillivray, L.R. (2004) Supramolecular construction of molecular ladders in the solid state. *Angew. Chem. Int. Ed.*, **43**, 232–236.

59 MacGillivray, L.R., Papaefstahiou, G.S., Friščić, T., Varshney, D.B. and Hamilton, T.D. (2004) Template-controlled synthesis in the solid state. *Top. Curr. Chem.*, **248**, 201–221.

60 Mori, Y. and Matsumoto, A. (2007) Molecular stacking and photoreactions of fluorine-substituted benzyl muconates in the crystals. *Cryst. Growth Des.*, **7**, 377–385.

61 Gnanaguru, K., Ramasubbu, N., Venkatesan, K. and Ramamurthy, V. (1985) A study on the photochemical dimerization of coumarins in the solid state. *J. Org. Chem.*, **50**, 2337–2346.

62 Ramamurthy, K. (ed.) (1991) *Photochemistry in Organized & Constrained Media*, VCH, New York.

63 Mori, Y. and Matsumoto, A. (2007) Photodimerization mechanism of bis (3,4,5-trifluorobenzyl) (E,E)-muconate in a columnar assembly in the crystalline state. *Chem. Lett.*, **36**, 510–511.

64 Matsumoto, A., Yokoi, K., Aoki, S., Tashiro, K., Kamae, T. and Kobayashi, M. (1998) Crystalline-state polymerization of diethyl (Z,Z)-2,4-hexadienedioate via a radical chain reaction mechanism to yield an ultrahigh molecular weight and stereoregular polymer. *Macromolecules*, **31**, 2129–2136.

65 Matsumoto, A., Odani, T., Chikada, M., Sada, K. and Miyata, M. (1999) Crystal-lattice controlled photopolymerization of di(benzylammonium) (Z,Z)-muconates. *J. Am. Chem. Soc.*, **121**, 11122–11129.

66 Tashiro, K., Kamae, T., Kobayashi, M., Matsumoto, A., Yokoi, K. and Aoki, S. (1999) Structural change in the topochemical solid-state polymerization process of diethyl cis,cis-muconate crystal. 1. Investigation of polymerization process by means of X-ray diffraction, infrared/raman spectra, and DSC. *Macromolecules*, **32**, 2449–2454.

67 Matsumoto, A., Katayama, K., Odani, T., Oka, K., Tashiro, K., Saragai, S. and Nakamoto, S. (2000) Feature of γ-radiation polymerization of muconic acid derivative in the crystalline state. *Macromolecules*, **33**, 7786–7792.

68 Nagahama, S. and Matsumoto, A. (2002) Thermally induced topochemical polymerization of 1,3-diene monomers. *Chem. Lett.*, **31**, 1026–1027.

69 Matsumoto, A., Chiba, T. and Oka, K. (2003) Topochemical polymerization of N-substituted sorbamides to provide thermally stable and crystalline polymers. *Macromolecules*, **36**, 2573–2575.

70 Mori, Y., Chiba, T., Odani, T. and Matsumoto, A. (2007) Molecular arrangement and photoreaction of sorbamides and hexadienyl carbamates with various N-substituents in the solid state. *Cryst. Growth Des.*, **7**, 1356–1364.

71 Bloor, D. (1982) *Developments in Crystalline Polymers II* (ed. D.C. Bassett), Applied Science Publishers, London, pp. 151–193.

72 Enkelmann, V. (1984) Structural aspects of the topochemical polymerization of diacetylenes. *Adv. Polym. Sci.*, **63**, 91–136.

73 Sun, A.W., Lauher, J.W. and Goroff, N.S. (2006) Preparation of poly (diiododiacetylene), an ordered conjugated polymer of carbon and iodine. *Science*, **312**, 1030–1034.

74 Matsumoto, A., Nagahama, S. and Odani, T. (2000) Molecular design and polymer structure control based on polymer crystal engineering. Topochemical polymerization of 1,3-diene mono- and dicarboxylic acid derivatives bearing a naphthylmethylammonium group as the countercation. *J. Am. Chem. Soc.*, **122**, 9109–9119.

75 Matsumoto, A., Sada, K., Tashiro, K., Miyata, M., Tsubouchi, T., Tanaka, T., Odani, T., Nagahama, S., Tanaka, T., Inoue, K., Saragai, S. and Nakamoto, S. (2002) Reaction principles and crystal structure design for topochemical polymerization of 1,3-diene monomers. *Angew. Chem. Int. Ed.*, **41**, 2502–2505.

76 Matsumoto, A., Tanaka, T., Tsubouchi, T., Tashiro, K., Saragai, S. and Nakamoto, S. (2002) Crystal engineering for topochemical polymerization of muconic esters using halogen-halogen and CH/π interactions as weak intermolecular interactions. *J. Am. Chem. Soc.*, **124**, 8891–8902.

77 Matsumoto, A. and Nakazawa, H. (2004) Two-step and reversible phase transitions of organic polymer crystals produced by topochemical polymerization. *Macromolecules*, **37**, 8538–8547.

78 Matsumoto, A., Tanaka, T. and Oka, K. (2005) Stereospecific radical polymerization of substituted benzyl muconates in the solid state under topochemical control. *Synthesis*, 1479–1489.

79 Matsumoto, A., Furukawa, D., Mori, Y., Tanaka, T. and Oka, K. (2007) Change in crystal structure and polymerization reactivity for the solid-state polymerization of muconate esters. *Cryst. Growth Des.*, **7**, 1078–1085.

80 Hasegawa, M., Saigo, K., Mori, T., Uno, H., Nohara, M. and Nakanishi, H. (1985) Topochemical double photocyclodimerization of the 1,4-dicinnamoylbenzene crystal. *J. Am. Chem. Soc.*, **107**, 2788–2793.

81 Ueno, T., Furukawa, D. and Matsumoto, A. (2008) Thermally-induced polymerization of muconic esters in the solid state studied by infrared microscope spectroscopy under temperature control. *Macromol. Chem. Phys.*, **209**, 357–365 See also the cover picture for the issue of Vol. 209, No. 4.

82 Tanaka, T. and Matsumoto, A. (2002) First disyndiotactic polymer from a 1,4-disubstituted butadiene by alternate molecular stacking in the crystalline state. *J. Am. Chem. Soc.*, **124**, 9676–9677.

83 Nagahama, S., Tanaka, T. and Matsumoto, A. (2004) Supramolecular control over the stereochemistry of diene polymers. *Angew. Chem. Int. Ed.*, **43**, 3811–3814.

84 Matsumoto, A., Furukawa, D. and Nakazawa, H. (2006) Stereocontrol of diene polymers by topochemical polymerization of substituted benzyl muconates and their crystallization properties. *J. Polym. Sci. Part A Polym. Chem.*, **44**, 4952–4965.

85 Furukawa, D. and Matsumoto, A. (2007) Reaction mechanism based on X-ray crystal structure analysis during the solid-state polymerization of muconic esters. *Macromolecules*, **40**, 6048–6056.

86 Oshita, S., Tanaka, T. and Matsumoto, A. (2005) Synthesis of new stereoregular host polymers for organic intercalation by solid-state hydrolysis using layered syndiotactic polymer crystals. *Chem. Lett.*, **34**, 1442–1443.

87 Oshita, S. and Matsumoto, A. (2006) Orientational control of guest molecules in organic intercalation system by host polymer tacticity. *Eur. J. Chem.*, **12**, 2139–2146.

88 Oshita, S. and Matsumoto, A. (2006) Fluorescence from aromatic compounds isolated in the solid state by double intercalation using layered polymer crystals as the host solid fluorescence from aromatic compounds isolated in the solid state by double intercalation using layered polymer crystals as the host solid. *Langmuir*, **22**, 1943–1945.

89 Odani, T., Okada, S., Kabuto, C., Kimura, T., Matsuda, H., Matsumoto, A. and Nakanishi, H. (2004) Single-crystal-to-single-crystal polymerization of 4,4′-butadiynedibenzylammonium disorbate. *Chem. Lett.*, **33**, 1312–1313.

90 Matsumoto, A. and Odani, T. (2004) Fabrication of polymer crystals/Ag nanocomposite by intercalation. *Chem. Lett.*, **33**, 42–43.

91 Matsumoto, A., Ishikawa, T., Odani, T., Oikawa, H., Okada, S. and Nakanishi, H. (2006) An organic/inorganic nanocomposite consisting of polymuconate and silver nanoparticles. *Macromol. Chem. Phys.*, **207**, 361–369.

92 Matsumoto, A., Matsumoto, A., Kunisue, T., Tanaka, A., Tohnai, N., Sada, K. and Miyata, M. (2004) Solid- state photopolymerization of diacetylene-containing carboxylates with naphthylmethyl- ammonium as the countercation in a two-dimensional array. *Chem. Lett.*, **33**, 96–97.

93 Sada, K., Inoue, K., Tanaka, T., Tanaka, A., Epergyes, A., Nagahama, S., Matsumoto, A. and Miyata, M. (2004) Organic layered crystals with adjustable interlayer distances of 1-naphthylmethyl-ammonium *n*-alkanoates and isomerism of hydrogen bond networks by steric dimension. *J. Am. Chem. Soc.*, **126**, 1764–1771.

94 Sada, K., Inoue, K., Tanaka, T., Epergyes, A., Tanaka, A., Tohnai, N., Matsumoto, A. and Miyata, M. (2005) Multicomponent organic alloys based on organic layered

crystals. *Angew. Chem. Int. Ed.*, **44**, 7059–7062.

95 Tanaka, A., Inoue, K., Hisaki, I., Tohnai, N., Miyata, M. and Matsumoto, A. (2006) Supramolecular chirality in lyered crystals of achiral ammonium salts and fatty acids: a hierarchical interpretation. *Angew. Chem. Int. Ed.*, **45**, 4142–4145.

96 Dei, S. and Matsumoto, A. (2007) Thermochromism of polydiacetylene with a hysteresis loop in the solid state and in solution. *Chem. Lett.*, **36**, 784–785.

97 Dei, S., Matsumoto, A. and Matsumoto, A. (2008) Thermochromism of polydiacetylene in the solid state and in solution by the self-organization of polymer chains containing no polar group. *Macromolecules*, **41**, 2467–2473.

98 Dei, S., Shimogaki, S. and Matsumoto, A. (2008) Thermochromism of polydiacetylenes containing robust 2D hydrogen bond network of naphthylmethylammonium carboxylates. *Macromolecules*, **41**, 6055–6065.

99 Dei, S. and Matsumoto, A. (2009) Synthesis, structure, chromatic property, and induced circular dichromism of polydiacetylenes including an extended conjugating system in the side chain. *Macromol. Chem. Phys.*, **210**, 11–20

25
Reaction Dynamics Studies on Crystalline-State Photochromism of Rhodium Dithionite Complexes

Hidetaka Nakai and Kiyoshi Isobe

25.1
Introduction

A photochromic reaction is a reversible isomerization reaction induced by photo-irradiation [1]. Crystalline-state photochromic reactions, especially, are becoming an important topic because of their potential advantages for construction of novel switching devices and materials [2]. Although many photochromic compounds have been reported so far, compounds that undergo photochromic reactions in the crystalline state are rare and their reaction dynamics are not well characterized as a result of the low degree of interconversion ratios and/or instability of the photo-generated isomers in the solid phase [3, 4]. We have recently found that a new class of photochromic compounds, rhodium dinuclear complex [(RhCp*)$_2$(μ-CH$_2$)$_2$(μ-O$_2$SSO$_2$)] (**1**) (Cp* = η^5-C$_5$Me$_5$) with a photo-responsive dithionite group (μ-O$_2$SSO$_2$), shows the reversible crystalline-state photochromic reaction between **1** and [(RhCp*)$_2$(μ-CH$_2$)$_2$(μ-O$_2$SOSO)] (**2**) with essentially 100% interconversion ratio (Figure 25.1) [5]. Taking advantage of the full reversibility of **1**, we started the unexplored reaction dynamics study of crystalline-state photochromism [6]. This chapter presents the following three reaction dynamics related to single crystals of the dithionite complexes: (i) dynamics of molecular structural changes in single crystals, (ii) dynamics of reaction cavities in the crystalline-state reaction, and (iii) dynamics of surface morphology changes of photochromic single crystals.

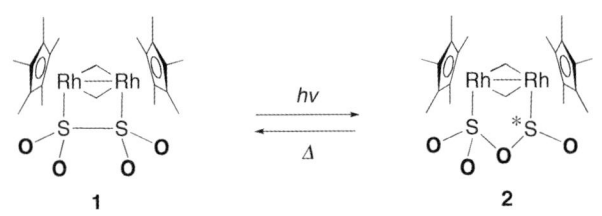

Figure 25.1 Crystalline-state photochromism between **1** and **2**.

Molecular Nano Dynamics, Volume II: Active Surfaces, Single Crystals and Single Biocells
Edited by H. Fukumura, M. Irie, Y. Iwasawa, H. Masuhara, and K. Uosaki
Copyright © 2009 WILEY-VCH Verlag GmbH & Co. KGaA, Weinheim
ISBN: 978-3-527-32017-2

25.2
Photochromism of Rhodium Dithionite Complexes

Recently some metal complexes showing a photochromic reaction have been prepared [7, 8]. However, almost all of these compounds possess an organic photochromic moiety as a ligand. For instance, the diarylethene derivatives have been widely used as powerful ligands to synthesize the photo-functional metal complexes [7]. On the other hand, the preparation of a metal complex having photochromic properties by itself, "transition-metal based photochromic compound" (Figure 25.2) [4–6, 8], is still difficult because of a lack of systematic synthetic strategies. Further, finding new transition-metal based photochromic compounds poses a challenge because of the lack of understanding of the photochromic mechanism in these complexes. The rhodium dithionite complex **1** is one of the rare examples of transition-metal based photochromic compounds, as described in the following discussion.

The rhodium dithionite complex **1** is composed of the rhodium dinuclear moiety $(RhCp^*)_2(\mu\text{-}CH_2)_2$ and the dithionite ion $S_2O_4^{2-}$. The dithionite ion is very attractive as the external stimuli-responsive ligand because the ion has a weak covalent S–S bond (2.389 ± 10 Å), which is easily cleaved to form radical species by stimuli [9].

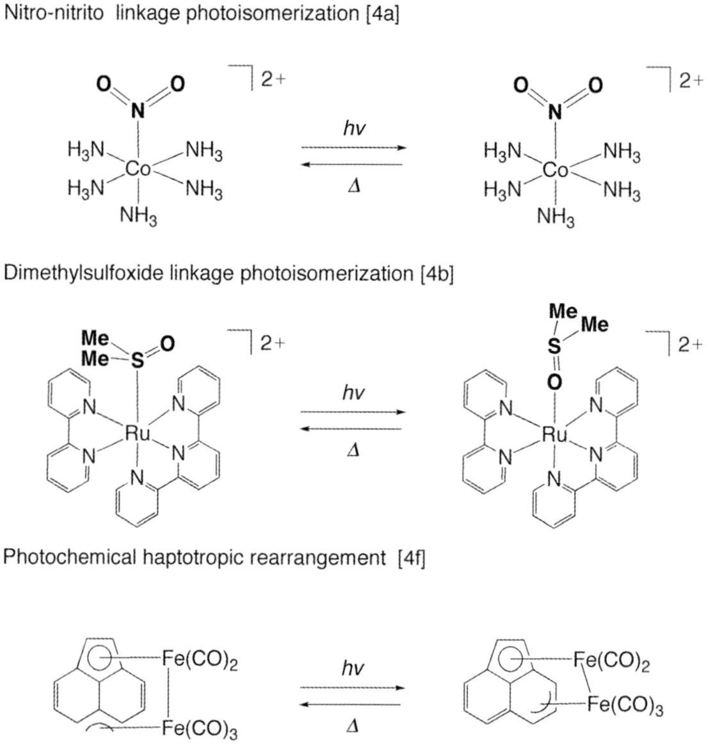

Nitro-nitrito linkage photoisomerization [4a]

Dimethylsulfoxide linkage photoisomerization [4b]

Photochemical haptotropic rearrangement [4f]

Figure 25.2 Some examples of transition-metal based photochromic compounds.

However, it has not been well established whether $S_2O_4^{2-}$ has an ability to coordinate [10]. The organorhodium dinuclear moiety is a soft acidic center and can accept a soft donor atom like sulfur. Indeed, the dinuclear moiety forms the complexes [(RhCp*)$_2$(μ-CH$_2$)$_2$(μ-S$_2$)] and [(RhCp*)$_2$(μ-CH$_2$)$_2$(μ-SSO$_2$)] where the μ-S$_2$ and μ-SSO$_2$ ligands are coordinated parallel to the Rh–Rh bond through two S atoms (side-on fashion) [11]. Treatment of trans-[(RhCp*)$_2$(μ-CH$_2$)$_2$Cl$_2$] [12] with $S_2O_4^{2-}$ affords the first novel side-on type dithionite complex **1**.

The photochromism of **1** is achieved by atom rearrangement into the intramolecular oxygen-atom insertion product **2**, in contrast to the frequently reported organic photochromic systems based on photo-induced cyclization, cis/trans isomerization, or H atom transfer [1]. The quantum yield of the photoreaction from **1** to **2** at 509 nm in acetonitrile without O$_2$ is 0.14 ± 0.01. In solution, photoreaction of **1** causes the oxidation reaction by atmospheric oxygen, resulting in a mixture of **2** and further oxidation products such as [(RhCp*)$_2$(μ-CH$_2$)$_2$(μ-SO$_3$)] and [(RhCp*)$_2$(μ-CH$_2$)$_2$(μ-SO$_4$)]. In contrast, in the crystalline state, the photochromic system between **1** and **2** is stable and repeatable with essentially 100% interconversion ratio.

Crystalline-state photochromism usually proceeds with considerably lower interconversion ratios of less than 15% because the light penetration into the bulk crystal is prohibited by the absorption of the photo-generated isomer (inner-filter effects) [3, 4]. The fully reversible crystalline-state photochromism of **1** can be partly attributed to its photochromic property. The rhodium dithionite complex **1** belongs to a unique class of photochromic compounds, which exhibits a unimolecular type T inverse photochromism [13]. The type T inverse photochromism means that the back reaction occurs thermally and the λ_{max} of the absorption spectrum of **1** is longer than that of **2**. If the back reaction occurs photochemically and the λ_{max} of the initial absorption spectrum is shorter than that of the photo-generated isomer, it is called type P positive photochromism and is known as a common photochromic system.

Figure 25.3 shows the UV–vis spectral change from **1** to **2** in a microcrystalline powder film. The hypsochromic (blue) shift of λ_{max} from 510 to 475 nm was

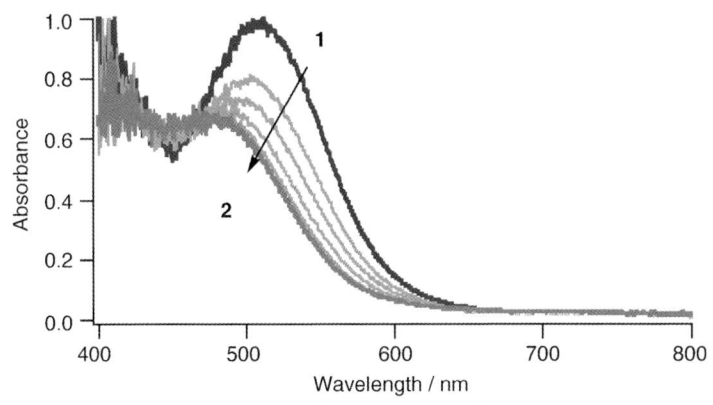

Figure 25.3 Irradiation time-resolved UV–vis spectra of **1** to **2** in a micro-crystalline powder film.

observed as a characteristic of inverse photochromism. The B3LYP hybrid DFT and the time-dependent DFT calculations support the assignment of the absorption band at 510 nm for **1** as the charge transfer band from $\sigma(S-S)$ to $\sigma^*(S-S)$ and $\sigma^*(Rh-Rh)$ orbitals and the assignment of the absorption band at 475 nm for **2** as the charge transfer band from $\sigma(Rh-SO)$ to $\sigma^*(Rh-Rh)$ orbital. More importantly, the absorption coefficient of λ_{max} in **2** is smaller than that in **1** by about one third. Thus, the light is able to pass more easily through the crystals when the photoreaction of **1** proceeds. This is one of the reasons why the photoreaction of crystals of **1** proceeds with almost 100% interconversion ratio. The back reaction from **2** to **1** is an exothermic reaction with liberation of about 4.5 kcal mol^{-1} of heat. On leaving crystals of **2** in the dark for three weeks at room temperature, the back reaction takes place leading to complete regeneration of crystals of **1**. The unique full reversibility of the crystalline-state photochromic system between **1** and **2** provides us with an opportunity to clarify the reaction dynamics by X-ray diffraction analysis and spectroscopic methods, as described in the next section.

25.3
Reaction Dynamics of Crystalline-State Photochromism

25.3.1
Dynamics of Molecular Structural Changes in Single Crystals

When single-crystal integrity is preserved during the reaction, the reaction is called a "crystalline-state reaction" and the process of the crystalline-state reaction can be observed by stepwise crystal structure analysis [4d, 14]. Although some crystalline-state reactions, including the photochromic reaction, have been investigated based on crystal structures before and after reactions, little is known about the dynamics of the molecular structural changes in the crystalline-state reaction [3d, 4a, 15]. Intriguingly, the crystalline-state photochromic process between **1** and **2**, including the stereoselective oxygen-atom transfer and thermodynamically controlled photo-isomerization, can be directly followed by conventional single-crystal X-ray diffraction analyses.

The solid-state molecular structures of **1** and **2** are shown in Figure 25.4. The μ-O_2SSO_2 ligand in **1** is coordinated parallel to the Rh–Rh bond and has a weak S–S bond of 2.330(2) Å. Complex **2** contains a new type of oxysulfur species, O_2SOSO, having one oxo-bridge between the two S atoms: S1 has two terminal O atoms, while S2 has only one terminal O atom and is asymmetric. Although **2** has an asymmetric sulfur atom, S2, there are pairs of enantiomers in one single crystal of **2** because of the centrosymmetric space group $P2_1/n$. The O_2SOSO type of oxysulfur compounds has been considered as an important unstable intermediate for the oxidation of disulfide compounds to SO_3^{2-} and SO_4^{2-} and has only been characterized theoretically [16]. The bridging S1–O5 and S2–O5 bonds in **2** differ significantly in length (S1–O5: 1.709(5), S2–O5: 1.636(6) Å), and the latter has double-bond character.

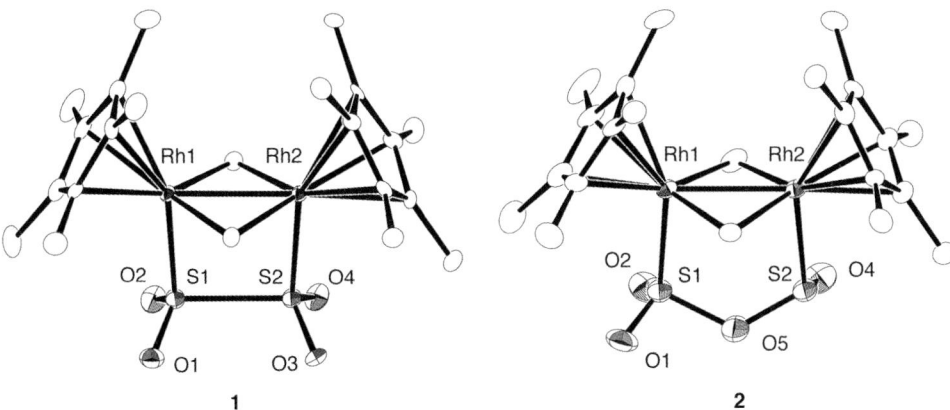

Figure 25.4 ORTEP drawings of **1** and **2** with 50% probability ellipsoids. Hydrogen atoms are omitted for clarity. Selected bond lengths (Å) and angles (°) for **1**: Rh1−Rh2 2.6224(5), Rh1−S1 2.279(1), Rh2−S2 2.277(1), S1−S2 2.330(2), S1−O1 1.462(5), S1−O2 1.459(4), S2−O3 1.467(5), S2−O4 1.464(4), Rh2−Rh1−S1 86.63(4), Rh1−Rh2−S2 86.00(3), Rh1−S1−S2 93.29(6) Rh2−S2−S1 94.06(5). Selected bond lengths (Å) and angles (°) for **2**: Rh1−Rh2 2.6257(6), Rh1−S1 2.270(2), Rh2−S2 2.285(2), S1−O1 1.445(7), S1−O2 1.440(6), S1−O5 1.709(5), S2−O5 1.636(6), S2−O4 1.486(6), S1···S2 2.964(3), Rh2−Rh1−S1 94.15(5), Rh1−Rh2−S2 94.35(5), Rh1−S1−O5 107.9(2), Rh2−S2−O5 107.8(2), S1−O5−S2 124.7(3).

Since the crystals of **1** and **2** possess the same space group and molecular arrangement, by comparison of a certain molecule in the unit cell of **1** with that of **2**, it can be readily seen which terminal oxygen in the μ-O_2SSO_2 ligand of **1** is transferred to the bridging oxygen in **2** (Figure 25.5). It is the O3 atom. In the μ-O_2SSO_2 ligand, the four oxygen atoms are stereochemically nonequivalent in the cavity formed by six Cp* ligands. As shown in Figure 25.6, the O3 atom is the most congested by methyl groups of the Cp* ligands. Based on this situation in the crystal of **1**, what kind of selectivity and specificity is observed through the photoreaction?

During the photoreaction from **1** to **2**, positional disorder and occupancy changes of the oxygen atoms with irradiation time are observed in X-ray diffraction analyses (Figure 25.7). A careful analysis of the time dependence of the positional disorder and occupancy changes of the oxygen atoms during photoreaction at 20 °C indicates that species **2a–d** are formed in the initial and middle stages of the reaction (Figure 25.8a). The species **2a–d** that cause the positional disorder are the stereoisomers of **2** in the crystal. Although the crystal has mirror image species of **2a–d** as a set, in the present treatment only one asymmetric unit in the crystal is considered. Isomers **2a** and **2c** are a pair of enantiomers, **2a** and **2b** are identical species but differ orientationally in the cavity, and **2c** and **2d** are also identical but differ in orientation. Surprisingly, in the final stage of the reaction, only **2a** is formed and the positional disorder almost disappears. This disappearance means that the thermodynamically unstable μ-O_2SOSO species **2b–d** generated in the crystal convert under irradiation to the most stable species **2a**. Thus, the specificity of oxygen transfer is observed during the photoreaction at 20 °C (Figure 25.7).

(a)

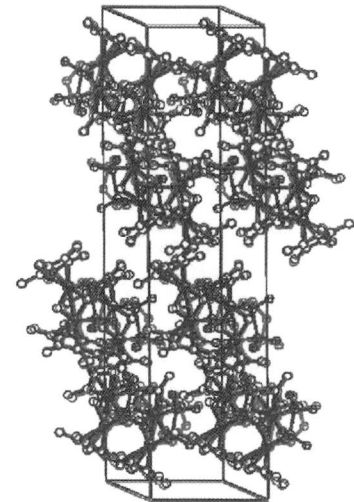

Lattice parameters

$P2_1/n$	$P2_1/n$
$a = 8.394(2)$ Å	$a = 8.339(2)$ Å
$b = 36.532(8)$ Å	$b = 36.750(9)$ Å
$c = 8.430(2)$ Å	$c = 8.447(2)$ Å
$\beta = 116.213(10)$ °	$\beta = 115.594(10)$ °
$V = 2319.2(10)$ Å^3	$V = 2334(1)$ Å^3
$Z = 4$	$Z = 4$

(b)

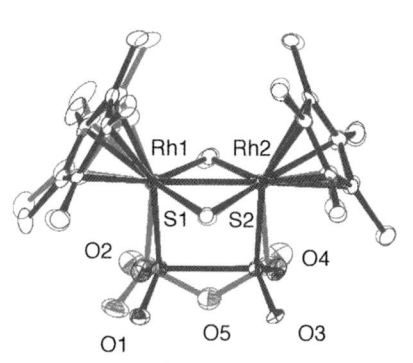

Figure 25.5 Superimposition of (a) unit cells and (b) crystal structures of **1** (black) and **2** (gray).

The solid-state dynamics from **1** to **2a** is disclosed by using low temperature experiments. The photoreaction at −163 °C shows that, in the initial stage of the reaction, the highest population product is not the most stable product **2a** but **2b** (Figure 25.8b). As described above, the O3 atom migration, which forms **2a**, has much difficulty compared with other O atoms. The initial population of the products at −163 °C reflects the selectivity based on the shape of the cavity in the crystal of **1** (topochemical principle [17]). It must be emphasized that no obvious conversions between the isomers are observed through the photoreaction at −163 °C. These results strongly indicate that the kinetically controlled reaction takes place predominantly at −163 °C. On the other hand, in the dark, the time dependence of the population of the isomers generated at −163 °C after 190 h of irradiation shows that isomer **2c** is directly transformed into **2a** without irradiation (Figure 25.9). At the same time, the unchanged **2b** and **2d** populations confirm that the direct thermal

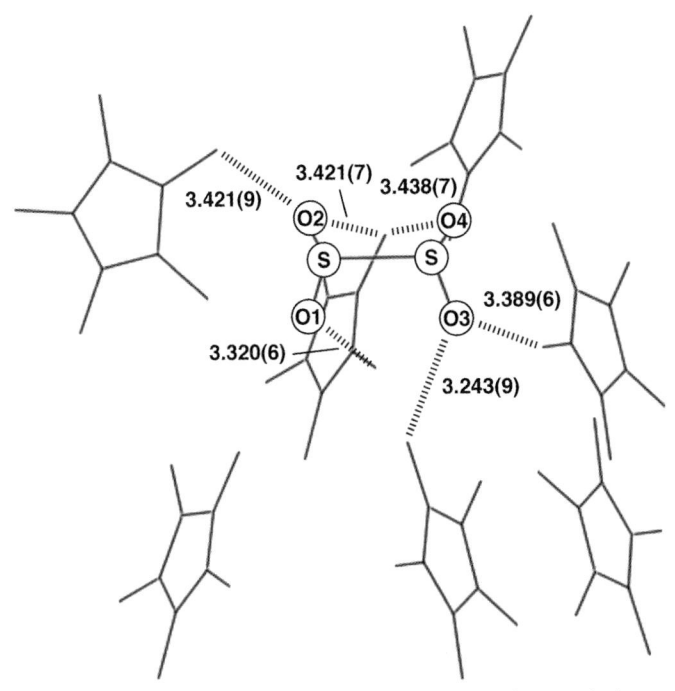

Figure 25.6 Reaction cavity around the μ-O_2SSO_2 unit in the crystal of **1**.

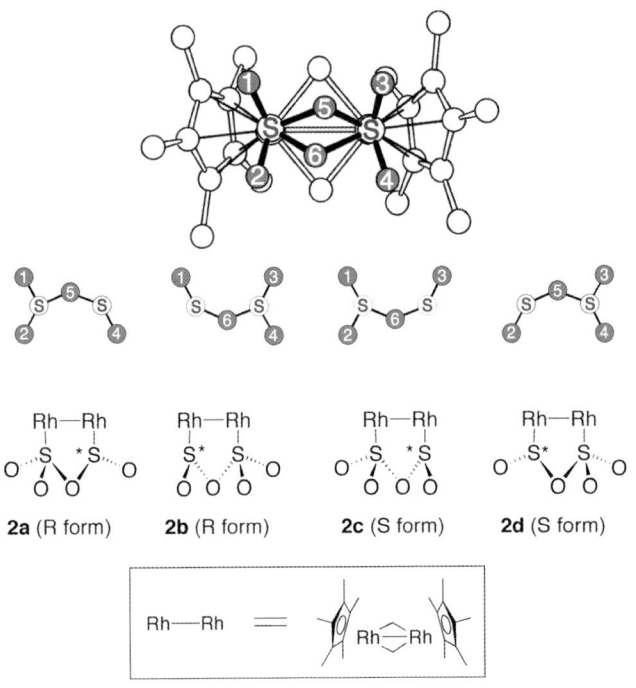

2a (R form)   2b (R form)   2c (S form)   2d (S form)

Figure 25.7 Positional disorder of oxygen atoms during the photoreaction.

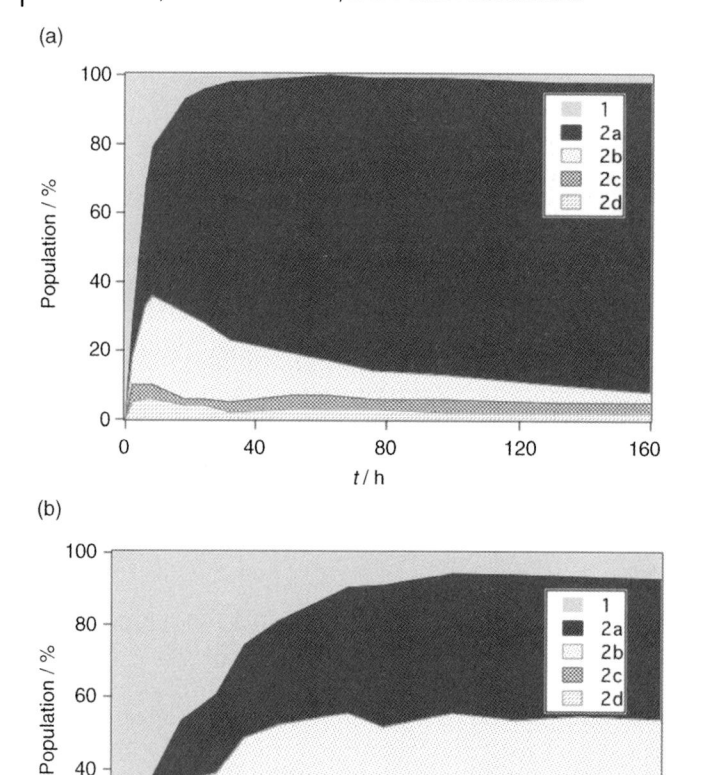

Figure 25.8 Change in population of the photochemically generated isomers with irradiation time at (a) 20 °C and (b) −163 °C.

conversion process of **2b** (or **2d**) to **2a** is not present at 20 °C. Based on the facts that the photoreactions from **2** to **1** and between the isomers of **2** do not proceed within photoirradiation, we conclude that the conversion of **2b** (or **2d**) into **2a** contains a photochromic process of **2b** (or **2d**) → **1** → **2a**. Thus, the specific photoreaction from **1** to **2a** at 20 °C was observed because thermodynamically favorable **2a** accumulates through the direct thermal conversion from **2c** to **2a** and the repetitive photochromic process (**2b** (or **2d**) → **1** → **2a**) (Figure 25.10).

In contrast, the thermal back reaction from **2a** to **1** does not show any disorder phenomena except for that due to the formation of **1**. The bonding character in **2a** supports that specific cleavage of the weak S1–O5 bond is the lowest-energy path for isomerization to **1** and may be the only path for isomerization to **1**.

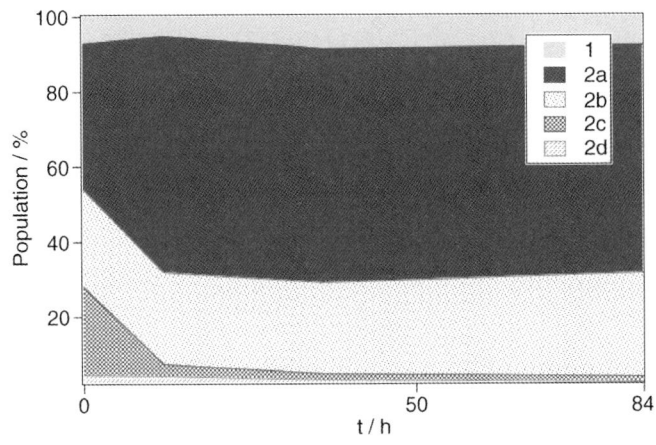

Figure 25.9 Time dependence of the population of the light-induced isomers generated at −163 °C for 190 h irradiation (during the experiment, the backward reaction scarcely takes place).

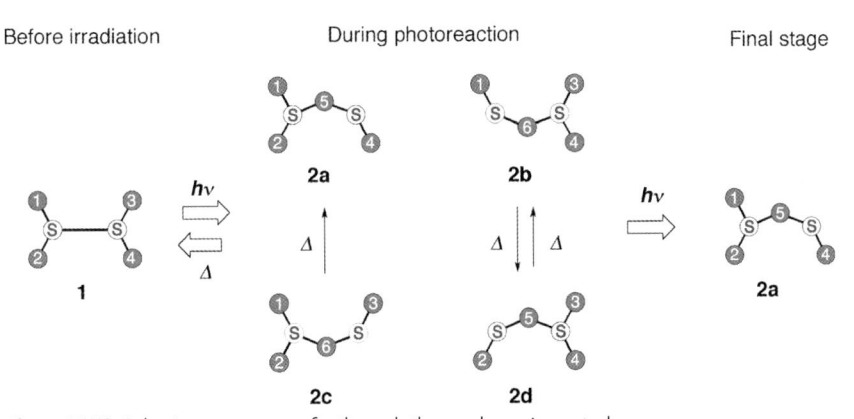

Figure 25.10 Selective oxygen transfer through thermodynamic control.

25.3.2
Dynamics of Reaction Cavities in a Crystalline-State Reaction

It is well known that flat disk-like ligands such as Cp (η^5-C$_5$H$_5$) and Cp* can undergo $2\pi/5$ jumping motions around the ligand-metal coordination C$_5$ axis in the crystal [18]. The crystalline-state photochromic reaction between **1** and **2** proceeds in the reaction cavities formed by the Cp* ligands. Figure 25.11 shows the packing diagram of **1** and thermal ellipsoids (50% probability) of C-atoms of two crystallographically independent Cp* rings in **1** at −163 °C. In Figure 25.11, the following three features can be pointed out.

1. The crystallographically independent Cp* ligands are in the parallel and perpendicular Cp* ring plane arrangements. Thus, the chemical environment of the

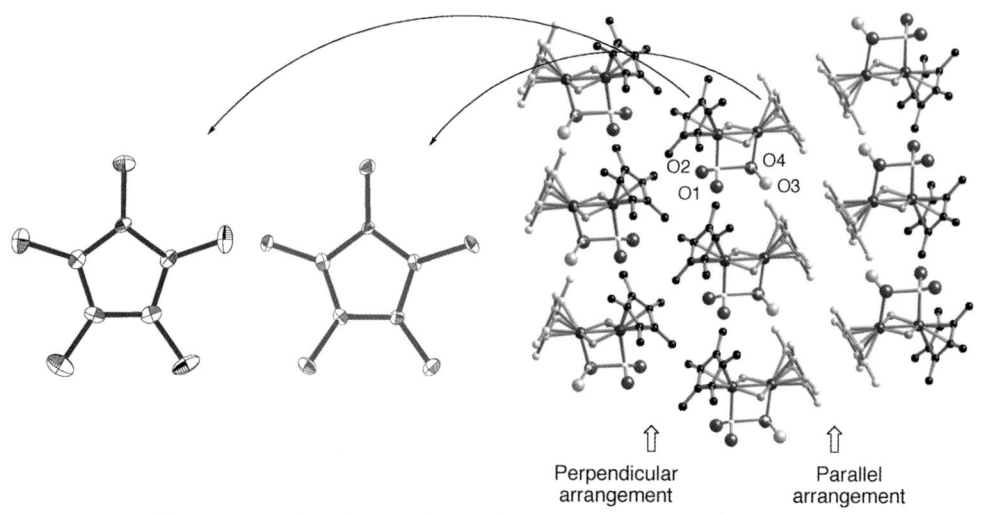

Figure 25.11 Packing diagram of **1** and thermal ellipsoids (50% probability) of C-atoms of two crystallographically independent Cp* rings in **1**.

two Cp* ligands in one molecule in the crystal is different. Indeed, the solid-state ^{13}C CPMAS NMR spectrum of the Cp* ligands in the crystals of **1** at 21 °C shows two methyl carbon signals at 10.5 and 10.8 ppm and two ring carbon signals at 104.0 and 106.0 ppm, reflecting the chemical environment in the crystal. Of course, in solution, only one methyl and one ring carbon signal appear at 9.6 and 104.1 ppm, respectively.

2. The shapes of the thermal ellipsoids of the C-atoms, which contain information on the dynamic process in the solid state [18–20], are quite different between the parallel and perpendicular Cp* rings. The shape difference indicates that an activation energy for the jumping motion of parallel Cp* rings is higher than that for perpendicular Cp* rings.

3. The parallel Cp* ligands are the closest to the terminal O3 atoms (indicated pale gray in Figure 25.11) while the parallel Cp* rings in two adjacent molecules are staggered (two parallel Cp* rings form an intermolecular staggered arrangement). These crystallographic situations indicate that the motion of the parallel Cp* ligands in **1** is restricted by the steric hindrance of the O3 atom, in addition to the intermolecular staggered form itself. In other words, complex **1** has a molecular stress in the crystal.

Figure 25.12 shows the packing diagram and thermal ellipsoids of complex **2** similar to Figure 25.11 of **1**. Two features can be noted by comparison between Figures 25.11 and 25.12.

First, the increase in packing distance for the parallel Cp* rings is observed after irradiation: 3.5793(20) and 3.6323(25) Å in crystal **1** and **2**, respectively. Secondly, the shapes of the thermal ellipsoids of the C-atoms in **2** are different from those

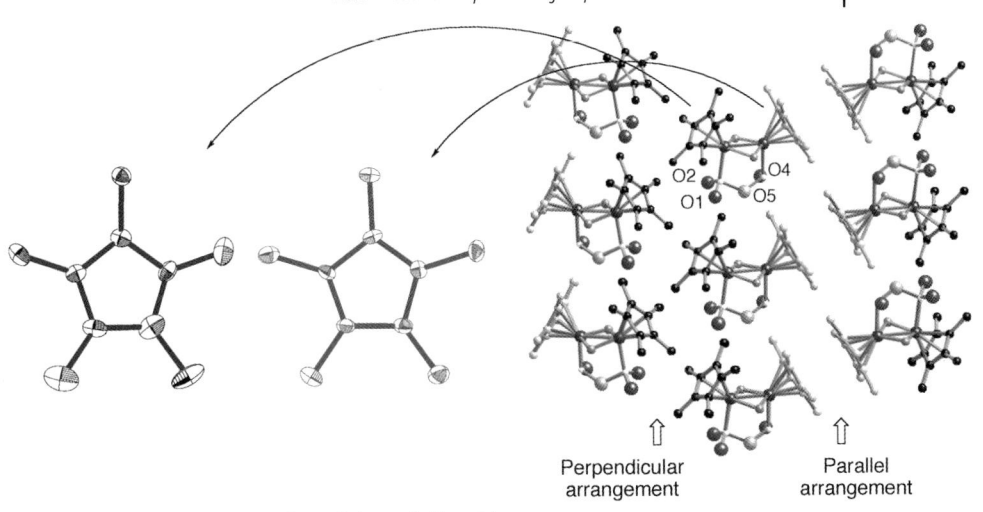

Figure 25.12 Packing diagram of **2** and thermal ellipsoids (50% probability) of C-atoms of two crystallographically independent Cp* rings in **2**.

in **1**. The shape differences indicate that the activation energies for the jumping motion of the parallel and perpendicular Cp* rings in **2** are lower than those in **1**. These crystallographic features strongly suggest that the molecular motions of the Cp* ligands, which form the reaction cavities of the μ-O$_2$SSO$_2$ and μ-O$_2$SOSO groups in the crystals of **1** and **2**, are closely connected to the photochromic reaction. In order to elucidate the relationship between the molecular motion of the Cp* ligands and the photochromic reaction, the dynamic behavior of the Cp* in the crystals of **1** and **2** was examined by the variable temperature (VT) solid-state NMR analyses [6d].

The activation energies for the Cp* ligand motions in **1** and **2** were unequivocally determined by the VT solid-state ^{2}H NMR analyses of deuterated analogues **1**-d_{30} and **2**-d_{30}, in which the Cp*^{d15} ligands (η^5-C$_5$(CD$_3$)$_5$) are used instead of Cp* ligands in **1** and **2**. The resulting activation energies (kJ mol^{-1}) are as follows: 33 ± 3 and 7.8 ± 1 for the parallel and perpendicular Cp* ligand in **1**, respectively; 21 ± 2 and 4.7 ± 0.5 for the parallel and perpendicular Cp* ligand in **2**, respectively. The results show that the molecular motion of the Cp* ligands couples to the photochromic reaction (atom rearrangement of the dithionite ligand). This strongly indicates that the dynamic behavior of the Cp* ligands assists the crystalline-state reaction to proceed, maintaining the single-crystal integrity and forming only one enantiomeric pair of the μ-O$_2$SOSO complex in the final stage of the photoreaction at 20 °C. The activation energy change of 12 kcal mol^{-1} (subtract 21 from 33) for the parallel Cp* ligand motion with the photoreaction from **1** to **2** is much greater than that of 3.1 kcal mol^{-1} (subtract 4.7 from 7.8) for the perpendicular Cp* ligand motion. The results indicate that the relaxation of the above-mentioned molecular stress by the O3 atom and between the parallel Cp* rings in the crystal of **1** mainly contribute to a reduction in the activation energy for the parallel Cp* ligand motion.

In order to find the role of the dynamic reaction cavity in **1**, we are currently preparing various derivatives of the dithionite complex by chemical modification of the Cp* ligands of **1**. Very recently, we have successfully synthesized a new dithionite complex [(RhCp^{Et})$_2$(μ-CH$_2$)$_2$(μ-O$_2$SSO$_2$)] (**1^{Et}**) with Cp^{Et} (η^5-C$_5$Me$_4$Et) ligands instead of Cp*. The crystal structure of the Cp^{Et} derivative **1^{Et}** was determined by X-ray diffraction analysis. The three-dimensional crystal packing of **1^{Et}** is similar to that of the Cp* complex **1** (The Cp^{Et} ligands also form the reaction cavities in the crystal). However, the flexibility of the cavities related to the rotational motion of the Cp^{Et} ligands, which is disclosed by variable temperature CP/MAS ^{13}C NMR measurements, is substantially less than that of **1**. Intriguingly, in the case of **1^{Et}**, the single crystal integrity is sometimes not preserved during the photoreaction. Thus, the dynamic behavior of the Cp* ligands in **1** plays an important role in keeping the single-crystal integrity during the crystalline-state reaction.

25.3.3
Dynamics of Surface Morphology Changes of Photochromic Single Crystals

Surface nano-morphology changes of photoreactive molecular crystals are an attractive area of research, because the phenomena could potentially be applied to photodriven nanometer-scale devices and provide important information on crystalline-state reaction mechanisms and dynamics [2a, 21]. As described in Section 25.3.2, the single crystal of **1^{Et}**, in which the Cp^{Et} rings have no reorientation freedom in the crystal, tends to collapse and degrade as the reaction proceeds. This observation for the crystal of **1^{Et}** can be explained by the local stress induced by the photoreaction that is not suitably released by the crystal lattice. In such a crystal, does the surface morphology of the crystal change?

To observe the changes on the nanometer-scale, we have examined the surface morphology changes by using atomic force microscopy (AFM). When a crystal of **1^{Et}** is mildly irradiated such that macroscopically it does not collapse, the light-induced surface nano-morphology changes on the crystal of **1^{Et}** could be reproducibly observed on the nanometer-scale. Careful AFM experiments show two kinds of stepwise surface morphology changes triggered by the photoreaction, as described in the following.

First, a crystal of **1^{Et}** was irradiated for 2 min with visible light (>500 nm, intensity 5.0 mW cm^{-2}). The conversion ratio from **1^{Et}** to **2^{Et}** in the crystal was about 5%. This roughly corresponds to the photoisomerization reactions of the molecules in 3000 layers. Continuous AFM measurements of the photo-irradiated crystal were then performed at certain intervals under dark conditions and the AFM image changes were observed, as shown in Figure 25.13 (A prominent surface was used for the AFM experiments).

Figure 25.13a is the schematic representation of the AFM image at 5 min after stopping the photoirradiation. No morphological change of the surface was discerned during this initial 5 min interval. After this induction period, steps gradually appeared on the surface and formed a ripplelike pattern (Figure 25.13b). The height of the steps is 6.6 ± 2.4 nm, which corresponds to 13 molecular layers. The steps are nearly parallel to a line X in Figure 25.13b. When the observation time was prolonged,

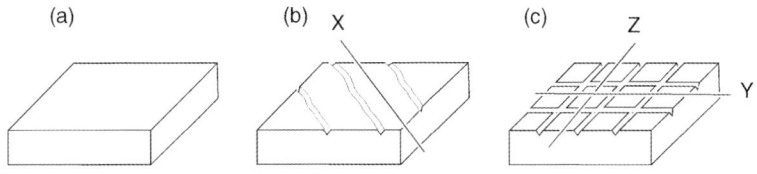

Figure 25.13 Schematic-representation for AFM image changes of the (11-1) crystal surface of 1^{Et} at the following intervals: (a) 5 min. (b) 30 min (c) 120 min after irradiation.

a new cross-stripe pattern with regularity appeared and the initial ripplelike pattern gradually disappeared (Figure 25.13c). The heights of the steps along lines Y and Z in Figure 25.13c were determined to be 8.1 ± 5.8 and 7.9 ± 8.4 nm, respectively. When a long photoirradiation time (>5 min) was employed, the height of the cross-stripe steps became higher though this pattern remains. Of course, without photoirradiation, no morphological changes were observed after more than 3 h (blank experiments). The key to the observation of the pure ripplelike pattern lies in the choice of a suitable light intensity and irradiation time for the photoisomerization. If the light intensity is high and/or the irradiation time is prolonged, only the cross-stripe pattern is observed. Thus, the present experiment using the devised AFM equipment adduces clear evidence to show surface morphology dynamics along with the formation of the ripplelike pattern. The results indicate that the development of surface structures is a consequence of the relaxation of strain. Further work including the elucidation of this dynamic behavior is currently in progress.

25.4 Summary

In this chapter, crystalline-state photochromic dynamics of rhodium dithionite complexes are reviewed. The chemistries described here have been achieved not only by recent developments of the analytical technique but also by discovery of a new class of transition-metal based photochromic compounds. One of the advantages of transition-metal complexes is structural diversity. In order to find the rule of an exquisite combination of metal ions and ligands, we are currently synthesizing various dithionite derivatives with other metal ions and/or modified Cp* ligands. As shown in this chapter, dithionite complexes are a very useful photochromic system to investigate crystalline-state reaction dynamics. We believe that dynamics studies of newly synthesized dithionite derivatives provide useful insight into the construction of sophisticated molecular switches. A dithionite complex may appear in a practical application field in the near future.

Acknowledgments

This work was financially supported by the Grant-in-Aids for Scientific Research (KAKENHI) in Priority Area "Molecular Nano Dynamics, No. 17034018",

"Chemistry of Coordination Space, No. 180330", "Synergy of Elements, No. 190270", and "Photochromism, No. 20044010" and by the Grant-in-Aids "No. 1635002" and "Nanotechnology Support Project" from Ministry of Education, Culture, Sports, Science and Technology (MEXT), Japan. This work was also supported in part by funds from the "Tokuyama Science Foundation".

References

1 (a) Brown, G.H. (ed.) (1971) *Photochromism*, Wiley-Interscience, New York; (b) Irie, M. (2000) Photochromism: memories and switches. *Chem. Rev.*, **100** (5), Special thematic issue; (c) Dürr, H. and Bouas-Laurent, H. (eds) (2003) *Photochromism: Molecules and Systems*, Elsevier, Amsterdam.

2 (a) Irie, M., Kobatake, S. and Horichi, M. (2001) Reversible surface morphology changes of a photochromic diarylethene single crystal by photoirradiation. *Science*, **291**, 1769–1772; (b) Kobatake, S., Takami, S., Muto, H., Ishikawa, T. and Irie, M. (2007) Rapid and reversible shape changes of molecular crystals on photoirradiation. *Nature*, **446**, 778–781; (c) Garcia-Garibay, M.A. (2007) Molecular crystals on the move: from single-crystal-to-single-crystal photoreactions to molecular machinery. *Angew. Chem. Int. Ed.*, **46**, 8945–8947.

3 (a) Harada, J., Uekusa, H. and Ohashi, Y. (1999) X-ray analysis of structural changes in photochromic salicylideneaniline crystals. Solid-state reaction induced by two-photon excitation. *J. Am. Chem. Soc.*, **121**, 5809–5810; (b) Yamada, T., Kobatake, S., Muto, K. and Irie, M. (2000) X-ray crystallographic study on single-crystalline photochromism of bis(2,5-dimethyl-3-thienyl)perfluorocyclopentene. *J. Am. Chem. Soc.*, **122**, 1589–1592; (c) Yamada, T., Kobatake, S. and Irie, M. (2000) X-ray crystallographic study on single-crystalline Photochromism of 1,2-Bis(2,5-dimethyl-3-thienyl)perfluorocyclopentene. *Bull. Chem. Soc. Jpn.*, **73**, 2179–2184; (d) Morimoto, M. and Irie, M. (2005) Photochromism of diarylethene single crystals: Crystal structures and photochromic performance. *Chem. Commun.*, 3895–3905.

4 (a) Boldyreva, E.V. (1994) Intramolecular linkage isomerization in the crystals of some Co(III) – ammine complexes – a link between inorganic and organic solid state chemistry. *Mol. Cryst. Liq. Cryst.*, **242**, 17–52; (b) Rack, J.J., Winkler, J.R. and Gray, H.B. (2001) Phototriggered Ru(II)–dimethylsulfoxide linkage isomerization in crystals and films. *J. Am. Chem. Soc.*, **123**, 2432–2433; (c) Nishimura, H. and Matsushita, N. (2002) Single-crystalline photochromism of bis[2-(aminomethyl)pyridine] platinum(II) chloride monohydrate. *Chem. Lett.*, 930–931; (d) Coppens, P., Novozhilova, I. and Kovalevsky, A. (2002) Photoinduced linkage isomers of transition-metal nitrosyl compounds and related complexes. *Chem. Rev.*, **102**, 861–883; (e) Kovalevsky, A.Y., Bagley, K.A., Cole, J.M. and Coppens, P. (2003) Light-induced metastable linkage isomers of ruthenium sulfur dioxide complexes. *Inorg. Chem.*, **42**, 140–147; (f) Niibayashi, S., Matsubara, K., Haga, M. and Nagashima, H. (2004) Thermally reversible photochemical haptotropic rearrangement of diiron carbonyl complexes bearing a bridging acenaphthylene or aceanthrylene ligand. *Organometallics*, **23**, 635–646.

5 Nakai, H., Mizuno, M., Nishioka, T., Koga, N., Shiomi, K., Miyano, Y., Irie, M., Breedlove, B.K., Kinoshita, I., Hayashi, Y.,

Ozawa, Y., Yonezawa, T., Toriumi, K. and Isobe, K. (2006) Direct observation of photochromic dynamics in the crystalline state of an organorhodium dithionite complex. *Angew. Chem. Int. Ed.*, **45**, 6473–6476.

6 (a) Miyano, Y., Nakai, H., Hayashi, Y. and Isobe, K. (2007) Synthesis and structural characterization of a photoresponsive organodirhodium complex with active S–S bonds: [(Cp^{Ph}Rh)$_2$(μ-CH$_2$)$_2$ (μ-O$_2$SSO$_2$)] (Cp^{Ph} = η^5-C$_5$Me$_4$Ph). *J. Organomet. Chem.*, **692**, 122–128; (b) Nakai, H., Uemura, S., Irie, M., Miyano, Y., Hayashi, Y. and Isobe, K. (2007) Exploration of dynamics of photoreaction and morphology change of dithionite complex in crystalline-state. 1st Asian Conferece on Coordination Chemistry, Okazaki, Japan, July 29-August 2, Abstr. OA-14; (c) Miyano, Y., Nakai, H., Mizuno, M. and Isobe, K. (2008) Substitution effects of Cp ring benzyl groups on photoisomerization of a rhodium dithionite complex in the crystalline state. *Chem. Lett.*, 826–827; (d) Nakai, H., Nonaka, T., Miyano, Y., Mizuno, M., Ozawa, Y., Toriumi, K., Koga, N., Nishioka, T., Jrie, M. and Jsobe, K., (2008) Photochromism of an organorhodium dithionite complex in the crystalline-state: molecular motion of pentamethylcyclo-pentadienyl ligands coupled to atom rearrangement in a dithionite ligand. *J. Am. Chem. Soc.*, **130**, 17836–17845.

7 (a) Han, J., Maekawa, M., Suenaga, Y., Ebisu, H., Nabei, A., Kuroda-Sowa, T. and Munakata, M. (2007) Photochromism of novel metal coordination polymers with 1,2-bis(2′-methyl-5′-(carboxylic acid)-3′-thienyl)perfluorocyclopentene in the crystalline phase. *Inorg. Chem.*, **46**, 3313–3321; (b) Lee, P.H.-M., Ko, C.-C., Zhu, N. and Yam, V.W.-W. (2007) Metal coordination-assisted near-infrared photochromic behavior: a large perturbation on absorption wavelength properties of N,N-donor ligands containing diarylethene derivatives by coordination to the rhenium(I) metal center. *J. Am. Chem. Soc.*, **129**, 6058–6059; (c) Uchida, K., Inagaki, A. and Akita, M. (2007) Preparation and photochemical behavior of organoruthenium derivatives of photochromic dithienylethene (DTE): DTE–(RRuL$_m$)$_n$ (RRuL$_m$ = (η^6-C$_6$H$_5$)Ru (η^5-C$_5$Me$_5$), (η^6-C$_6$H$_5$)RuCl$_2$(PPh$_3$), (η^5-C$_5$Me$_4$)Ru(CO)$_2$; $n = 1, 2$). *Organometallics*, **26**, 5030–5041.

8 (a) Kawano, Y., Tobita, H. and Ogino, H. (1992) Reversible thermal and photochemical interconversion between cis and trans isomers of the silylene-bridged diiron complex Cp′$_2$Fe$_2$(CO)$_3$ (μ-SiHTol) (Cp′ = η-C$_5$Me$_5$, Tol = p-CH$_3$C$_6$H$_4$). *Organometallics*, **11**, 499–500; (b) Adams, R.D., Cortopassi, J.E., Aust, J. and Myrick, M. (1993) Energy capture by a tetranuclear metal cluster complex. Synthesis and characterization of PtOs$_3$(CO)$_{10}$(μ-η^2-dppm)[Si(OMe)$_3$](μ-H) and its metastable photoisomer. *J. Am. Chem. Soc.*, **115**, 8877–8878; (c) Boese, R., Cammack, J.K., Matzger, A.J., Pflug, K., Tolman, W.B., Vollhardt, K.P.C. and Weidman, T.W. (1997) Photochemistry of (fulvalene)tetracarbonyldiruthenium and its derivatives: efficient light energy storage devices. *J. Am. Chem. Soc.*, **119**, 6757–6773; (d) Yuki, M., Okazaki, M., Inomata, S. and Ogino, H. (1998) Formation of isomeric tetrathiotungstate clusters [{Cp*Ru(CO)}$_2$(WS$_4$){W(CO)$_4$}] by the reaction of [Cp*$_2$Ru$_2$S$_4$] with [W(CO)$_3$(MeCN)$_3$]. *Angew. Chem. Int. Ed.*, **37**, 2126–2128; (e) Burger, P. (2001) [Me$_2$C (η^5-C$_5$H$_4$)$_2$Ru$_2$(CO)$_4$]–An organometallic thermo-optical switch. *Angew. Chem. Int. Ed.*, **40**, 1917–1919; (f) Shibahara, T., Tsuboi, M., Nakaoka, S. and Ide, Y. (2003) Photochromic isomerization of a dinuclear molybdenum complex with ethylene-1,2-dithiolate and disulfur ligands: X-ray structures of the two isomers. *Inorg. Chem.*, **42**, 935–937.

9 (a) Pearson, W.B. (ed.) (1956) *Structure Reports*, vol. 20, International Union of

Crystallography, p. 331; (b) Cotton, F.A. and Wilkinson, G. (1988) *Advanced Inorganic Chemistry*, 5th edn, Wiley-Interscience, p. 522.

10 (a) Poffenberger, C.A., Tennent, N.H. and Wojcicki, A. (1980) Synthesis, characterization and some reactions of organometallic complexes containing bridging dithionite ligand. *J. Organomet. Chem.*, **191**, 107–121; (b) Kubas, G.J., Wasserman, H.J. and Ryan, R.R. (1985) Reduction of SO_2 by $(C_5R_5)M(CO)_3H$ (M=Mo, W; R=H, Me). Chemistry and structures of $(C_5H_5)Mo(CO)_3(SO_2H)$, the first example of insertion of SO_2 into a M—H bond, and $[(C_5Me_5)Mo(CO)_3]_2$ (μ-S_2O_4), an S-bonded dithionite complex. *Organometallics*, **4**, 2012–2021; (c) Bitterwolf, T.E. (1987) Reaction of sulfur dioxide with new dinuclear rhodium compounds. *J. Organomet. Chem.*, **320**, 121–127; (d) Matsumoto, K., Koyama, T. and Koide, Y. (1999) Oxidation of the sulfide ligands to SO_4^{2-} in the dinuclear complex $\{RuCl[P(OMe)_3]_2\}_2(\mu$-$S_2)(\mu$-Cl$)(\mu$-$N_2H_4)$: synthesis and characterization of $\{RuCl[P(OMe)_3]_2\}_2(\mu$-$S_2)(\mu$-Cl$)(\mu$-$N_2H_4)^+$ HSO_4^-, $\{RuCl_2[P(OMe)_3]_2\}_2(\mu$-S$)$ (μ-N_2H_4), and $\{RuCl[P(OMe)_3]_2\}_2(\mu$-$S_2O_5)$ (μ-N_2H_4). *J. Am. Chem. Soc.*, **121**, 10913–10923.

11 (a) Nishioka, T., Kitayama, H., Breedlove, B.K., Shiomi, K., Kinoshita, I. and Isobe, K. (2004) Novel nucleophilic reactivity of disulfido ligands coordinated parallel to M—M (M=Rh, Ir) bonds. *Inorg. Chem.*, **43**, 5688–5697.

12 (a) Isobe, K., Okeya, S., Meanwell, N.J., Smith, A.J., Adams, H. and Maitlis, P.M. (1984) Synthesis and characterisation of disubstituted Di-μ-methylene-bis (μ-pentamethylcyclopentadienyl) dirhodium(IV) complexes; X-ray structure of $[\{(C_5Me_5)Rh\}_2(\mu$-$CH_2)_2(CO)_2]^{2+}$. *J. Chem. Soc., Dalton Trans.*, 1215–1221.

13 Bouas-Laurent, H. and Dürr, H. (2001) Organic photochromism. *Pure Appl. Chem.*, **73**, 639–665.

14 Ohashi, Y. (1998) Real-time *in situ* observation of chemical reactions. *Acta. Crystallogr.*, **A54**, 842–849.

15 (a) Ramamurthy, V. and Venkatesan, K. (1987) Photochemical reactions of organic crystals. *Chem. Rev.*, **87**, 433–481; (b) Desiraju, G.R. (1989) *Crystal Engineering: The Design of Organic Solids*, Elsevier, Amsterdam; (c) Braga, D. and Grepioni, F. (2004) Reactions between or within molecular crystals. *Angew. Chem. Int. Ed.*, **43**, 4002–4011; (d) Kawano, M. and Fujita, M. (2007) Direct observation of crystalline-state guest exchange in coordination networks. *Coord. Chem. Rev.*, **251**, 2592–2605.

16 Lacombe, S., Loudet, M., Dargelos, A. and Robert-Banchereau, E. (1998) Oxysulfur compounds derived from dimethyl disulfide: an ab initio study. *J. Org. Chem.*, **63**, 2281–2291.

17 (a) Cohen, M.D. and Schmidt, G.M.J. (1964) Topochemistry. Part I. A survey. *J. Chem. Soc.*, 1996–2000; (b) Schmidt, G.M.J. (1971) Photodimerization in the solid state. *Pure Appl. Chem.*, **27**, 647–678.

18 Braga, D. (1992) Dynamical processes in crystalline organometallic complexes. *Chem. Rev.*, **92**, 633–665.

19 (a) Garcia-Garibay, M.A. (2005) Crystalline molecular machines: encoding supramolecular dynamics into molecular structure. *Proc. Natl. Acad. Sci. U. S. A.*, **102**, 10771–10776; (b) Khuong, T.-A.V., Nuñez, J.E., Godinez, C.E. and Garcia-Garibay, M.A. (2006) Crystalline molecular machines: a quest toward solid-state dynamics and function. *Acc. Chem. Res.*, **39**, 413–422; (c) Horike, S., Matsuda, R., Tanaka, D., Matsubara, S., Mizuno, M., Endo, K. and Kitagawa, S. (2006) Dynamic motion of building blocks in porous coordination polymers. *Angew. Chem. Int. Ed.*, **45**, 7226–7230.

20 Bryan, R.F., Greene, P.T., Newlands, M.J. and Field, D.S. (1970) Metal-metal bonding in coordination complexes. Part X. Preparation, spectroscopic properties,

and crystal structure of the cis-isomer of di-μ-carbonyl-dicarbonyldi-*n*-cyclopentadienyldi-iron. *J. Chem. Soc. A*, 3068–3074.

21 (a) Kaupp, G. (1992) Photodimerization of cinnamic acid in the solid state: new insights on application of atomic force microscopy. *Angew. Chem. Int. Ed.*, **31**, 592–595; (b) Kaupp, G. (1992) Photodimerization of anthracenes in the solid state: new results from atomic force microscopy. *Angew. Chem. Int. Ed.*, **31**, 595–598; (c) Kaupp, G. (1995) *Advances in Photochemistry*, vol. **19** (eds Neckers, D.C., Volman, D.H. and von Bünau, G.), Wiley-Interscience, New York, pp. 119–178; (d) Kaupp, G. (2002) Solid-state reactions, dynamics in molecular crystals. *Curr. Opin. Solid State Mater. Sci.*, **6**, 131–138.

26
Dynamics in Organic Inclusion Crystals of Steroids and Primary Ammonium Salts

Mikiji Miyata, Norimitsu Tohnai, and Ichiro Hisaki

26.1
Introduction

Organic inclusion crystals are attractive substances from the viewpoint of the dynamics of molecular assemblies. The inclusion crystals comprise nanocomposites between host and guest molecules and exhibit supramolecular properties through non-covalent bonds [1]. The host molecules provide molecular-level cavities where the guest molecules undergo intercalations and reactions. We have elucidated such dynamics by using the inclusion crystals of steroids and primary ammonium salts (Figure 26.1).

This chapter consists of three sections. The first is an overview of comprehensive steroidal crystals on the basis of our lengthy research [2]. The steroidal hosts yield diverse host–guest assemblies which display a variety of dynamical behaviors. For example, classical studies of inclusion polymerization established a dynamical process of monomeric guests in channel-type cavities, leading to recognition of molecular-level space effects therein [3]. Such polymerization research gave us the chance to observe intercalation of the guest molecules with retention of the host assemblies in the crystalline state, compared to working with inorganic compounds such as clays and graphite [4]. Intercalation with sandwich-type crystals and layer-reversion crystals is highlighted, followed by consideration of the fit of the guest molecules in the cavities.

The second section deals with the dynamics of organic primary ammonium salts of carboxylic acids, sulfonic acids, and phosphonic acids. Inspired by the steroidal crystals described above, our research has been extended to the supramolecular properties of organic salts. We focus on the latest research on solid-state fluorescence emission and pseudo-cubic hydrogen bonding clusters. It is noteworthy that the salts have practical advantages, including the relatively simple preparation of their crystals,

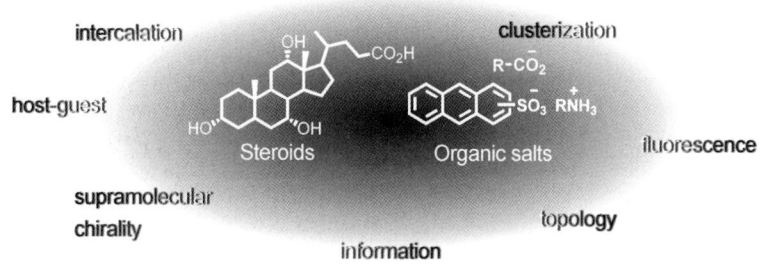

Figure 26.1 Dynamical properties related to organic inclusion crystals of steroids and primary ammonium salts.

the enormous number of combinations of acids and bases, and diverse crystal structures with different hydrogen bonding networks. Such organic salts could serve as useful materials for exploring supramolecular functions toward crystal engineering.

The third section describes the hierarchy and supramolecular chirality of molecular assemblies in the crystalline state. The steroidal molecules construct hierarchical assemblies on the basis of sequential information, as in the case of proteins. The notable feature is that each hierarchical assembly exhibits supramolecular chirality, such as three-axial, tilt, helical, and bundle chirality. On the other hand, the primary ammonium salts construct hierarchical hydrogen bonding networks which, in some cases, create supramolecular chirality from achiral components. The creation of chirality can be interpreted from a topological viewpoint, leading us to define the handedness of the supramolecular chirality. At the end of this section we present the general concept that molecular-level information on organic substances can be expressed by their assemblies through non-covalent interactions.

Finally, perspectives are briefly provided from the viewpoint of information and expression of organic molecules through their supramolecular architectures.

26.2
Dynamics of Steroidal Inclusion Crystals

26.2.1
Guest-Responsive Molecular Assemblies

Naturally-occurring steroidal bile acids involve cholic acid, deoxycholic acid, chenodeoxycholic acid, and lithocholic acid, as shown in Figure 26.2. These acids consist of skeletons and side-chains, and can be modified to more than one hundred derivatives. For example, their side-chains may have different chain length and hydrogen bonding groups in the form of carboxylic acids, amides, esters, alcohols and so on. Their highly asymmetric skeletons have one or more hydroxy groups with different

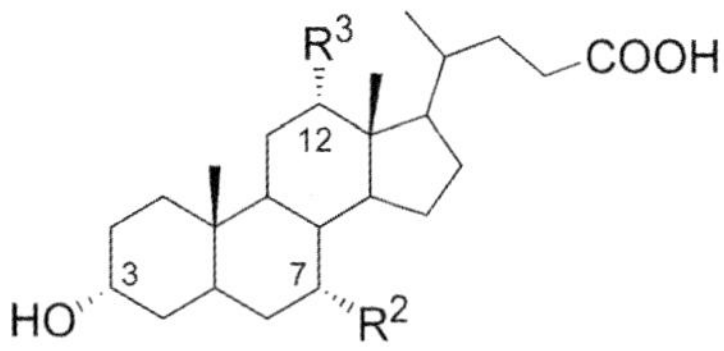

cholic acid (R^2 = OH, R^3 = OH)
deoxycholic acid (R^2 = H, R^3 = OH)
chenodeoxycholic acid (R^2 = OH, R^3 = H)
lithocholic acid (R^2 = H, R^3 = H)

Figure 26.2 Naturally-occurring steroidal bile acids involving cholic acid and its related derivatives.

locations and directions. Such unique molecular structures cause extensive formation of their guest-responsive assemblies through cooperative non-covalent bonds.

These steroidal derivatives yield inclusion crystals with a variety of organic substances. The enormous range of inclusion experiments indicated that each host exhibits unique inclusion ability. In other words, comparative studies with a pair or set of steroidal hosts show slight or drastic differences in their inclusion behavior. For example, both cholic acid and deoxycholic acid include a wide range of organic substances, while chenodeoxycholic acid includes only a little and lithocholic acid not at all. Their amide derivatives can include many aliphatic alcohols, except for the amide of chenodeoxycholic acid. The alcohol derivatives of cholic acid and deoxycholic acid include aromatic guests, while those of chenodeoxycholic acid and lithocholic acid do not. Bishomocholic acid with two additional methylene units has larger inclusion spaces than cholic acid, while bisnorcholic acid with two decreased methylene units has no inclusion spaces. Epimers of cholic acid and deoxycholic acid include various aliphatic alcohols, in contrast to their original acids.

Such diverse inclusion behavior is interpreted on the basis of X-ray crystallographic data and molecular graphics. Nowadays, more than three hundred sets of crystal data give us valuable information about non-covalent bonds between the host and guest molecules. The dominant interactions are attributed to conventional hydrogen bonds among oxygen and nitrogen atoms. Moreover, we can confirm a variety of weak hydrogen bonds, such as $CH \cdots O$, $CH \cdots \pi$, $NH \cdots \pi$, $\pi \cdots \pi$ and so on [5]. These hosts have multiple hydrogen bonding groups, which combine in many ways among four host molecules to make linear, cyclic, branched or arched networks. These networks are often accompanied by additional guest groups and are modified by slightly different hydrogen bonding groups. For example, cholic acid forms a cyclic network with four host molecules, and includes over a hundred aromatic molecules through $CH \cdots \pi$ interactions. The additional hydrogen of its amide induces a cyclic network with an extra hydrogen bonding donor which catches over fifty kinds of aliphatic alcohols. On the other hand, an epimer of cholic acid produces a branched network instead of the cyclic one, allowing aliphatic alcohol guests to insert between two hydrogen bonding groups of the host.

bilayer

herring-bone

triangular prism

honeycomb

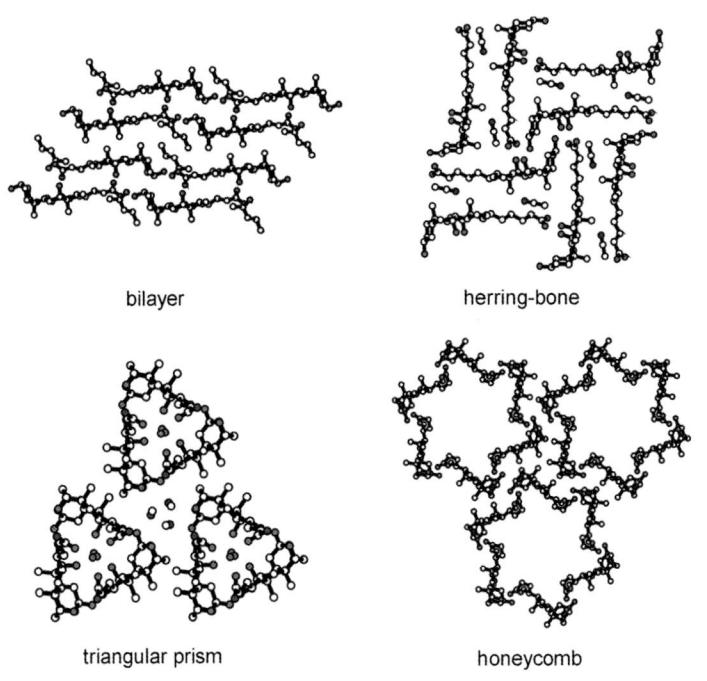

Figure 26.3 Representative packing diagrams for inclusion crystals of cholic acid and its derivatives. The crystals exhibit guest-responsive structures.

The steroidal acids and their derivatives form various crystal structures. Figure 26.3 illustrates four kinds of crystal structures, such as a bilayer, a herringbone, a triangular prism, and a honeycomb. The representative bilayer structures have diverse guest-dependent assembly modes. This diversity comes from facial molecular structures with three-axial chirality, resulting in four kinds of combinations of hydrophilic and lipophilic sides in a parallel- or anti-parallel manner. In addition, the cumulated bilayers can slide on the lipophilic sides, followed by conformational changes of the side-chains. Such flexibilities of the bilayers can explain the versatile inclusion behavior mentioned above. Another feature is that the bilayers are constructed by connection of 2_1 helical assemblies of the host molecules, indicating that their side-chain length decisively influences the assembly modes of the helices [6].

26.2.2
Intercalation in Steroidal Bilayer Crystals

The amphiphilic structures of steroidal molecules are closely related to the dynamical properties of their crystals. The hydrophilic sides are used to make the host framework through strong hydrogen bonds, while the lipophilic sides are used to make the cavities through weak hydrogen bonds and van der Waals forces. The latter weak interactions cause intercalation phenomena which correspond to reversible

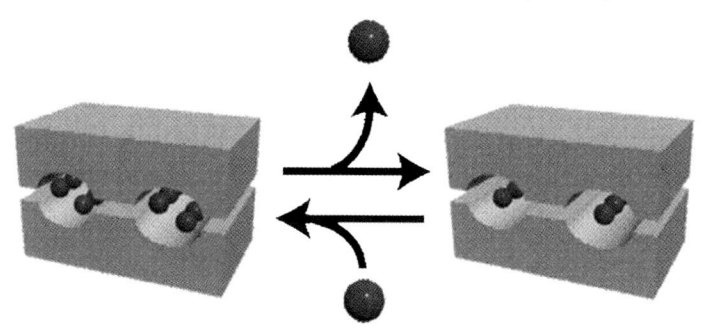

Figure 26.4 Schematic representation of intercalation phenomenon in a layer material.

insertion, release, and exchange of guest molecules accompanying any fluctuation of the host framework (Figure 26.4). An important problem lies in whether the host frameworks are preserved during the intercalation or not. This is the reason why the intercalation of steroidal crystals was quite rare among organic crystalline materials about twenty years ago [4a]. Recently, the intercalation phenomena have become ubiquitous for organic and inorganic layered materials as well as nano-porous crystalline materials based on three-dimensional organic or organometallic frameworks [7]. Hereafter, we introduce the latest two examples of the attractive intercalation of steroidal bilayer crystals.

One is the intercalation and deintercalation by using sandwich-type inclusion crystals of cholic acid, as shown in Figure 26.5 [8]. The sandwich-type structure has the same host bilayers as the bilayer-type, but involves additional sandwiched guest molecules between the bilayers. The intercalation and deintercalation of the guest molecules take place smoothly with retention of the crystalline state, resulting in reversible changes of the interlayer distances. The sandwich-type crystals in a 1 : 2 host-to-guest ratio are formed with five disubstituted benzenes, such as o-toluidine, m-fluoroaniline, o-chlorotoluene, o-bromotoluene and indene. Their thermogravimetric analyses revealed that the guest molecules, except o-toluidine, are released in two steps, indicating formation of intermediate crystals in a 1 : 1 or 2 : 1 host-to-guest ratio. X-ray powder diffraction patterns of the intermediate crystals revealed that the crystals have the same bilayer structures as those of the common inclusion crystals. Furthermore, reverse structural changes were achieved with absorption of the guest

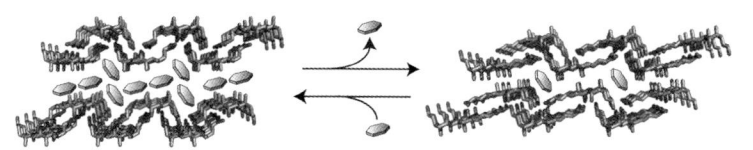

Figure 26.5 Intercalation in sandwich-type inclusion crystals of steroidal cholic acid. Benzene derivatives undergo reversible desorption and absorption, accompanied by changes in bilayer distances.

molecules to regenerate the starting sandwich-type inclusion crystals. The host bilayers expand and shrink by 1.8 to 4.5 Å perpendicular to the layer direction. Such notable changes serve as a novel example of organic intercalation materials.

Moreover, the intercalation method was applied to the enantioresolution of racemic alcohols. The amide derivative of cholic acid, called cholamide, serves as an excellent host for enantioresolution of 2,2-dimethyl-3-hexanol among secondary aliphatic alcohols. The intercalation method gave highly effective enantioresolution of the alcohol, as in the case of the recrystallization method. For example, dioxane molecules were exchanged with the alcohol molecules in channels of the bilayers with a high enantioselectivity of 96% ee, which was close to the value (98% ee) obtained by the recrystallization method. X-ray diffraction studies proved that the host bilayer framework has an anti-parallel arrangement on the lipophilic side in the case of dioxane, while the framework has a parallel one after intercalation of the alcohol. This indicates that the intercalation accompanies a layer-reversion. Such dynamical enantioresolution confirms the efficiency of the intercalation method.

26.2.3
Guest Fit Through Weak Non-Covalent Bonds

The guest molecules are accommodated in molecular cavities with different strengths, relating to the dynamical behavior of organic inclusion crystals. In this section we introduce three related examples about guest fit in the host cavities through weak non-covalent bonds.

The first example is concerned with the packing coefficient of the host cavity, termed the *PC*cavity, which corresponds to the ratio of the volume of the guest molecule to the volume of the host cavity [9]. The *PC*cavity values of the inclusion crystals of cholic acid and cholamide were evaluated for a wide range of aromatic compounds and aliphatic alcohols. The values are approximately in the range 45–75%, meaning inhibition of too close or loose packing. In the case of the same host frameworks the values tend to increase with increasing guest volumes, indicating that the guest molecules exceeding this range force the host to adopt different assembly modes or molar ratios. Otherwise, they may form guest-free crystals or not form any crystals. The values in the inclusion state (45–75%) are intermediate between those in the liquid state (44–56%) and in the crystalline state (66–77%).

The second example deals with weak hydrogen bonds such as $N-H\cdots\pi$, $C-H\cdots\pi$, $C-H\cdots O$ and so on [5, 10]. We compared these bonds between the inclusion crystals of cholamide and cholic acid, where weak hydrogen bonds play a key role in linking the host and guest molecules. The steroidal side-chains involving methyl, methylene, and amide groups serve as the hydrogen bond donors, and aromatic guest molecules serve as the acceptors. The inclusion crystals are categorized into two host frameworks, β-trans- and *DCA*-type. The former involves four weak hydrogen bonds; one $N-H\cdots\pi$, two $C-H\cdots\pi$, and one $C-H\cdots O$ interactions, whereas the latter involves only $C-H\cdots\pi$ interactions. Figure 26.6 displays these interactions and the Hirshfeld surface of the guest molecule. It can be seen that

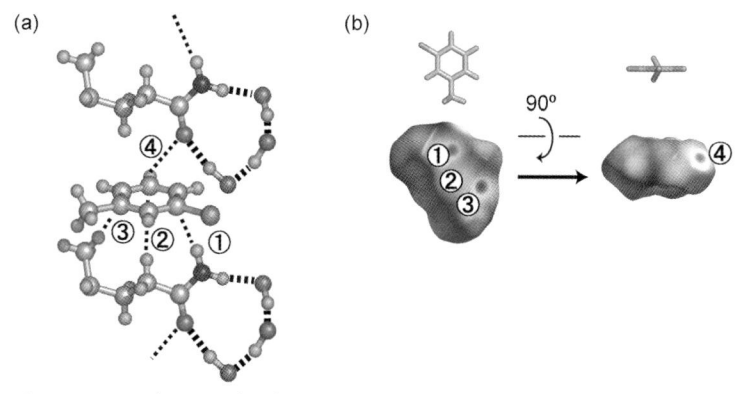

Figure 26.6 Weak intermolecular interactions among two cholamide host and one aromatic guest molecules. (a) Four kinds of the interactions around the guest molecule. (b) Hirshfeld surface of the guest molecule, which shows four spots corresponding to the interactions shown in (a).

the N−H···π weak hydrogen bond forces cholamide to choose the β-trans-type crystal rather than the α-gauche-type. This exemplifies that the weak hydrogen bond enables the host to select the framework having low steric matching. In this way our systematic investigation provides effective examples for displaying weak hydrogen bonds among the host and guest molecules.

The third example deals with efficient enantioresolution due to the highest guest fit in the steroidal crystals. When the host assemblies have guest-dependent crystal structures, it is not so easy to obtain resolution greater than 90% ee. This is because the assembly modes can change to different modes before the highest PCcavity values. So far we have obtained two successful enantioresolutions of secondary alcohols. One is the remarkable recognition of the (2R,3S)-isomer among four isomers of 3-methyl-2-pentanol by epicholic acid [11]. The other is the enantioresolution of (S)-2,2-dimethyl-3-hexanol by cholamide [12]. As for the recognition mechanism, we proposed that a four-location model [13] is more suitable than the conventional three-point attachment model. As shown in Figure 26.7, the former is based on a deformed hole, while the latter on a planar surface. As the fourth location of the former model, the smallest hydrogen atom together with a chiral carbon or methyl group have to be recognized in the case of secondary alcohols. In principle, at least, two kinds of disordered structures of guest components may be observed in the concave cavity. One is that in which a stereogenic carbon of the guest is disordered together with the fourth site (D, hydrogen). Such a disordered structure was confirmed in the case of the inclusion crystals of cholamide with 2-methyl-3-hexanol. An additional methyl group brought about highly efficient enantioresolution of (S)-2,2-dimethyl-3-hexanol. The other is that in which the stereogenic carbon is almost fixed while two neighboring substituents (C, D), such as a methyl group and hydrogen, are disordered.

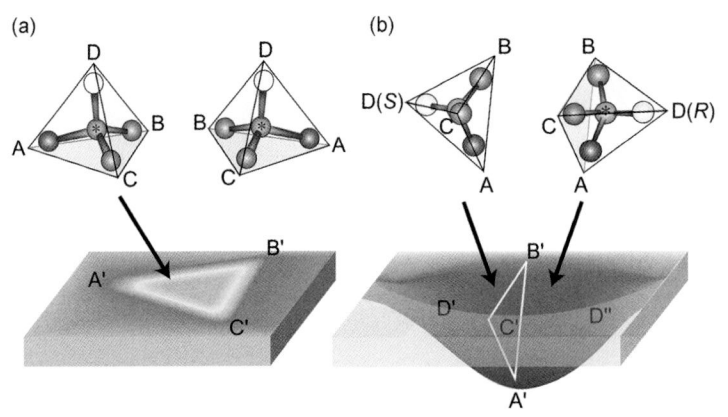

Figure 26.7 Three- (left) and four- (right) location models for chiral recognition. The former is based on a planar surface, while the latter on a deformed hole where the fourth substituent (D) together with stereogenic carbon has to be recognized.

26.3
Dynamics of Organic Crystals of Primary Ammonium Salts

26.3.1
Solid-State Fluorescence Emission

Solid-state luminescent properties of organic materials have attracted much attention because of their direct and widespread applications in organic light-emitting diodes, organic dye laser photonic and photoelectronic devices. Currently, research on the luminescent properties is faced with the serious problem that identical organic fluorophores emit strongly in solution but weakly in the solid state. This is probably because the luminescent properties depend not only on the molecular structures but also on the molecular arrangements of the fluorophores. In order to develop sophisticated crystalline materials, it is essential to screen the arrangements necessary for solid-state fluorescence emission as well as to regulate slight movements of the fluorophores through intermolecular interactions.

We employed organic salts of anthracene-2,6-disulfonic acid (ADS) with various aliphatic primary amines (Scheme 26.1), and investigated their solid-state luminescent properties as well as their arrangements of anthracene moieties. The ADS salts exhibited amine-dependent properties, such as emitting color and intensity (Figure 26.8). Single X-ray crystallographic studies revealed that the salts have amine-dependent arrangements, which are classified on the basis of hydrogen bonding networks between sulfonate anions and ammonium cations. These results indicate that the luminescent properties depend closely on the molecular arrangements.

We have found that the ADS salt with n-heptylamine, which has discrete anthracene moieties in the crystal structure, exhibits the highest quantum yield in

Scheme 26.1 Preparation of organic salts of anthracene-2,6-disulfonic acid (ADS) and various primary amines.

a series of ADS salts and luminesces more strongly than intact anthracene [14]. Moreover, we have demonstrated that a powerful strategy for luminescent enhancement lies in the rigidity of the luminant packing in the solid-state [15]. X-ray crystallographic studies confirmed the formation of a rigid one-dimensional arrangement by inserting benzylamine molecules into tubulate spaces of an anthracene host framework. This result shows that the strong fluorescence intensity observed is due to the increase in rigidity around the fluorophores, and not to the decrease in intermolecular interactions, as previously thought.

A ternary system was found to afford different molecular arrangements and excimer emission from a binary system. The former system involves guest molecules as the third modulators which control the solid-state emission modes by chemical or physical stimuli. In our sophisticated system the modulators act as molecular information sources which are transcribed to the arrangements, and then are translated into the emission mode.

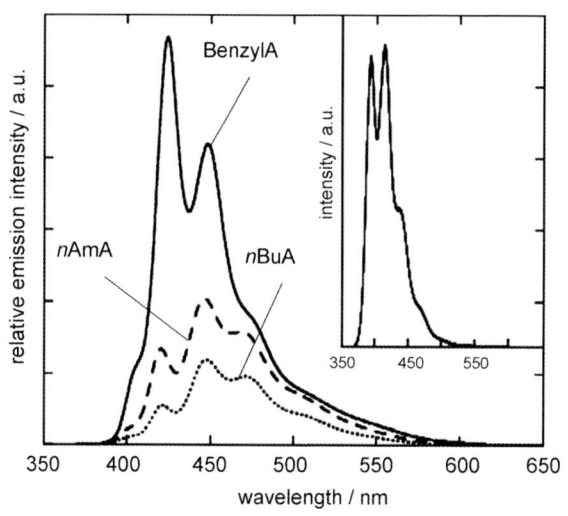

Figure 26.8 Solid-state fluorescence spectra of the ADS salts with benzyl amine (solid line), n-amyl amine (dashed line), and n-butyl amine (dotted line). Their spectra in solution (inset).

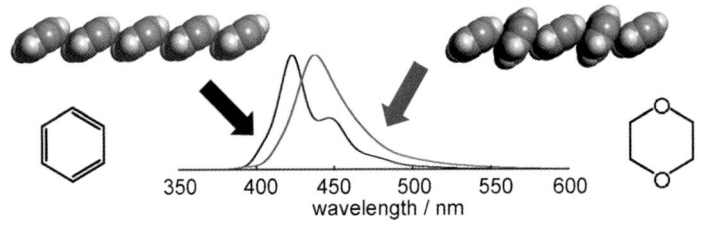

Figure 26.9 Solid-state fluorescence spectra in the ADS salts with rac-sec-butyl amine involving dioxane or benzene as the third modulators.

For instance, 1,4-dioxane as the modulator yields a repetitive cycle consisting of desorption, absorption, and cooling processes among three different crystal forms. We observed switching of the corresponding solid-state emission modes between monomer and excimer states (Figure 26.9) [16]. In contrast, benzene as the modulator did not show such switching, although benzene was absorbed into the crystals and desorbed from them. The former crystals showed a change from a monomer emitting arrangement to an excimer emitting one of anthracene moieties under the cooling process, while the latter crystals kept the monomer emitting arrangement. Moreover, the switching of the solid-state emission modes can be repeated with good reproducibility. This programmed switching system is based on a dynamical change of arrangements of the anthracene moieties. These observations led to the idea that these crystal dynamics depend on molecular information of the guest molecules. Such programmed dynamics may lead to a novel paradigm of organic crystalline materials.

26.3.2
Hydrogen Bond Clusters

Recently we have found that triphenylmethylamine and various sulfonic acids form supramolecular cubic-type clusters in the crystalline and solution states (Scheme 26.2) [17]. This finding originates from a combinatorial study for screening new host compounds other than steroidal compounds. The research gave us the idea that the bulkiness of the components can regulate the dimensionalities of hydrogen bonding networks. Such an idea has been applied to the preparation of novel materials such as artificial clusters, clays, zeolites and so on.

The hydrogen bonded [4 + 4] clusters have a cubic hydrogen bonding network in a core-shell, which is covered by bulky triphenylmethyl groups. Figure 26.10 shows three kinds of [4 + 4] ion-pair clusters of organic salts composed of four triphenylmethylamines and four organic carboxylic acids, sulfonic acids, and phosphonic acids in the crystalline state. Such clusters involving various carboxylates are obtained

$$4\,R^1\text{-}SO_3H + 4\,R^2\text{-}NH_2 \xrightarrow{\text{clusterization}} \left[R^1\text{-}\overline{SO_3}\right]_4 \left[R^2\text{-}\overset{+}{NH_3}\right]_4$$

Scheme 26.2 Formation of [4 + 4] clusters from salts of organic sulfonic acids and primary amines.

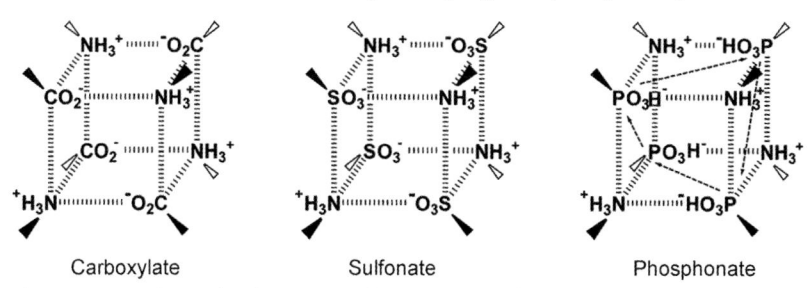

Figure 26.10 Hydrogen bonding [4 + 4] clusters composed of primary ammonium salts of triphenylmethylamine with organic acids, such as carboxylate, sulfonate and phosphonate.

only in a limited range of combinations, whereas those involving sulfonates are obtained in a wide range. The retention of the sulfonate clusters in solution was confirmed by ^{1}H NMR and mass spectroscopy. It is attractive that the cubic clusters exhibit supramolecular chirality like dice in our daily lives, since the clusters have two vortex-like patterns of the hydrogen bonding network; in a clockwise or anticlockwise direction.

A wide range of organic sulfonic acids yield clusters with diverse sizes and shapes (Figure 26.11). The clusters have potency which derives from: the topology of the hydrogen bond networks, steric effects of the substituents and the acidities of the sulfonic acids. A sulfonate is a tridentate hydrogen bond acceptor and a primary ammonium is a tridentate hydrogen bond donor. Since each direction of the hydrogen bond is at the same tetrahedral angle, complementary pairs among primary ammoniums and sulfonates assemble into a cubic hydrogen bonding cluster without strain. The closed hydrogen bond network can be considered as zero dimensional. Bulky triphenylmethyl groups cover a core-shell which consists of a cubic hydrogen bonding network. High acidity overcomes other hydrogen bond acceptors and solvent molecules. These factors enable the cubic hydrogen bond networks to be tightly held together. The twelve hydrophobic phenyls of four triphenylmethyl groups

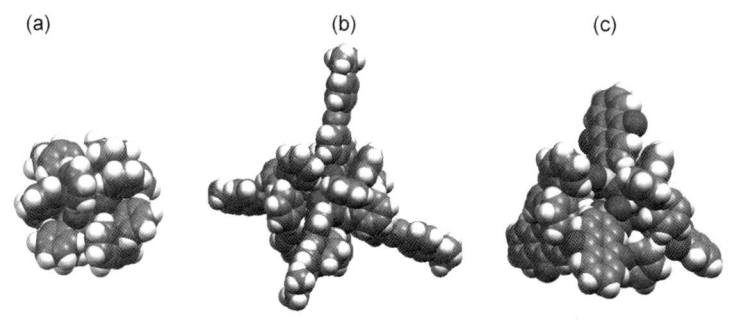

Figure 26.11 Different sizes and shapes of the [4 + 4] clusters composed of various sulfonates: methane sulfonate (a), benzene sulfonate (b), and 2-anthraquinone sulfonate (c).

shield the hydrophilic core, and segregate it from the outside. This unique feature, like a reverse micelle, gives high solubilities of the clusters in nonpolar solvents.

Moreover, organic salts of triphenylmethylamine with some phosphonic acids yield more complicated [4 + 4] clusters than those with sulfonates. The second hydrogen atoms in the ammonium phosphonates remain as OH groups without deprotonation. Hence, the additional hydrogen atoms serve as hydrogen-bond donors to yield four OH $\cdots$ O hydrogen bonds in addition to the cubic-type hydrogen bonds. As a result, a total of sixteen hydrogen bonds produces a novel pseudo-decahedral hydrogen-bond network with different topology. Since these complicated clusters form topological isomers, dynamical control of the topology may contribute to molecular informatics.

26.4
Dynamical Expression of Molecular Information in Organic Crystals

26.4.1
Hierarchical Structures with Supramolecular Chirality

At present there are no general principles for explaining crystal growth processes of organic molecules. In addition, it still remains very difficult to observe the process with the latest microscopes. These situations prompted us to overview the above-mentioned studies which clarified various relations between the molecular structures and the crystal structures. Such an overview gave us an analogy between the crystals and proteins, resulting in an interpretation of hierarchy and three-dimensional chirality of the crystalline assemblies. This interpretation comes from the brief hypothesis that various small assemblies can be formed before total crystal structures. Figure 26.12 shows schematically that the steroidal molecules construct a hierarchical structure, as in the case of proteins. Namely, the molecules serve as primary structures (Figure 26.12a) and associate with their hydrogen-bonding groups of the skeletons and side-chains. The resulting secondary structures are naturally bimolecular assemblies (Figure 26.12b), followed by helical assemblies (Figure 26.12c). The helices are tied up in a bundle (Figure 26.12d), which corresponds to the tertiary structures and leaves cavities for accommodating guest components (Figure 26.12e).

Since the steroidal molecules are highly asymmetric, the resulting hierarchical assemblies may have the corresponding three-dimensional structures with supramolecular chirality. Starting from molecular chirality, each assembly must be chiral. In this context, we encountered a new problem, how we describe such molecular and supramolecular chirality. The first idea is that the steroidal molecules are analogous to a vertebrate animal which has three-axial chirality based on three directions such as head–leg, right–left, and belly–back. The three-axial chirality enables us to determine the three-axes of the hierarchical assemblies, as in the case of the helices of proteins and DNA.

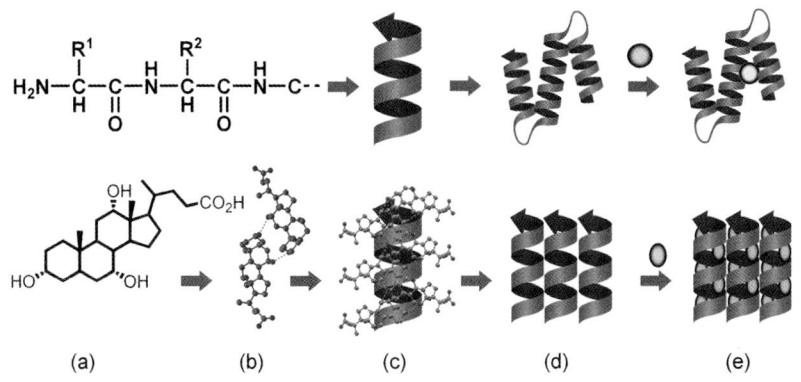

Figure 26.12 Hierarchical structures with supramolecular chirality for a folded polypeptide chain (top) and a cholic acid crystal (bottom). (a) A single molecule, (b) a bimolecular assembly, (c) a helical assembly, (d) a bundle of the helices, and (e) a host–guest system.

Kitaigorodskii pointed out the fact that molecules without symmetry elements tend to form 2_1 helical assemblies predominantly and induce close packing modes with chiral space groups, such as $P2_1$, $P2_12_12_1$, and so on [18]. It was strange for us that there are no general rules to determine handedness of the 2_1 helical assemblies, and therefore we cannot determine their handedness [19]. In our daily lives, however, we have experiences of going up and down right- or left-handed stairs with 2_1 helical arrangements, indicating that the key structures, 2_1 helical assemblies, may have their handedness. We have introduced the term, three-axial and tilt chirality to define the handedness. This definition has proved to be powerful in the elucidation of many structural problems of hierarchical assemblies, prompting further research not only for steroidal molecules but also the surrounding organic molecules. The next section deals with our research directed toward this problem.

26.4.2
Expression of Supramolecular Chirality in Hierarchical Assemblies

26.4.2.1 Three-Axial Chirality
In order to distinguish enantiomorphic structures, molecular chirality has conventionally been expressed in terms of center-, axis-, and plane-chirality. In the case of compounds involving several stereogenic centers, however, these terms seem to be insufficient to express their whole molecular chirality or anisotropy, although their local chirality is well confirmed in the conventional manner. Indeed, the steroidal molecules (Figure 26.13a), which have asymmetric, amphiphilic, facial structures with multiple stereogenic carbon atoms, are saddled with such a structural complexity. In order to solve this problem, we introduced a simple but unique concept, "three-axial chirality", as shown in Figure 26.13b. Such three-axial chirality is based on the orthorhombic three axes applied in a molecular structure and is expressed by

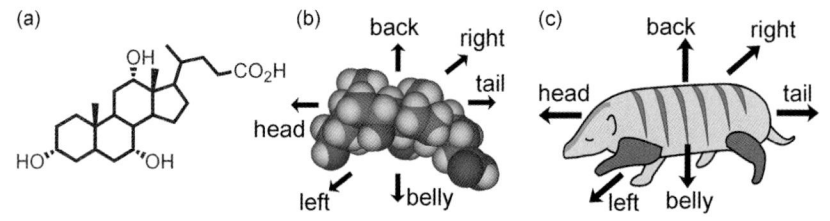

Figure 26.13 Expression for three-axial chirality of cholic acid molecule. (a) A molecular structure of cholic acid, (b) a space-filling model of cholic acid with the three-axial chirality expression, as in the case of (c) a vertebrate amimal.

the following terms: head and tail (leg), right and left, as well as belly and back, just like the direction of vertebrate animals in our daily lives (Figure 26.13c).

26.4.2.2 Tilt Chirality

Generally, a symmetrical object is naturally achiral; its mirror image is identical with the original. On the other hand, an assembly composed of two objects can exhibit chirality and have distinguishable enantioisostructures, as shown in an illustration of polyhedrons (Figure 26.14a). This example is classical and must be well understood. The important thing here is that the appearance of supramolecular chirality requires a tilt between the two objects [20]. Thus, we have emphasized the use of the term, tilt chirality, to express such chirality [21], although the conventional term "axis-chirality" essentially comes from the same origin as the tilt chirality.

Right- or left-handedness of the 2_1 helical molecular assembly cannot be determined from a mathematical and crytallographical viewpoint, because the assembly is produced by a twofold screw axis operation which includes 180° rotation and translation. However, since the assembly obviously gives a distinguishable enantiomorphic pair, it is reasonable to aspire to define the handedness of the 2_1 helical assembly. Indeed, we have proposed how to determine the handedness on the basis of the tilt chirality. As shown schematically in Figure 26.14b, given that the molecules are inclined to the right or left in front of a 2_1 screw axis, the assembly is defined to be right- or left-handed, respectively.

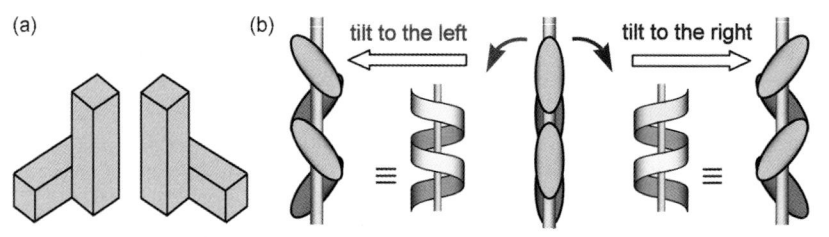

Figure 26.14 Supramolecular chirality provided by the tilt of the elements. (a) An enantiomorphic pair of polyhedrons, (b) definition of tilt chirality of a 2_1 helical assembly.

26.4.2.3 Helical and Bundle Chirality in a 2_1 Assembly

In the case of facial and asymmetric molecules such as steroids, combination of the three-axial and tilt chirality enables us to express supramolecular chirality of the 2_1 helical assembly. The first step directed toward the molecular assembly lies in the formation of a bimolecular assembly. We employ a simple example to describe the helical chirality, as shown in Figure 26.15. When the molecules with three-axial chirality (Figure 26.15a) align in the same head-to-tail direction, they have the following three association modes; belly-to-back, belly-to-belly as well as back-to-back (Figure 26.15b). In the case of the latter two modes, the bimolecular assemblies with a right- or left-tilt are expanded to the corresponding asymmetric 2_1 helical assemblies by twofold screw axis operation. In addition, an asymmetric helix is generally designated by the following three axes; right and left, up and down, and in and out (Figure 26.15c). Combination of these three-axial and tilt chirality enables us to define the helical chirality of the 2_1 helical assemblies, as shown in Figure 26.15d. In this figure, the up- and in-side of the helix may correspond to the head- and belly-side of the asymmetric molecule, respectively. The molecules are stacked with combinations of the belly-to-belly or back-to-back mode on the in-side, accompanied by the tilt, to give four kinds of bimolecular assemblies (a_1)–(d_1). They are extended to the corresponding 2_1 helical assemblies (a_2)–(d_2) [22].

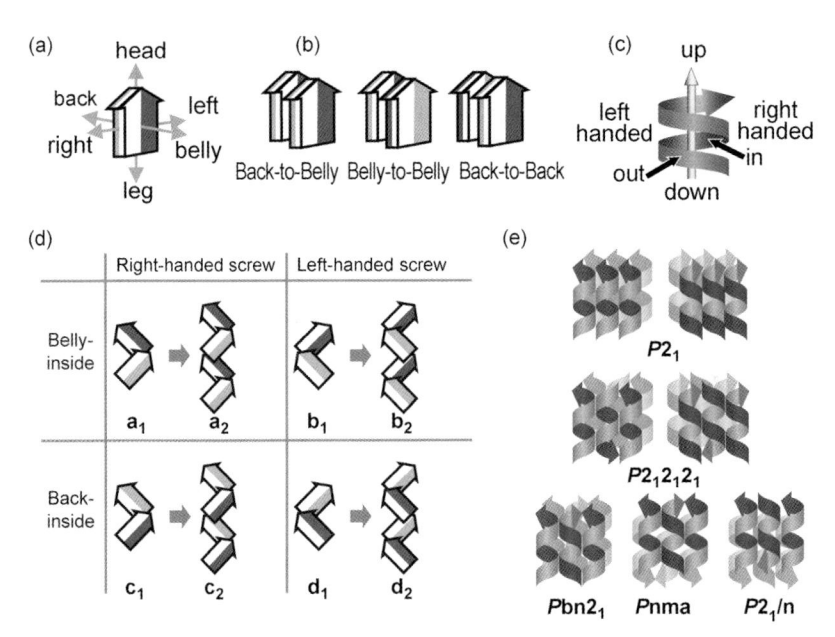

Figure 26.15 Schematic representation of supramolecular chirality in the 2_1 helices and their bundles. (a) Three-axial chirality in an asymmetric molecule, (b) assembly modes of the two molecules, (c) terms expressing the chirality of the helix: right- or left-handed, up or down, and in or out, (d) schematic representation of chirality in four kinds of bimolecular and 2_1 helical assemblies, and (e) chirality of the bundle of the helices, where the space groups of crystals are represented by stacking modes of the helices.

The 2_1 helical assemblies are then tied up to yield bundles as tertiary structures. Considering the way to assembly the helices with helical chirality, the most popular and fundamental aggregation pattern in crystals is a parallel or anti-parallel alignment. In Figure 26.15e, the typical seven kinds of bundles of helices are shown. A uniform alignment of either left- or right-handed helices gives a chiral structure which is observed in the crystal with space group $P2_1$. Another bundle of either left- or right-handed helices with the reverse up-down directions also gives the corresponding chiral crystal with space group $P2_12_12_1$. On the other hand, bundles composed of both left- and right-handed helices in a parallel or antiparallel fashion give achiral crystals with space groups such as $Pbn2_1$, $Pnma$, and $P2_1/n$.

26.4.3
Supramolecular Chirality of Hydrogen Bonding Networks

It is natural that steroidal crystals belong to chiral space groups, such as $P2_1$, $P2_12_12_1$, and so on. But it was surprising to us that primary ammonium salts sometimes gave chiral crystals, even when achiral amines and carboxylic acids were employed [23]. Such chirality creation was ascertained in various crystalline assemblies with zero-, one-, and two-dimensionalities. This suggests that the assemblies involve supramolecular chirality through non-covalent bonds in contrast to molecular chirality through covalent bonds. This idea prompted us to thoroughly analyze the hydrogen bonding networks of the ammonium salts.

Mathematical interpretation led us to the conclusion that we should recognize a topological aspect of the hydrogen bond networks in the crystals of primary ammonium carboxylates. This comes from the following two facts. First, that each cation and anion acts as a tridentate hydrogen bond donor and acceptor, respectively. Secondly, that one of two oxygen atoms of the carboxylate anion ($O(\alpha)$) acts as a two hydrogen bond acceptor, while another ($O(\beta)$) acts as a one hydrogen bond acceptor (Scheme 26.3). This discrimination of the two oxygen atoms enables us to recognize absolute configurations of the primary ammonium cations, leading to the interpretation of supramolecular chirality of the zero-, one-, and two-dimensional networks [24].

Such a consideration enabled us to clarify the topological diversity of pseudo-cubic hydrogen bond networks composed of primary ammonium triphenylacetates. The carboxylates involve three kinds of hydrogen bond which display different combinations of four carboxylates on the vertices of the cube to generate nine different topologies. Figure 26.16a exemplifies an enantiomeric pseudo-cubic network with C2 symmetry. Such a topological issue may appear in one-dimensional hydrogen bond networks of the carboxylates. Absolute configurations of the ammonium cations play a decisive role for discriminating supramolecular chirality of one-dimensional ladder-type hydrogen bond networks. As shown in Figure 26.16b, the ladder network involving the 2_1-axis is not superimposable on its mirror image, leading to the first definition of right- or left-handedness of its 2_1 helicity on the basis of supramolecular tilt chirality. Moreover, the 2_1 helical assemblies with three-axial chirality can be bundled in various ways to yield chiral crystals.

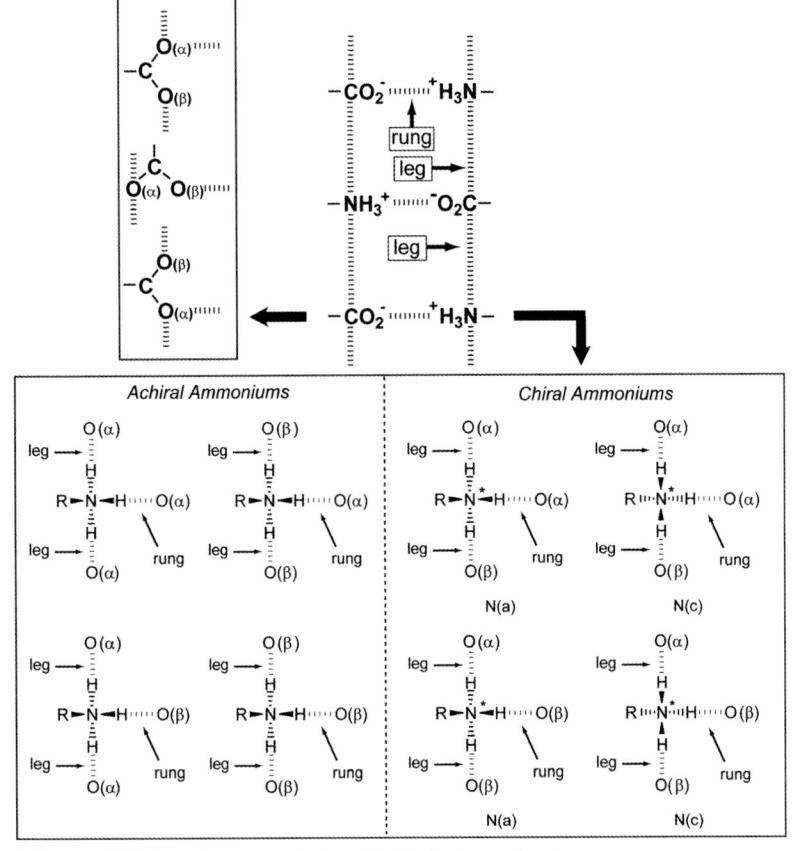

26.4 Dynamical Expression of Molecular Information in Organic Crystals

Scheme 26.3 Schematic representation of ladder hydrogen bond networks composed of primary ammonium cations and carboxylate anions with three possible alignments. The nitrogen atoms are classified into four achiral and four chiral isostructures.

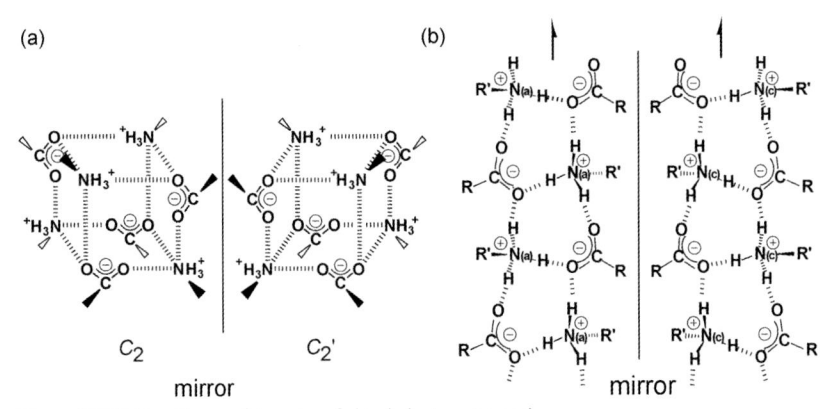

Figure 26.16 Enantiomorphic pairs of chiral clusters (a) and ladders (b) which are composed of primary ammonium carboxylates.

On the other hand, two-dimensional networks of the carboxylates involve a more complicated problem to discriminate the configurations of the nitrogen cations and the supramolecular chirality.

26.4.4
Expression of Molecular Information

For the past century, various relations between organic molecules and their assemblies have been made clear. It is well-known that spherical, axial, and facial molecules form hexagonal close packing, layered, and stacked assemblies, respectively. These relations have currently been extended to supramolecular synthons towards crystal engineering [25]. In this context, steroidal molecules with chiral, facial, curved, and amphiphilic features correspond to chiral supramolecular synthons for self-organization. In addition, our long-term study on the steroidal assemblies made clear a relation between organic molecules and their assemblies, bridging the gap between small molecules and biopolymers. We noticed that the steroidal assemblies bear a resemblance to proteins and gave us a novel concept for elucidating the relation between organic molecules and their assemblies.

As mentioned above, steroidal assemblies consist of hierarchical structures with supramolecular chirality, leading us to find an analogy on the basis of the concept; molecular information and their expression of biopolymers. The information originates from sequential arrangements of α-amino acids. Since it is considered that the steroidal molecules consist of chains of methylene units with various substituents, the concept may be applied to steroidal molecules. Such sequential chains may be considered to hold for other related organic molecules, leading to the idea that chiral methylene chains with various substituents function as universal molecular storage. The methylene chains can be chemically modified to various sequential chains, such as polypetides, polynucleotides, polysaccharides, and ster-

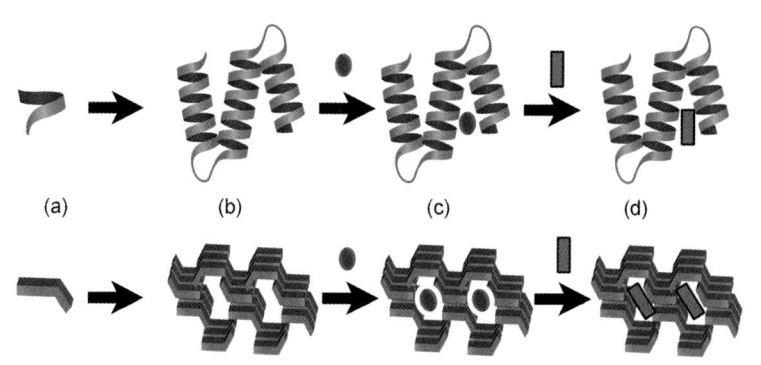

Figure 26.17 Dynamical processes for molecular information and expression in the cases of protein (upper) and steroidal molecules (lower). (a) A molecule as an information carrier, (b) a molecular architecture based on folding or assembly, (c) a host-guest complexation, and (d) a guest exchange reaction.

oids. Their information of the chains can be expressed through three-dimensional architectures by the best use of non-covalent bonds. The architectures are folding structures of proteins as well as assemblies of the steroidal molecules, as shown schematically in Figure 26.17.

26.5
Conclusion and Perspectives

We have demonstrated dynamics in organic inclusion crystals of steroids and primary ammonium salts from various viewpoints. We started such a dynamical study with inclusion polymerization in channel-type cavities. Subsequently, we directed our attention toward host–guest assemblies of the steroids. Although we suffered from diverse crystal structures of steroids at the beginning, the dynamics of the host and guest molecules in the assemblies always encouraged us. During the long-term studies we acquired valuable knowledge about diversity, hierarchy and supramolecular chirality. Analogy with proteins played a decisive role in reaching the unique concept that chiral carbon chains of the steroidal molecules are similar to sequential peptide chains. The chains involve enormous isomers. For example, when twenty different amino acids make a sequential chain having ten amino acids, the chain has over a trillion isomers. This reminds us of a form of Japanese literature, thirty one- or seventeen-syllable short poems, called Tanka (Waka) or Haiku, respectively. The short poems have infinite expression, much as the steroidal molecules do. We hope such analogy plays a significant role in performing various research studies towards supramolecular crystal engineering as well as the prediction of crystal structures.

References

1 (a) MacNicol, D.D., McKendrick, J.J. and Wilson, D.R. (1978) Clathrates and molecular inclusion phenomena. *Chem. Soc. Rev.*, **7**, 66–87; (b) Davies, J.E.D., Kemula, W., Powell, H.M. and Smith, N.O. (1983) Inclusion compounds–past, present, and future. *J. Inclusion Phenom.*, **1**, 3–44; (c) Weber, E.(ed.) (1987) *Top. Curr. Chem.*, **140**; (1988) *Top. Curr. Chem.*, **149**; (d) Atwood, J.L., Davies, J.E.D. and MacNicol, D.D. (eds) (1984) *Inclusion Compounds*, Academic Press, London, vols 1–3;(1991) Oxford Press, Oxford, vols 4, 5; (e) MacNicol, D.D., Toda, F. and Bishop, R. (eds) (1996) *Comprehensive Supramolecular Chemistry, Solid-State Supramolecular Chemistry: Crystal Engineering*, Vol. 6, Pergamon, Oxford; (f) Bishop, R. (1996) Designing new lattice inclusion hosts. *Chem. Soc. Rev.*, **25**, 311–319; (g) Herbstein, F.F. (2005) *Crystalline Molecular Complexes and Compounds*, vols. 1, 2, Oxford University Press, New York.

2 (a) Miyata, M., Tohnai, N. and Hisaki, I. (2007) Crystalline host–guest assemblies of steroidal and related molecules: diversity, hierarchy, and supramolecular chirality. *Acc. Chem. Res.*, **40**, 694–702; (b) Miyata, M., Tohnai, N. and Hisaki, I. (2007) Supramolecular chirality in crystalline assemblies of bile acids and

their derivatives; three-axial, tilt, helical, and bundle chirality. *Molecules*, **12**, 1973–2000.

3 (a) Miyata, M. (1996) in *Comprehensive Supramolecular Chemistry, Supramolecular Technology,* vol. 10 (ed. D.N. Reinhoudt,), Pergamon, Oxford, pp. 557–582; (b) Miyata, M. (2004) in *Encyclopedia of Supramolecular Chemistry*, vol. 1 (eds J.L. Atwood and J.W. Steed), Marcel Dekker, New York, pp. 705–711.

4 (a) Miyata, M., Shibakami, M., Chirachanchai, S., Takemoto, K., Kasai, N. and Miki, K. (1990) Guest-responsive structural changes in cholic acid intercalation crystals. *Nature*, **343**, 446–447; (b) Matsumoto, A., Odani, T., Sada, K., Miyata, M. and Tashiro, K. (2000) Intercalation of alkylamines into an organic polymer crystal. *Nature*, **405**, 328–330.

5 (a) Yoswathananont, N., Sada, K., Nakano, K., Aburaya, K., Shigesato, M., Hishikawa, Y., Tani, K., Tohnai, N. and Miyata, M. (2005) The effect of a host-guest hydrogen bond on the inclusion of alcoholic guests in the host cavities of cholamide. *Eur. J. Org. Chem.*, 5330–5338; (b) Aburaya, K., Nakano, K., Sada, K., Yoswathananont, N., Shigesato, M., Hisaki, I., Tohnai, N. and Miyata, M. (2008) Importance of weak hydrogen bonds in the formation of cholamide inclusion crystals with aromatic guests. *Cryst. Growth Des.*, **8**, 1013–1022.

6 Kato, K., Sugahara, M., Tohnai, N., Sada, K. and Miyata, M. (2004) Drastic increase of flexibility of open host frameworks of a steroidal host compound by the shortened spacer. *Eur. J. Org. Chem.*, 981–994.

7 Müller, A., Reuter, H. and Dillinger, S. (1995) Supramolecular inorganic chemistry: small guests in small and large hosts. *Angew. Chem.*, **107**, 2505–2539; (1995); *Angew. Chem., Int. Ed. Engl.*, **34**, 2328–2361.

8 Nakano, K., Sada, K., Nakagawa, K., Aburaya, K., Yoswathananont, N., Tohnai, N. and Miyata, M. (2005) Organic intercalation material: reversible change in interlayer distances by guest release and insertion in sandwich-type inclusion crystals of cholic acid. *Chem. Eur. J.*, **11**, 1725–1733.

9 Nakano, K., Sada, K., Kurozumi, Y. and Miyata, M. (2001) Importance of packing coefficients of host cavities in the isomerization of open host frameworks: guest-size-dependent isomerization in cholic acid inclusion crystals with monosubstituted benzenes. *Chem. Eur. J.*, **7**, 209–220.

10 (a) McKinnon, J.J., Mitchell, A.S. and Spackman, M.A. (1998) Hirshfeld surfaces: a new tool for visualising and exploring molecular crystals. *Chem. Eur. J.*, **4**, 2136–2141; (b) McKinnon, J.J., Spackman, M.A. and Mitchell, A.S. (2004) Novel tools for visualizing and exploring intermolecular interactions in molecular crystals. *Acta Crystallogr. B*, **60**, 627–668.

11 Kato, K., Aburaya, K., Miyake, Y., Sada, K., Tohnai, N. and Miyata, M. (2003) Excellent enantio-selective enclathration of (2R, 3S)-3-methyl-2-pentanol in channel-like cavity of 3-epideoxycholic acid, interpreted by the four-location model for chiral recognition. *Chem. Commun.*, 2872–2873.

12 Aburaya, K., Hisaki, I., Tohnai, N. and Miyata, M. (2007) Dependence of the enantioselectivity on reversion of layer directions in cholamide inclusion compounds. *Chem. Commun.*, 4257–4259.

13 Mesecar, A.D. and Koshland, D.E. Jr, (2000) Structural biology: a new model for protein stereospecificity. *Nature*, **403**, 614–615.

14 Mizobe, Y., Tohnai, N., Miyata, M. and Hasegawa, Y. (2005) Tunable solid-state fluorescence system consisted of organic salts of anthracene-2, 6-disulfonic acid with primary amines. *Chem. Commun.*, 1839–1841.

15 Mizobe, Y., Ito, H., Hisaki, I., Miyata, M., Hasegawa, Y. and Tohnai, N. (2006) A novel strategy for fluorescence enhancement in the solid-state: affording rigidity to fluorophores packing. *Chem. Commun.*, 2126–2128.

16 Mizobe, Y., Miyata, M., Hisaki, I., Hasegawa, Y. and Tohnai, N. (2006) Anomalous anthracene arrangement and the corresponding rare excimer emission in the solid-state by transcription and translation of molecular information. *Org. Lett.*, **8**, 4295–4298.

17 Tohnai, N., Mizobe, Y., Doi, M., Sukata, S., Hinoue, T., Yuge, T., Hisaki, I., Matsukawa, Y. and Miyata, M. (2007) Well-designed Supramolecular Clusters Comprising Triphenylmethylamine and Various Sulfonic Acids. *Angew. Chem.*, **119**, 2270–2273; (2007) *Angew. Chem. Int. Ed.*, **46**, 2220–2223.

18 Kitaigorodskii, A.I. (1973) *Molecular Crystals and Molecules*, Academic Press, London.

19 Hahn, T. (ed.) (1983) *International Tables for Crystallography, Vol. A, Space-Group Symmetry*, Kluwer Academic Publishers, London, pp. 6048–6055.

20 Tanaka, A., Hisaki, I., Tohnai, N. and Miyata, M. (2007) Supramolecular tilt-chirality derived from symmetric benzene molecules: handedness of the 2_1 helical assembly. *Chem. Asian J.*, **2**, 230–238.

21 Hisaki, I., Tohnai, N. and Miyata, M. (2008) Supramolecular tilt chirality in crystals of steroids and alkaloids. *Chirality*, **20**, 330–336.

22 Watabe, T., Hisaki, I., Tohnai, N. and Miyata, M. (2007) Four kinds of 2_1 helical assemblies with the molecular tilt as well as three-directional and facial chirality. *Chem. Lett.*, **36**, 234–235.

23 Tanaka, A., Inoue, K., Hisaki, I., Tohnai, N., Miyata, M. and Matsumoto, A. (2006) Supramolecular chirality in layered crystals of achiral ammonium salts and fatty acids: a hierarchical interpretation. *Angew. Chem.*, **118**, 4248–4251; (2006) *Angew. Chem. Int. Ed.*, **45**, 4142–4145.

24 (a) Yuge, T., Tohnai, N., Fukuda, T., Hisaki, I. and Miyata, M. (2007) Topological study of pseudo-cubic hydrogen bond networks in a binary system composed of primary ammonium carboxylates: an analogue of ice cube. *Chem. Eur. J.*, **13**, 4163–4168; (b) Yuge, T., Sakai, T., Kai, N., Hisaki, I., Miyata, M. and Tohnai, N. (2008) Topological classification and supramolecular chirality of 2_1-helical ladder-type hydrogen-bond networks composed of primary ammonium carboxylates: bundle controls in 2_1-helical assemblies. *Chem. Eur. J.*, **14**, 2984–2993.

25 Desiraju, G.R. (1995) Supramolecular synthons in crystal engineering—a new organic synthesis. *Angew. Chem., Int. Ed. Engl.*, **34**, 2311–2327.

27
Morphology Changes of Organic Crystals by Single-Crystal-to-Single-Crystal Photocyclization
Hideko Koshima

27.1
Introduction

Since the study of [2 + 2] photocyclization of *trans*-cinnamic acids in 1964 [1], a large number of crystalline state reactions have been reported [2–6]. The relationship between the crystal structures and the reactions has been elucidated. Crystalline state photoreactions should give rise to molecular motion in the crystal lattice and thereby changes in surface morphology. However, the morphological changes in crystals during the reactions have never been intensively studied. Recently, actuators based on molecular crystals that change morphology in response to light have attracted interest. For the development of crystal actuators, it is necessary to elucidate the correlation between the structural changes and the morphological changes.

It was first reported that the topochemical photopolymerization of diolefin crystals gave rise to cracks and deformation [7]. An atomic force microscopic (AFM) study made possible the observation that the photodimerizations of *trans*-cinnamic acids and anthracenes in the crystalline state induced surface morphological changes at the tens and hundreds of nanometers level by the transportation and rebuilding of the surface molecules [8]. The appearance of a surface relief grating on the single crystal of 4-(dimethylamino)azobenzene was demonstrated by repeated irradiation with two coherent laser beams [9].

A recent study of the mechanical motion of photochromic diarylethene crystals is outstanding [10–14]. New steps appeared on the crystal surface on irradiation with UV light and disappeared with visible light [10]. The square single crystals changed reversibly to the lozenge shape upon irradiation and the rectangular crystals reversibly contracted and expanded [11]. The rapid and reversible bending of the crystalline rod on irradiation gave us the visual imagination for the development of crystal actuators [11]. The rolling thin crystals and the formation of microfibrils were reported [12, 13]. Furthermore, the diarylethene microcrystals made directional jumps upon UV irradiation [14]. Other jumping crystals induced by thermal phase transition are also known, *myo*-inositol [15, 16], hexadecahydropyrene [17, 18], and inorganic spinels [19].

Isopropylbenzophenone derivatives are well known compounds to undergo Norrish type II photocyclization in the crystalline state on UV irradiation [20]. Asymmetric induction is also possible by using the salt crystals of carboxylic acids with the chiral amines [21–24]. Furthermore, the absolute asymmetric photocyclization was achieved using the chiral salt crystals self-assembled from diisopropylbenzophenone derivative and achiral amine [25]. Some of the crystals reacted without any cracks or break due to small changes in the crystal structures, which involved single-crystal-to-single-crystal transformation [22–27]. For correlation between morphological changes and structural changes, single-crystal-to-single-crystal reactions are indispensable because the changes in morphology can be continuously determined throughout the reactions and the reaction processes can be traced by X-ray crystallographic analysis of the single crystals.

We would like to introduce our recent study of the morphological changes in isopropylbenzophenone crystals via single-crystal-to-single-crystal photocyclization. First, the salt crystals gave rise upon photoirradiation to unevenness on the surfaces and returned to the initial smooth surfaces after completion of the reaction [28, 29]. Second, the bulk crystals of the triisoproylbenzophenone derivative broke on irradiation despite the single-crystal-to-single-crystal transformation, but the microcrystals maintained the single-crystalline morphology during the course of the photocyclization [26].

27.2
Surface Morphology Changes in the Salt Crystals of a Diisopropylbenzophenone Derivative with Amines via Single-Crystal-to-Single-Crystal Photocyclization

27.2.1
Solid-State Photocylization

The salt crystals of **1**·(*S*)-**2** of 4-(2,5-diisopropylbenzoyl)benzoic acid **1** with (*S*)-phenylethylamine (*S*)-**2** were prepared by recrystallization from a methanol solution of both components (Scheme 27.1a). The crystals underwent enantiospecific photocyclization on irradiation at >290 nm with a high-pressure mercury lamp through Pyrex glass under argon at 288 K to give a cyclobutenol (*S*)-**4** in almost quantitative optical yield and 100% chemical yield [23].

The chiral crystals were spontaneously obtained from the solutions of **1** and 2,4-dichlorobenzylamine **3** (Scheme 27.1b). This kind of chiral crystallization of achiral molecules leads neccessarily to left- and right-handed crystals. The enantiomeric crystals of *M*- and *P*-**1**·**3** were selectively prepared by seeding. Irradiation of the crystals of *M*-**1**·**3** at around 350 nm through a UV filter gave a cyclopentenol (*R,R*)-**5** as almost the sole product, achieving absolute asymmetric photocyclization [25]. In contrast, irradiation at >290 nm afforded (*R,R*)-**5**, (*R*)-**4**, and (*R*)-hydrol in a 6:3:1 molar ratio; the enantiomeric excesses of the three products were higher than

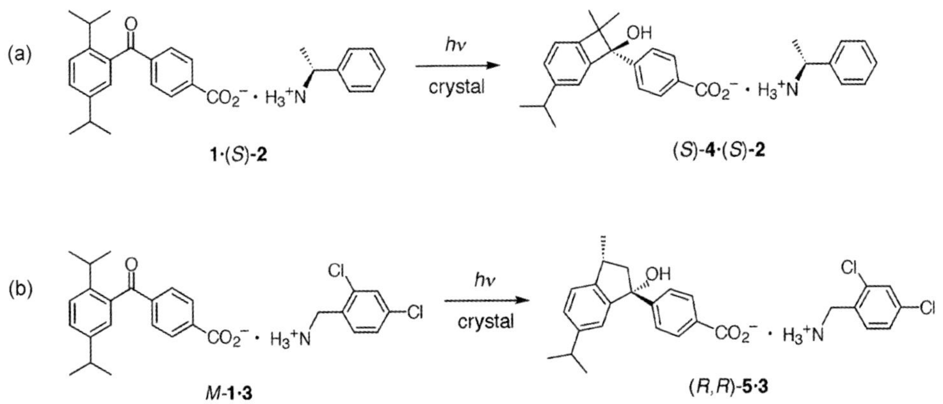

Scheme 27.1 (a) Enantiospecific and (b) absolute asymmetric photocyclization via single-crystal-to-single-crystal transformation.

80% ee at about 90% conversion, that is, showing lower product selectivity than the irradiation at around 350 nm.

27.2.2
Crystal Structures and the Reaction Mechanism

Photoirradiation of a single crystal of **1·(S)-2** maintained the initial transparency, confirming the single-crystal-to-single-crystal transformation. Finally, a piece of single crystal (1.28 × 0.21 × 0.09 mm^3) of **1·(S)-2** was submitted to X-ray crystallographic analysis at 293 K before and after successive irradiation at >290 nm at 293 K [23]. The absolute structure was determined on the basis of the S configuration of the phenylethylamine molecule **2**. The reaction proceeded smoothly and was completed after irradiation for 45 min. The crystal data are summarized in Table 27.1.

Table 27.1 Unit cell constants before and after photoirradiation.

	1·(S)-2			M-1·3		
	Before irrad.	Irrad. for 45 min (S)-4·(S)-2		Before irrad.	Irrad. for 712 h[a]	
			(change, %)			(change, %)
Meas. temp./K	293	293		293	293	
Space group	$P2_12_12_1$	$P2_12_12_1$		$P2_12_12_1$	$P2_12_12_1$	
a/Å	6.1890(5)	6.3312(3)	(+2.29)	6.3028(5)	6.235	(−1.08)
b/Å	14.411(1)	13.776(1)	(−4.41)	12.7985(9)	13.146	(+2.71)
c/Å	28.732(2)	28.510(2)	(−0.77)	31.838(4)	31.913	(+0.24)
V/Å^3	2562.6(4)	2486.6(3)	(−2.97)	2568.2(4)	2616.0	(+1.84)
D_c/g cm^{-1}	1.119	1.153	(+3.04)	1.258	1.235	(−1.83)

[a]Product: reactant = (R,R)-5: 1 = 53: 47.

The sizes of the unit cells changed slightly, increasing (+2.29%) along the a-axis, decreasing (−4.41%) along the b-axis, and decreasing (−0.77%) along the c-axis, resulting overall in a decrease (−2.97%) in total cell volume.

The ORTEP drawing of a salt bond pair arranged in the reactant **1**·(S)-**2** is shown in Figure 27.1a. Irradiation of the reactant **1**·(S)-**2** causes n–π* excitation of the C1=O1 carbonyl group of the benzophenone unit. The O1 atom abstracts the methine H1 hydrogen atom of the o-isoproyl group in the highest priority due to the shortest O1···H1 distance (3.15 Å) to give the ketyl radical •C1 and the methine radical •C2. Another hydrogen abstraction from the m-isoproyl group does not occur due to the

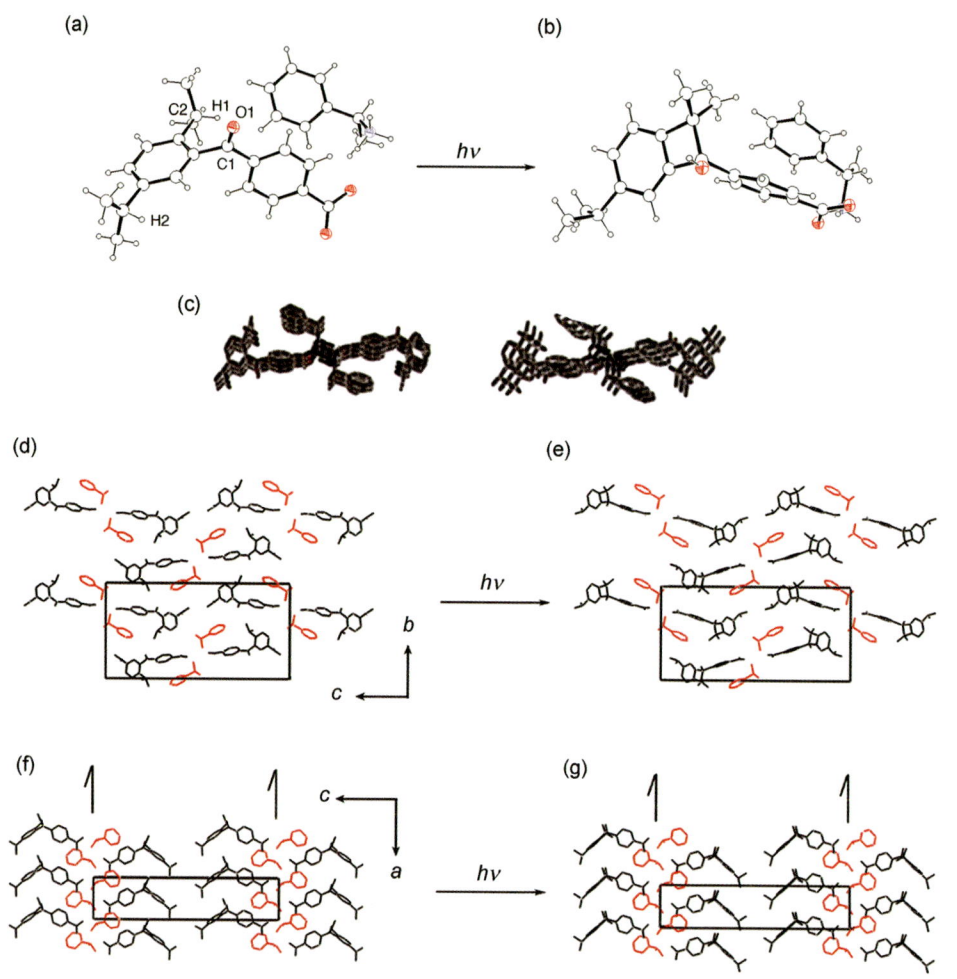

Figure 27.1 ORTEP drawings of (a) the reactant **1**·(S)-**2** and (b) the product (S)-**4**·(S)-**2** at 25% probability level, (c) stereoview of **1**·(S)-**2**, and the molecular arrangements on the (100) (d and e) and (010) faces (f and g) in **1**·(S)-**2** and (S)-**4**· (S)-**2**, respectively. Note that the methine H1 hydrogen atom in (a) is projected towards the carbonyl O1 oxygen atom for abstraction to yield the cyclobutenol (S)-**4**. (From Ref. [28] with permission. © 2008 Am. Chem. Soc.)

long O1···H2 distance (5.19 Å). Subsequently, the •C1 and •C2 biradicals are enantiospecifically coupled to produce the cyclobutenol (S)-4 (Figure 27.1b).

Such a photocyclization, however, did not greatly change the molecular conformation. The ionic bridge between the carboxylate anion of **1** and the ammonium cation of (S)-**2** forms a twofold helical chain in the reactant crystal along the a-axis. The stereoview and the molecular arrangements in the (100) and (010) planes are shown in Figure 27.1c, d, and f, respectively. The helical salt bridges remain after the reaction and are arranged in a similar manner in the product crystal (Figure 27.1e and g). The strong salt bonding is believed to fix the molecules and promote complete reaction without destroying the crystal, leading to single-crystal-to-single-crystal transformation.

The single crystals of **1·3** were also transparent after irradiation, confirming the single crystalline phase reaction. The single crystal of M-**1·3** was successively irradiated at around 350 nm with a superhigh-pressure mercury lamp through a UV transparent filter. The reaction proceeded very slowly and was not complete even after irradiation for 712 h (Table 27.1). The structure was solved as a disordered structure of the product (R,R)-**5** and the remaining reactant **1** in 53 : 47 occupancy. The sizes of the unit cells changed slightly, decreasing (-1.08%) along the a-axis, increasing ($+2.71\%$) along the b-axis, and increasing ($+0.24\%$) along the c-axis, resulting overall in an increase ($+1.84\%$) in total cell volume. The molecular arrangement was similar to that of **1·(S)-2**; a twofold helical chain through the salt bridge between the carboxylate anion of **1** and the ammonium cation of **3**. The reaction path to give (R,R)-**5** could be explained based on the crystal structure [26].

27.2.3
Morphology Changes in Bulk Crystals

The morphological changes in the salt crystals of **1·(S)-2** on photoirradiation were determined by AFM [28]. The single crystals prepared by evaporating the methanol solutions of both components at room temperature gave needle-shaped crystals grown along the a-axis (Figure 27.2A). The (001) face was the most developed, followed by the (010) face. The (100) face was cut perpendicular to the (001) and (010) faces with a razor blade to afford a piece of single crystal ($210 \times 209 \times 68 \, \mu m^3$). The crystal was then irradiated at >290 nm with a high-pressure mercury lamp through Pyrex glass under argon at 288 K.

Figure 27.2B shows the AFM images of the changes in surface morphology during the reaction. Before irradiation, the (001) face was flat (a). Upon UV irradiation, numbers of hemispheric unevennesses appeared on the (001) surface and these gradually grew to heights of tens of nanometers after irradiation for 20 min (b). The relative unevenness was less than 0.01% of the crystal thickness (68 μm), indicating that it was limited to the crystal surface. Prolonged irradiation for 100 min resulted in merger of the hemispheres and a return to the initial smooth surface (c). Similar morphological changes appeared on the (010) face on photoirradiation (d–f). In contrast, the (100) face was very rough with heights of several nanometers due

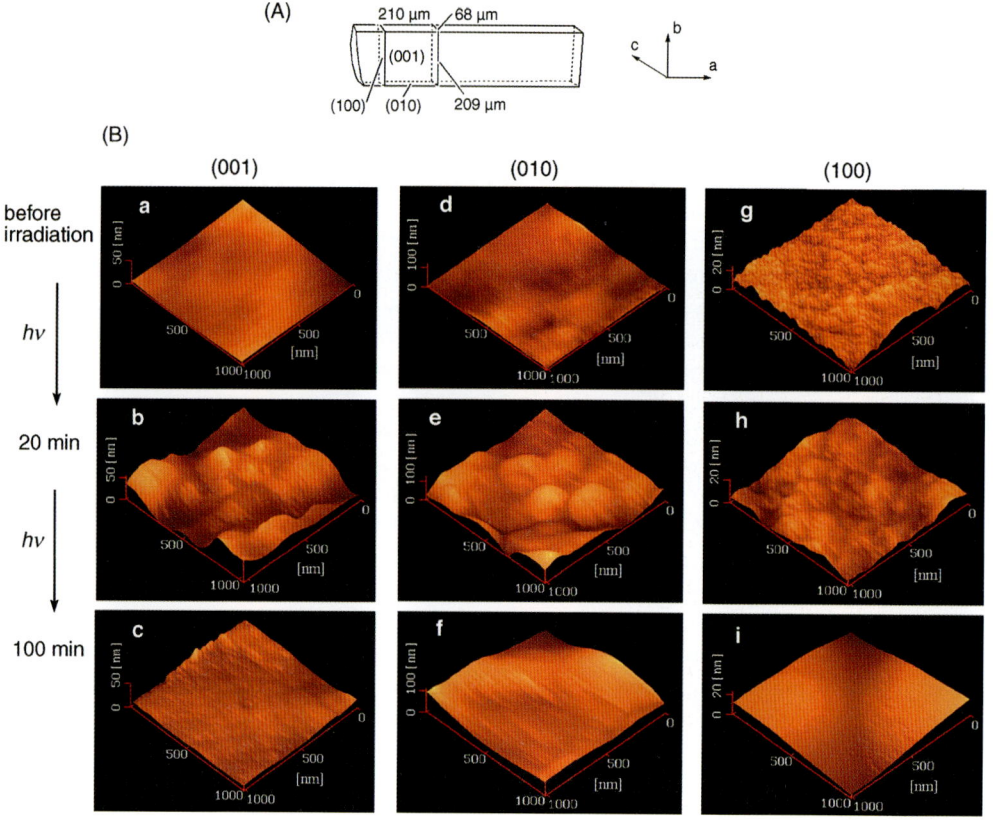

Figure 27.2 (A) Crystal habit of **1·(S)-2** and (B) the AFM images of the morphological changes on UV irradiation for 0, 20, and 100 min: (a–c) (001), (d–f) (010), and (g–i) (100) faces. (From Ref. [28] with permission. © 2008 Am. Chem. Soc.)

to the cut with the razor blade (g). UV irradiation decreased the roughness (h) and finally gave a very smooth surface (i).

The completion of reaction after irradiation for 100 min was confirmed by HPLC analysis. The smooth progress of the reaction is consistent with the result obtained in the X-ray crystallographic analysis of a single crystal of **1·(S)-2** (1.28 × 0.21 0.09 mm) in which irradiation for 45 min gave the product crystal (S)-**4**·(S)-**2** (Table 27.1).

27.2.4
Morphology Changes in Microcrystals

Next, the microcrystals of **1·(S)-2** were submitted to UV irradiation [29]. The microcrystals of **1·(S)-2** were prepared by dropping equimolar solutions of both components in methanol onto quartz plates and evaporating the solvents to give the long needle-shaped crystals (Figure 27.3A). The top surface was determined to be

27.2 Surface Morphology Changes in the Salt Crystals | 533

A

10 μm

B

X
Y

C: (001) surface

a
b
c
d
e

D: X–Y section

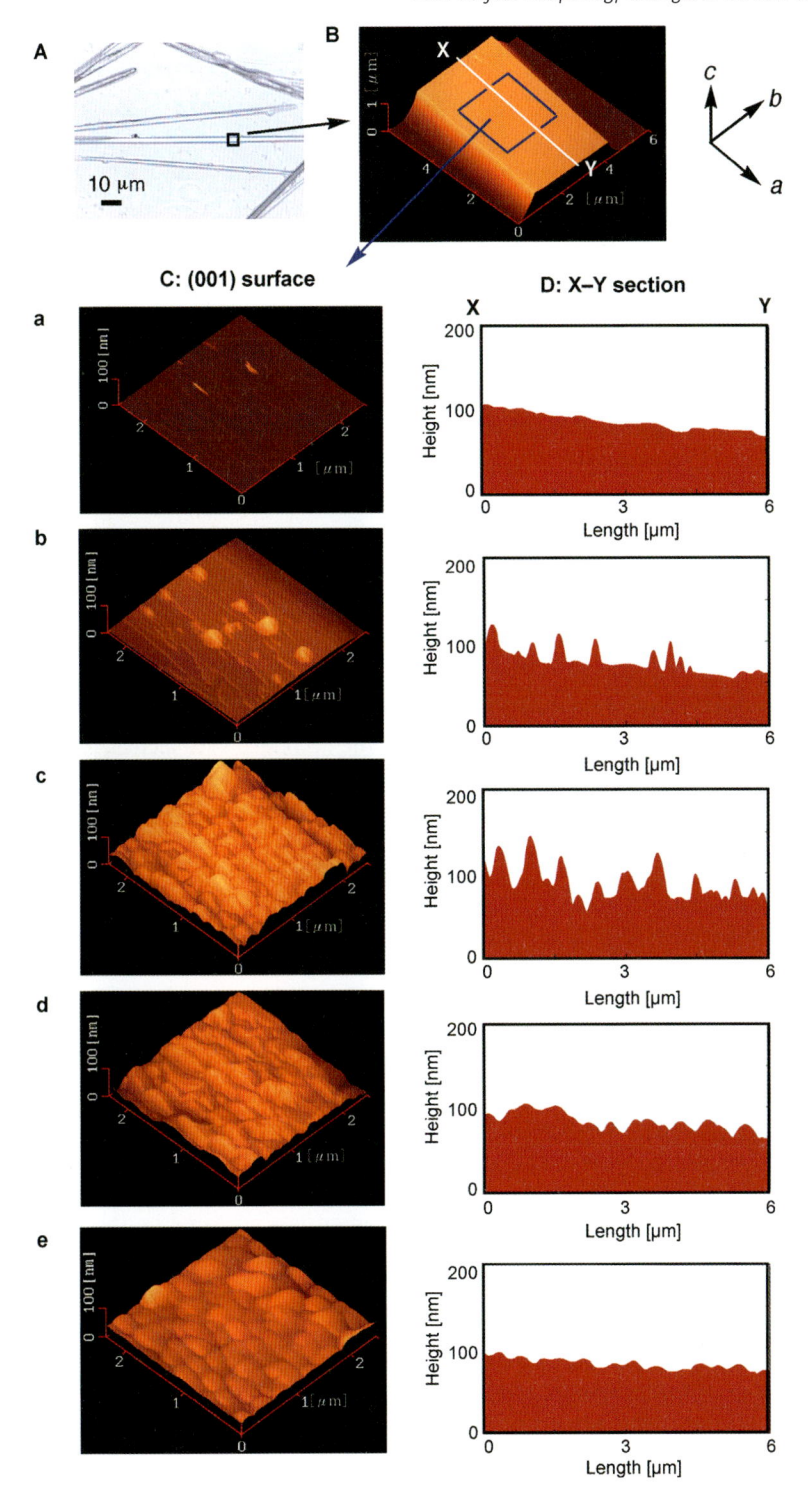

a
b
c
d
e

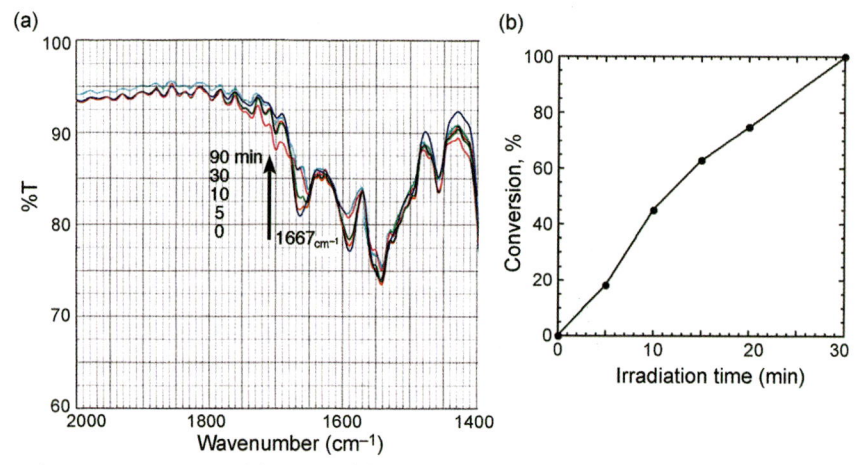

Figure 27.4 (a) IR spectral changes and (b) the conversion of the photocyclization of **1** in the microcrystals of **1**·(S)-**2** on UV irradiation.

the (001) face from comparison with the needle-shaped bulk crystals grown along the a-axis (Figure 27.2A). The photocyclization process was monitored by FTIR spectroscopy. Figure 27.4a shows the change in the IR spectrum under UV irradiation at >290 nm. The intensities of the reactant **1**·(S)-**2** band at 1667 cm^{-1}, the stretching vibration due to the benzophenone carbonyl group, decreased with increasing UV irradiation time. Irradiation for longer than 30 min did not change the IR spectrum, showing that the photocyclization was complete within 30 min. The conversion was calculated based on the intensity change in the absorption at 1667 cm^{-1} to give an almost linear relationship with the irradiation time (Figure 27.4b).

The morphological changes in the microcrystal of **1**·(S)-**2** during the photocyclization were observed by AFM (Figure 27.3C and D). Before irradiation, the (001) surface was flat (a). Upon UV irradiation at >290 nm, unevennesses appeared on the surface (b), and these gradually grew to a height of 50–70 nm at 75% conversion after 20 min (c); the relative unevenness was about 7–10% for the crystal thickness (680 nm). Irradiation for 30 min completed the reaction and decreased the height of unevennesses to about 20 nm (3% relative unevenness) (d). Prolonged irradiation for 90 min further decreased the unevenneses to several nanometers (1% relative unevenness), returning to a nearly smooth surface (e). The whole crystal shape did not change much before and after irradiation. In the case of the bulk crystal (68 μm thickness), similar unevennesses with heights of tens of nanometers appeared on the (001) surface (<0.01% relative unevenness) (Figure 27.2b). The similar magnitude of

Figure 27.3 (A) Microcrystals of **1**·(S)-**2**, (B) the AFM image before irradiation, and the changes of (C) the (001) surface morphology and (D) the X–Y section on photoirradiation for (a) 0 min, (b) 10 min (45% conversion), (c) 20 min (75% conversion), (d) 30 min (100% conversion), and (e) 90 min (100% conversion).

unevennesses between the microcrystal and the bulk crystal revealed that the morphological changes occurred only near the crystal surfaces.

The dependence on the light wavelength of the surface morphological changes of the (001) face of the microcrystal of **1·(S)-2** was also checked. Irradiation at around 350 nm with a high-pressure mercury lamp through a UV filter gave a slower reaction than that at >290 nm due to the decrease in light intensity. Similar unevennesses appeared on the (001) surface and these gradually grew to a height of 15 nm at 67% conversion after 70 min; about 2% relative unevenness for the crystal thickness (830 nm). Irradiation for 110 min completed the reaction, increasing the height of the unevennesses to about 25 nm (3% relative unevenness). On irradiation for 180 min the unevennesses grew large,r to about 40 nm (5% relative unevenness). Prolonged irradiation for 300 min at last decreased the unevenneses to about 30 nm (4% relative unevenness). Thus, irradiation at around 350 nm delayed the appearance and disappearance of unevennesses.

The morphological changes in the microcrystals of **1·3** during the photocyclization were also determined by AFM. The microcrystals of **1·3** prepared by dropping equimolar solutions of both components in ethanol onto quartz plates and evaporating the solvents gave long needle-shaped crystals, similar to those of **1·(S)-2** in Figure 27.3A. The top surface was determined to be the (001) face from comparison with the needle-shaped bulk crystals grown along the a-axis. Figure 27.5 shows the morphological changes in the microcrystals of **1·3** during photocyclization. Before irradiation, the (001) face was nearly flat (a). Upon UV irradiation at >290 nm, unevennesses appeared on the (001) surface and these gradually grew to a height of about 15 nm at 66% conversion after 12 h (b); the relative unevenness was about 2% of the crystal thickness (700 nm). Irradiation for 25 h completed the reaction and slightly decreased the height of unevennesses to about 10 nm (1.5% relative unevenness) (c). Prolonged irradiation for a further 40 h decreased the unnevennesses to 2 nm (0.7% relative unevenness) (d). The whole crystal shape did not change much before and after irradiation.

27.2.5
Correlation between the Morphology Changes and the Crystal Strucural Changes

The crystalline state reaction proceeds heterogeneously from the surface to the inside on UV irradiation. The X-ray crystal data (Table 27.1) of **1·(S)-2** before and after irradiation reveal that the crystal expands slightly along the a-axis and contracts along the b- and c-axes, without destroying the crystal. Therefore, some stress should be induced within the crystal lattice. The helical salt bond chains are strong, like molecular springs. In contrast, the intermolecular interaction among the neighboring salt bond chains is weak due to van der Waals forces alone, as shown in Figure 27.1d and f, suggesting that the (001) plane can be cleaved in a thin layer. Each molecular chain near the (001) surface can move along the a-axis, which leads to the appearance of hemispheric unevennesses on the (001) surface shown in Figures 27.2b and 27.3b–d. When the reaction is complete, the inner stress disappears, and an essentially smooth surface is restored (Figures 27.2c and 27.3e).

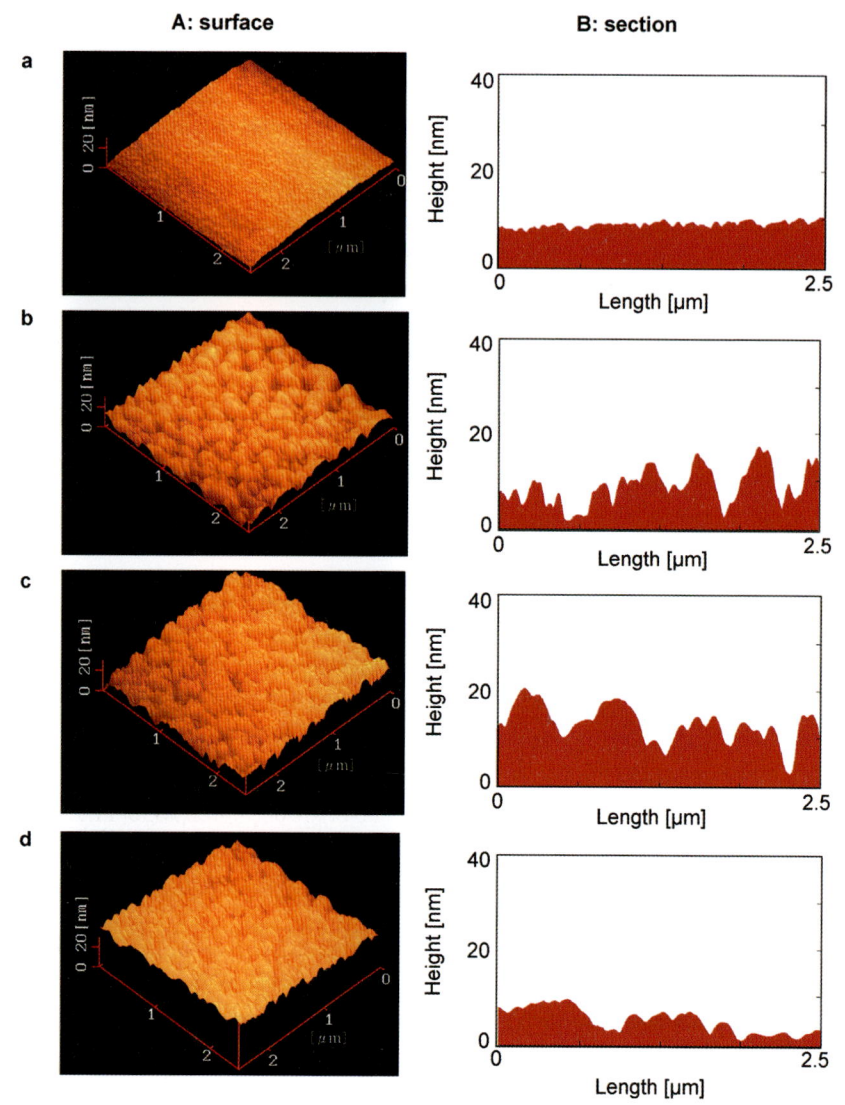

Figure 27.5 (A) AFM images of the changes in surface morphology of the microcrystal of **1·3** and (B) the section on photoirradiation for (a) 0 h, (b) 12 h (66% conversion), (c) 25 h (100% conversion), and (d) 40 h (100% conversion).

A similar explanation can be applied for the morphological changes of the (010) surface (Figure 27.2d–f).

In contrast, the roughness of the (100) face of **1·(S)-2** decreased on irradiation and the surface ultimately became flat (Figure 27.2g–i). The experimental results provide the visual evidence that the molecules on the (100) surface moved inside or outside the crystal leading to a decrease in the surface energy to a minimum, resulting in the formation of a very smooth surface. This is like photochemical annealing of a crystal.

To evaluate the molecular mobility during the morphological change, the melting point change was measured. The melting point of the microcrystals of **1**·(S)-**2** before photoirradiation was 163–164 °C, decreasing to a minimum of 92–106 °C on irradiation due to the coexistence of the reactant **1**·(S)-**2** and the product (S)-**4**·(S)-**2** in the crystals. The melting point then reached an almost constant value of 102–107 °C due to the formation of (S)-**4**·(S)-**2** alone at the completion of photocyclization. The considerable decrease in the melting point should facilitate the movement of molecules in the photochemical process.

The same explanation for the morphological changes in the salt crystal of **1**·**3** is applicable because the molecular arrangement is similar to that of **1**·(S)-**2**.

27.3
Morphology Changes in Triisobenzophenone Crystals via Diastereospecific Single-Crystal-to-Single-Crystal Photocyclization

27.3.1
Solid-State Photocyclization and the Crystal Structures

Crystals of 2-(2,4,6-triisopropylbenzoyl)((S)-1-phenylethyl)benzamide (S)-**6** were irradiated at >290 nm with a high-pressure mercury lamp through Pyrex glass under argon at 293 K for 8 h to afford a cyclobutenol (R, S)-**7** in 100% chemical yield and 100% de, revealing the diastereospecific photocyclization in the crystalline state (Scheme 27.2) [26]. In contrast, the solution photolysis of (S)-**6** in acetonitrile gave (R, S)- and (S, S)-**7** in 24% and 21% chemical yield, respectively, revealing the low diastereoselectivity of only 6% de.

X-Ray crystallographic analyses of the reactant and the product were carried out to elucidate the reaction mechanism for the diastereospecific photocyclization. Irradiation of the single crystal of (S)-**6** at >290 nm with a high-pressure mercury lamp led to breakage of the crystal. Therefore, the reaction was not thought to be single-crystal-to-single-crystal transformation. However, careful irradiation at around 350 nm through a UV filter for 180 h completed the reaction without any cracking so that the X-ray structure analysis was successful [27]. Table 27.2 summarizes the crystal data. The size of the unit cells changed slightly, increasing

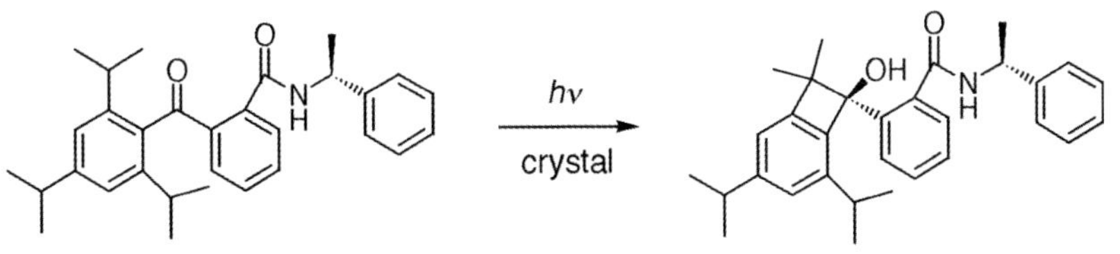

Scheme 27.2 Diastereospecific single-crystal-to-single-crystal photocyclization.

27 Morphology Changes of Organic Crystals by Single-Crystal-to-Single-Crystal Photocyclization

Table 27.2 Selected crystal data before and after photoirradiation.

	(S)-6		
	Before irrad.	Irrad. for 180 h (R,S)-7	
			(change, %)
Meas. temp./K	173	173	
Space group	$P2_12_12_1$	$P2_12_12_1$	
a/Å	9.5739(16)	9.742(2)	(+1.76)
b/Å	13.154(3)	13.735(4)	(+4.42)
c/Å	21.267(5)	20.009(6)	(−5.92)
V/Å^3	2678.3(10)	2677.4(13)	(−0.03)
D_c/g cm^{-3}	1.130	1.130	(0.00)

(+1.76%) along the a-axis, increasing (+4.42%) along the b-axis, and decreasing (−5.92%) along the c-axis, resulting overall in no change (0.0%) in the total cell volume.

ORTEP drawings of the reactant (S)-6 and the product (R, S)-7 are shown in Figure 27.6a and b, respectively. The molecular conformations are very similar. In the

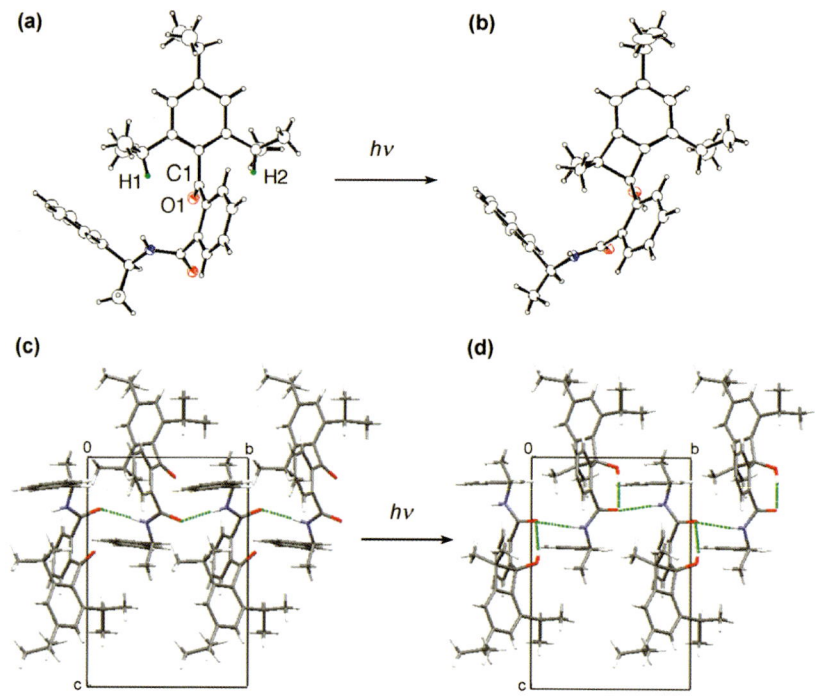

Figure 27.6 ORTEP drawings of (a) (S)-6 and (b) (R,S)-7 at 15% probability level, and the molecular arrangements in (c) (S)-6 and (d) (R,S)-7.

crystal of (S)-**6**, a twofold helical chain is formed among the amide groups through the N−H···O=C hydrogen bond (2.12 Å) along the b axis (Figure 27.6c). A similar twofold helical chain is also formed in the product crystal of (R, S)-**7** among the amide groups through the N−H···O=C hydrogen bond (2.32 Å) along the b axis, and a further intramolecular O−H···O=C hydrogen bond (1.86 Å) is formed between the hydroxy group and the amide group.

Irradiation of the crystals of (S)-**6** at >290 nm with a high-pressure mercury lamp through Pyrex glass causes n−π^* excitation of the carbonyl group of the (S)-**6** molecule (Figure 27.6a). The O1 atom has the possibility of abstraction of both the methine H1 and H2 hydrogen atoms of two o-isopropyl groups to produce the ketyl radical and corresponding methine radicals, because both distances of O1···H1 (2.84 Å) and O1···H2 (2.97 Å) are short enough and similar, and the angles of C1−O1−H1 (57.1°) and C1−O1−H2 (53.2°) are also similar. Then, the radicals should approach each other and finally couple to afford (R)- and (S)-cyclobutenol, that is, both the diastereomeric products (R,S)-**7** (Figure 27.6b) and (S,S)-**7**. However, only (R,S)-**7** was obtained by the crystalline state photocyclization. This means that the occurrence of the diastereospecific O1···H1 hydrogen abstraction could not be explained from the distance and angle parameters between the carbonyl oxygen atom and the two γ-hydrogen atoms. Recent intensive studies have revealed that crystalline state reactions proceed generally with the minimum molecular motion in the crystal lattice because the molecules are arranged at close positions in three-dimensional regularity and the motion is very restricted [2–6]. As shown in Figure 27.6, the similarity of the molecular shapes as well as the twofold helical arrangements in the crystals suggests that the transformation from (S)-**6** to (R, S)-**7** can proceed smoothly with minimum motion within the helical chains, leading to the single-crystal-to-single-crystal reaction.

27.3.2
Morphology Changes

Despite the single-crystal-to-single-crystal transformation of (S)-**6**, photoirradiation of the bulk crystals led to cracking after several minutes (Figure 27.7). The following

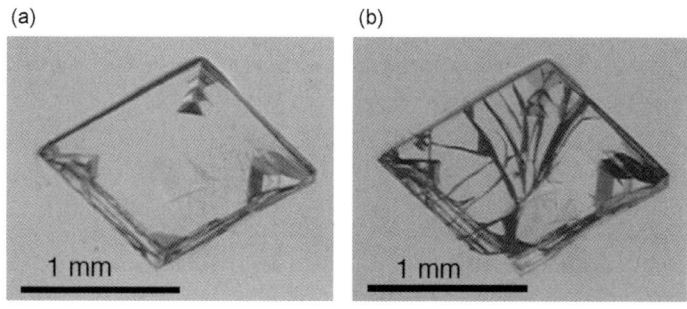

Figure 27.7 Morphology change in the bulk crystal of (S)-**6** (a) before and (b) after UV irradiation. The top surface is the (100) plane.

explanation is possible. The photocyclization proceeds from the surface to the inside of the bulk crystal. Therefore, upon UV irrradiation, the (100) surface expands slightly along the b-axis and contracts along the c-axis, as understood from the crystal data in Table 27.2, and the strain is generated within the crystal lattice leading to cracking and breaking into polycrystals.

A recent report encouraged us that in the topochemical polymerization by successive [2 + 2] photocyclization of diolefin derivatives, the bulk crystals were broken into fragments during polymerization but the nanocrystals maintained the single-crystalline phase in the course of polymerization [30]. We tried UV irradiation of the microcrystals of (S)-**6**. The microcrystals were prepared by vaporizing the powder samples on a heater at slightly lower temperature than the melting point, collecting the vapor on a quartz plate, and crystallizing at room temperature (Figure 27.8a). The photocyclization process was monitored by FTIR spectroscopy. The intensities of the reactant (S)-**6** band at 1678 and 1640 cm^{-1} for the stretching vibration due to the benzophenone carbonyl group and the amide carbonyl group, respectively, decreased with increasing irradiation time. Conversely, the intensity of the product (R,S)-**7** band at 1609 cm^{-1} for the stretching vibration due to the amide carbonyl group increased. Irradiation for longer than 10 min did not change the IR spectrum, revealing that the photocyclization was complete within 10 min. The conversion was calculated based on the intensity change in the absorption at 1678 cm^{-1} to give an almost linear relationship with irradiation time.

Figure 27.8b shows the AFM image of a piece of microcrystal before irradiation. The top surface is the (100) plane. In contrast to the bulk crystals, the microcrystal maintained the single-crystalline phase during photocyclization (Figure 27.8c and d).

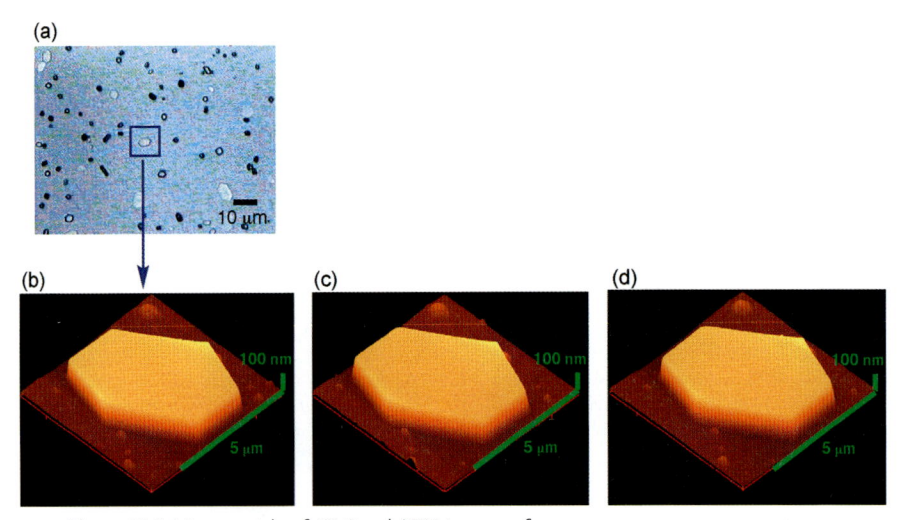

Figure 27.8 Microcrystals of (S)-**6** and AFM images of the morphological change on UV irradiation: (a and b) before irradiation, (c) irradiation for 4 min (50% conversion), and (d) 10 min (100% conversion).

No cracks were found during or after the completion of reaction. The surface morphology and the shape of the microcrystal were the same as the reactant crystal. The crystal should expand slightly along the a- and b-axis and contract along the c-axis, but not change the volume (Table 27.2). Therefore, the inner strain accumulated in the microcrystals is so small that the reaction can be completed while keeping the initial single-crystalline shape.

27.4
Concluding Remarks

During single-crystal-to-single-crystal phototransformation, the salt crystals of the diisopropylbenzophenone derivative with amines changed their surface morphology on photoirradiation but returned to the initial smooth surfaces after completion of the reaction. We also found that the bulk crystals of the triisoproylbenzophenone derivative broke on UV irradiation, but the microcrystals maintained the single-crystalline morphology during photocyclization. The morphological changes could be explained based on the crystal structure changes. However, many subjects remain for the elucidation of the correlation between the motion of molecules and the morphological changes in bulk crystals. More intensive and basic research is necessary for the development of mechanical crystal devices.

Acknowledgments

This study was supported by the Grant-in-Aid for Scientific Research in Priority Area (Area No. 432, No. 17034047) from the Ministry of Education, Culture, Sports, Science and Technology of Japan and the Asahi Glass Foundation.

References

1 Cohen, M.D. and Schmidt, G.M. (1964) Topochemistry. Part VIII. The photochemisry of *trans*-cinnamic acids. *J. Chem. Soc.*, 1996–2000.
2 Desiraju, G.R. (ed.) (1987) *Organic Solid State Chemistry*, Elsevier, Amsterdam.
3 Ohashi, Y. (ed.) (1993) *Reactivity in Molecular Crystals*, Kodansha, VCH, Tokyo.
4 Tanaka, K. and Toda, F. (2000) Solvent-free organic synthesis. *Chem. Rev.*, **100**, 1025–1074.
5 Toda, F. (ed.) (2002) *Organic Solid-State Reactions*, Kluwer Academic Publishers, Dordrecht.
6 Inoue, Y. and Ramamurthy, V. (eds) (2004) *Chiral Photochemistry*, Marcel Dekker Inc., New York.
7 (a) Nakanishi, H., Hasegawa, M. and Sasada, Y. (1972) Four-center type photopolymerization in the crystalline state. V. X-Ray crystallographic study of the polymerization of 2,5-distyrylpyrazine. *J. Polym. Sci.*, A-2, **10**, 1537–1553; (b) Nakanishi, H., Hasegawa, M., Kirihara, H. and Yurugi, T. (1977) Various morphological changes in the solid-state photopolymerization of diolefinic compounds. *Nippon. Kagaku. Zasshi.*, 1046–1050.

8 (a) Kaupp, G. (1992) Photodimerization of cinnamic acid in the solid state: new insights on application of atomic force microscopy. *Angew. Chem. Int. Ed. Engl.*, **31**, 592–595; (b) Kaupp, G. (1992) Photodimerization of anthracenes in the solid state: new results from atomic force microscopy. *Angew. Chem. Int. Ed. Engl.*, **31**, 595–598; (c) Kaupp, G. and Plagmann, M. (1994) Atomic force microscopy and solid state photolyses: phase rebuilding. *J. Photochem. Photobiol. A: Chem.*, **80**, 399–407.

9 Nakano, H., Tanino, T. and Shirota, Y. (2005) Surface relief grating formation on a single crystal of 4-(dimethylamino) azobenzene. *Appl. Phys. Lett.*, **87**, 061910.

10 Irie, M. Kobatake, S. and Horichi, M. (2001) Reversible surface morphology changes of a photochromic diarylethene single crystal by photoirradiation. *Science*, **291**, 188–191.

11 Kobatake, S., Takami, S., Muto, H., Ishikawa, T. and Irie, M. (2007) Rapid and reversible shape changes of molecular crystals on photoirradiation. *Nature*, **446**, 778–781.

12 Uchida, K., Sukata, S., Matsuzawa, U., Akazawa, M., de Jong, J.J.D., Katsonis, N., Kojima, Y., Nakamura, S., Areephong, J., Meetsma, A. and Feringa, B.L. (2008) Photoresponsive rolling and bending of thin crystals of chiral diarylethenes. *Chem. Comm.*, 326–328.

13 Uchida, K., Izumi, N., Sukata, S., Kojima, Y., Nakamura, S. and Irie, M. (2006) Photoinduced reversible formation of microfibrils on a photochromic diarylethene microcrystalline surface. *Angew. Chem. Int. Ed.*, **45**, 6470–6473.

14 (a) Colombier, I., Spagnoli, S., Corval, A., Baldeck, P.L., Giraud, M., Leaustic, A. and Yu, P. (2005) Strong photomechanical effects in photochromic organic microcrystals. *Mol. Cryst. Liq. Cryst.*, **431**, 195–199; (b) Colombier, I., Spagnoli, S., Corval, A., Baldeck, P.L., Giraud, M., Leaustic, A., Yu, P. and Irie, M. (2007) Diarylethene microcrystals make directional jumps upon ultraviolet irradiation. *J. Chem. Phys.*, **126**, 011101/1–011101/3.

15 Gigg, J., Gigg, R., Payne, S. and Conant, R. (1987) The allyl group for protection in carbohydrate chemistry. Part 21. (±)-1,2:5,6- and (±)-1,2:3,4-Di-*O*-isopropylidene-*myo*-inositol. The unusual behaviour of crystals of (±)-3,4-Di-*O*-acetyl-1,2,5,6-tetra-*O*-benzyl-*myo*-inositol on heating and cooling: a 'thermosalient solid'. *J. Chem. Soc., Perkin Trans. I*, 2411–2414.

16 Steiner, T., Hnfried, H. and Saenger, W. (1993) 'Jumping crystals': X-ray structures of the three crystalline phases of (±)-3,4-Di-*O*-acetyl-1,2,5,6-tetra-*O*-benzyl-*myo*-inositol. *Acta Crystallogr. B*, **49**, 708–718.

17 Kohne, B., Praefcke, K. and Mann, G. (1988) Perhydropyren, ein beim phasenübergang hüpfender kohlenwasserstoff. *Chimia*, **42**, 139–141.

18 Ding, J., Herbst, R., Praefcke, K., Kohne, B. and Saenger, W. (1991) A crystal that hops in phase transition, the structure of *trans,trans,anti,trans,trans*-perhydropyrene. *Acta. Crystallogr. B*, **47**, 739–742.

19 Crottaz, O., Kubel, F. and Schmid, H. (1997) Jumping crystals of the spinels $NiCr_2O_4$ and $CuCr_2O_4$. *J. Mater. Chem.*, **7**, 143–146.

20 Ito, Y., Nishimura, H., Umehara, Y., Yamada, Y., Tone, M. and Matsuura, T. (1983) Photoinduced reaction. 142. Intramolecular hydrogen abstraction from triplet states of 2,4,6-triisopropylbenzophenones: importance of hindered rotation in excited states. *J. Am. Chem. Soc.*, **105**, 1590–1597.

21 Koshima, H., Maeda, A., Matsuura, T., Hirotsu, K., Okada, K., Mizutani, H., Ito, Y., Fu, T.Y., Scheffer, J.R. and Trotter, J. (1994) Ionic chiral handle-induced asymmetric synthesis in a solid state norrish type II photoreaction. *Tetrahedron: Asym.*, **5**, 1415–1418.

22 Hirotsu, K., Okada, K., Mizutani, H., Koshima, H. and Matsuura, T. (1996) X-ray study of asymmetric photoreaction of 2,4,6-triisopropyl-4′-benzophenone. *Mol. Cryst. Liq. Cryst.*, **277**, 99–106.

23 Koshima, H., Matsushige, D. and Miyauchi, M. (2001) Enantiospecific single crystal-to-single crystal photocyclization of 2,5-diisopropyl-4′-carboxybenzophenone in the salt crystals with (S)- and (R)-phenylethylamine. *Cryst.Eng.Comm.*, **33**, 1–3.

24 Koshima, H., Ide, Y., Fukano, M., Fujii, K. and Uekusa, H. (2008) Single-crystal-to-single-crystal photocyclization of 4-(2,4,6-triisopropylbenzoyl)benzoic acid in the salt crystal with (S)-phenylethylamine. *Tetrahedron Lett.*, **49**, 4346–4348.

25 Koshima, H., Kawanishi, H., Nagano, M., Yu, H., Shiro, M., Hosoya, T., Uekusa, H. and Ohashi, Y. (2005) Absolute asymmetric photocyclization of isopropylbenzophenone derivatives using cocrystal approach involving single-crystal-to-single-crystal transformation. *J. Org. Chem.*, **70**, 4490–4497.

26 Koshima, H., Fukano, M. and Uekusa, H. (2007) Diastereospecific photocyclization of isopropylbenzophenone derivative in crystals and the morphological changes. *J. Org. Chem.*, **72**, 6786–6791.

27 Fujii, K., Uekusa, H., Fukano, M. and Koshima, H., unpublished.

28 Koshima, H. Yuya, I. and Ojima, N. (2008) Surface morphology changes of a salt crystal of 4-(2,5-diisopropylbenzoyl)benzoic acid with (S)-phenylethylamine via single-crystal-to-single-crystal photocyclization. *Cryst. Growth Des.*, **8**, 2058–2060.

29 Koshima, H., Ide, Y., Yamazaki, S. and Ojima, N. (2009) Changes in the surface morphology of salt crystals of 4-(2,5-diisopropylbenzoyl)benzoic acid with amines via single-crystal-to-single-crystal photocyclization. *J. Phys. Chem. C*, **113**, 111683–111688.

30 Takahashi, S., Miura, S., Okada, S., Oikawa, H. and Nakanishi, H. (2002) Single-crystal-to-single-crystal transformation of diolefin derivatives in nanocrystals. *J. Am. Chem. Soc.*, **124**, 10944–10945.

Part Five
Single Biocells

28
Femtosecond Laser Tsunami Processing and Light Scattering Spectroscopic Imaging of Single Animal Cells

Hiroshi Masuhara, Yoichiroh Hosokawa, Takayuki Uwada, Guillaume Louit, and Tsuyoshi Asahi

28.1
Introduction

Spectroscopy and imaging have received much attention, not only for fundamental research in physics and chemistry but also for analyzing structures and functions of materials, devices, biological systems, and so on. Nowadays such studies utilizing spectroscopy and imaging are oriented toward bio and medical applications, and many researchers in life science, biotechnology, and medicine have introduced advances in laser and microscope technologies to their specific fields and proposed new measurement, processing, molecular and nanoparticle probes, and so on [1–5]. In general *in situ* analysis of dynamic structures, properties, and functionalities of living cells and tissues is performed by fluorescence and Raman spectroscopies, coherent anti-Stokes Raman spectroscopy(CARS), second harmonic generation, fluorescence lifetime measurement, and their imaging [6–11]. Fluorescent, Raman-active, and/or centro-symmetric molecules are necessary, but they are not always incorporated in living systems. Therefore new spectroscopic and imaging methods with higher spatial and temporal resolutions are needed for unraveling novel functionality.

Spectroscopy and imaging need suitable molecules and nanoparticles which can probe the properties and functions of components and organs in cells and tissues. When we do not add such probes, we consequently detect signals from the component molecules constituting living systems, for example, autofluorescence. The analysis is usually not easy as the photophysical and photochemical natures of the molecules in the systems are not clarified, and even molecular structures are mostly beyond our knowledge. Note that so many molecules, maybe more than 1000 kinds of molecules, are included in a single cell, meaning that identification of molecular structure should be the first step for understanding biological function in terms of molecules and molecular interactions. Thus for probing and understanding biological activity in living cells and tissues, many fluorescent molecules have been designed and synthesized [12] and, recently, metallic nanoparticles, gold nanorods, and semiconductor dots have been developed and applied [13]. Inorganic nanoparticles have the advantages of high emission and scattering efficiencies and of no bleaching upon long illumination and measurement. It is assumed that the physical and chemical properties in homogeneous media of all these probe molecules and particles are elucidated well and their studies in living systems give information on the properties and functions at the molecular or single nanoparticle level.

Small molecules can be easily pinocytosed by living cells, but nanoparticles larger than 100 nm are not easily transferred to their interiors. It sometimes takes a few hours for such nanoparticles to enter inside the cells. Therefore manipulation of cells and injection of probe molecules and nanoparticles are extremely important and their development is strongly needed to fulfill the promise of potential applications of spectroscopy and imaging in life science, biotechnology, medicine, and so on.

Based on our systematic studies on nano-spectroscopy and nano-photochemistry, nano-manipulation and chemistry of photon pressure, and nano-ablation and its biological application [14–16], we have planned to develop new methodologies for answering some of the above described problems. One area is to develop novel spectroscopy and imaging which makes it possible to interrogate non-fluorescent nanomaterials incorporated in living cells. If this is realized, we will have more potential probes and elucidate a wider range of functionalities. Rayleigh light scattering spectroscopy has given one answer, namely, electronic transitions of single polymer nanocrystals were sensed by the scattering spectroscopy by us, although their size was too small for absorption spectroscopy [14–18]. Now the high potential of scattering spectroscopic measurement has been widely demonstrated for surface plasmon resonance of gold and silver nanoparticles [13, 19–23]. Indeed scattering spectroscopy and imaging are complementary to fluorescence spectroscopy and its imaging.

The second area is to develop new manipulation and modification of living cells and injection of nanoparticles into living cells. Our approach is based on femtosecond laser processing, but the laser pulse is not introduced to target cells and tissues. That makes it possible to introduce artificial nanoparticles, whose physical and chemical nature is well known, into the targets. By combining new spectroscopy/imaging and injection methods, we will be able to reach one of the milestones in the understanding of molecular level dynamics and the functionality of single living cells. Some of our novel results obtained between April 2004 and March 2007 are summarized and considered here.

28.2
Femtosecond Laser Ablation and Generated Impulsive Force in Water: Laser Tsunami

Laser ablation is the base on which many laser processing technologies for living systems have been developed, where the targets which should be fabricated are, in general, directly exposed to intense laser pulse irradiation. Thus laser processing is always accompanied by some damage to their surface and internal structures, which is indeed critical in application, especially to living systems. By tuning laser fluence and pulse width, better solutions to achieve the manipulation and injection with less damage have been sought. Among the trials, femtosecond laser processing has been receiving much attention, as three-dimensional processing is made possible in aqueous solution, culture medium, and even inside living cells and tissues. The conventional Ti:Sapphire femtosecond laser oscillates around 800 nm, at which wavelength most biological materials are transparent. By focusing it, multi-photon absorption is induced selectively at the central part of the focal point, where the ablation takes place. Namely, ablation can be caused freely in the three-dimensional space of transparent media, which is indeed one advantage of the femtosecond laser compared to the nanosecond one.

It had been believed that intense femtosecond laser excitation results in plasma formation, causing ablation phenomena [24]. This explanation can be applied to metals and semiconductors but we doubted whether this plasma is responsible for the ablation of molecular materials. We applied femtosecond absorption spectroscopy and femtosecond surface light scattering imaging to phthalocyanine films and liquid benzenes and revealed the primary electronic and morphological processes [25–30]. We could not detect ionic species of the corresponding molecules as the main transient species just at and moderately above the ablation threshold, although some ionic radicals should be observed, at least in the initial stage of ablation involving plasma formation. Thus we pointed out the negative possibility of the plasma mechanism, at least for molecular solids and liquids in the fluence range not far above the ablation threshold.

Here, we briefly describe the laser ablation dynamics of phthalocyanine film where the primary species was confirmed, by femtosecond transient absorption spectroscopy [25], to be its electronically excited state. The exciton absorption band was replaced in 20 ps by a hot band of the ground electronic state. This rapid decay was ascribed to mutual interactions between densely formed excitons leading to sudden temperature elevation. The elevated temperature was estimated by comparing transient absorption spectra with the temperature difference ones. It is worth noting that quite normal dynamics of the excited states are detected, even under ablation conditions, which is the reason why we do not accept the plasma mechanism. Then we considered how to correlate these electronic processes of phthalocyanine films with their fragmentation.

The evolution of the surface roughening just before the ejection of fragments was evaluated by time-resolved surface light scattering imaging. The root-mean-squared roughness of the phthalocyanine film increased up to a few tens of nm around a few ns after excitation, although the film was excited by a 170 fs laser pulse. The temperature elevation was completed in a few tens of ps after excitation, while appreciable surface roughening was not clearly detected before a few ns. These dynamic processes are depicted schematically in Figure 28.1. Vigorous molecular and lattice vibrations are enhanced in the irradiated area for 20 ps – a few ns delay time, which may be represented as a transient pressure [31]. This idea of laser-induced transient high pressure was already proposed by Dlott by using a picosecond CARS experiment on an anthracene single crystal [32]. The irradiated part is surrounded by an unexcited area and a substrate glass, while thermal conduction

Femtosecond Laser Excitation

Nanosecond Laser Excitation

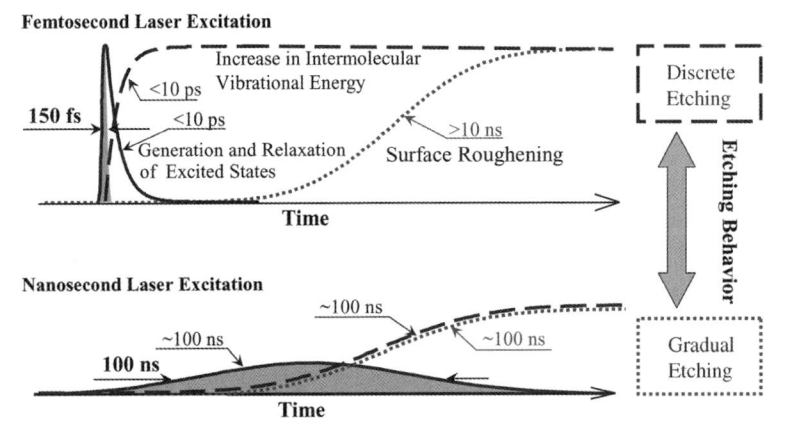

Figure 28.1 Molecular electronic and vibrational relaxation processes and following morphological changes under laser ablation condition.

does not take place rapidly in the organic film. Therefore the transient pressure can be released only by fragmentation of the film to open air. This explanation of femtosecond laser ablation is consistent with the experimental result of etching behavior (discrete etching) which is quite different from that in nanosecond laser ablation (gradual etching). In the latter case, electronic excitation, conversion to molecular motions, and surface roughening take place simultaneously in hundreds of ns, and the melting phenomenon is surely involved. On the basis of the spectroscopic consideration, we proposed that a photomechanical mechanism operates for femtosecond laser ablation of molecular materials, which is contradictory to the conventional plasma mechanism of femtosecond laser ablation [24].

During these laser ablation studies to elucidate the relevant mechanism, we have come to the idea that the photomechanical force generated by laser ablation can be utilized as a useful perturbation to manipulate and fabricate living systems in aqueous solution. Namely, multi-photon excitation of water which is induced by a focused near-IR femtosecond pulse under a microscope evolves to ablation of water at the focal point, which is followed by shockwave propagation, cavitation bubbling, and convection flow. The phenomenon is already well known [33], and we have utilized this as an impulsive force and demonstrated how the force is useful to manipulate, modify, and pattern various nanomaterials and single living cells in solution [34]. Here we name this phenomenon laser-induced micro/nano tsunami (laser tsunami) in view of the impulsive water flow and describe some recent advances in our studies. Ablation is induced near the target, and the generated laser tsunami attacks the target. Of course, the laser pulse is not irradiated onto the target so that photochemical decomposition leading to damage is fully avoided.

A schematic illustration of the laser tsunami and laser power dependence of shockwave generation, plasma emission, and cavitation bubbling is given in Figure 28.2. It is noticeable that the threshold of the shockwave is lowest and that of the bubbling is highest [35]. Of course this tendency is rather qualitative, as direct

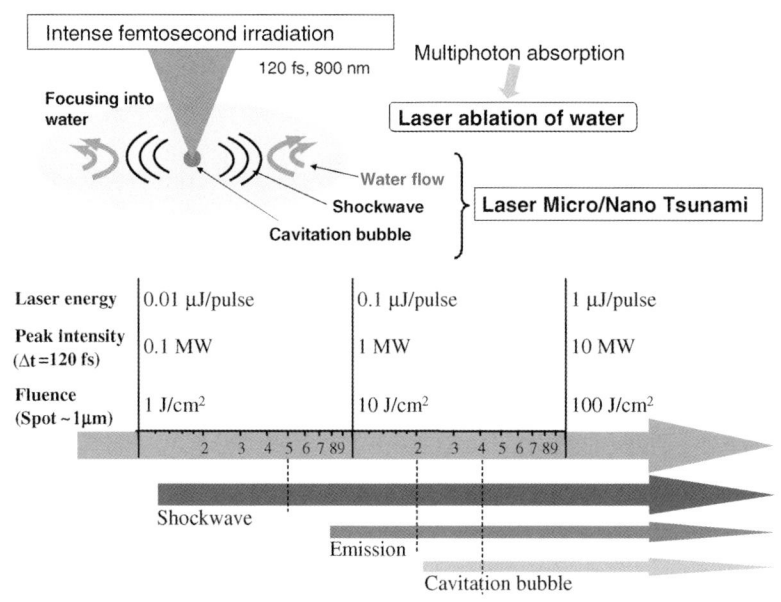

Figure 28.2 A schematic representation of the laser tsunami and laser energy necessary to induced shockwave, emission, and cavitation.

observation of shockwave, emission, and bubbling depends on the sensitivity/resolution of the time-resolved detection instruments and experimental conditions such as laser wavelength, fluence, pulse width, temperature, solvent and so on.

The threshold of the pulse energy to induce the laser tsunami is relatively low for a femtosecond laser compared with nanosecond and picosecond lasers. The laser tsunami expands to a volume of (sub µm)^3 around the focal point, when an intense laser pulse is focused into an aqueous solution by a high numerical aperture objective lens. When a culture medium containing living animal cells is irradiated, they could be manipulated by laser tsunami. Mouse NIH3T3 cells cultured on a substrate can be detached and patterned arbitrarily on substrates [36]. We have also demonstrated that the laser tsunami is strong enough to transfer objects with size of a few 100 µm [37], which is impossible by conventional optical tweezers because the force due to the optical pressure is too weak. In addition we demonstrated for the first time the crystallization of organic molecules and proteins in their super-saturated solution by laser tsunami [38–40].

28.2.1
Manipulation of a Single Polymer Bead by Laser Tsunami

The laser tsunami can push a small object in solution, which was clearly visualized by conducting a model experiment. As shown in Figure 28.3, a single polymer bead moved when a femtosecond laser pulse was focused near to it. Upon one shot of irradiation, the bead was pushed step-wise away from the focal point. More detailed

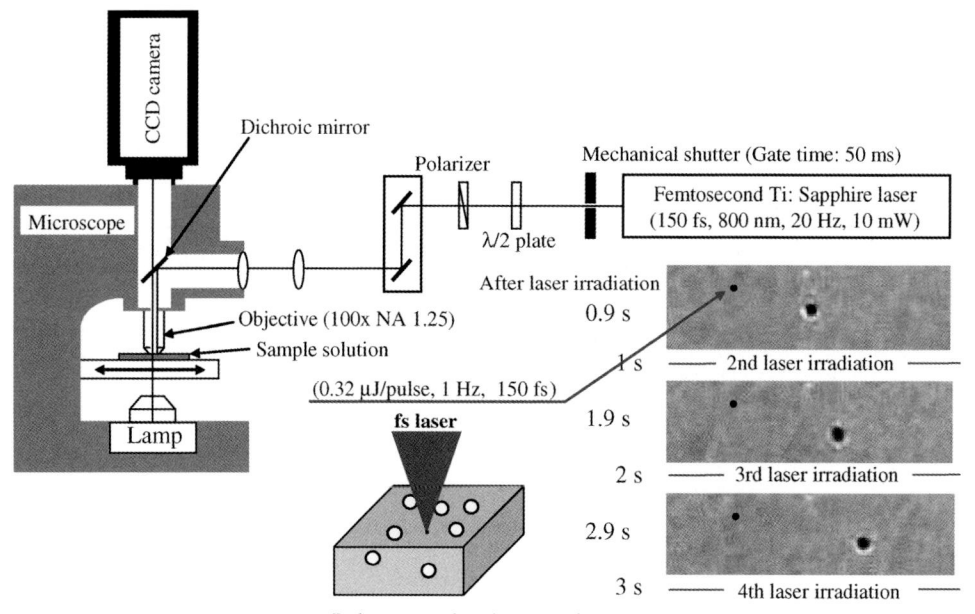

Figure 28.3 An experimental set-up for direct observation of the laser tsunami pushing a single polymer bead in solution.

observation was carried out by employing a high-speed camera to record the whole process of the motion of a larger bead after the femtosecond laser irradiation as shown in Figure 28.4 [41]. The phenomenon is due to the laser tsunami and can be divided into three stages. After femtosecond laser irradiation, a cavitation bubble is soon generated and the nearby bead is pushed away by bubble expansion, which is considered stage A. The bubble expands up to about 200 μm diameter and then collapses within 30 μs, and consequently the bead is pulled back to a certain extent, which is stage B. Following this, a jet water flow is induced, which carries small remaining gas bubbles to the surface of the solution. The jet flow is regarded as convection of water, which is stage C. During stages A and B, the maximum velocity was only $2.8\,\mathrm{m\,s^{-1}}$, but the maximum acceleration was about $3.6 \times 10^5\,\mathrm{m\,s^{-2}}$. As the mass density of the polystyrene spheres is $1.05\,\mathrm{g\,ml^{-1}}$, the force exerted on the bead can be estimated, $f_A = 1.2\,\mathrm{\mu N}$. This force is much larger, by a few orders of magnitude, than the trapping force of the conventional optical tweezers. However, this force itself makes little contribution to the total displacement of the bead. The final position of the bead is determined mainly by the jet flow, that is, stage C. The maximum acceleration of the bead induced by the jet flow was about $9.0 \times 10^4\,\mathrm{m\,s^{-2}}$, so the force exerted on the bead is estimated to be $f_C = ma = 0.29\,\mathrm{\mu N}$. This force is smaller than that in stages A and B, but still much larger than the conventional optical trapping force.

Now we describe a novel method to trap and manipulate a micro-object with the laser tsunami. As a demonstration, we chose polystyrene beads of 90 μm diameter and dispersed them in de-ionized water. As the bead is relatively large and heavy, it

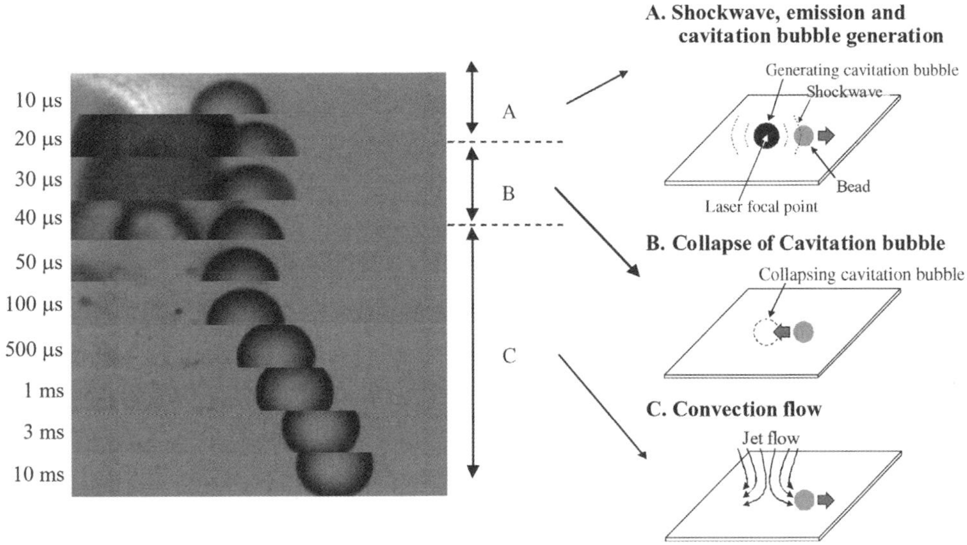

28.2 Femtosecond Laser Ablation and Generated Impulsive Force in Water: Laser Tsunami

Figure 28.4 Time-resolved observation of the laser tsunami pushing a single polymer bead in solution and its model explanation.

remained at the bottom of the specimen cell without visible Brownian motion. When the laser was introduced, the water near the laser focus clearly caused a laser tsunami. When the laser focal point was moved toward the target bead by adjusting the mortar-driven microscope stage, and the distance between them reached about 300 μm or so, the bead started to move. The force was so strong that even the beads adhering to the substrate could be moved easily. In order to manipulate the sample with the laser tsunami, the laser beam was scanned around the target bead by using a Galvano mirror to form a trapping circle [37], as shown in Figure 28.5. The trapping circle

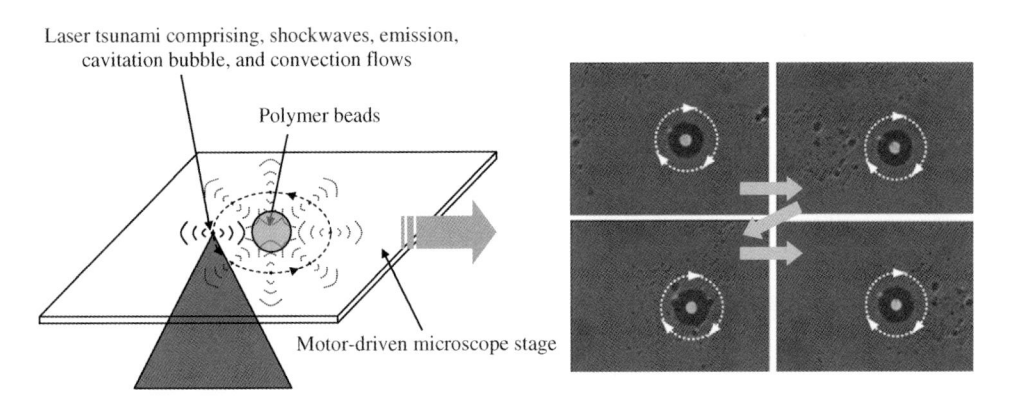

Figure 28.5 Manipulation of a single polymer bead by spatially controlling the laser tsunami.

radius was set to 150 μm, the scanning frequency was tuned to 10 cycles s^{-1}, the laser pulse energy was adjusted to 5 μJ, and its repetition rate was set to 100 Hz, so that one cycle consisted of ten laser shots. The femtosecond laser was irradiated and the trapping circle was brought close to the target bead from its upper position. When the circle plane was shifted to about 50 μm above the bead, it began to move to the center of the trapping circle. As shown in Figure 28.5, the dust on the substrate which looked like black dots were moved from left to right by driving the stage, so that we can say that the target bead was transferred from right to left, while still remaining in the center area of the trapping circle. Thus, it was obvious that the bead was successfully trapped inside the circle and manipulated by the laser tsunami. It is worth noting that even the bead attached to the substrate could be trapped and moved easily by the laser tsunami, which is usually impossible by conventional optical tweezers.

Optical trapping has been used as a powerful tool and it is being developed for exploring new molecular phenomena and clarifying novel molecule–light interactions. Nevertheless, optical trapping has some limitations preventing practical application to living systems. The first is that the trapping force by the CW laser is not strong, as pointed out above. The second is due to the damage caused by intense CW laser irradiation. Manipulation by the laser tsunami overcomes these limitations, as its mechanical force is strong despite its transient nature and the laser tsunami is caused by irradiation of the surrounding medium, not the target. The advantages of manipulation by laser tsunami will receive much attention as an alternative to optical trapping in the near future.

28.2.2
Manipulation of Single Animal Cells by Laser Tsunami

Usually the living animal cells are cultured on a substrate and its removal is realized by chemical treatments, while laser tweezers cannot be applied as the cells are strongly adhered to the substrate. As described above, the impulsive force of the laser tsunami reaches μN order, so that its one-by-one removal is now possible. The first demonstration of non-destructive isolation was given for an animal cell, mouse NIH3T3, having lamellipodia and filopodia, with which the cells were firmly attached in the collagen matrix [36]. By just applying the laser tsunami, we found that nothing happened, which is of course because they were connected to the substrate using filopodia. Therefore we tried to cut each filopodium by direct femtosecond irradiation. The filopodia contracted soon upon one shot irradiation and the cell changed its shape to spherical. As Brownian motion of the cell was not observed, the cell should still be, at least partly, adhered to the matrix. Then a laser with pulse energy 0.51 μJ pulse was focused at 20 μm, far from the ventral side of the cell, to cause laser tsunami. As a result the cell was detached from the matrix and pushed away. The observed behavior is summarized in Figure 28.6, where a schematic illustration is also given. The isolated cell showed Brownian motion, then it stopped on the substrate again and the filopodia were confirmed to be regenerated, namely, the cell adhered again to the matrix. This means that the cells were not killed. The viability of the cells was also confirmed by examining the conventional dye exclusion test

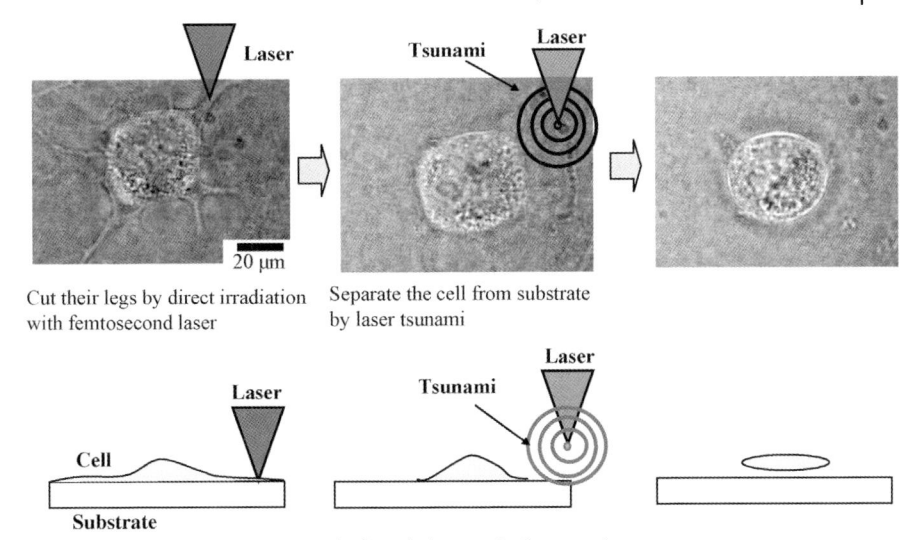

28.2 Femtosecond Laser Ablation and Generated Impulsive Force in Water: Laser Tsunami | 555

Figure 28.6 Non-destructive removal of single living cells from a substrate.

before and after laser irradiation. These results indicate that the one-by-one removal of cells utilizing the laser tsunami is a method with high potential.

A successful example is summarized above, while the result of the non-destructive removal should be statistical as the cell has distributions in shape, cell cycle, viability, and so on. To estimate the physical damage to the cells, unspecified 80 NIH3T3 cells were selected and examined. First, their filopodia were cut directly, as described above, and then the laser irradiation was started at a position where the distance between the focal point and the end of the cell was 60 µm. The distance was then reduced until the cell was detached by laser tsunami, which was examined as a function of laser fluence. Of course, as the distance becomes shorter, the necessary energy to remove the cells is smaller. When the distance is typically less than 35 µm, however, the laser probably irradiates and kills a cell. Note that the height of the present mouse cell is usually about 10 µm and a lens with high NA giving large incident angle was used. The probability of direct laser irradiation should increase with the decrease in distance, although the laser energy is decreased. One result is given here as a concrete example. 80% of detached cells kept their viability when a 0.72 µJ pulse^{-1} was applied at a distance greater than 15 µm, while 80% of cells were killed even at 0.41 µJ pulse^{-1} upon decreasing the distance to less than that. We consider that the higher probability at a greater distance promises a practical extension of the laser tsunami method in the near future. For example, once a cell is detached from the collagen matrix by laser tsunami, it can be successfully manipulated by employing conventional laser trapping with a CW Nd^{3+}:YAG laser [35], which should realize fine manipulation and arrangement of the cell. The combination of laser tsunami and conventional laser tweezers will open new horizons in the studies on single cell analysis of differentiation and issue formation.

The removal by laser tsunami has important advantages which are not available by conventional removal and patterning methods for living cells: ink jet printing, micro-contact printing, laser-induced forward transfer. In printing methods, liquid droplets with a volume less than 1 nl containing cells, proteins, and so on, are deposited on a substrate. These small droplets are very volatile due to the relatively large surface-to-volume. The laser transfer technique also involves vaporization. Drying of cells and proteins may lead to death and de-naturarization, respectively, which is critical for cell manipulation. On the other hand our laser tsunami method never exposes the cells to air during the manipulation. Moreover, the cells are never irradiated by the laser, so that photochemical and photothermal damage are not, in principle, involved. These features are very important and promise high potential for our laser tsunami method in bio-related research involving cell manipulation.

The present removal technique by laser tsunami was developed as a non-destructive micro-patterning method for living animal cells [41]. To conduct the whole process in solution, we set two substrates at a distance of 100 µm; a source substrate where NIH3T3 cells were cultured and a target substrate onto which the cells are transferred. Immediately after the treatment the cells kept their spherical shape and adhered rather weakly to the source substrate. The femtosecond laser pulse was focused at 5 µm below the source substrate. The resultant tsunami attacked the cells and they fell down to the target substrate. By motor-driving the microscope stage laterally, we could demonstrate *in situ* micro-patterning of living cells. Similarly, we applied the laser tsunami to micro-patterning of protein cubes a few µm in size, and demonstrated how easily two kinds of cubes with and without green fluorescence protein are alternately arranged [42, 43]. The whole process was performed in water, so that the cells were never exposed to air and de-naturalization of proteins would not be expected. The spatial resolution of the micro-pattern of the protein crystal was determined to be 80 µm, which is sufficient for practical applications.

28.2.3
Modification and Regeneration Process in Single Animal Cells by Laser Tsunami

By decreasing laser fluence, the force caused by the laser tsunami decreases, so that the cell cannot be detached. However, we found a fluence region where the cell position was shifted by the effect of the laser tsunami [34]. High-speed imaging of the culture medium containing NIH3T3 cells revealed how the cell is shifted by the laser tsunami. Focusing the femtosecond pulse at 20 µm from one side of the cell, bubbling was clearly observed and the cell was shifted. It is interesting to see that the cell still adhered to the substrate. The distance of the shift and its velocity were estimated to be 5 µm and 10 mm s^{-1}, respectively, under some conditions. We consider that the cell movement is induced by mechanical shock due to the laser tsunami, while another possible explanation is that the shift is due to cellular biological processes such as de-polymerization, synthesis, and polymerization of cytoskeletons which are stimulated by the laser tsunami. The latter seems important, but such biological and chemical processes will not take place in the present time scale of milliseconds.

The shift of the living cell is accompanied by modification of the cell shape. The elucidation of this phenomenon should be important for the further development of the laser tsunami as a general tool for manipulation of living cells. As one of such approaches we studied the dynamic change of actin stress fibers induced by the laser tsunami. It is well known that the fibers are the inside framework giving the cell shape, while actin is a cytoskeleton protein and forms actin filament. The three-dimensional network of actin stress fiber, which is an association of actin filaments, provides mechanical support for the cell, determines the cell shape, and enables cell movement. Thus the shape change in the cell due to the laser tsunami can be examined by observing the laser-induced dynamics of fibers. The actin stress fiber was visualized by binding it with enhanced green fluorescence protein (EGFP), and monitored by total internal reflection fluorescence (TIRF) imaging [34].

Epi-fluorescence and TIRF images of a single NIH3T3 cell before and after laser irradiation are shown in Figure 28.7. Although the EGFP-actin is expressed in the whole cell, only the actin fibers at the interface between the cell and substrate were selectively detected as a TIRF image. This is one of the suitable conditions to interrogate the initial process of how the laser tsunami detaches the cell. The bright lines in the TIRF image can be attributed to actin stress fibers which are located close to or adhered to the substrate. The time evolution of the cell when the laser with $0.3\,\mu\text{J}\,\text{pulse}^{-1}$ was focused at the ventral side of the cell with the 60× objective lens is

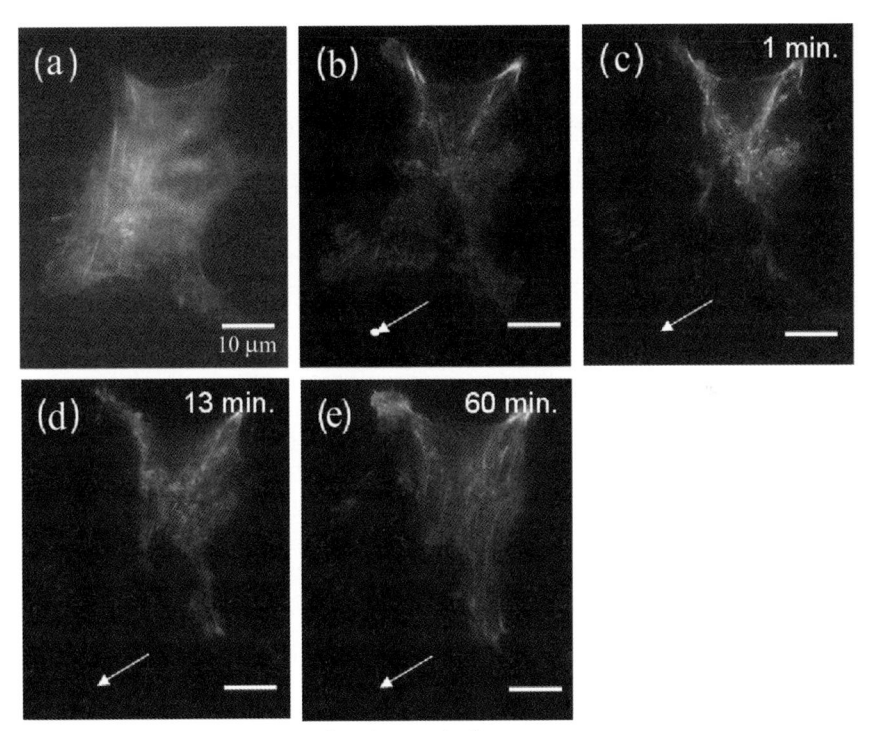

Figure 28.7 Epi-fluorescence (a) and total internal reflection fluorescence (b–e) images of a single NIH3T3 living cell before (a, b) and after (c–e) inducing a laser tsunami.

summarized in Figure 28.7. Immediately after the laser shot, fluorescence of actin filaments near the laser focal point was lost, while the intensity of the filaments increased at the opposite side to the laser focal point.

The loss of the actin filaments near the laser focal point will be due to the detachment of the cytoskeleton from the culture substrate. The detachment will induce relaxation of tension of the cytoskeleton, which may be the one of the origins of cell adhesion and migration processes. On the other hand, in the area far from the laser focal point, there is a possibility that the generation of cytoskeleton in the cell is enhanced, because the increase in the fluorescence intensity means that the actin molecules accumulate at the adhesion area. A few tens of minutes after laser irradiation, the actin filaments were regenerated near the substrate. The direct cutting of actin fibers by focussing a femtosecond laser pulse has already been reported by Mazur et al. [44], however, the regeneration process of the cell had not been described in the literature. Our result deals with the regeneration of the fiber cut by the laser tsunami, which is a unique result on regeneration dynamics.

28.2.4
Injection of Nanoparticles into Single Animal Cells by the Laser Tsunami

The laser tsunami changes the shape and morphology of the membrane of living NIH3T3 cells without their death, while the change does not take place uniformly. The membrane structure should be affected mechanically and recover spontaneously due to the self-assembling nature of lipid molecules. We considered that this deformation may facilitate the penetration of external nanoparticles into the cells. If the nanoparticles with several kinds of functionalities can be freely injected into the living cells, probing and sensing the cell from its inside will be much advanced. Therefore, there is a strong need to develop a new injection technique for nanoparticles with size ranging from 10 to a few hundreds of nm into a specific cell. Recently, we have demonstrated for the first time that a fluorescent polymer latex bead located on the surface of an animal cell can be injected by applying the laser tsunami [45]. It is well known that those nanoparticles cannot enter quickly into living cells, although small molecules can be pinocytosed spontaneously.

A schematic illustration and our experimental set-up for the laser tsunami injection are shown in Figure 28.8. Mouse NIH3T3 fibroblasts were cultured on a glass bottom dish, and polystyrene nanoparticles of 200 nm diameter containing a fluorescent dye were dispersed in the culture medium. Without the laser tsunami the particles can penetrate into the cell by themselves, but this endocytosis takes 2 h after the nanoparticles were dispersed in the culture medium. Since the laser was irradiated within 10 min after the dispersion of the nanoparticles, the penetration through spontaneous endocytosis was negligible in the experiment. In order to observe simultaneously the fluorescent nanoparticles and the cell membranes and to know their relative geometry, the membrane was stained with a styryl dye, N-(3-triethylammoniumpropyl)-4-(4-(dibutylamino) styryl) pyridinium dibromide for its imaging. The three-dimensional fluorescence image was obtained by employ-

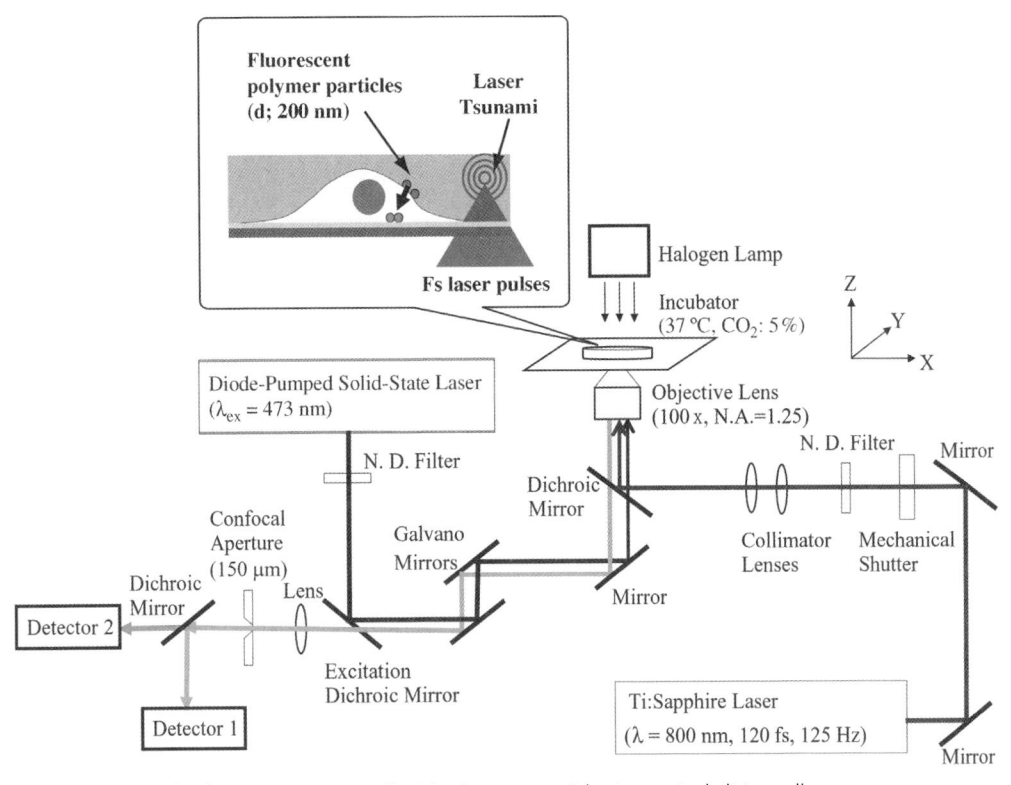

Figure 28.8 A confocal microscopy system for injecting nanoparticles into a single living cell.

ing a confocal fluorescence microscope system, where the fluorescence through a confocal aperture was split into two wavelength regions, giving two fluorescence distributions together.

After dispersion of the nanoparticles in the sample solution, Brownian motion of the nanoparticle was observed in the medium. Once the nanoparticles were attached to the cell, the Brownian motion stopped. Here, we present the transmission and confocal fluorescence images at 10 min after adding the polystyrene nanoparticles to the culture medium. Initially, fluorescent nanoparticles and stained cell membranes were observed as white spots and a red wide silhouette, respectively, and the particles were observed only on and near the cell membrane. However, after 20 pulses of the focused femtosecond laser with energy 10 nJ pulse^{-1} were shot into the culture medium at 20 μm from the edge of the cell, the nanoparticles were observed inside the cell. This is clearly shown in Figure 28.9, where the fluorescence images at different heights are included. White spots were observed at heights of 2 and 4 μm in the upper part inside the cell, confirming the successful injection of the 200 nm particles. The injection behavior strongly depended on the shape and size of the cells, penetration was confirmed for ∼10% of the targeted cells. We consider that the injection of the nanoparticles is achieved not only by pushing the nanoparticle onto the cell membrane but also by enhancing its fluidity with the laser tsunami.

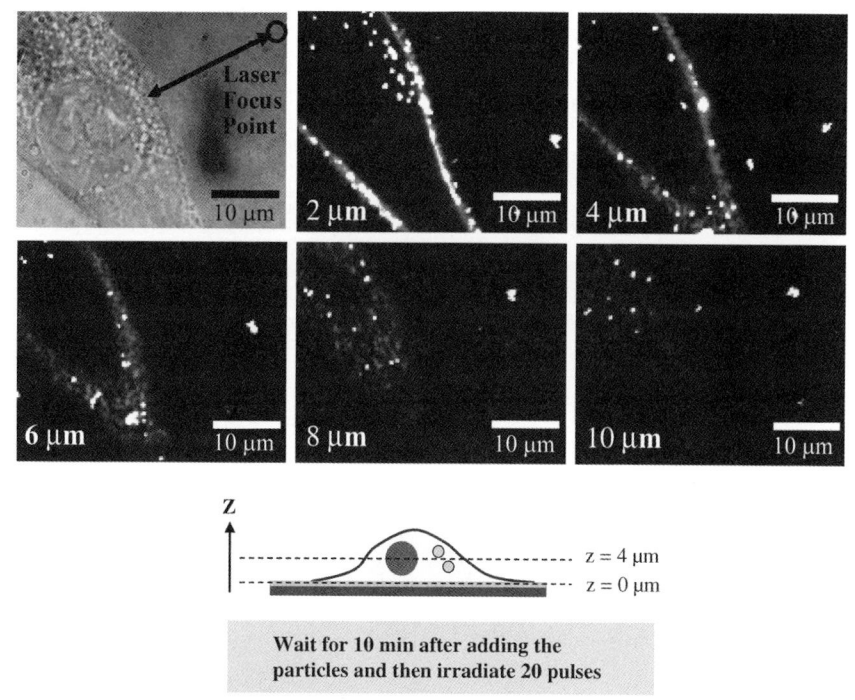

Figure 28.9 Fluorescence images of the polymer particles at different heights on/in a single NIH3T3 cell in culture medium. The laser was focused at 20 mm from the right edge of the cell, and some nanoparticles were injected at the upper right part inside the cell, which is shown at 2 and 4 mm.

How the single cell is affected by the laser tsunami was interrogated by time-resolved imaging, and indeed the cell facing the laser focal point was pushed and the nucleus shape was bent to some extent, corresponding to the morphological change in the cell. It is worth noting that the size, shape, and distribution of cellular organelles, giving white micro-objects inside the cell, were modified. Such observation suggests that the cell morphology is much affected by the impact caused by the laser tsunami and penetration through the cell membrane is greatly enhanced.

Direct irradiation of a single cell by a femtosecond laser pulse is an alternative and more conventional method to inject nanoparticles, which has already been reported by some groups [46–48]. A small ablated hole in the cell membrane may easily allow the nanoparticles to penetrate into the cell. However, some photochemical damage should occur, so that viability of the injected cells may be lower than after processing with the laser tsunami. As a more conventional injection method, high pressure has been employed as an impactor, using a "particle gun" [49], but selective injection into a specific single cell is not possible. Mechanical injection using a micropipette is another popular tool [50], but it suffers from clogging problems. The present laser tsunami injection is, in principle, a non-invasive, non-clogging, and three-dimensionally controllable method, and directional injection giving specific selective penetration of nanoparticles is realized, which is indeed one of the advantages of this method.

28.3 Development of Rayleigh Light Scattering Spectroscopy/Imaging System and its Application to Single Animal Cells

Most spectroscopy and imaging methods for single-cell analysis are based on fluorescence measurement of molecules and nanoparticles. The non-luminescent molecules and materials have not received attention in the relevant research fields, while electronic absorption spectroscopy is another important tool for the general identification and understanding of the optical properties and the molecular electronic structure. However, one can hardly measure the visible absorption spectrum under a microscope, because much of the illumination light cannot interact with the sample due to the short pass length and the large illumination spot compared to molecules, aggregates, and particles in small domains, resulting from the diffraction limit of light. This results in low sensitivity, therefore, conventional absorption spectroscopy is not a useful technique for investigating single cells. Instead we have developed a light scattering microspectroscopic system using darkfield illumination and investigated single gold nanoparticles [22, 23, 51–54]. The dark background realized high contrast imagings like fluorescence detection leading to spectroscopic measurements at the single particle level. In particular, the light scattering spectra are closely related to electronic absorption ones, so that this spectral measurement is an alternative approach to understanding molecular electronic structures. On the other hand, the spatial resolution of light scattering microspectroscopy remains lower than conventional confocal laser scanning microscopy because a halogen lamp is applied as a light source. Light scattering microspectroscopy with high three-dimensional resolution is strongly required for spectroscopy and imaging of living cells.

Recently, we have developed a confocal light scattering microspectroscopy and imaging system for three-dimensional spectroscopic investigation and imaging with submicron resolution [55, 56]. Femtosecond laser-induced supercontinuum light obtained by a photonic crystal fiber [57] has been applied to microspectroscopy, and indeed the spectral coverage of supercontinuum light makes it possible to achieve cellular imaging of multicolor two-photon fluorescence [58–60] and multiplex CARS [10, 11, 61]. Here the supercontinuum allows us to obtain light scattering spectra covering the whole visible region. This means that we are able to investigate electronic spectral properties without measuring auto-fluorescence and without adding probe molecules. Some groups have reported light scattering microscopic imaging by means of supercontinuum light [62–64], however, no one has achieved simultaneous three-dimensional spectroscopy and imaging yet, and, as far as we know no one has examined living cells. By combining the supercontinuum light with the confocal microscope technique, we have developed a light

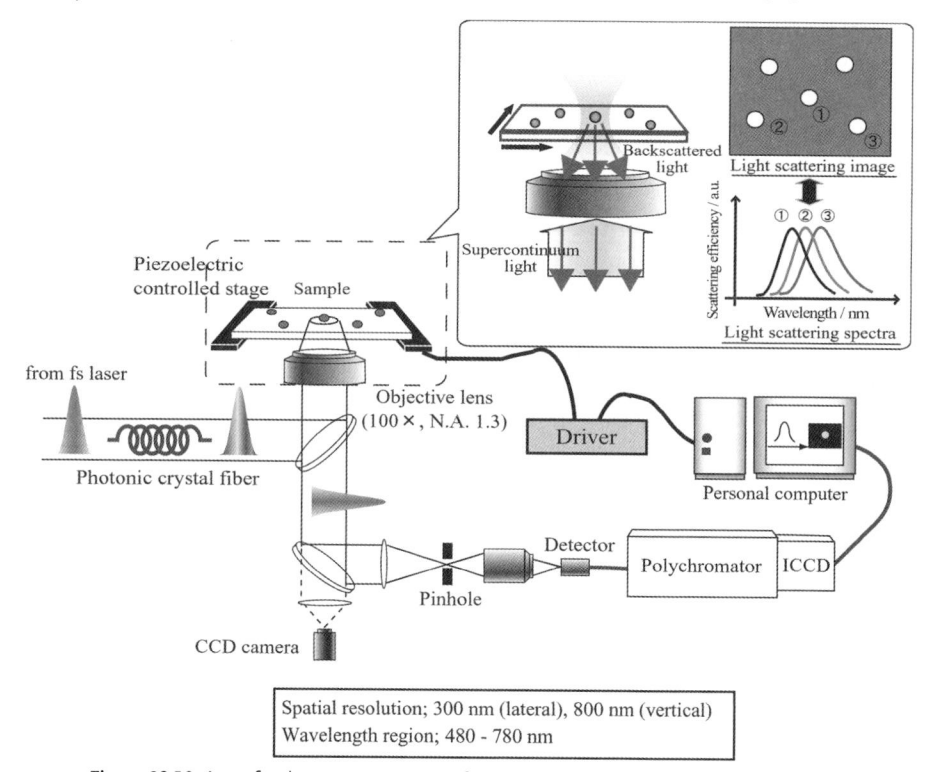

Figure 28.10 A confocal microscopy set-up for light scattering spectroscopy and imaging using femtosecond white light continuum as a probe light.

scattering spectroscopy/imaging system, whose spatial and spectral resolutions were evaluated on the basis of single gold nanoparticle measurements. Here, we describe its performance and application to the three-dimensional spectroscopic imaging of single unstained cells with and without gold nanoparticles and consider its promising future.

Figure 28.10 shows a schematic diagram of the experimental set-up of our confocal light scattering microspectroscopy/imaging system. A femtosecnd laser pulse train of 780 nm, 200 fs, 800 mW, and 80 MHz is introduced into a photonic crystal fiber by a microscope objective. The out-out pulse was spectrally broadened in the wavelength region from 500 nm to 800 nm, and its output power was several tens of megawatts. The supercontinuum light was coupled into an inverted microscope, focused to a sample with an objective lens (100×, N.A. = 1.30). The sample was set on a piezoelectric controlled-stage which was mounted on an XY motorized microscope stage. The back scattered light from the sample was collected with the same objective lens and introduced to an imaging pinhole, its spectrum was then measured with a polychromator coupled with an image-intensified charge-coupled device (ICCD). The sample was raster scanned by the piezo stage, which was synchronized with ICCD detection. Thus the light scattering spectra were measured at each focused

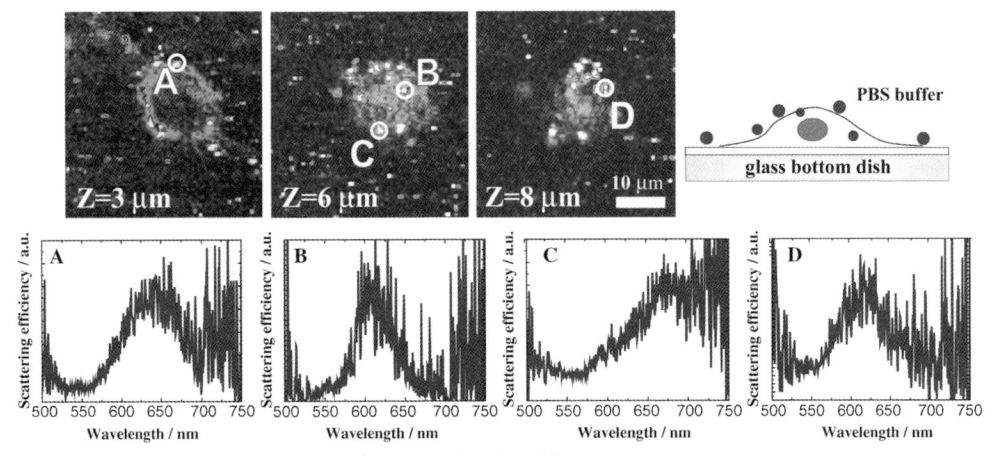

Figure 28.11 Light scattering spectra and images of single gold nanoparticles dispersed on/in a single living NIH3T3 cell.

point. By sampling an arbitrary wavelength region of the light scattering spectrum at each point of the target, a spectroscopic image was obtained.

Confocal light scattering spectroscopic imaging of mouse NIH-3T3 living cells at different depths are shown in Figure 28.11, which was obtained after incubation with dispersed gold nanoparticles. Images are roughly similar to that before incubation with nanoparticles, while some bright points are interestingly overlapped. These features suggest that the points seem to be gold nanoparticles. The light scattering spectra of the bright points A-D in Figure 28.11 are also given. As the particles A and C were located at the end of the cell, we can conclude that they are adsorbed on the surface of the cellular membrane, while the particle B is clearly included in the cell. The latter is surrounded in the scattering medium with relatively high refractive index. On the other hand the particle D at 8 μm should be near the top of the cell. Each bright point shows a spectral peak in the wavelength range 600–700 nm, which is clearly ascribed to the surface plasmon resonance band of single gold nanoparticles [13]. Without adding gold nanoparticles we could not find any bright point showing such a spectral character. The spectral peak and intensity differ from particle to particle, which could indicate not only variation in particle size or shape but also the change in intercellular environmental condition, depending on the particle position [22]. The spectrum of the particle labeled C strongly indicates the aggregation of particles, as the peak is much shifted to the red compared with others [13]. The present results clearly demonstrate that the wide spectroscopic wavelength and the submicron spatial resolution, which are characteristics of our system, allow us to distinguish precisely single gold nanoparticles on or in the single cell. It is important to note that the incident supercontinuum light into the cell is bright enough to penetrate into the cell and to give reliable spectra of gold nanoparticles.

Since light scattering spectra of gold nanoparticles reflect the environmental condition, simultaneous monitoring of the position and spectrum give us a new way to understand dynamic aspects of single living cells. When they are dissolved in homogeneous solution or fixed in a matrix no spectral change is expected. If nanoparticles move in and on a single cell, the scattering spectrum changing from position to position can be followed in the time scale determined by the employed instruments. In this case spatial and spectral tracking of a single nanoparticle can be elucidated and real time observation of how nanoparticles behave will receive much attention.

One example is shown in Figure 28.12, where the nanoparticle near the top of the single mouse NIH3T3 cell started its migration at the position of 12.5 μm from the bottom of the dish. During the measurement the particle migrated laterally keeping constant height for a while, giving constant peak wavelength of the scattering spectra. At about 28 min after the start the particle sank down by 3 μm and the peak showed a quick change. After that the position and the peak recovered to their original values in a few minutes. Later the particle rose to a higher position in the cell, accompanied by a small spectral shift.

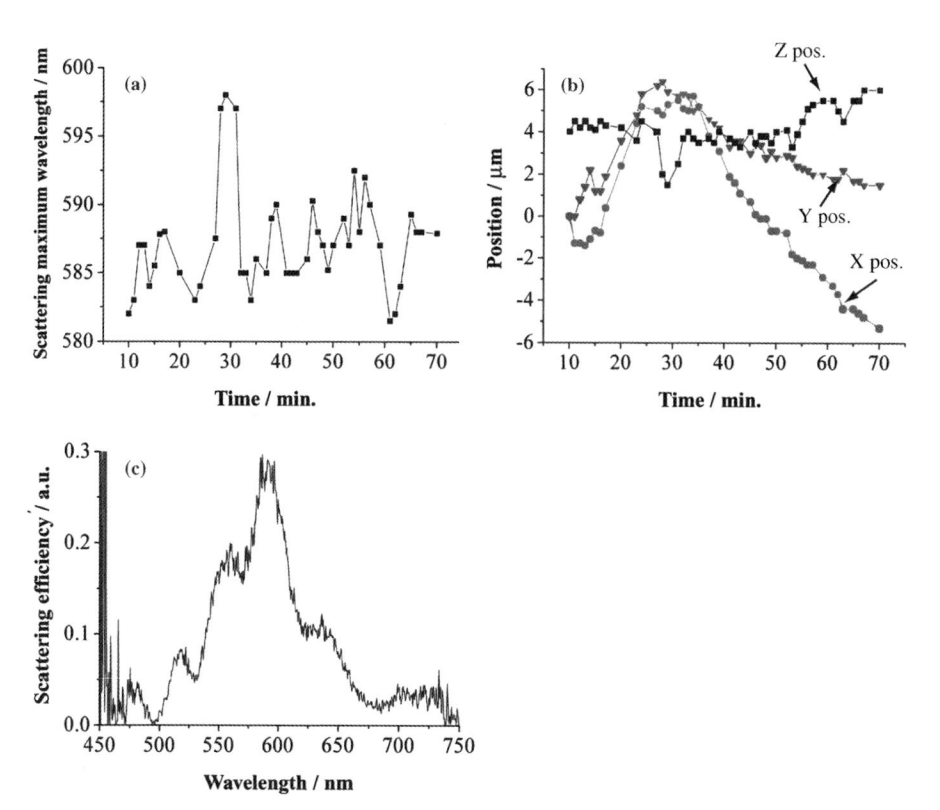

Figure 28.12 Single gold nanoparticle tracking on/in a single living NIH-3T3 cell by light scattering spectroscopy and imaging. (a) Scattering spectrum maximum wavelength evolution with time. (b) Particle position evolution with time (x, y and z positions in light gray circles, gray triangles and black squares, respectively); x and y positions are measured starting from the initial tracking position, while z values are the distance from the upper surface of the dish. (c) Typical spectrum (at 28 min).

A possible explanation of this interesting behavior is as follows. After an initial contact of the gold nanoparticle with the cell, it may migrate laterallly and then may be included in the cell either through pinocytosis or by entering through a small channel [65]. At this point the particle would be tightly surrounded by a membrane layer, and this process would be followed by rapid efflux out of the cell in the next few minute because of its exogenous character. Short endocytosis and exocytosis cycles have indeed been reported by Xu et al. [66]. The shift of 12 nm in the scattering spectral maximum may represent an increase in the local dielectric constant upon penetration of the particle into the cell. The aparent return to the initial maximum could be explained by the opposite process, possibly escape of the particle from the cell before the endocytosis is completed.

28.4
Summary

In this chapter we have described two methodologies of femtosecond laser tsunami processing and light scattering spectroscopic imaging, and have shown their application to single animal cells. The first is a new processing method based on femtosecond laser ablation in solution. The femtosecond pulse from a Ti:Sapphire laser (800 nm, ~150 fs) is focused into water, culture medium, and so on, under a microscope, leading to multiphoton excitation at the focal point. Above a certain threshold shockwave propagation, plasma emission, and cavitation bubbling are induced, which is a well known phenomenon. We have named the phenomenon "laser tsunami" and have paid attention to the resultant high pressure and convection flow. When the femtosecond laser is focused near a target animal cell, the high pressure and convection flow affect the cell transiently and locally, which is regarded as an impulsive force and should be a soft mechanical perturbation. As the targets are not irradiated directly, photochemical damage can be avoided. Three-dimensional processing is possible even inside a single living cell, which is ascribed to multiphoton excitation by femtosecond pulse. The force is of the order of μN, although it is transient, and extremely stronger than that produced by conventional optical trapping with CW laser beam irradiation which is of the order of pN.

By taking advantage of the laser tsunami, we have developed new methods which enable us to manipulate and to process single living animal cells in solution without damaging them photochemically and exposing them to air. Manipulation of single large polymer beads and living cells, modification of actin stress fibers in a single cell, and injection of nanoparticles into a single cell are presented here to show the high potential of laser tsunami processing.

The second methodology is a novel spectroscopy and imaging method which makes it possible to interrogate non-fluorescent nanomaterials incorporated in living cells. This is very promising, since we will be able, in principle, to probe more materials and elucidate a wider range of functionalities. Rayleigh light scattering spectroscopy using a supercontinuum which is generated by focusing a femtosecond laser pulse to a photonic crystal fiber is combined with a confocal microscope.

Electronic transitions of single molecular aggregates, nanocrystals, polymer beads, metal nanoparticles, and so on can be followed three-dimensionally. As an example, gold nanoparticles were dispersed in a culture medium containing living cells and their tracking behavior was monitored. Spectral measurement of the surface plasmon resonance of gold nanoparticles gives information on the surrounding environment, so we can consider where the nanoparticle is located: in water, on the membrane surface, or in the membrane.

These two methodologies are novel and complementary to conventional laser manipulation, fabrication, spectroscopy, and imaging methods. Processing by laser tsunami is clearly different from laser processing which is based on direct laser irradiation. Manipulation by laser tsunami is due to an impulsive force, which is in contrast to the laser tweezer by continuous laser irradiation. The present light scattering spectroscopy and imaging are complementary to fluorescence spectroscopy and its imaging. Our approaches are new and thus expected to contribute to systematic exploration of new phenomena in living cells, particularly at molecular and nanoparticle levels.

Acknowledgments

The preset work is supported by KAKENHI (the Grant-in-Aids for Scientific Research) on Priority Area "Molecular Nano Dynamics" (April 2004-March 2007) from the Ministry of Education, Culture, Sports, Science and Technology of Japan (MEXT). HM also thanks MOE-ATU (the Ministry of Education-Aiming for Top University) Project (National Chiao Tung University) by the Ministry of Education, Taiwan and the National Science Council, Taiwan (0970027441) for their support.

References

1 Kawata, S. **and** Masuhara, H. (eds) (2004) *Nanophotonics: Integrating Photochemistry, Optics, NanoBio Materials Studies*, Elsevier, Amsterdam.
2 Kawata, S. and Masuhara, H. (2006) *Nanoplasmonics: From Fundamentals to Applications*, Elsevier, Amsterdam.
3 Kawata, S., Masuhara, H. and Tokunaga, F. (eds) (2007) *Nano Biophotonics: Science and Technology*, Elsevier, Amsterdam.
4 Mycek, M.A. and Pogue, B.W. (eds) (2003) *Handbook of Biomedical Fluorescence*, Marcel Dekker, New York.
5 Bouma, B.E. and Tearney, G.J. (2001) *Handbook of Optical Coherence Tomography*, Marcel Dekker, New York.
6 Wang, X.F. and Herman, B. (1996) *Fluorescence Imaging Spectroscopy and Microscopy*, John Wiley & Sons, New York.
7 Gremlich, H-.U. and Yan, B. (eds) (2001) *Infrared and Raman Spectroscopy of Biological Materials*, Marcel Dekker, New York.
8 Prasad, P.N. (2004) *Nanophotonics*, Wiley-VCH, Weinheim.
9 Huang, Y.-S., Karashima, T., Yamamoto, M. and Hamaguchi, H. (2003) Molecular-level pursuit of yeast mitosis by time- and space-resolved Raman spectroscopy. *J. Raman Spectrosc.*, **34**, 1–3.
10 Kano, H. and Hamaguchi, H. (2005) Ultrabroadband (>2500 cm^{-1}) multiplex

coherent anti-Stokes Raman scattering microspectroscopy using a supercontinuum generated from a photonic crystal fiber. *Appl. Phys. Lett.*, **86**, 121113–121115.

11 Kano, H. and Hamaguchi, H. (2006) In-vivo multi-nonlinear optical imaging of a living cell using a supercontinuum light source generated from a photonic crystal fiber. *Opt. Express*, **14**, 2798–2804.

12 Schulman, S.G. (ed.) (1985) *Molecular Luminescence Spectroscopy Part I*, John Wiley & Sons, New York.

13 Kreibig, U. and Vollmer, M. (1995) *Optical Properties of Metal Clusters, Springer Series in Materials Science*, Springer, Berlin.

14 Masuhara, H., Asahi, T. and Hosokawa, Y. (2006) Laser nanochemistry. *Pure Appl. Chem.*, **78**, 2205–2226.

15 Asahi, T., Sugiyama, T. and Masuhara, H. (2008) Laser fabrication and spectroscopy of organic nanoparticles. *Acc. Chem. Res.*, **41**, 1790–1798.

16 Masuhara, H., Nakanishi, H. and Sasaki, K. (eds) (2002) *Single Organic Nanoparticles*, Springer, Berlin.

17 Asahi, T., Matsune, H., Yamashita, K., Masuhara, H., Kasai, H. and Nakanishi, H. (2007) Size-dependent fluorescence spectra of individual perylene nanocrystals studied by far-field fluorescence microspectroscopy coupled with atomic force microscope observation. *Polish J. Chem.*, **84**, 687–699.

18 Volkov, V.V., Asahi, T., Masuhara, H., Masuhara, A., Kasai, H., Oikawa, H. and Nakanishi, H. (2004) Size-dependent optical properties of polydiacetylene nanocrystal. *J. Phys. Chem. B*, **108**, 7674–7680.

19 El-Sayed, M.A. (2001) Some interesting properties of metals confined in time and nanometer space of different shapes. *Acc. Chem. Res.*, **34**, 257–264.

20 Link, S. and El-Sayed, M.A. (2000) Shape and size dependence of radiative, non-radiative and photothermal properties of gold nanocrystals. *Int. Rev. Phys. Chem.*, **19**, 409–453.

21 West, J.L. and Halas, N.J. (2003) Engineered nanomaterials for biophotonics applications: improving sensing, imaging, and therapeutics. *Annu. Rev. Biomed. Eng.*, **5**, 285–292.

22 Itoh, T., Asahi, T. and Masuhara, H. (2002) Direct Demonstration of environment-sensitive surface plasmon resonance band in single gold nanoparticles. *Jpn. J. Appl. Phys.*, **41**, L76–L78.

23 Itoh, T., Asahi, T. and Masuhara, H. (2001) Femtosecond light scattering spectroscopy of single gold nanoparticles. *Appl. Phys. Lett.*, **79**, 1667–1669.

24 Baurelere, D. (2000) *Laser Processing and Chemistry*, 3rd edn, Springer, Berlin.

25 Hosokawa, Y., Yashiro, M., Asahi, T. and Masuhara, H. (2001) Photothermal conversion dynamics in femtosecond and picosecond discrete laser etching of Cu-phthalocyanine amorphous film analysed by ultrafast UV-VIS absorption spectroscopy. *J. Photochem. Photobiol., A*, **142**, 197–207.

26 Hosokawa, Y., Yashiro, M., Asahi, T., Fukumura, H. and Masuhara, H. (2000) Femtosecond laser ablation dynamics of amorphous film of a substituted Cu-phthalocyanine. *Appl. Surf. Sci.*, **154–155**, 192–195.

27 Hatanaka, K., Itoh, T., Asahi, T., Ichinose, N., Kawanishi, S., Sasuga, T., Fukumura, H. and Masuhara, H. (1999) Time-resolved ultraviolet-visible absorption spectroscopic study on femtosecond KrF laser ablation of liquid benzyl chloride. *Chem. Phys. Lett.*, **300**, 727–733.

28 Hatanaka, K., Itoh, T., Asahi, T., Ichinose, N., Kawanishi, S., Sasuga, T., Fukumura, H. and Masuhara, H. (1998) Time-resolved surface scattering imaging of organic liquids under femtosecond KrF laser pulse excitation. *Appl. Phys. Lett.*, **73**, 3498–3500.

29 Hatanaka, K., Itoh, T., Asahi, T., Ichinose, N., Kawanishi, S., Sasuga, T., Fukumura, H. and Masuhara, H. (1999) Femtosecond laser ablation of liquid toluene: Molecular mechanism studied by time-resolved

absorption spectroscopy. *J. Phys. Chem. A*, **103**, 11257–11263.

30 Hatanaka, K., Tsuboi, Y., Fukumura, H. and Masuhara, H. (2002) Nanosecond and femtosecond laser photochemistry and ablation dynamics of neat liquid benzenes. *J. Phys. Chem. B*, **106**, 3049–3060.

31 Zhigilei, L.V. and Garrison, B.J. (1998) Computer simulation study of damage and ablation of submicron particles from short-pulse laser irradiation. *Appl. Surf. Sci.*, **127–129**, 142–150.

32 Dlott, D.D., Hambir, S. and Franken, J. (1998) The new wave in shock waves. *J. Phys. Chem. B*, **102**, 2121–2130.

33 Vogel, A. and Venugopalan, V. (2003) Mechanisms of pulsed laser ablation of tissue. *Chem. Rev.*, **103**, 577–644.

34 Hosokawa, H., Yasukuni, R., Spitz, J.-A., Tada, T., Negishi, T., Shukunami, C., Hiraki, Y., Asahi, T. and Masuhara, H. (2007) Single living cell processing in water medium using focused femtosecond laser-induced shockwave and cavitation bubble, in *Nano Biophotonics: Science and Technology* (eds S. Kawata, H. Masuhara and F. Tokunaga), Elsevier, Amsterdam, p. 245.

35 Hosokawa, Y., Takabayashi, J., Shukunami, C., Hiraki, Y. and Masuhara, H. (2004) "Single cell manipulation using femtosecond laser induced shockwave"(Japanese). *Rev. Laser Eng.*, **32**, 94–98.

36 Hosokawa, Y., Takabayashi, H., Miura, S., Shukunami, C., Hiraki, Y. and Masuhara, H. (2004) Nondestructive isolation of single cultured animal cells by femtosecond laser-induced shockwave. *Appl. Phys. A*, **79**, 795–798.

37 Jiang, Y., Matsumoto, Y., Hosokawa, Y., Masuhara, H. and Oh, I. (2007) Trapping and manipulation of a single micro-object in solution with femtosecond laser-induced mechanical force. *Appl. Phys. Lett.*, **90**, 061107–061103.

38 Adachi, H., Hosokawa, Y., Takano, K., Tsunesada, F., Masuhara, H., Yoshimura, M., Mori, Y. and Sasaki, T. (2002) "Effect of short pulse laser on organic and protein" (Japanese). *J. Jpn. Assoc. Cryst. Growth*, **29**, 35–39.

39 Adachi, H., Takano, K., Hosokawa, Y., Inoue, T., Mori, Y., Matsumura, H., Yoshimura, M., Tsunaka, Y., Morikawa, M., Kanaya, S., Masuhara, H., Kai, Y. and Sasaki, T. (2003) Laser irradiated growth of protein crystal. *Jpn. J. Appl. Phys.*, **42**, L798–L800.

40 Hosokawa, Y., Adachi, H., Yoshimura, M., Mori, Y., Sasaki, T. and Masuhara, H. (2005) Femtosecond laser-induced crystallization of 4-(Dimethylamino)-N-methyl-4- stilbazolium tosylate. *Cryst. Growth Des.*, **5**, 861–863.

41 Kaji, T., Ito, S., Miyasaka, H., Hosokawa, Y., Masuhara, H., Shukunami, C. and Hiraki, Y. (2007) Nondestructive micropatterning of living animal cells using focused femtosecond laser-induced impulsive force. *Appl. Phys. Lett.*, **91**, 023904_1–023904_3.

42 Hosokawa, Y., Kaji, T., Shukunami, C., Hiraki, Y., Kotani, E., Mori, H. and Masuhara, H. (2007) Nondestructive micro-patterning of proteinous occlusion bodies in water by femtosecond laser-induced mechanical force. *Biomed. Microdevices*, **9**, 105–111.

43 Masuhara, H., Hosokawa, Y., Yoshikawa, H.Y., Nakamura, K., Sora, Y., Mori, Y., Jiang, Y.Q., Oh, I., Kaji, T., Mori, H., Hiraki, Y., Yamaguchi, A. and Asahi, T. (2007) Femtosecond nonlinear processing in solution: from crystallization to manipulation and patterning, in *Nano Biophotonics: Science and Technology* (eds S. Kawata, H. Masuhara and F. Tokunaga), Elsevier, Amsterdam, p. 227.

44 Kumar, S., Maxwell, I.Z., Heisterkamp, A., Polte, T.R., Lele, T.P., Salanga, M., Mazur, E. and Ingber, D.E. (2006) Viscoelastic retraction of single living stress fibers and its impact on cell shape, cytoskeletal organization, and extracellular matrix mechanics. *Biophys. J.*, **90**, 3762–3773.

45 Yamaguchi, A., Hosokawa, Y., Louit, G., Asahi, T., Shukunami, C., Hiraki, Y. and Masuhara, H. (2008) Nanoparticle

injection to single animal cells using femtosecond laser-induced impulsive force. *Appl. Phys. A*, **93**, 39–43.

46 Tirlapur, U.K. and Konig, K. (2002) Cell biology: targeted transfection by femtosecond laser. *Nature*, **418**, 290–291.

47 Stracke, F., Rieman, I. and König, K., (2005) Optical nanoinjection of macromolecules into vital cells. *J. Photochem. Photobiol., B*, **81**, 136–142.

48 Peng, C., Palazzo, R.E. and Wilke, I. (2007) Laser intensity dependence of femtosecond near-infrared optoinjection. *Phys. Rev. E*, **75**, 041903–041908.

49 Boynton, J.E., Gillham, N.W., Harris, E.H., Hosler, J.P., Johnson, A.M., Jones, A.R., Randolph-Anderson, B.L., Robertson, D., Klein, T.M., Shark, K.B. et al. (1988) Chloroplast transformation in Chlamydomonas with high velocity microprojectiles. *Science*, **240**, 1534–1538.

50 Wolf, D., Wu, R. and Sanford, J.C. (1987) High-velocity microprojectiles for delivering nucleic acids into living cells. *Nature*, **327**, 70–73.

51 Uwada, T., Toyota, R., Masuhara, H. and Asahi, T. (2007) Single Particle Spectroscopic Investigation on the interaction between exciton transition of cyanine dye J-aggregates and localized surface plasmon polarization of gold nanoparticles. *J. Phys. Chem. C*, **111**, 1549–1552.

52 Asahi, T., Uwada, T. and Masuhara, H. (2006) Single particle spectroscopic study on surface plasmon resonance of ion-adsorbed gold nanoparticles, in *Nanoplasmonics: From Fundamentals to Applications* (eds S. Kawata and H. Masuhara), Elsevier, Amsterdam, pp. 219–228.

53 Uwada, T., Asahi, T., Masuhara, H., Ibano, D., Fujishiro, M. and Tominaga, T. (2007) Multipole resonance modes in localized surface plasmon of single hexagonal/triangular gold nanoplates. *Chem. Lett.*, **36**, 318–319.

54 Itoh, T., Uwada, T., Asahi, T., Ozaki, Y. and Masuhara, H. (2007) Analysis of localized surface plasmon resonance by elastic light-scattering spectroscopy of individual Au nanoparticles for surface-enhanced Raman scattering. *Can. J. Anal. Sci. Spectrosc.*, **52**, 130–141.

55 Asahi, T., Uwada, T., Louit, G. and Masuhara, H. (2008) Single particle spectroscopy and tracking of gold nanospheres in living cells by conforcal light scattering microscopy. Digest of the IEEE/LEOS Summer Topical Meetings, pp. 67–68.

56 Uwada, T., Asahi, T. and Masuahra, H. (2008) "Microspectroscopic analysis of nanostructures by femtosecond laser induced supercontinuum light beam" (Japanese), in *Nanoimaging* (ed. A. Wada), NTS, Tokyo, p. 184.

57 Ranka, J.K., Windeler, R.S. and Stentz, A.J. (2000) Visible continuum generation in air silica microstructure optical fibers with anomalous dispersion at 800 nm. *Opt. Lett.*, **25**, 25–27.

58 Ogilvie, J.P., Débarre, D., Solinas, X., Martin, J.-L., Beaurepaire, E. and Joffre, M. (2006) Use of coherent control for selective two-photon fluorescence microscopy in live organisms. *Opt. Express*, **14**, 759–766.

59 Isobe, K., Watanabe, W., Matsunaga, S., Higashi, T., Fukui, K. and Itoh, K. (2005) Multi-spectral two-photon excited fluorescence microscopy using supercontinuum light source. *Jpn. J. Appl. Phys.*, **44**, L167–L169.

60 Wildanger, D., Rittweger, E., Kastrup, L. and Hell, S.W. (2008) STED microscopy with a supercontinuum laser source. *Opt. Express*, **16**, 9614–9621.

61 Konorov, S.O., Akimov, D.A., Serebryannikov, E.E., Ivanov, A.A., Alfimov, M.V. and Zheltikov, A.M. (2004) Cross-correlation frequency-resolved optical gating coherent anti-Stokes Raman scattering with frequency-converting photonic-crystal fibers. *Phys. Rev. E*, **70**, 57601–57604.

62 Lindfors, K., Kalkbrenner, T., Stoller, P. and Sandoghdar, V. (2004) Detection and

spectroscopy of gold nanoparticles using supercontinuum white light confocal microscopy. *Phys. Rev. Lett.*, **93**, 37401–37404.

63 Li, P., Shi, K. and Liu, Z. (2005) Optical scattering spectroscopy by using tightly focused supercontinuum. *Opt. Express*, **13**, 9039–9044.

64 Povazay, B., Bizheva, K., Unterhuber, A., Hermann, B., Sattmann, H., Fercher, A.F., Drexler, W., Apolonski, A., Wadsworth, W.J., Knight, J.C., Russell, P.S.J., Vetterlein, M. and Scherzer, E. (2002) Submicrometer axial resolution optical coherence tomography. *Opt. Lett.*, **27**, 1800–1802.

65 Chithrani, B.D., Ghazani, A.A. and Chan, W.C.W. (2006) Determining the size and shape dependence of gold nanoparticle uptake into mammalian cells. *Nano Lett.*, **6**, 662–668.

66 Xu, X.H.N., Chen, J., Jeffers, R.B. and Kyriacou, S. (2002) Direct Measurement of Sizes and Dynamics of single living membrane transporters using nanooptics. *Nano Lett.*, **2**, 175–182.

29
Super-Resolution Infrared Microspectroscopy for Single Cells

Makoto Sakai, Keiichi Inoue, and Masaaki Fujii

29.1
Introduction

29.1.1
Infrared Microscopy

Molecular vibration is often referred to as a "fingerprint" of molecules, because it sensitively reflects their geometry and environment. The technique of infrared microscopy, combining optical microscopes with IR spectroscopy, enables one to visualize the "molecular fingerprint" in various materials including biological samples. The infrared microscope has been used widely for the following reasons: (i) The IR absorption spectrum of small samples and domains can be easily measured, (ii) without causing any damage, and (iii) spatial mapping of specific materials by IR is possible [1–8]. However, the spatial resolution of infrared microscopes is very poor relative to other optical microscopes, meaning that other analytical methods are used, such as Raman microscopy, in order to achieve subcellular resolution [9–13]. This is because the spatial resolution of a conventional infrared microscope is restricted by the diffraction limit, which is almost the same as the wavelength of IR light. This diffraction limit prevents conventional infrared microscopes from achieving a better spatial resolution, because the IR wavelength is very long, ranging from 3 to 25 µm in the mid-infrared region. If a specific IR absorption band can be mapped with sub-micron spatial resolution, visualization of the molecular structure and reaction dynamics in a non-uniform environment such as a cell becomes a possibility.

29.1.2
Super-Resolution Microscopy by Two-Color Double Resonance Spectroscopy

In recent years, we have developed a two-color super-resolution laser scanning fluorescence microscope based on the up-conversion fluorescence depletion

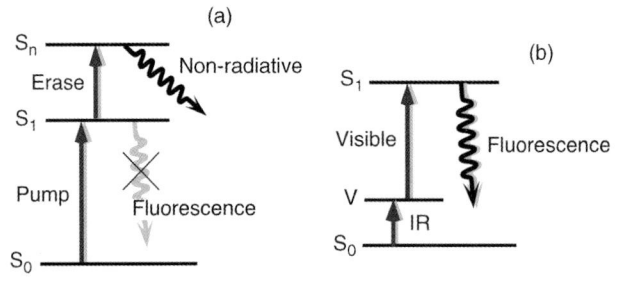

Figure 29.1 (a) Scheme of two-color fluorescence dip spectroscopy. The pump beam excites a molecule from the ground state (S_0) to the S_1 state. Then, the erase beam further excites the S_1 molecule to a higher excited state, S_n. Due to various relaxation processes from S_n states, such as internal conversion to the ground state, the molecule can decay from S_n without fluorescence. (b) Excitation scheme of transient fluorescence detected IR (TFD-IR) spectroscopy. The IR light excites a molecule to a vibrationally excited level. The visible light further excites it to the electronically excited S_1 state, where the transient fluorescence is emitted.

technique, that is a form of two-color double resonance spectroscopy [14–17]. The excitation scheme is shown in Figure 29.1a. In this technique, when a molecule is excited to the S_1 state by the first beam of laser light (pump light), the fluorescence intensity from S_1 is monitored. Here, the fluorescence intensity is proportional to the population in S_1. The S_1 molecule is further excited to a higher excited state, S_n by irradiating with a second beam of laser light (erase light). Because non-radiative relaxation processes, such as internal conversion or inter-system crossing, are generally accelerated in the S_n state, the fluorescence intensity from S_1 is depleted (fluorescence depletion effect). Therefore, excitation to the S_n state can be detected by fluorescence depletion. By applying this fluorescence depletion process to fluorescence microscopy, the size of the fluorescent region becomes smaller than the diffraction limit of the focused pump beam because the up-conversion fluorescence depletion process occurs only in the overlapping area of the two laser beams. This approach, which we have developed for over 20 years [18–20], proves the principle that super-resolution optical microscopes can be achieved by combining optical microscopes with two-color double resonance spectroscopies. In fact, super-resolution optical microscopes employing two-color double resonance spectroscopy are a hot topic, chosen by *Science* as one of the top 10 breakthroughs of 2006, and many scientists and biologists have noted the significance of this technique [21–24].

29.1.3
Transient Fluorescence Detected IR Spectroscopy

Figure 29.1b shows the principle of transient fluorescence detected IR (TFD-IR) spectroscopy, which was developed about 30 years ago by Kaiser and Laubereau [25, 26]. This technique is also a form of two-color double resonance spectroscopy, and allows operation in the IR region. Briefly, tunable IR light is introduced together with visible light of which the wavelength is fixed to be slightly longer than the visible absorption band. If the frequency of the IR light is not resonant with a vibrational level, no fluorescence will appear because the wavelength of visible light does not match the absorption band. When the IR frequency is resonant with an excited vibrational level, the vibrationally excited molecule generated by the IR absorption can absorb the visible light, and generate fluorescence (transient fluorescence). Thus the vibrational transition can be detected as this transient fluorescence. TFD-IR spectroscopy has high sensitivity because it is zero-background, and the IR absorption can be detected by the electronic transition, which has a large absorption cross-section. Moreover, it is also possible to observe the population dynamics of vibrational cooling by adjusting the time delay between the IR and visible light.

29.1.4
Application to Super-Resolution Infrared Microscopy

Super-resolution infrared microscopy is achieved by applying TFD-IR spectroscopy to a fluorescence microscope. A sample is irradiated with IR and visible light. At the focal point, IR and visible light are focused to diameters restricted by their respective diffraction limits. The transient fluorescence due to IR excitation appears only in the spatial region where both IR and visible beams overlap. This overlap region can be smaller than the diffraction limit of IR or visible light (see Figure 29.2a). Thus it is possible to obtain the IR absorption spectrum of a region smaller than the diffraction limit.

More realistically, samples are typically irradiated with co-linear IR and visible light beams. In this case, super-resolution is also achieved in the IR light. When the same objective lens is used, the diffraction limit is proportional to the wavelength. The IR

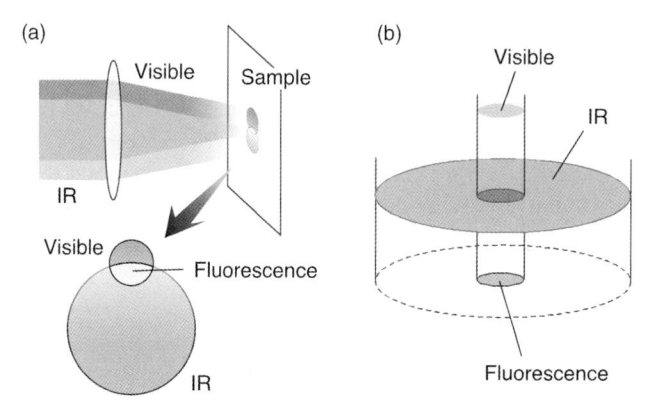

Figure 29.2 Concept of super-resolution infrared microscopy based on TFD-IR spectroscopy. (a) Principle of super-resolution infrared microscopy. The transient fluorescence due to IR excitation takes place in the area where the IR and visible lights overlap. This overlapped region can be smaller than the diffraction limit of IR or visible light. (b) Co-linear visible and IR light beams are introduced to the sample. The size of the overlapping region is identical to the diffraction limit of the visible light employed.

wavelength is much longer than the visible wavelength; visible light can be focused much more tightly than IR light. Transient fluorescence appears only in the overlapping region of IR and visible light, and the size of the overlapping region is precisely the same as the diffraction limit of visible light (see Figure 29.2b). This means that IR information is obtained with the spatial resolution of visible light, that is, the IR is super-resolved. Furthermore, picosecond time-resolved TFD-IR spectroscopy gives access to the time-evolution of the vibrationally excited molecule. This means that the measurement of picosecond dynamics is possible with IR super-resolution in this form of microscopy.

29.2
Experimental Set-Up for Super-Resolution Infrared Microscopy

29.2.1
Picosecond Laser System

In super-resolution infrared microscopy, we monitor the transient fluorescence pumped from a vibrationally excited level. In a vibrationally excited level, vibrational relaxation occurs on a picosecond timescale, which means that the use a picosecond laser system is required. The laser set-up for infrared microscopy is essentially the same as that for the transient fluorescence detected IR spectroscopy [27, 28]. Seed pulses from a cw mode-locked Ti:sapphire laser (Spectra Physics, Tsunami) were introduced into a regenerative Ti:sapphire amplifier (Quanta Ray, TSA-10). The amplified output was composed of 2–3 ps pulses with energies of 6 mJ at 800 nm. 20% of the output pulse was used to pump a traveling-wave optical parametric amplifier system (Light conversion, TOPAS 400) after frequency doubling in BBO, to provide tunable visible light. The remaining 80% of the output pulse was split into two beams, and one was introduced into another OPA system (Light conversion, TOPAS 800). The second harmonic of the idler wave from the OPA and the remaining output pulse were differentially mixed in a KTA crystal to generate tunable IR light in the 3 μm region. For the generation of mid-IR light around the 6 μm region, the signal and idler waves from the OPA were differentially mixed in a $AgGaS_2$ crystal. Both visible and IR light have a spectral bandwidth of ≈15 cm^{-1} and a pulse width of ≈3 ps.

29.2.2
Fluorescence Detection System

29.2.2.1 Optical Layout for the Solution and Fluorescent Beads
The fluorescence detection system employed is shown in Figure 29.3. Both the IR and visible light (∼3 ps, ∼15 cm^{-1}) generated by the picosecond laser system were introduced into a home-made laser fluorescence microscope [29, 30]. For the measurement of both solutions and fluorescent beads, both beams were adjusted onto a co-linear path by a beam-combiner and focused into the sample by an objective

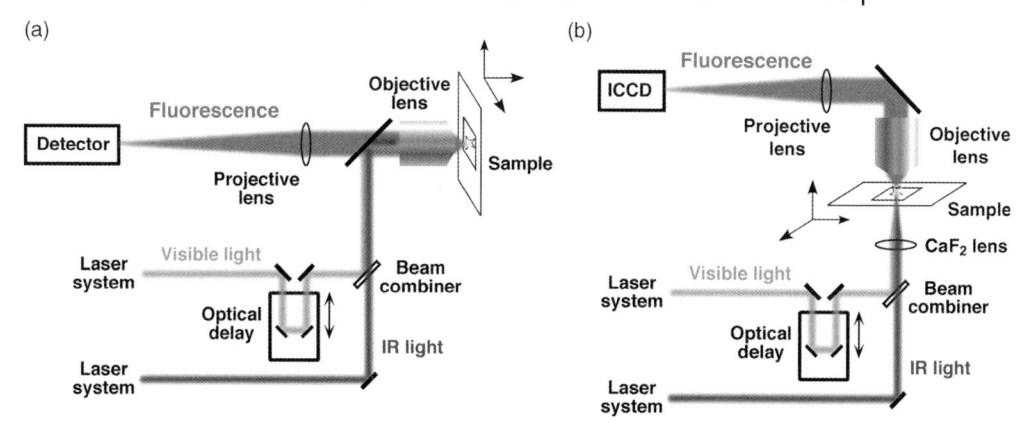

Figure 29.3 (a) Optical layout for experiments on solutions and fluorescent beads. Picosecond IR and visible light beams were coaxially combined by a beam combiner, and were focused on the sample through an objective lens. The transient fluorescence at the focal point was collected by the same objective lens and was focused onto a photodetector. (b) Optical layout for the cells. Both IR and visible light beams are used as the illumination light, and the transient fluorescence is collected from the opposite side by an objective lens.

reflection lens (Sigma, OBLR-20, NA = 0.38). Here, the sample used was Rhodamine-6G dissolved in chloroform-d_1 to give a concentration of 5×10^{-4} mol dm^{-3} or fluorescent beads of 15 μm diameter.

The transient fluorescence from the sample at the focal point was collected by the same objective lens and was projected onto a photodetector. For the measurement of the transient fluorescence image, the signal was recorded by a CCD camera with an image-intensifier (Princeton Instruments Inc., PI-MAX-512) and stored on a personal computer as a fluorescence image. For the measurement of IR spectra, the transient fluorescence signal was detected by a photomultiplier (Hamamatsu, 1P28) and integrated by a digital boxcar (EG&G PARC, 4420/4422) before being recorded by a personal computer as a function of the IR laser frequency. The delay time between the IR and visible light was varied by an optical delay system (Sigma, LTS-400X). The optical layout is shown in Figure 29.3a.

29.2.2.2 Optical Layout for Biological Samples

The optical layout for the measurement of biological samples (cells) is shown in Figure 29.3b. The sample was irradiated with co-linear IR and visible light beams. The transient fluorescence from the sample was collected from the opposite side by an objective lens. In this optical layout, the spatial resolution was determined by the objective numerical aperture (NA) and the visible fluorescence wavelength; IR super-resolution smaller than the diffraction limit of IR light was achieved. Here, *Arabidopsis thaliana* roots stained with Rhodamine-6G were used as a sample. We applied this super-resolution infrared microscope to the *Arabidopsis thaliana* root cells, and also report the results of time-resolved measurements.

IR and visible light beams were superposed onto a co-linear path by a beam-combiner, and focused into the sample by a CaF$_2$ lens ($f = 100$ mm). The focal spot sizes of the IR and visible light beams were adjusted to about 200 μm at the sample position. The transient fluorescence from the sample was collected from the opposite side by a NA = 0.25 objective lens (Newport, M-10X), and was projected onto a CCD camera with an image-intensifier (Princeton Instruments Inc., PI-MAX-1K-HB) and recorded by a personal computer as a fluorescence image. For measuring the time-resolved transient fluorescence detected IR image, the delay time between the IR and visible light beams was varied by an optical delay system (Sigma, LTS-400X). In this optical layout, IR and visible light beams were used as the illumination light for the microscope, and the spatial resolution at ≈570 nm, which is the fluorescence wavelength of Rhodamine-6G, was 1.4 μm from the Rayleigh diffraction limit. This is much smaller than the diffraction limit (8.1 μm) of IR light.

29.2.3
Sample

Rhodamine-6G, a typical dye for fluorescence probes, shows S_1–S_0 absorption around the 500–550 nm region, with the maximum at 530 nm [31]. The visible light was fixed to around 610 nm, which is longer than the S_1–S_0 absorption. The IR light was scanned in the region of the CH and NH stretching vibrations of Rhodamine-6G (2700–3700 nm). Under these conditions, the total energy of visible plus IR reaches somewhere around the absorption maximum at 530 nm. On the other hand, the fluorescence region of Rhodamine-6G is from 550 to 590 nm, with the maximum at 565 nm [32]. Therefore, the transient fluorescence was monitored through a 555–575 nm band-pass filter. The visible light around 610 nm was cut off by a notch filter.

Dye laser grade Rhodamine-6G was purchased from Exciton (R590) and was used without any further purification. Fluorescent beads of 15 μm diameter were purchased from Molecular Probes (Orange, excitation 540 nm/fluorescence 560 nm) and were used after spreading on a glass plate.

Arabidopsis thaliana seedlings were grown using a 1×10^{-4} M Rhodamine-6G aqueous solution on a filter paper of glass fiber. For the preparation of microscopic specimens, *Arabidopsis thaliana* roots were crushed between two glass slides after soaking in 1 M HCl in a hot-water bath for 10 min, and washed with 45% acetic acid and 3:1 methanol/acetic acid. The preparation was sealed in nujol mull with a cover slip.

29.3
Results and Discussion

29.3.1
Transient Fluorescence Image with IR Super-Resolution in Solution

Figure 29.4 shows the fluorescence images of Rhodamine-6G in a chloroform-d_1 solution (concentration, 5×10^{-4} mol dm^{-3}) observed by introducing (a) only IR

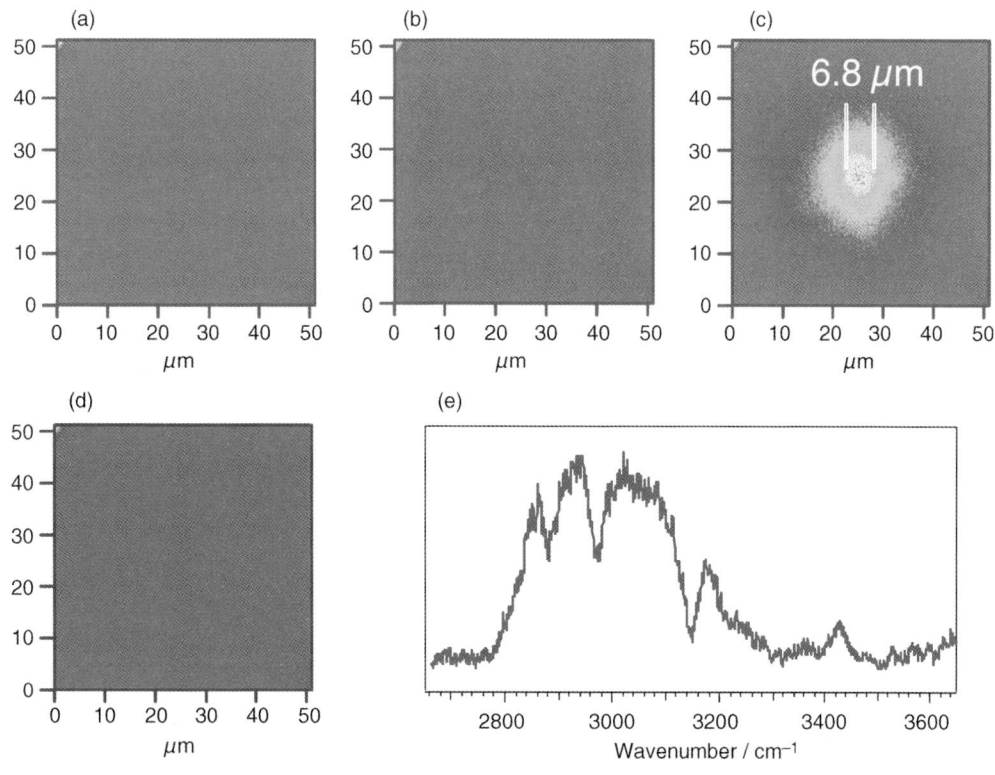

Figure 29.4 The fluorescence images of Rhodamine-6G in a chloroform-d_1 solution (concentration, 5×10^{-4} mol dm^{-3}) observed by introducing (a) only IR (IR wavelength, 3400 nm; diameter, 2.5 mmϕ), (b) only visible (visible wavelength, 621 nm; diameter, 7.0 mmϕ), (c) both IR (IR wavelength, 3400 nm) and visible light, and (d) both IR (IR wavelength, 2750 nm) and visible light into a quartz cell. (e) TFD-IR spectrum of Rhodamine-6G. The IR wavelength was scanned from 2.7 μm (3700 cm^{-1}) to 3.8 μm (2650 cm^{-1}) while monitoring the transient fluorescence signal of (c).

light (IR wavelength, 3400 nm; diameter, 2.5 mmϕ), (b) only visible light (visible wavelength, 621 nm; diameter, 7 mmϕ), (c) both IR (IR wavelength = 3400 nm) and visible light, and (d) both IR (IR wavelength = 2750 nm) and visible light into a quartz cell. No fluorescence appears with either (a) (only IR) or (b) (only visible light). In contrast, the transient fluorescence image is clearly observed when the sample is irradiated with both infrared and visible light (c). Moreover, the transient fluorescence image disappears when the IR wavelength is changed from 3400 nm to 2750 nm (d), which is out of the range of IR absorption of Rhodamine-6G. This clearly demonstrates that the transient fluorescence images originate from an IR–visible double resonance signal that appears via the vibrationally excited level.

To confirm that the transient fluorescence is indeed an action signal due to resonant infrared absorption, the fluorescence intensity in the image was measured as a function of IR wavelength (TFD-IR spectroscopy, [27]). Figure 29.4e shows the TFD-IR spectrum of Rhodamine-6G. When the IR wavelength was scanned over the

CH/NH/OH stretching vibration region from 2650 to 3650 cm^{-1}, the transient fluorescence signal clearly varied as a function of the IR wavelength. This spectrum matches well the IR spectrum of Rhodamine-6G obtained using a conventional IR absorption spectrometer [33]. Therefore, it can be concluded that the observed fluorescence image shown in Figure 29.4c does indeed correspond to the infrared image at 3400 nm.

The observed diameter of the transient fluorescence image, that is the TFD-IR image, is 6.8 μm FWHM (see Figure 29.4c). The theoretical diffraction limit for the IR light (3400 nm, 2.5 mmϕ, $f = 10$ mm; effective NA = 0.125) is about 16 μm FWHM. Thus it is clear that the TFD-IR image obtained by this technique gives an infrared image smaller than the diffraction limit. The expected spatial resolution of the infrared microscopy is at least two times higher than the observed diffraction limit. Furthermore, the resolution can be improved because the TFD-IR image resolution of 6.8 μm is still larger than the diffraction limit of the visible light, which is 1 μm under these particular experimental conditions (621 nm, 7.0 mmϕ, $f = 10$ mm; effective NA = 0.35). In principle, it should be possible to reach a resolution of about 1 μm. The loss of resolution observed here may be due to low axial resolution and we intend to improve it significantly by means of a confocal optical system. This is significant for the application of this microscopy and microspectroscopy to realistic systems such as biological tissues and so on.

One of the important functions of this infrared microscope is the measurement of the IR spectrum from a spatial region smaller than the diffraction limit. This possibility is already illustrated in Figure 29.4e. The TFD-IR spectrum, that corresponds to the IR absorption spectrum, was measured from a fluorescence region smaller than the IR diffraction limit. Infrared spectroscopy in a sub-micron region will be possible by using a high NA objective lens with the confocal optical system.

29.3.2
Picosecond Time-Resolved Measurement

Another important aspect of this method is picosecond time-resolved IR imaging. Originally TFD-IR spectroscopy was used to investigate vibrational relaxation processes [25–27]. In this form of spectroscopy, the population dynamics can be observed by the time-evolution of transient fluorescence images as the delay time between IR and visible light is adjusted. Figure 29.5 shows picosecond time-resolved transient fluorescence images when the IR and visible lights are fixed to 3400 nm and 621 nm, respectively. At a delay time of −3 ps, when the visible is applied before the IR, no signal is observed. At 0 ps delay time when the IR and visible beams are applied at the same time, a transient fluorescence image clearly appears. The transient fluorescence increases in intensity up to 3 ps and subsequently decays with time. The observed time-evolution of the fluorescence image represents the population decay from the vibrational level prepared by the IR together with the spatial mapping of the hot molecules at a specific time. This time-resolved IR imaging shows excellent promise for space and time-resolved spectroscopy, which is necessary

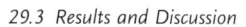

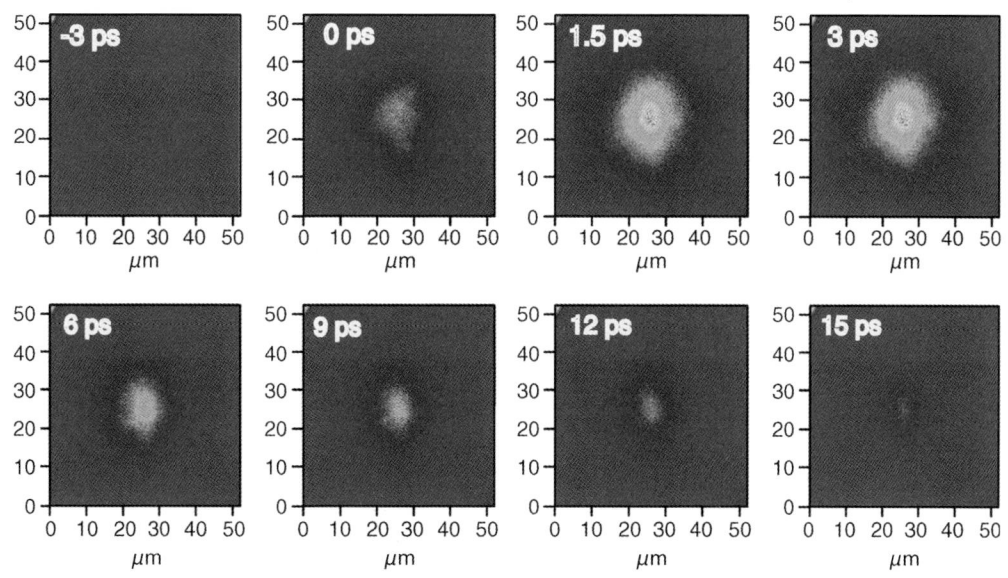

Figure 29.5 Picosecond time-resolved transient fluorescence images at several delay times, when the IR (diameter, 2.5 mmϕ) and visible (diameter, 7 mmϕ) lights are fixed to 3400 nm and 621 nm, respectively.

for the study of non-uniform systems such as catalysts, or biological systems like whole cells.

29.3.3
Application to Fluorescent Beads

In this section, we report the application of super-resolution infrared microscopy to a Rhodamine-6G doped fluorescence bead; a minute sample. Figure 29.6a shows a fluorescence image of a 15 μm diameter fluorescent bead observed by using 539 nm pumping light. This corresponds to the fluorescence image taken by a conventional fluorescent microscope. An image of about 15 μm diameter is observed. The images of the fluorescent bead measured by our super-resolution infrared microscope are shown in Figure 29.6b–d. No fluorescence is observed when only visible light at 621 nm (Figure 29.6b) or only IR light at 3300 nm (Figure 29.6c) are applied to the bead. However, when both IR and visible lights are introduced, the transient fluorescence is clearly observed, as shown in Figure 29.6d. The fluorescence intensity decreased remarkably when the IR wavelength was changed from 3300 nm to 2810 nm. This shows that transient fluorescence also depends on IR absorption in the fluorescent bead sample.

The spatial resolution of the observed TFD-IR image can be evaluated from Figure 29.7. Figure 29.7a is a normal fluorescence image obtained by 539 nm pumping, and Figure 29.7b is a TFD-IR image measured using both IR and visible pumping lasers. Cross-section profiles along the white lines drawn in Figure 29.7a

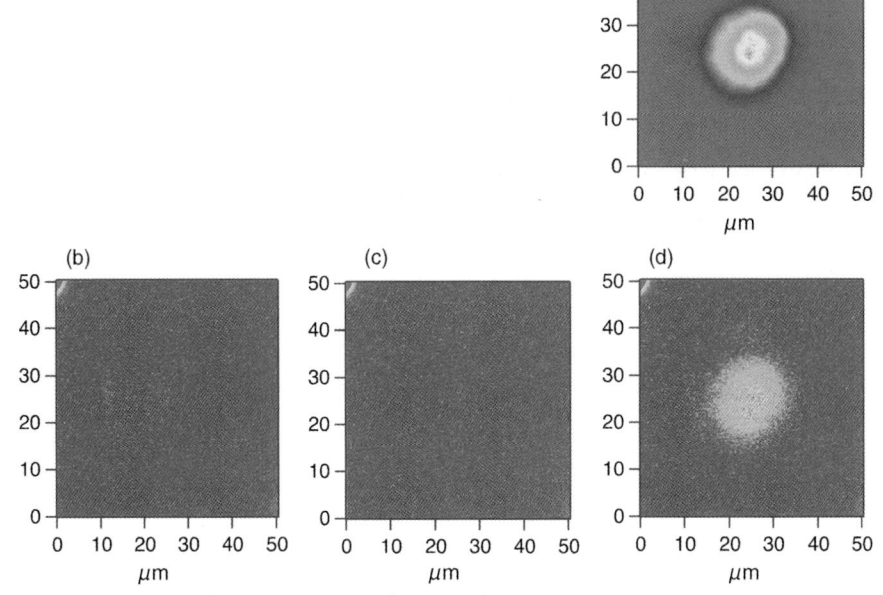

Figure 29.6 (a) A fluorescence image of a 15 μm diameter fluorescent bead observed by using 539 nm pumping light. The image of a fluorescent bead measured by applying (b) only IR (IR wavelength, 3300 nm; diameter, 2.5 mmϕ), (c) only visible (visible wavelength, 621 nm; diameter, 7 mmϕ), (d) both IR and visible light.

and b are shown in c and d. The simulated fluorescence intensity for the 15 μm diameter bead with the diffraction limit (16 μm FWHM) of IR light (IR wavelength, 3300 nm; effective NA = 0.125) is shown in Figure 29.7d as a solid curve. It is clear that the TFD-IR image is smaller than the IR diffraction limit, and super-resolution is achieved for IR microscopy. At a glance, both the normal fluorescence image and the TFD-IR image are similar. This suggests that the spatial resolution of the TFD-IR image is close to the visible fluorescence resolution. The dashed curve shows the simulated fluorescence intensity for the 15 μm diameter bead with the assumption that the effective spatial resolution of this microscope is 6.8 μm FWHM, which was measured by the TFD-IR image in the Rhodamine-6G solution (see Figure 29.4c). Here, the size of the bead is large enough relative to the calculated focal spot size (1 μm) and depth of focus (2.5 μm) that we used the size of the image in the dye solution as the effective spatial resolution. The simulated profile matches well both the normal fluorescence and the TFD-IR images. This means that the spatial resolution is determined only by visible light, not IR light, and provides evidence of super-resolution in the TFD-IR image.

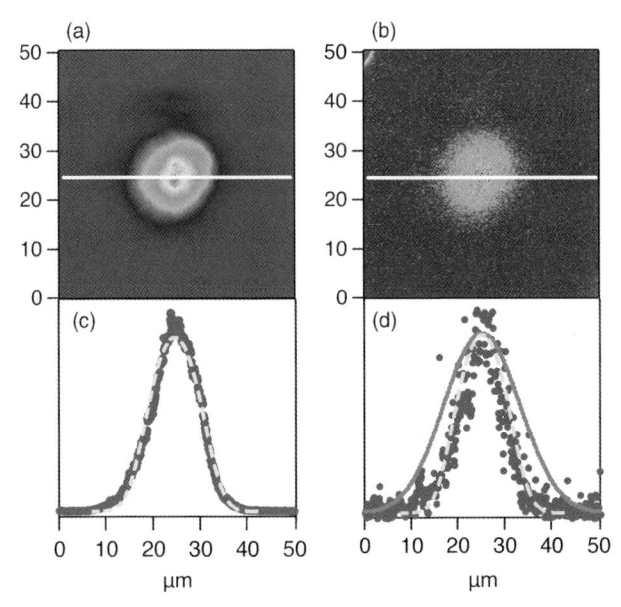

Figure 29.7 (a) A normal fluorescence image by 539 nm pumping. (b) TFD-IR image measured by IR and visible pumping lasers. (c), (d) Cross-section profiles along the white lines of (a) and (b). The simulated fluorescence intensity for the 15 μm diameter bead with 6.8 μm FWHM is shown by a dashed white curve, and with 16 μm FWHM that is the diffraction limit of IR light (IR wavelength, 3300 nm; diameter, 2.5 mmϕ; effective NA = 0.125) is shown by a gray curve.

29.3.4
Application to Whole Cells

29.3.4.1 Super-Resolution IR Imaging of *Arabidopsis thaliana* Roots

Figure 29.8a shows a fluorescence image of *Arabidopsis thaliana* roots labeled with Rhodamine-6G, observed by introducing 539 nm light. This corresponds to a fluorescence image taken by a conventional fluorescent microscope. As can be seen from the figure, Rhodamine-6G stains the inside of the cell uniformly. Figure 29.8b–d show transient fluorescence detected IR (TFD-IR) images of *Arabidopsis thaliana* roots labeled by Rhodamine-6G, observed by introducing: (b) only visible light (wavelength, 607 nm), (c) only IR light (wavelength, 3300 nm), and (d) both IR and visible light. No fluorescence appears in (b) only visible and in (c) only IR light. On the other hand, transient fluorescence clearly appears by introducing in (d) both IR and visible light, and a TFD-IR image that is almost the same as (a) is observed with a spatial resolution higher than the diffraction limit of IR light. TFD-IR images disappear when the visible light is applied before the IR light.

We also measured picosecond time-resolved TFD-IR images from −10 to 50 ps. Figure 29.9 shows the picosecond time-resolved TFD-IR images obtained around a 0 ps delay time. For these TFD-IR images, the population decay of the vibrationally excited Rhodamine-6G molecule in a cell is demonstrated as the delay-time dependent fluorescence. At a −5 ps delay time, when visible light is applied before IR light,

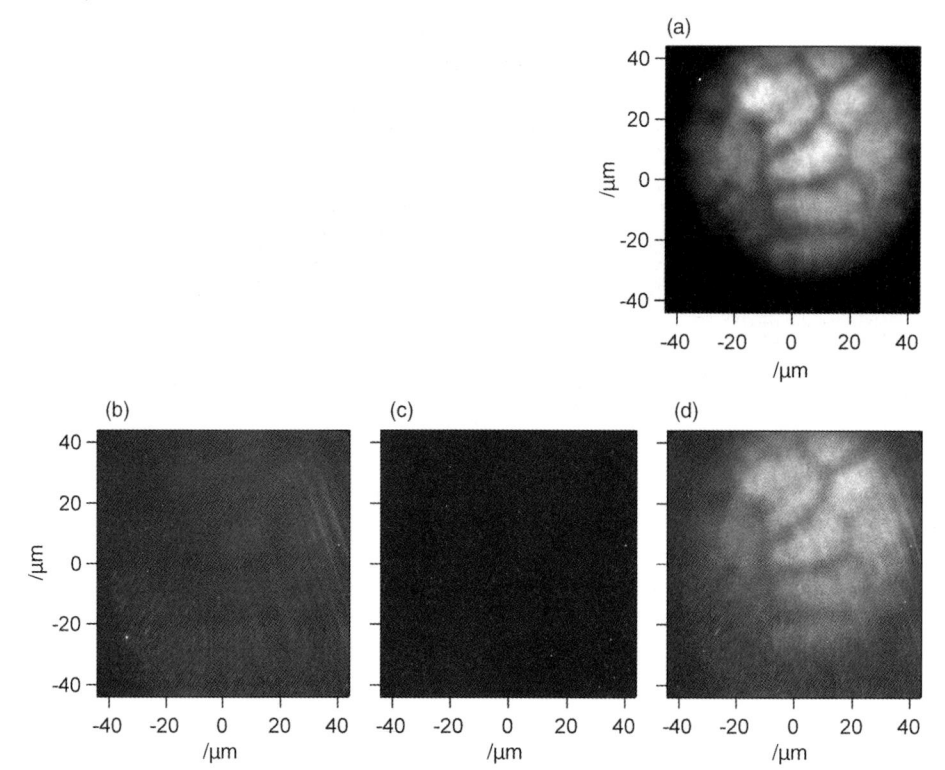

Figure 29.8 (a) Fluorescence image of *Arabidopsis thaliana* roots labeled by Rhodamine-6G, observed by introducing 539 nm light. (b)–(d) TFD-IR images of *Arabidopsis thaliana* roots by introducing (b) only visible (wavelength, 607 nm), (c) only IR (wavelength, 3300 nm), and (d) both IR and visible light. The spatial resolution observed from the cross-section of the image in (d) is almost the same as the diffraction limit of 1.4 μm.

the TFD-IR image is not observed at all. However, at 0 ps when visible and IR light are applied simultaneously, a TFD-IR image clearly appears, and subsequently decays with time.

29.3.4.2 Vibrational Relaxation Dynamics in the Cells

Figure 29.10 shows time-profiles of the TFD-IR signal intensity at positions A–C on the TFD-IR image in Figure 29.9. As can be seen, all of the time-profiles show the same behavior. An interesting feature is that the vibrational energy of Rhodamine-6G is not completely lost, even at a 50 ps delay time, which is sufficiently longer than the fast vibrational energy flow with an exponential decay constant of 1.5 ps.

Many previous reports [34–44] on the relaxation dynamics of vibrational energy in molecules of a similar size have shown that vibrational energy flows from a solute to solvent molecules on a picosecond time scale, and is completely lost after typically $\approx$20 ps in most solutions. This suggests that the vibrational relaxation dynamics in

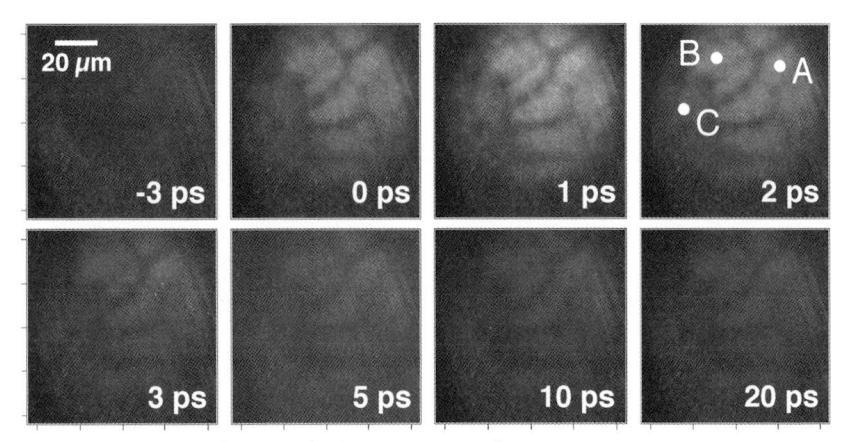

Figure 29.9 Picosecond time-resolved TFD-IR images of *Arabidopsis thaliana* roots labeled with Rhodamine-6G at delay times around 0 ps. The IR and visible light are fixed to 3300 nm and 607 nm, respectively.

the cell is quite different from that of the solute–solution system. We assume that the vibrational relaxation of Rhodamine-6G in the cell is greatly influenced by the neighboring environment. Inside the cell is a non-uniform environment, consisting of many parts, including a nucleus, cytoplasm, and a cell membrane, which interact

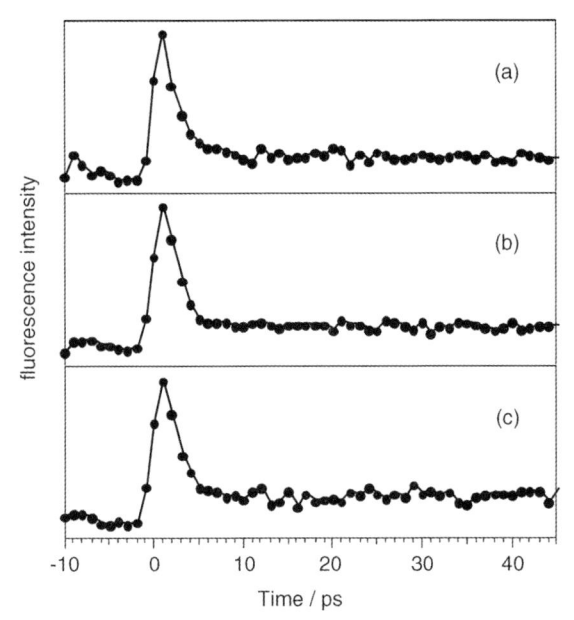

Figure 29.10 Time-profiles of the TFD-IR signal intensity at positions A–C on the TFD-IR image of Figure 29.9. All of the time-profiles show that the vibrational energy of Rhodamine-6G is not completely lost, even at a 50 ps delay time. The IR and visible lights are fixed to 3300 nm and 607 nm, respectively.

with each other or water molecules in the cell. This may cause the site dependence of vibrational relaxation in a cell. In order to understand the more detailed characteristic dynamics occurring within cells, vibrational relaxation of probe molecules in specific parts of cells, such as the nucleus, cytoplasm, and cell membrane, will be measured by using other probe fluorescent dyes in the next stage of our investigations.

29.4
Summary

We have performed super-resolution infrared microscopy by combining a laser fluorescence microscope with picosecond time-resolved TFD-IR spectroscopy. In this chapter, we have demonstrated that the spatial resolution of the infrared microscope improved to more than twice the diffraction limit of IR light. It should be relatively straightforward to improve the spatial resolution to less than 1 μm by building a confocal optical system. Thus, in the near future, the spatial resolution of our infrared microscope will be improved to a sub-micron scale.

Furthermore, this new super-resolution infrared microscopy is capable of capitalizing on TFD-IR spectroscopy's ability to study vibrational relaxation processes.

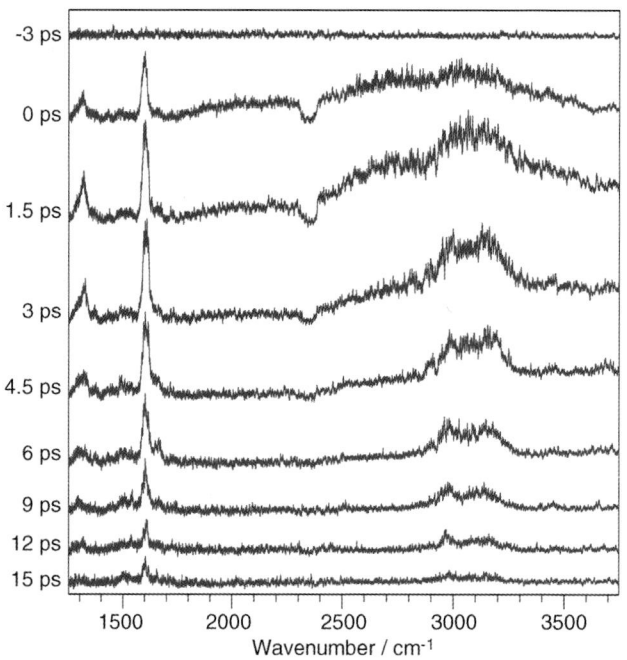

Figure 29.11 Picosecond time-resolved TFD-IR spectra of Rhodamine-6G at several delay times. The visible light is fixed to 585 nm and the IR wavelength was scanned from 2.7 μm (3700 cm^{-1}) to 8 μm (1250 cm^{-1}).

Therefore, by using this super-resolution infrared microscope, we will be able to carry out space- and time-resolved vibrational microspectroscopy in the IR super-resolved region. Given that IR absorption is regarded as the "fingerprint" of a molecule, the new super-resolution infrared "microspectroscopy" will become an extremely important tool, not only in microscopy but also in spectroscopy.

Recent tunable picosecond IR lasers can generate IR light up to 10 µm. This means we can easily extend this imaging and microspectroscopic method into the mid-IR region, where the strong IR absorption of water around the 3 µm region can be avoided. Figure 29.11 shows a demonstration of picosecond time-resolved TFD-IR spectra of Rhodamine-6G/chloroform-d_1 solution in the 1250–3800 cm^{-1} (8–2.7 µm) mid-IR region. The visible light is fixed to 585 nm. As can be seen, we are able to perform infrared microspectroscopy at least up to the 8 µm (1250 cm^{-1}) region. The resolution of this infrared microscopy is determined by the diffraction limit of visible light, not by IR. Consequently, sub-micron resolution should be retained, even in the mid-IR region for which the diffraction limit can be as high as a few tens of micrometers.

We have also demonstrated picosecond time-resolved TFD-IR imaging of the vibration relaxation of Rhodamine-6G in *Arabidopsis thaliana* roots, and found an abnormally long-lived component of vibrational relaxation in a cell. This may result from a site dependence of vibrational relaxation within whole cells. These results indicate the possible utility of the two-color super-resolution infrared microscope in mapping specific IR absorptions with high spatial resolution, and the observation of dynamics in a non-uniform environment, such as a cell. By using this infrared super-resolution microscope, we will be able to visualize the structure and reaction dynamics of molecules in a wide range of non-uniform environments.

Acknowledgments

The present work was financially supported in part by a Grants-in-Aid for Scientific Research (KAKENHI) on Priority Areas (Area No. [432] and Area No. [477]) from the Ministry of Education, Culture, Sports, Science and Technology (MEXT) of Japan. The authors gratefully thank Professor M. Kinjo and Professor N. Ohta in Hokkaido University for providing the *Arabidopsis thaliana* sample, and Dr T. Ohmori for stimulating discussion. The authors also thank Dr T. Watanabe and Dr Y. Iketaki of the Olympus Company for providing technical support for the laser fluorescence microscope, and Professor J. R. Woodward for valuable comments on the manuscript.

References

1 Kidder, L.H., Haka, A.S. and Lewis, E.N. (2002) *Handbook of Vibrational Spectroscopy* (eds J.M. Chalmers and P.R. Griffiths), John Wiley and Sons, Chichester.

2 Chan, K.L.A. and Kazarian, S.G. (2003) New opportunities in micro- and macro-attenuated total reflection infrared spectroscopic imaging: Spatial resolution

and sampling versatility. *Appl. Spectrosc.*, **57**, 381–389.

3 Yano, K., Ohoshima, S., Gotou, Y., Kumaido, K., Moriguchi, T. and Katayama, H. (2000) Direct measurement of human lung cancerous and noncancerous tissues by Fourier transform infrared microscopy: Can an infrared microscope be used as a clinical tool? *Anal. Biochem.*, **287**, 218–225.

4 Bailey, J.A., Dyer, R.B., Graff, D.K. and Schoonover, J.R. (2000) High spatial resolution for IR imaging using an IR diode laser. *Appl. Spectrosc.*, **54**, 159–163.

5 Aoki, K., Nakagawa, M. and Ichimura, K. (2000) Self-assembly of amphoteric azopyridine carboxylic acids: Organized structures and macroscopic organized morphology influenced by heat, pH change, and light. *J. Am. Chem. Soc.*, **122**, 10997–11004.

6 Rammelsberg, R., Boulas, S., Chorongiewski, H. and Gerwert, K. (1999) Set-up for time-resolved step-scan FTIR spectroscopy of noncyclic reactions. *Vib. Spectrosc.*, **19**, 143–149.

7 Kidder, L.H., Kalasinsky, V.F., Luke, J.L., Levin, I.W. and Lewis, E.N. (1997) Visualization of silicone gel in human breast tissue using new infrared imaging spectroscopy. *Nat. Med.*, **3**, 235–237.

8 Marcott, C., Reeder, R.C., Paschalis, E.P., Tatakis, D.N., Boskey, A.L. and Mendelsohn, R. (1998) *Cell. Mol. Biol.*, **44**, 109–115.

9 Born, M. and Wolf, W. (1997) *Principle of Optics*, 7th edn, Cambridge University Press, Cambridge.

10 Wilson, T. and Sheppard, C.J.R. (1984) *Theory and Practice of Scanning Optical Microscopy*, Academic Press, London.

11 Puppels, G.J., De Mul, F.F.M., Otto, C., Greve, J., Robert-Nicoud, M., Arndt-Jovin, D.J. and Jovin, T.M. (1990) Studying single living cells and chromosomes by confocal Raman microspectroscopy. *Nature*, **347**, 301–303.

12 Huang, Y.-S., Karashima, T., Yamamoto, M., Ogura, T. and Hamaguchi, H. (2004) Raman spectroscopic signature of life in a living yeast cell. *J. Raman Spectrosc.*, **35**, 525–526.

13 Kano, H. and Hamaguchi, H. (2006) Vibrational imaging of a single pollen grain by ultrabroadband multiplex coherent anti-stokes Raman scattering microspectroscopy. *Chem. Lett.*, **35**, 1124–1125.

14 Watanabe, T., Iketaki, Y., Omatsu, T., Yamamoto, K., Ishiuchi, S., Sakai, M. and Fujii, M. (2003) Two-color far-field super-resolution microscope using a doughnut beam. *Chem. Phys. Lett.*, **371**, 634–639.

15 Iketaki, Y., Watanabe, T., Ishiuchi, S., Sakai, M., Omatsu, T., Yamamoto, K. and Watanabe, T. (2003) Investigation of the fluorescence depletion process in the condensed phase; application to a tryptophan aqueous solution. *Chem. Phys. Lett.*, **372**, 773–778.

16 Watanabe, T., Iketaki, Y., Omatsu, T., Yamamoto, K., Sakai, M. and Fujii, M. (2003) Two-point-separation in super-resolution fluorescence microscope based on up-conversion fluorescence depletion technique. *Opt. Express*, **11**, 3271–3276.

17 Watanabe, T., Iketaki, Y., Omatsu, T., Yamamoto, K. and Fujii, M. (2005) Two-point separation in far-field super-resolution fluorescence microscopy based on two-color fluorescence dip spectroscopy, Part I: Experimental evaluation. *Appl. Spectrosc.*, **59**, 868–872.

18 Fujii, M., Ebata, T., Mikami, N. and Ito, M. (1983) 2-Color multiphoton ionization of diazabicyclooctane in a supersonic free jet. *Chem. Phys. Lett.*, **101**, 578–582.

19 Ito, M. and Fujii, M. (1988) *Advances in Multiphoton Processes and Spectroscopy*, vol. 4 (ed. S.H. Lin), World Scientific, Singapore.

20 Endo, Y. and Fujii, M. (1998) *Nonlinear Spectroscopy for Molecular Structure Determination* (eds R.W. Field, E. Hirota, J.P. Maier and S. Tsuchiya), Blackwell Science, Oxford.

21 Willig, K.I., Rizzoli, S.O., Westphal, V., Jahn, R. and Hell, S.W. (2006) STED microscopy reveals that synaptotagmin

remains clustered after synaptic vesicle exocytosis. *Nature*, **440**, 935–939.

22 Betzig, E., Patterson, G.H., Sougrat, R., Lindwasser, O.W., Olenych, S., Bonifacino, J.S., Davidson, M.W., Lippincott-Schwartz, J. and Hess, H.F. (2006) Imaging intracellular fluorescent proteins at nanometer resolution. *Science*, **313**, 1642–1645.

23 Donnert, G., Keller, J., Medda, R., Andrei, M.A., Rizzoli, S.O., Lührmann, R., Jahn, R., Eggeling, C. and Hell, S.W. (2006) Macromolecular-scale resolution in biological fluorescence microscopy. *Proc. Natl. Acad. Sci. USA*, **103**, 11440–11445.

24 Kittel, R.J., Wichmann, C., Rasse, T.M., Fouquet, W., Schmidt, M., Schmid, A., Wagh, D.A., Pawlu, C., Kellner, R.R., Willig, K.I., Hell, S.W., Buchner, E., Heckmann, M. and Sigrist, S.J. (2006) Bruchpilot promotes active zone assembly, Ca^{2+} channel clustering, and vesicle release. *Science*, **312**, 1051–1054.

25 Seilmeier, A., Kaiser, W., Laubereau, A. and Fischer, S.F. (1978) Novel spectroscopy using ultrafast 2-pulse excitation of large polyatomic-molecules. *Chem. Phys. Lett.*, **58**, 225–229.

26 Gottfried, N.H., Seilmeier, A. and Kaiser, W. (1984) Transient internal temperature of anthracene after picosecond infrared excitation. *Chem. Phys. Lett.*, **111**, 326–332.

27 Sakai, M. and Fujii, M. (2004) Vibrational energy relaxation of the 7-azaindole dimer in CCl_4 solution studied by picosecond time-resolved transient fluorescence detected IR spectroscopy. *Chem. Phys. Lett.*, **396**, 298–302.

28 Sakai, M., Ishiuchi, S. and Fujii, M. (2002) Picosecond time-resolved nonresonant ionization detected IR spectroscopy on 7-azaindole dimer. *Eur. Phys. J. D*, **20**, 399–402.

29 Sakai, M., Kawashima, Y., Takeda, A., Ohmori, T. and Fujii, M. (2007) Far-field infrared super-resolution microscopy using picosecond time-resolved transient fluorescence detected IR spectroscopy. *Chem. Phys. Lett.*, **439**, 171–176.

30 Sakai, M., Ohmori, T. and Fujii, M. (2007) *Nano Biophotonics* (eds H. Masuhara, S. Kawata and F. Tokunaga), Elsevier, Oxford, pp. 189–195.

31 Sahar, E. and Treves, D. (1997) Excited singlet-state absorption in dyes and their effect on dye-lasers. *IEEE. J. Quantum Electron.*, **13**, 962–967.

32 Soan, P.J., Case, A.D., Damzen, M.J. and Hutchinson, M.H.R. (1992) High-reflectivity 4-wave-mixing by saturable gain in rhodamine-6G dye. *Opt. Lett.*, **17**, 781–783.

33 SDBSWeb: http://www.aist.go.jp/RIODB/SDBS/ (National Institute of Advanced Industrial Science and Technology, March 2004).

34 Elsaesser, T. and Kaiser, W. (1991) Vibrational and vibronic relaxation of large polyatomic-molecules in liquids. *Annu. Rev. Phys. Chem.*, **42**, 83–107.

35 Hamaguchi, H. and Gustafson, T.L. (1994) Ultrafast time-resolved spontaneous and coherent Raman-spectroscopy – the structure and dynamics of photogenerated transient species. *Annu. Rev. Phys. Chem.*, **45**, 593–622.

36 Iwata, K. and Hamaguchi, H. (1994) Picosecond time-resolved Raman-spectroscopy of S1 p-terphenyl and p-terphenyl-d_{14} in solution – time-dependent changes of Raman band shapes. *J. Raman Spectrosc.*, **25**, 615–621.

37 Sension, R.J., Szarka, A.Z. and Hochstrasser, R.M. (1992) Vibrational-energy redistribution and relaxation in the photoisomerization of *cis*-stilbene. *J. Chem. Phys.*, **97**, 5239–5242.

38 Mizutani, Y. and Kitagawa, T. (1997) Direct observation of cooling of heme upon photodissociation of carbonmonoxy myoglobin. *Science*, **278**, 443–446.

39 Okamoto, H., Nakabayashi, T. and Tasumi, M. (1997) Analysis of anti-Stokes resonance Raman excitation profiles as a method for studying vibrationally excited molecules. *J. Phys. Chem. A*, **101**, 3488–3493.

40 Sakai, M., Mizuno, M. and Takahashi, H. (1998) Picosecond-nanosecond time-resolved resonance Raman study of the structure and dynamics of the excited states of 5-dibenzosuberene derivatives. *J. Raman Spectrosc.*, **29**, 919–926.

41 Dlott, D.D. (2001) Vibrational energy redistribution in polyatomic liquids: 3D infrared-Raman spectroscopy. *Chem. Phys.*, **266**, 149–166.

42 Laenen, R. and Simeonidis, K. (1999) Energy relaxation and reorientation of the OH mode of simple alcohol molecules in different solvents monitored by transient IR spectroscopy. *Chem. Phys. Lett.*, **299**, 589–596.

43 Cheatum, C.M., Heckscher, M.M., Bingemann, D. and Crim, F.F. (2001) CH_2I_2 fundamental vibrational relaxation in solution studied by transient electronic absorption spectroscopy. *J. Chem. Phys.*, **115**, 7086–7093.

44 Dashevskaya, E.I., Litvin, I., Nikitin, E.E. and Troe, J. (2001) Classical diffusion model of vibrational predissociation of van der Waals complexes - Part II. Comparison with trajectory calculations and analytical approximations. *Phys. Chem. Chem. Phys.*, **3**, 2315–2324.

30
Three-Dimensional High-Resolution Microspectroscopic Study of Environment-Sensitive Photosynthetic Membranes

Shigeichi Kumazaki, Makotoh Hasegawa, Mohammad Ghoneim, Takahiko Yoshida, Masahide Terazima, Takashi Shiina, and Isamu Ikegami

30.1
Introduction

30.1.1
Thylakoid Membranes of Oxygenic Photosynthesis

In nearly all biological systems, synthesis of organic molecules directly or indirectly requires energy from the sun. The light-to-chemical energy conversion is called photosynthesis, which occurs in various forms around the globe. Oxygenic photosynthesis in plants takes place in chloroplasts, which are believed to be descendants of intracellular symbionts, probably cyanobacteria. The thylakoid membrane in both chloroplasts and cyanobacteria is the most essential part of the light-to-chemical energy conversion for oxygenic photosynthesis. The primary photochemical reactions of oxygenic photosynthesis are essentially achieved by two types of pigment–protein complexes photosystem I and photosystem II (PSI and PSII), which form a serial photoinduced electron-transfer chain [1]. The ideal condition for the serial electron-transfer system is that the two photosystems are equally excited within their maximum reaction rates. The two photosystems possess slightly different spectral features, and the spectral features and the numbers of light-harvesting systems (antenna) interacting with each of the two photosystems varies. If only PSII is highly and continuously excited, the acceptor side of PSII is fully reduced and some of the extra chlorophylls in the singlet excited state are converted into the lowest triplet state. The triplet chlorophyll can generate highly reactive and damaging singlet oxygen. In natural environments, both the intensity and spectral composition of solar light fluctuate with time, which necessitates an active balancing of excitation energy distribution between the two photosystems according to the light conditions. Such adaptation is achieved through changes in the distance relation and number of pigment–protein complexes, and through the morphology of the thylakoid membrane on a time scale from seconds to months [2]. The adaptation mechanism is a subject of active research, but this active balancing has not yet been well visualized for either cyanobacteria or chloroplasts.

30.1.2
Thylakoid Membranes in Chloroplasts

In higher plants, thylakoid membranes in chloroplasts form a flattened sac and are impermeable to most molecules and ions. There are two major light-harvesting chlorophyll–protein complexes LHCI and LHCII, which work as the main antenna for PSI and PSII, respectively. The light-harvesting complexes, LHCI and LHCII, absorb light and transfer the excitation energy to the two photosystems. The fine structure of the thylakoid membrane has been studied mainly using electron microscopy. A well accepted model of the structure suggests that it consists of two types of membranes [3]. One type is a stacked form of membranes known as grana membranes. The other form is unstacked or stroma-exposed membranes. It has been established that PSI and PSII are segregated. About 85% of PSII and 90% of its light-harvesting complexes, LHCII, are localized in the grana thylakoid. About 90% of PSI is localized in the unstacked part of the membrane, the stroma-exposed thylakoid [4]. The dimensions of a single grana region of the thylakoid membrane are of the order of submicrometers. Since PSI and PSII show different colors in fluorescence, red and far-red, respectively [5, 6], it could be expected that color-selected fluorescence imaging may resolve the fine structures of the thylakoid. Optical microscopy, including fluorescence imaging, has the potential to trace physiological changes of the thylakoid membrane during various adaptation processes. Whilst the electron microscopy is necessary to reveal the fine structure on a nanometer scale in solidified samples, such techniques may not provide details of all physiological aspects of the thylakoid membrane.

30.1.3
Thylakoid Membrane of Cyanobacteria

One unique feature of the thylakoid membrane in cyanobacteria is that the main light-harvesting antenna for PSII is phycobilisome (PBS), which is a pigment–protein complex containing phycobilins with absorption peaks at about 530–620 nm, between carotenoids and chlorophylls [1]. PBS is not a transmembrane complex, but is attached to the stromal side of the thylakoid membrane. Another unique feature of the cyanobacterial thylakoid is that PSI and PSII are not segregated, unlike in higher plants [7]. This raises a question: why are grana necessary in higher plants and some algae? To answer this question is not simple, but it may be addressed if we understand the detailed responses of the thylakoid membrane of various organisms to oxygenic photosynthesis: higher plants, alga and cyanobacteria.

30.1.4
Applications of Fluorescence Microscopy to a Thylakoid Membrane

The last two decades have seen tremendous progress in optical microscopy, especially, confocal fluorescence microscopy. Confocal laser-scanning fluorescence microscopy (CLSM) has been applied to chloroplasts. Chloroplasts in green algae were found to show different levels of complexity, when they were studied through PSII fluorescence [9]. Grana regions in the thylakoid membrane of chloroplasts have been successfully observed as intense fluorescence spots of chlorophyll autofluorescence inside chloroplasts [8–12]. The quantum yield of the PSII fluorescence is known to be far higher than that of PSI at physiological temperatures, such that chlorophyll autofluorescence usually infers fluorescence from PSII and its closely associated antenna, LHCII. For example, time-lapse imaging of chloroplast division was achieved using CLSM based on two-photon-absorption induced chlorophyll autofluorescence [13]. Two-photon excitation of pigments based on a near-infrared short pulsed laser is an attractive technique since it can suppress out-of-focus excitation and has high penetration ability through thick samples, such as a leaf [14, 15]. In studies of thylakoid membranes using fluorescence microscopy, the excitation light necessary for generating fluorescence inevitably causes some adaptive changes of the thylakoid membrane, depending on the wavelength and power. Suppression of the out-of-focus excitation, as realized with nonlinear excitation, is thus favorable for minimizing the disturbance to the physiological state of the thylakoid membrane.

Images with spectral information have been frequently obtained from leaves of plants [16]. However, fluorescence microscopic studies on chloroplasts to resolve the spectrum have been very limited. To resolve the spectrum a decrease in the number of photons per single detector channel is required, which is problematic in highly spatially resolved microscopic studies requiring photons. The weak fluorescent signal may be compensated by using a relatively long accumulation time for the imaging, but slow image acquisition may not enable us to study relatively fast physiological processes. In spite of such difficulties in spectral microscopy, several works have recently been published, focusing on chloroplasts [12, 17, 18] or cyanobacteria [19–21]. Although there are several types of commercial microscopic systems that can measure fluorescence spectra with a reasonably high wavelength resolution (≈ 2 nm), most record spectra using step-by-step tuning of detectable wavelength regions and/or sacrifice the wavelength resolution and/or number of channels. The most serious drawback in the choice of a limited number of spectral windows and detectors is a significant rejection of fluorescence in the other spectral regions that could be physically detectable and scientifically informative.

30.1.5
Simultaneous Spectral Imaging and its Merits

A line-scan of the illumination laser light has been used to enhance the scan rate, at least compared to the point-by-point scan [18, 21–25]. It remains compatible with a spectral acquisition of high resolution, since total fluorescence from the linear region can be projected onto the slit of an imaging polychromator. A CCD camera at the exit port records the fluorescence intensity as a function of the spatial coordinate and the wavelength in the two-dimensional sensitive area of the camera. This method sustains the advantage of nearly parallel illumination and simultaneous detection of fluorescence, as far as the linear region is concerned.

A microscopic fluorescence spectrum is potentially very informative, since it reflects stoichiometric ratios, efficiencies of electronic excitation transfers among the pigment–protein complexes, and quenching mechanisms inherent in the photosynthetic reactions. Since the multiple fluorescence bands are overlapping, spectral detection based on a polychromator and multichannel detector is more informative than detection using a few channels based on dichroic mirrors and band-pass filters.

Cyanobacterial thylakoid membranes show fluorescence not only from chlorophylls but also from phycobilins. Even in the case of plant and algal thylakoid membranes, in which chlorophyll fluorescence is predominant in the red to far-red region, it is possible that multiple fluorescence components arise. The light-harvesting complexes, LHCI and LHCII, are characterized by slightly different chlorophyll fluorescence peaks, the distinction of which is usually feasible only at cryogenic temperatures [1, 6]. However, the ratios of their contributions certainly change during some physiological changes, and these may potentially be detectable at room temperature if the fluorescence spectrum is recorded at a high resolution (≈ 2 nm).

30.2
Spectral Fluorescence Imaging of Thylakoid Membrane

30.2.1
Realization of Fast Broadband Spectral Acquisition in Two-Photon Excitation Fluorescence Imaging

Full details of our-home-made spectromicroscope system have been described previously [21]. The set-up is shown in Figure 30.1a. A near-infrared femtosecond pulse train at 800 nm is illuminated on a line (one lateral axis, denoted as the X axis) on a specimen using a resonant scanning mirror oscillating at 7.9 kHz. Total multiphoton-induced fluorescence from the linear region was focused on the slit of an imaging polychromator.

An electron-multiplying CCD (EMCCD) camera is used to resolve fluorescence of different colors at different horizontal pixels and fluorescence of different spatial positions in the specimen at different vertical pixels. Scanning on the other two axes (Y and Z) is achieved by a closed-loop controlled sample scanning stage and a piezo-driven objective actuator. The full widths at half maximum of the point-spread function of the system have been estimated to be 0.39–0.40, 0.33 and 0.56–0.59 μm for the X (lateral axis along the line-scan), Y (the other lateral axis) and Z axes (the axial direction), respectively, at fluorescence wavelengths between 644 and 690 nm. The efficiency of the new line-scanning spectromicroscope has been estimated in comparison with our own point-by-point scanning spectromicroscope [21]. Under typical conditions for observations of cyanobacterial cells (cf. Section 30.2.2), the total

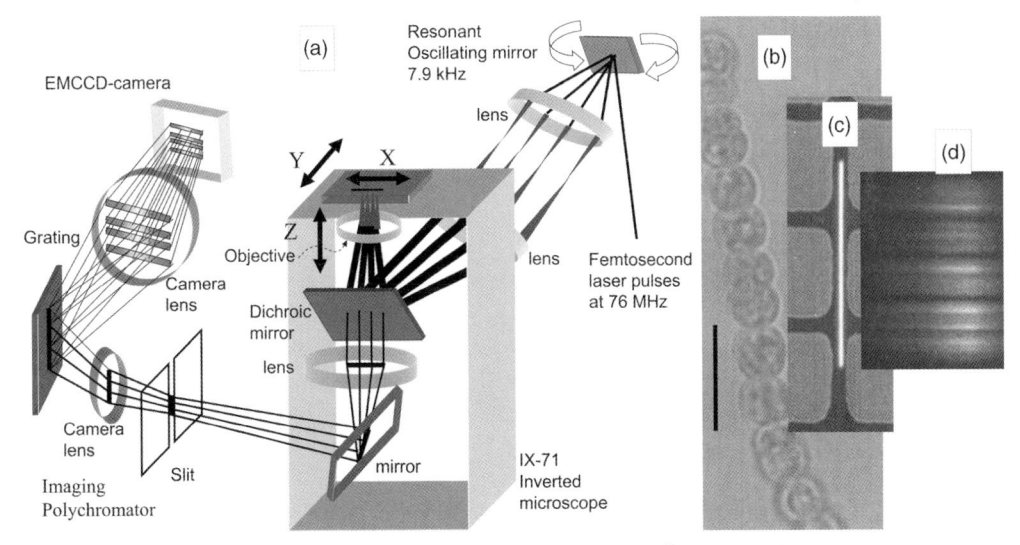

Figure 30.1 (a) Schematic of the optical set-up of the line-scanning spectromicroscope. (b) Typical wide-field microscopic image of the cyanobacterium *Anabaena* PCC7120 under the illumination of a halogen lamp. The scale bar is 10 μm. (c) Reflection image of the femtosecond pulsed laser from the surface of an objective scale with a spacing of 10 μm. (d) Typical image of fluorescence intensity distribution detected by the EMCCD camera. The horizontal and vertical axes correspond to the fluorescence wavelength and a lateral axis in the specimen, respectively.

exposure time was shortened by about 50 times for a constant average excitation density. The improvement factor was proportional to the length of the line-scanned region, as expected.

The idea of line-scanning or line-focus is not new. There is only one drawback which is that a confocal effect does not occur along the slit direction. Nevertheless, it is still adopted in a number of commercial fluorescence and Raman spectromicroscopes (LSM5-LIVE from Carl Zeiss, Raman 11 from Nanophoton, etc.) when spectral information is desirable. Our set-up seems to be unique in that multiphoton excitation with the rapidly moving resonant scan mirror is used for broadband spectral imaging. Linear illumination has been achieved either by expanding a laser beam along one axis with a cylindrical lens [22–24] or by scanning a focal point with a scan mirror [21, 25]. In the former case, the laser intensity ($W\,cm^{-2}$) is inversely proportional to the out-of-focus distance in the optic axial coordinate, as far as the region outside the beam waist is concerned. In the latter case, the laser intensity is inversely proportional to the square of the out-of-focus distance in the optic axial coordinate. Thus, in cases employing multiphoton excitations to suppress out-of-focus excitation and improve penetration ability, better depth sectioning is expected in the case of linear illumination by a rapid scan mirror than in the case of linear illumination by a cylindrical lens. We have actually confirmed that the depth resolution is equal to the diffraction limit [21]. However, the lateral resolution is not equal to the theoretical limit, which may be limited by the imaging polychromator.

30.2.2
Spectral Imaging of a Filamentous Cyanobacterium, *Anabaena*

30.2.2.1 Thylakoid Membrane of Cyanobacterium

Anabaena is a filamentous cyanobacterium (Figure 30.1b), which is well known for its ability to fix nitrogen. Under nitrogen-starved conditions, the nitrogen fixation is undertaken only in terminally differentiated heterocyst cells that are separated from each other at a constant spacing in the same filament [26, 27]. A great deal of attention has been paid to *Anabaena* by a broad range of biologists, since it is regarded as one of the simplest organisms exhibiting cell differentiation. In our study, intracellular spectral features of vegetative cells, which are normal cells without nitrogen-fixation ability, were studied using fluorescence spectromicroscopy.

30.2.2.2 Stability of the *Anabaena* Fluorescence Spectra Under Photoautotrophic Conditions

Before any active research to induce and observe responses of the thylakoid membrane to environmental conditions can take place, it is important for us to understand the stability in fluorescence properties of cells, especially under probing light. Total fluorescence intensity is equal to the wavelength-integrated fluorescence signal in the case of spectrally resolved data, which is shown in Figure 30.2a for photoautotrophically grown *Anabaena* cells. All cells typically show a dark region at the center of

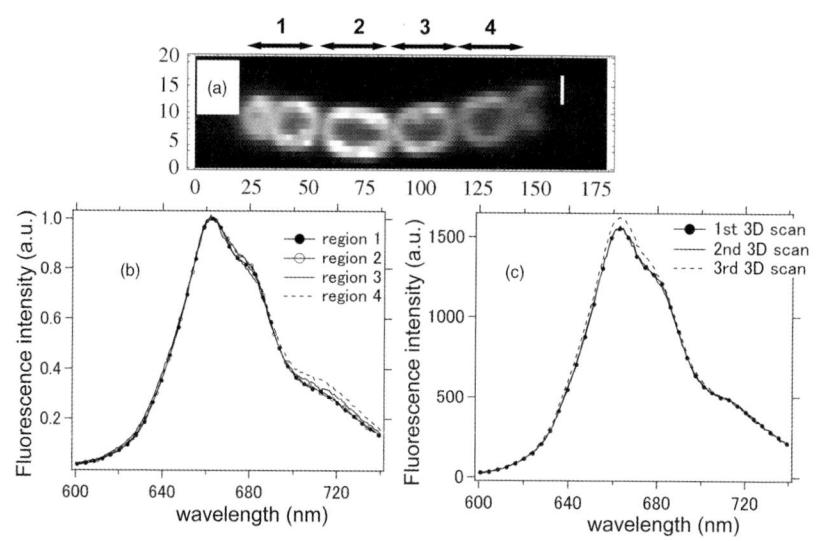

Figure 30.2 (a) A typical image of the total fluorescence intensity of filamentous *Anabaena* cells grown photoautotrophically. The scale bar on the right represents 2 μm. (b) Fluorescence spectra generated from the subregions numbered 1 to 4 in (a). The averaging for individual spectra was performed on all of the 15 Z-positions sharing the same *XY* coordinates. (c) Fluorescence spectra of the region 1 in (a) in consecutive three 3D scans.

cells, which correspond to regions free of thylakoid membranes, as shown by electron microscopic images [28]. In contrast to the thylakoid membrane in plants and some algae, grana-like regions do not occur in cyanobacterial cells [7, 28]. It is thus believed that PSI and PSII are not segregated.

Fluorescence spectra of individual cells in the same filament are compared in Figure 30.2b. These were generated from the subregions numbered from 1 to 4, as indicated in Figure 30.2a. The spectra are very similar. It should be noted that the spectra here are all generated based on the average of about 1000–2000 fluorescent points. In many articles studying fluorescence spectra of plants and cyanobacteria, cryogenic temperatures are used to decompose spectra. This technique is powerful and informative, but the cooling procedure is not compatible with real-time recording of physiological phenomena. It is noteworthy that the spectra, even at room temperature, show some shoulders corresponding to peaks of constituent fluorescent components (at around 680, 710 nm), which can be visualized only with a high spectral resolution, as used in the current set-up.

Stability or sensitivity of the fluorescence spectra under the current observation conditions has been investigated by repeating fluorescence spectral imaging in the same region. The spectra in Figure 30.2c show little change in terms of intensity and shape between three sets of consecutive 3D scans, one of which includes 15 z-sections with spacing of 0.90 μm. This suggests that the fluorescence properties of the cells are preserved even with the laser scanning.

30.2.2.3 Change of the *Anabaena* Fluorescence Spectra by Dark Conditions

Some *Anabaena* can grow heterotrophically under dark conditions if fructose is supplied as a respiratory substrate [29], although the growth rate is not so fast. It is interesting that some thylakoid membrane is preserved even in the dark. However, the morphology and response to light exposure or laser scanning is thought to be different from those grown photoautotrophically, as has been shown in part previously [29]. Under photoautotrophic growth conditions long filaments of cells (containing 10 cells or more) are frequent. In the case of *Anabaena* PCC 7120 that is unable to grow under dark conditions, we were unable to find long filaments of cells, reflecting stop of growth.

A typical image of total fluorescence intensity for a couple of *Anabaena* cells with dark treatment for about 30 days is shown in Figure 30.3a. The central dark regions in the cells were smaller than those in photoautotrophically grown cells. This is consistent with early electron microscopic observations [29]. Fluorescence spectra of individual cells are shown in Figure 30.3b and c for the left and right cells, respectively. Two sets of 9 z-sections of spectral images with a spacing of 1.00 μm were acquired consecutively from these cells. In comparison with photoautotrophically grown cells, several differences in the spectral properties were noted. First, the two cells clearly exhibit different spectral shapes, which are largely characterized by the ratio between the 660 and 740 nm peaks. Second, the spectra obtained in the second 3D scan were significantly weaker for the whole wavelength region than in the first scan. Third, the spectral shapes in the second 3D scans were different from those in the first scans. The overall decrease in fluorescence intensity in the

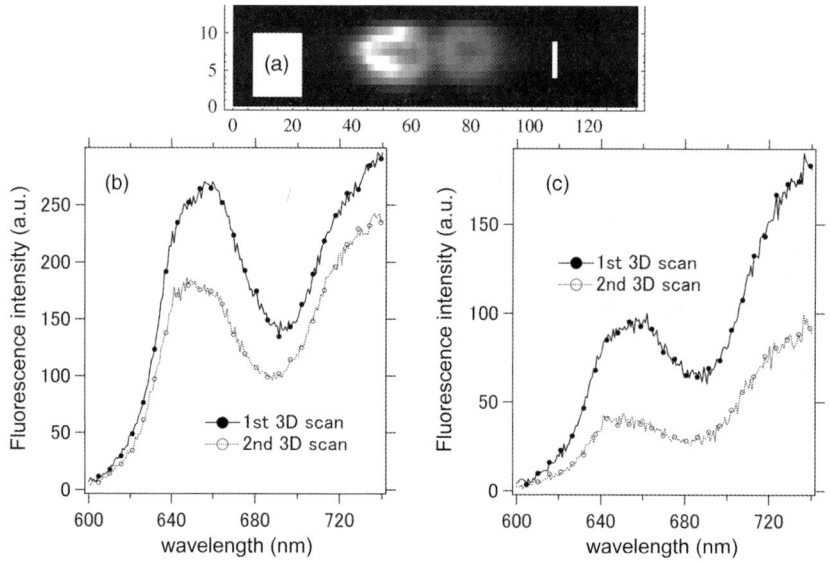

Figure 30.3 (a) A typical image of the total fluorescence intensity of a couple of *Anabaena* cells with dark treatment. The scale bar on the right represents 2 μm. (b) Fluorescence spectra averaged over the 3D space of the left cell in (a). (c) Fluorescence spectra averaged over the 3D space of the right cell in (a). The spectra in (b) and (c) were generated from two consecutive 3D scans.

consecutive scans is most likely due to photobleaching and/or photoinhibition of all types of photosynthetic pigment–protein complexes. This was not observed for photoautotrophically grown cells (Figure 30.2b), probably due to activation of photochemical and non-photochemical quenching (quenching for photosynthetic charge separation and protective quenching other than photosynthetic charge separation) only under photoautotrophic growth conditions.

30.2.2.4 Intracellular Spectral Gradient in *Anabaena* Cells

Our previous publication showed an intracellular gradient in the ratio of fluorescence intensities between two wavelength regions of 644–664 nm and 680–690 nm (F685/F654 ratio), for photoautotrophically grown cells [21]. The two wavelength regions centered at 654 and 685 nm are selected to focus probing on the fluorescence from phycobilisome (PBS) and PSII, respectively [19, 20]. An example is shown for the F685/F654 ratio map in Figure 30.4b, together with the total fluorescence intensity map in Figure 30.4a. These indicate that the ratio value is lower at the core than at the periphery of each cell. The locally averaged spectra in Figure 30.4c show that the difference at 680 nm fluorescence, after normalization using amplitudes at 645 nm, amounts to as much as 30%.

We have obtained such F685/F654 ratio maps for about 70 cells from 25 *Anabaena* filaments. Among them, about 60% of the cells showed substantially lower F685/F654 ratios (by 10–30%) in the central regions than in the peripheral regions. The

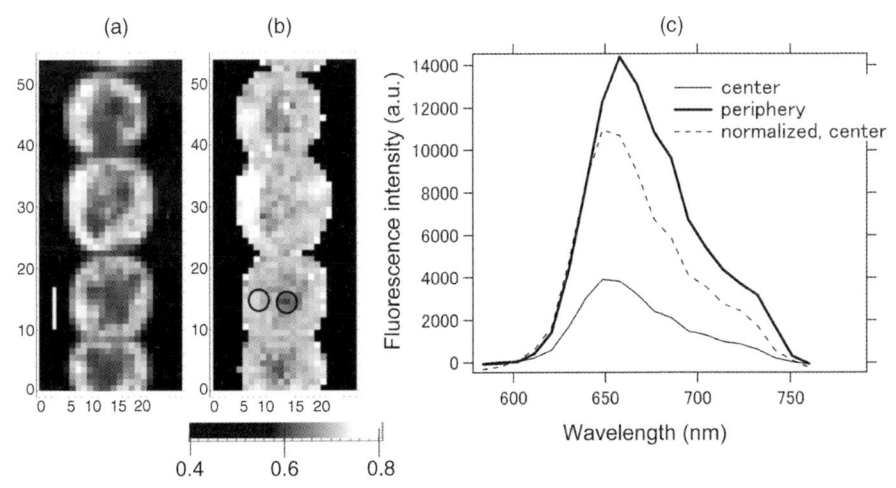

Figure 30.4 A typical case for the fluorescence ratio, F685/F654, being lower in the central parts of the cells than the periphery. (a) Gray-scale image of total fluorescence intensity. The scale bar represents 2 µm. (b) F685/F654 Ratio map. The corresponding ratio value is shown by the horizontal bar. The region outside the cell is also shown in black, where fluorescence intensity is too weak for the ratio value to be defined. (c) Local fluorescence spectra. They were calculated from the region demarcated by black circles with a diameter of 1.0 µm in (b). The spectral imaging data in this figure were obtained using a commercial point-scanning spectromicroscope (NanoFinder, Tokyo Instruments, Tokyo, Japan) modified to use two-photon excitation using a 820 nm femtosecond pulse train [21].

opposite trend has not been observed. The other 40% of cells did not show a clear gradient in the ratio maps. To the best of our knowledge, there has been no report of such a systematic modulation of the F685/F654 ratio on the same strain.

Even if the intrinsic fluorescence spectrum of the thylakoid membrane is constant, there are several possible causes for such a spectral gradient. Wavelength dependence of reabsorption and/or point-spread functions can, in principle, give rise to such a dependence of the spectra on the position of the probing fluorescence (e.g., whether the position is at the central part or near the edge). These two possible artifacts were ruled out in our previous work, based on measurements of wavelength dependence of transmission and point-spread function [21].

The intracellular gradient in the F685/F654 ratio map indicates that fluorescence of PSII is relatively lower in the central regions than in the peripheral regions. There are at least three interpretations of this phenomenon. The first and most straightforward interpretation is that the stoichiometric ratio of PBS is higher in the central regions than in the peripheral regions. It is easy to hypothesize a physiological meaning for this, since the thylakoid membrane at the deepest position in the cell receives less external light than that nearest to the outer membrane. The loss of light during transmission from the surface to the deepest part of the thylakoid membrane is estimated to be about 24% (assuming one half of the measured apparent absorbance of 0.25) [21]. This may be a small change in light intensity, but the increase in the effective absorption cross section would certainly benefit PSII in the deepest part of the thylakoid membrane. The second interpretation is that the efficiency of energy transfer from PBS to PSII is somehow lower in the central regions than in the peripheral regions. This interpretation may be relevant to the so-called state transition, in which the yield of the energy transfer from PBS to PSI and/or PSII is regulated by the intensity and spectrum of the incident light [2]. The third interpretation is a natural hybrid of the first and second interpretations. When there is more PBS per PSII in the central regions than in the peripheral regions, some of the extra PBS associated with the central thylakoid membrane cannot be closely connected with PSII. This, on average, results in higher weights of PBS fluorescence at the central thylakoid membrane than at the peripheral thylakoid membrane. It would be necessary to apply time-resolved fluorescence decay measurements and/or excitation wavelength dependence of the fluorescence ratio image to this system to determine which of the above-mentioned mechanisms is most appropriate.

Two recent works using spectral microscopy have reported dependence of fluorescence spectrum on radial positions in cyanobacterial cells [20, 30]. One is about variation between subunits of phycobilisomes in *Nostoc punctiforme* [20], the other is reporting that phycobilin-to-chlorophyll fluorescence ratio (phycobilin/chlorophyll) is higher along the periphery than inside the cells in *Synechocystis* sp. PCC6803 [30]. The latter phenomenon seems to be opposite to what we have found. How such different trends arise is interesting future subject to be addressed with the help of spectral microscopy.

30.2.3
Spectral Imaging of Chloroplasts

30.2.3.1 Chloroplasts from a Plant, *Zea mays*
In an effort to generate images of the morphology of thylakoid membranes inside plant chloroplasts, isolated chloroplasts resuspended in a buffer solution containing a herbicide, 3-(3,4-dichlorophenyl)-1,1-dimethylurea (DCMU), have often been used [10, 11]. The DCMU inhibits photosynthesis by blocking electron transport at the electron acceptor side of PSII and enhances chlorophyll fluorescence from PSII.

In our study, we tried to resolve PSI and PSII without introducing DCMU or any other chemical to enhance fluorescence. Chloroplasts were placed *in situ*, in a leaf, which was detached but not sliced. One prerequisite to acquire a fluorescence map of PSI and PSII is to determine the fluorescence spectra of the two photosystems. From careful examination of the sensitivity of the microscopic fluorescence spectra, the following rule was suggested. A volume-averaged fluorescence spectrum of a single chloroplast exhibits intensity changes at wavelengths between 655 and 715 nm, but the intensity between 715 and 740 nm remains more or less constant with repeated 3D scans for spectral imaging and illumination with a halogen lamp under the microscopic observation conditions. The volume-averaged fluorescence spectrum was obtained using 3D spectral imaging of a whole chloroplast, which gave chloroplast volume and spectra for every scanned point. The sum of the fluorescence spectra was divided by the number of scanned points with substantial chlorophyll fluorescence (relative volume). The two photosystems are known to show different dependence on redox-conditions and varying stress-sensitivity in fluorescence quantum yields [31].

Under our measurement conditions, unavoidable lamp illumination, laser scanning, and the enclosure of the leaf in a thin chamber seem to have induced changes in the fluorescence quantum yield of mainly PSII. This procedure was used to approximately decompose the observed fluorescence spectra into at least two components, which helped us to obtain fluorescence maps of PSI and PSII.

The wavelength for PSII imaging was thus selected to be between 665 and 681 nm (hereafter designated red image), and that for PSI was between 715 and 740 nm (far-red image), as shown in Figure 30.5. It should be noted that the fluorescence bands of PSII antenna are believed to be located at 683–685 nm at both a physiological temperature and 77 K [1, 6]. Our wavelength selection for the PSII fluorescence in Figure 30.5 was between 665 and 681 nm in order to minimize overlap with the fluorescence of PSI. The fluorescence images in Figure 30.5a and b are shown using a normal gray scale. In order to visualize small amplitude modulation clearly, pixels with intensities of at least 50% of the maximum intensity in each XY image are shown in Figure 30.5c and d.

There are distinctly bright spots at a sub-micrometer scale in the red fluorescence images in Figure 30.5c. The bright spots in Figure 30.5c are attributable to grana that are rich with PSII. Compared with the nearly circular grana regions obtained in previous works [8–12], the putative grana region in our red-fluorescence images is anisotropically elongated in the direction of the laser scan (X axis). This may simply reflect an artifact due to the lack of a confocal effect along the direction of the slit. According to the concept of a "string of grana", grana are not scattered at random, but tend to be arranged in strings of varying length, like beads on a necklace. This has been supported by previous CLSM images and electron microscopic images [3, 10]. With our resolution limitations, elongated spots may actually contain separate multiple grana in the same string. Closer inspection of Figure 30.5c also reveals that the putative grana regions are connected not only in the single XY planes, but also along the different Z positions. Our data seem to directly support the model of a string of grana rather than randomly scattered grana, based on the 3D red fluorescence distribution.

The PSI indicated by the far-red images in Figure 30.5d seems to be distributed almost homogeneously down to a sub-micrometer scale, which is in contrast to the case of the red fluorescence images. It seems that PSI is distributed not only in the non-appressed part (non-grana part) of the thylakoid membrane but also in the grana part of thylakoid membrane. This is possible if we consider the model in which PSI reside not only between neighboring grana regions but also in the stroma-exposed thylakoid closely surrounding the grana. Given the typical size of grana (0.3–0.6 µm) and the relatively weak far- red fluorescence than the red one [3, 10], it is probably hard for our optical microscopy to resolve local minima of PSI fluorescence intensity corresponding to the center of grana.

The fluorescence images attributed to PSI in a chloroplast of the plant *simplex var. metallica* in Ref. [10] show a clear intensity modulation with local maxima coincident with local minima of PSII fluorescence, in contrast to our far-red images that show rather homogeneous distribution of PSI. This difference may be attributable to the abnormally large size of the grana (up to 1.9 µm) in this special plant which is found in shady, dense and humid tropical forest.

Z = 0.0   0.5   1.0   1.5   1.5

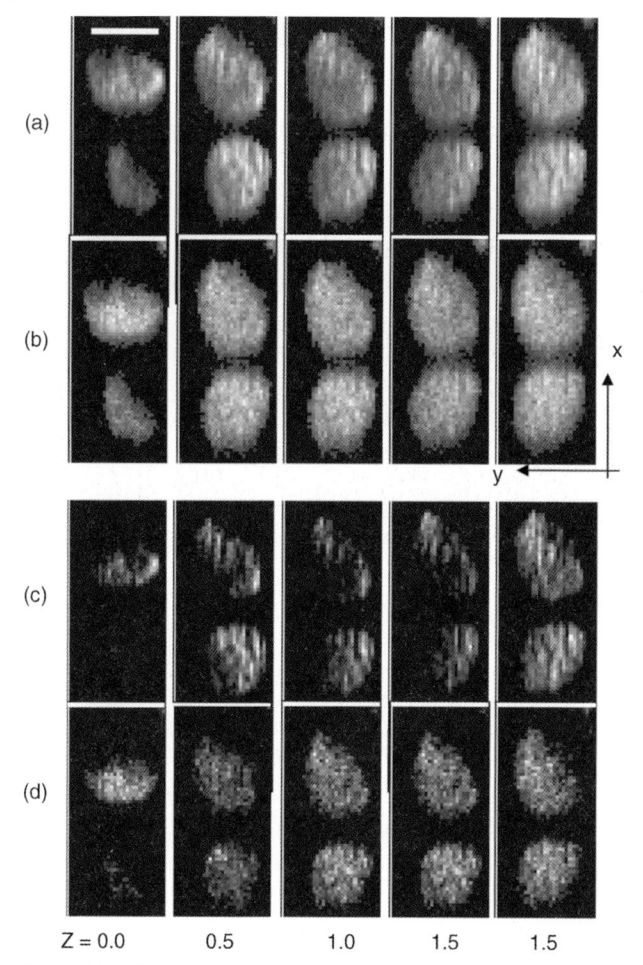

Figure 30.5 Fluorescence images of two neighboring chloroplasts in a mesophyll cell of Zea mays at different Z positions. (a) Integrated intensities between 665 and 681 nm (red image). The scale bar is 5.0 μm. The direction of the line scan is along the X axis. (b) Integrated intensities between 715 and 740 nm (far-red image). In (a) and (b), the maximum intensity in each image is indicated in white and a threshold intensity reasonably above noise level in black. (c) and (d) Generated from the same data set as in (a) and (b), respectively, but the threshold intensity was set at 50% of the maximum intensity in each image. Fluorescence points below the threshold are thus black. This figure is adapted from Figure 1 in Ref. [18] with kind permission of Springer Science and Business Media.

30.2.3.2 Chloroplast from the Green Alga, *Chlorella*

Figure 30.6 shows fluorescence images of a chloroplast from a cell of *Chlorella kessleri* using the same wavelength selection as for *Zea mays* (cf. Figure 30.5). The red and far-red fluorescence images are rather similar to each other. Although the red fluorescence is relatively localized near the edge of the cells compared to the

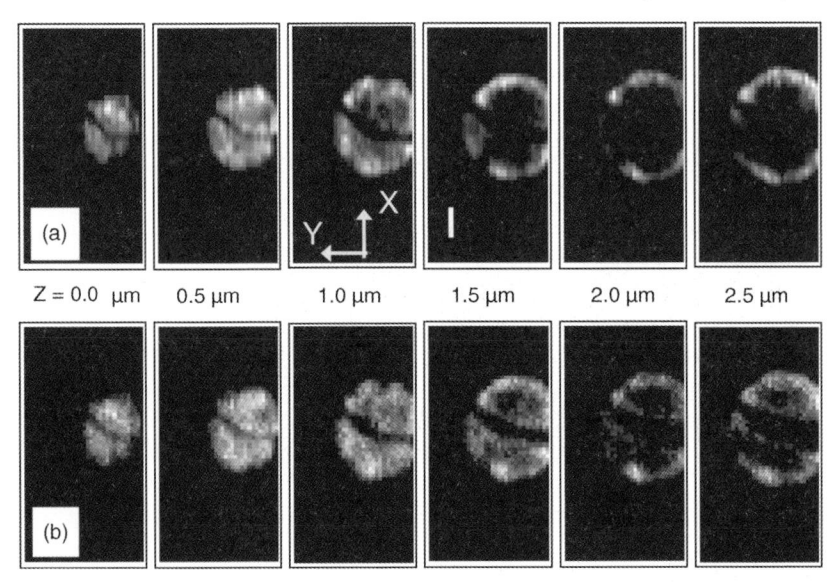

Figure 30.6 Fluorescence images of a *Chlorella* cell at different Z positions. (a) Integrated intensities between 665 and 681 nm (red image) The scale bar is 2.0 μm. The direction of the line scan is along the X axis. (b) Integrated intensities between 715 and 740 nm (far-red image). The maximum intensity in each XY image is indicated in white and the intensity at 50% of the maximum intensity in black. Fluorescent points below the threshold (50% of the maximum intensity) are thus black. This figure is adapted from Figure 2 in Ref. [18] with kind permission of Springer Science and Business Media.

far-red one, the overall difference between the red and far-red images is far less evident for *Chlorella* than for *Zea mays*.

The fluorescence spectrum of the *Chlorella* chloroplast is relatively flat between 715 and 740 nm compared with that of *Zea mays*, which often shows an increase in fluorescence intensity from 715 to 740 nm. The situation suggests that the far-red component of the *Zea mays* is substantially influenced by PSI whilst the far-red component of the *Chlorella* is dominated by a vibronic progression of the chlorophyll fluorescence of PSII. It is then reasonable that the red and far-red fluorescence images are relatively similar to one another.

These results seem to be consistent with the previous findings that segregation between PSI and PSII, or grana formation, is observed in only a limited number of taxa of green algae [9]. Electron microscopic images of *Chlorella* do not usually show typical grana stacking [32, 33], except when excess carbon dioxide is supplied [34].

30.3
Technical Verification and Perspective

One drawback of spectral microscopy is ambiguity in interpreting the observed spectral shape and intensity which can be affected by scattering and/or re-absorption (see Section 30.2.3.4). Lifetime is a more robust parameter which is not easily affected by reabsorption or scattering. Several types of fluorescence lifetime imaging microscopy (FLIM) are already available as commercial systems. Fluorescence decay rate can be used to record the extent of non-photochemical quenching and the effect of photosynthetic inhibitors. This has been studied for the green alga *Chlamydomonas reinhardtii* using FLIM [35]. However, time-resolved spectra are still difficult to generate, mainly due to the high cost of preparing multiple detector channels with time-resolving ability. Spectral imaging of time-integrated fluorescence will continue to be a basic microscopic tool, particularly for the research of the thylakoid membrane, because of the complexity of the spectra.

Fast confocal fluorescence imaging at a video rate or far better (1000–2000 Hz) is now available for several commercial systems. Imaging using a single focus requires higher peak power for faster imaging, which may lead to faster photobleaching and photodamage. Some of the systems thus employ a parallel illumination using multiple foci and parallel imaging by CCD cameras [36]. The responses of chloroplasts are not as fast, but the faster imaging capability is a benefit for observing slow phenomena. It enables fluorescence images to be obtained with a lower average and/or peak power by distributing the excitation photons over a number of scans. The number of photons necessary for fluorescence (spectral) imaging may be reduced as far as possible, ideally to the level of natural solar illumination.

There are some negative opinions on the use of multiphoton microscopy in botanical specimens, including chloroplasts [15]. In Ref. [15], it is reported that short-pulsed near-IR excitation (780 nm) leads to faster damage to the specimens than single photon excitation. However, our results show that at least light-adapted chloroplast and cyanobacteria grown photoautotrophically can preserve their fluorescence spectrum even after a few 3D scans. As mentioned previously, time-lapse imaging of chloroplast division was also demonstrated based on two-photon fluorescence [13]. In addition, we have estimated the relative densities of excited chlorophylls under the three experimental conditions detailed in Table 30.1. The rapid photodamage reported in Ref. [15] seems to be due to high peak power and high excitation density.

It should be also noted that so far we have only encountered fluorescence with a substantial intensity at wavelengths longer than 600 nm, in both *Anabaena* cells and chloroplasts. In contrast, significant autofluorescence was observed not only for chlorophylls but also for other shorter-wavelength emitting species in the multiphoton-induced fluorescence spectra of a chloroplast in Ref. [15]. The significantly different spectra are most plausibly explained by differences in the peak power of the laser and/or possible (near-) resonance effects. Excitation wavelength as short as 780 nm (in Ref. [15]) at the peak power level may lead to higher order nonlinear absorption by so many types of molecules in cells that it hampers non-invasive fluorescence imaging.

30.4
Summary

We have shown an application of fluorescence spectral imaging for the detailed study of thylakoid membranes in three types of organisms with oxygenic photosynthesis.

Table 30.1 Comparison of excitation density between the line scan system and two point-scan systems in application to plant leaves.

	Our line-scan System	CLSM, point-scan [15]	CLSM, point-scan [13]
Excitation wavelength: λ_{ex} (μm)	0.805	0.780	0.800
Average power of the laser at the sample position: P_A (mW)	0.03[a]	6.4	≈1[b]
Pulse repetition rate: R (Hz)	76×10^6	82×10^6	80×10^6
Single pulse energy: $E_P = P_A/R$ (pJ)	0.39[a]	78	4.5
Numerical aperture: NA	1.4	1.2	1.3
Spot Size: $S\varepsilon(0.61\lambda_{ex}/NA)^2$ (cm^2)	1.23×10^{-9}	1.57×10^{-9}	1.41×10^{-9}
Pulse duration: τ_p (ps)[c]	0.2	0.1	0.18
peak power density: $I\varepsilon E_P/\tau_p/S$ (W cm^{-2})	1.6×10^9	5.0×10^{11}	1.8×10^{10}
Dwell time or exposure time per pixel (s): τ_d	1	8.4×10^{-6}	30×10^{-6}
Number of pulses per pixel: $N_P = R\tau_d/6 \times (0.25/20)^d$ or $N_P = R\tau_d$	1.58×10^5	6.89×10^2	2.4×10^3
Relative two-photon absorption cross section: σ, Ref. [14][e]	≈0.9	8	1
pixel size in data acquisition (μm^2)	0.25×0.25	0.234×0.234	0.24×0.24
Relative two-photon excitation probability per single pulse: $P_S\varepsilon\sigma I^2\tau_p \times 10^{-5}$	4.50	1.97×10^6	5.7×10^2
Relative excitation density $D_E\varepsilon \, P_S N_P \times 10^{-5}$	7.13	1.36×10^4	14

[a] Single pulse energy previously reported in Ref. [21] was overestimated, since it was based on a power measured outside the microscope.
[b] In Ref. [13], the power used for the time-lapse imaging of a chloroplast division was not exactly described, but power as high as 2 mW was used without causing damage.
[c] In our study, pulse width at the sample position was not measured. In Ref. [15], it is not clear where the pulse width was obtained.
[d] In the case of the line scan, the number of pulses illuminated on the linear region to be imaged (20 μm) was about one-sixth of the total laser pulses, because of an angle-range selection of the scan mirror in which its angular speed is approximately constant.
[e] The pulse number per pixel was given by the number of pulses in the line divided by the number of pixels therein (80 = 20/0.25).

In the case of *Anabaena*, we found that the morphology and sensitivity of the thylakoid membrane, which are dependent on growth conditions, can be readily studied. Although the fluorescence spectra at room temperature are not so easily, or so well, resolved, they certainly show shoulders and dips, which reflect constituent pigment–protein complexes. It should be noted that the ratio of concentrations of the pigment–protein complexes and/or energy transfer efficiencies between them are position dependent in some cases.

PSI fluorescence in *Zea mays* was distributed rather homogeneously down to the submicrometer level. This was in contrast to the distribution of PSII which formed distinctly bright fluorescent spots. This supports the model in which PSI resides not only in the non-appressed regions of thylakoid membrane connecting neighboring grana, but also in the stroma-exposed part closely surrounding the grana.

Although high resolution of wavelengths is not necessarily regarded as important in all research areas, it is certainly advantageous in deriving the detailed spectral shape of fluorescence from microscopic regions. It is also necessary in order to select the best set of dichroic mirrors and filters for a limited number of detection channels. Nearly diffraction-limited and truly simultaneous fluorescence spectral imaging of chloroplasts *in situ* using the line-scanning spectromicroscope seems to be achieved with an excitation density comparable to that for commercial CLSM with only a few detector channels. The functionality of fluorescence spectral imaging needs to be further improved to study a wider range of physiological changes and developmental stages of chloroplasts and cyanobacteria.

Acknowledgments

The authors gratefully thank Professors H. Masuhara, K. Yoshihara and, S. Itoh for their encouragement of this study, and Drs M. Nishiyama, H. Oh-Oka and K. Okamoto for advice. This work was supported in part by Grants-in-Aid for Scientific Research (No.17750069 to SK) and for Scientific Research on Priority Areas (Area No. 432, No.16072209 to SK, and Area No. 477, No. 19056012 to SK) from the Ministry of Education, Culture, Sports, Science and Technology (MEXT) of Japan. This work was also supported in part by funds from the Kurata Memorial Hitachi Science and Technology Foundation (to SK), and the Asahi Glass Foundation (to SK).

References

1 Ke, B. (2001) *Photosynthesis*, Kluwer Adademic Publishers.
2 Allen, J.F. and Forsberg, J. (2001) Molecular recognition in thylakoid structure and function. *Trends Plant Sci.*, **6**, 317–326.
3 Staehelin, L.A. (2003) Chloroplast structure: from chlorophyll granules to supra-molecular architecture of thylakoid membranes. *Photosynth. Res.*, **76**, 185–196.
4 Buchanan, B.B., Gruissem, W. and Jones, R.L. (2000) *Biochemistry & Molecular Biology of Plants*, American Society of Plant Physiologists, Chapter 12.
5 Govindjee, (2004) in *Chlorophyll a Fluorescence a Signature of Photosynthesis* (eds C. Papageorgiou and Govindjee), Springer, Netherlands, pp. 1–42.
6 Itoh, S. and Sugiura, K. (2004) in *Chlorophyll a Fluorescence a Signature of Photosynthesis* (eds C. Papageorgiou and Govindjee), Springer, Netherlands, pp. 231–250.
7 Gantt, E. (1994) in *The Molecular Biology of Cyanobacteria*, (ed D.A. Bryant), Kluwer Academic Publisher, Netherlands, pp. 119–138.
8 van Spronsen, E.A., Sarafis, V., Brakenhoff, G.J., van der Voort, H.T.M. and Nanningga, N. (1989) Three-dimensional structure of living chloroplasts as visualized by confocal scanning laser microscopy. *Protoplasma*, 148, 8–14.
9 Gunning, B.E.S. and Schwartz, O.M. (1999) Confocal microscopy of thylakoid autofluorescence in relation to origin of

grana and phylogeny in the green algae. *Aust. J. Plant Physiol.*, **26**, 695–708.

10 Wildman, S.G., Hirsch, A.M., Kirchanski, S.J. and Spencer, D. (2004) Chloroplasts in living cells and the string-of-grana concept of chloroplast structure revisited. *Photosynth. Res.*, **80**, 345–352.

11 Garstka, M., Drożak, A., Rosiak, M., Venema, J.H., Kierdaszuk, B., Simeonova, E., van Hasselt, P.R., Dobrucki, J. and Mostowska, A. (2005) Light-dependent reversal of dark-chilling induced changes in chloroplast structure and arrangement of chlorophyll-protein complexes in bean thylakoid membranes. *Biochim. Biophys. Acta*, **1710**, 13–23.

12 Vácha, F., Sarafis, V., Benediktyová, B.L., Valenta, J., Vácha, M., Sheue, C.-R. and Nedbal, L. (2007) Identification of Photosystem I and Photosystem II enriched regions of thylakoid membrane by optical microimaging of cryo-fluorescence emission spectra and of variable fluorescence. *Micron*, **38**, 170–175.

13 Tirlapur, U.K. and König, K. (2001) Femtosecond near-infrared lasers as a novel tool for non-invasive real-time high-resolution time-lapse imaging of chloroplast division in living bundle sheath cells of *Arabidopsis*. *Planta*, **214**, 1–10.

14 Tirlapur, U.K. and König, K. (2002) in *Confocal and Two-Photon Microscopy* (ed. A. Diaspro), Wiley-Liss, New York, pp. 449–468.

15 Cheng, P.-C. (2006) in *Handbook of Biological Confocal Microscopy*, 3rd edn (ed. J.B. Pawley), Springer Science + Business Media, New York, pp. 414–441.

16 Lichtenthaler, H.K. and Miehé, J.A. (1997) Fluorescence imaging as a diagnostic tool for plant stress. *Trends Plant Sci.*, **2**, 316–319.

17 Lukins, P.B., Rehman, S., Stevens, G.B. and George, D. (2005) Time-resolved spectroscopic fluorescence imaging, transient absorption and vibrational spectroscopy of intact and photo-inhibited photosynthetic tissue. *Luminescence*, **20**, 143–151.

18 Kumazaki, S., Hasegawa, M., Yoshida, T., Taniguchi, T., Shiina, T. and Ikegami, I. (2008) in *Energy from the Sun*, vol. 1 (eds J. Allen, E. Gantt, J. Golbeck and B. Osmond), Springer, Netherlands, pp. 787–790.

19 Ying, L., Huang, X., Huang, B., Xie, J., Zhao, J. and Zhao, X.S. (2002) Fluorescence emission and absorption spectra of single *Anabaena* sp strain PCC7120 cells. *Photochem. Photobiol.*, **76**, 310–313.

20 Wolf, E. and Schüßler, A. (2005) Phycobiliprotein fluorescence of *Nostoc punctiforme* changes during the life cycle and chromatic adaptation: characterization by spectral confocal laser scanning microscopy and spectral unmixing. *Plant, Cell Environ.*, **28**, 480–491.

21 Kumazaki, S., Hasegawa, M., Ghoneim, M., Shimizu, Y., Okamoto, K., Nishiyama, M., Oh-oka, H. and Terazima, M. (2007) A line-scanning semi-confocal multi-photon fluorescence microscope with a simultaneous broadband spectral acquisition and its application to the study of the thylakoid membrane of a cyanobacterium *Anabaena* PCC7120. *J. Microsc.*, **228**, 240–254.

22 Brakenhoff, G.J., Squier, J., Norris, T., Bliton, A.C., Wade, M.H. and Athey, B. (1996) Real-time two-photon confocal microscopy using a femtosecond, amplified Ti:sapphire system. *J. Microsc.*, **181**, 253–259.

23 Lin, C.P. and Webb, R.H. (2000) Fiber-coupled multiplexed confocal microscope. *Opt. Lett.*, **25**, 954–956.

24 Kim, J., Kang, D.K. and Gweon, D.G. (2006) Spectrally encoded slit confocal microscopy. *Opt. Lett.*, **31**, 1687–1689.

25 de Grauw, C.J., Otto, C. and Greve, J. (1997) Line-scan Raman microspectrometry for biological applications. *Appl. Spectrosc.*, **51**, 1607–1612.

26 Thiel, T. and Pratte, B. (2001) Effect on heterocyst differentiation of nitrogen fixation in vegetative cells of the cyanobacterium *Anabaena variabilis* ATCC 29413. *J. Bacteriol.*, **183**, 280–286.

27 Meeks, J.C. and Elhai, J. (2002) Regulation of cellular differentiation in filamentous cyanobacteria in free-living and plant-associated symbiotic growth states. *Microbiol. Mol. Biol. Rev.*, **66**, 94–121.

28 Black, K., Buikema, W.J. and Haselkorn, R. (1995) The *hglK* gene is required for localization of heterocyst-specific glycolipids in the cyanobacterium *Anabaena* sp. Strain PCC7120. *J. Bacteriol.*, **177**, 6440–6448.

29 Peschek, G.A. and Sleytr, U.B. (1983) Thylakoid morphology of the cyanobacteria *Anabaena variabilis* and *Nostoc* MAC grown under light and dark conditions. *J. Ultrastruct. Res.*, **82**, 233–239.

30 Vermaas, W.F.J., Timlin, J.A., Jones, H.D.T., Sinclair, M.B., Nieman, L.T., Hamad, S.W., Melgaard, D.K. and Haaland, D.M. (2008) *In vivo* hyperspectral confocal fluorescence imaging to determine pigment localization and distribution in cyanobacterial cells. *Proc. Natl. Acad. Sci. USA*, **105**, 4050–4055.

31 Lichtenthaler, H.K. and Babani, F. (2004) in *Chlorophyll a Fluorescence a Signature of Photosynthesis* (eds C. Papageorgiou and Govindjee), Springer, Netherlands, pp. 713–736.

32 Reger, B.J. and Krauss, R.W. (1970) Photosynthetic response to a shift in chlorophyll-*a* to chlorophyll-*b* ratio of chlorella. *Plant Physiol.*, **46**, 568–575.

33 Hatano, S., Kabata, K., Yoshimoto, M. and Sadakane, H. (1982) Studies on frost hardiness in Chlorella-ellipsoidea. 8. Accumulation of free fatty-acids during hardening of Chlorella-ellipsoidea. *Plant Physiol.*, **70**, 1173–1177.

34 Gergis, M.S. (1972) Influence of carbon dioxide supply on chloroplast structure of Chlorella pyrenoidosa. *Arch. Mikrobiol.*, **83**, 321–327.

35 Holub, O., Seufferheld, M.J., Gohlke, C., Govindjee, Heiss, G.J. and Clegg, R.M. (2007) Fluorescence lifetime imaging microscopy of *Chlamydomonas reinhardtii*: non-photochemical quenching mutants and the effect of photosynthetic inhibitors on the slow chlorophyll fluorescence transient. *J. Microsc.*, **226**, 90–120.

36 Bewersdorf, J., Egner, A. and Hell, S.W. (2006) in *Handbook of Biological Confocal Microscopy* (ed. J. Pawley), Springer, New York, pp. 550–560.

31
Fluorescence Lifetime Imaging Study on Living Cells with Particular Regard to Electric Field Effects and pH Dependence

Nobuhiro Ohta and Takakazu Nakabayashi

31.1
Introduction

Fluorescence microscopy techniques have become invaluable tools for the study of biological systems because they can be applied to living cells under native, physiological conditions [1, 2]. The conventional fluorescence microscopy techniques generally use fluorescence intensity measurements to reveal chromophore concentration and location in cells. Numerous fluorescent chromophores are now available that enable selective imaging at the microscopic level. However, fluorescence intensity depends on excitation intensity fluctuations, absorption by the sample, and photobleaching of fluorescent chromophores, and thus is difficult to analyze quantitatively. To overcome these problems, either excitation ratio or emission ratio methods have been employed for quantitative imaging [3, 4]. However, the ratio methods are difficult to combine with a confocal microscope because of wavelength-dependent focal depth and absorption.

Measurements of fluorescence lifetime of a chromophore can enhance the potential of fluorescence microscopy [1, 2, 5–8]. Fluorescence lifetime is an inherent property of a chromophore, and thus is independent of chromophore concentration, photobleaching and, excitation intensity, but highly dependent on pH, ion concentration, and local environment that affects the non-radiative rate of a chromophore. This makes fluorescence lifetime imaging (FLIM) a powerful tool for quantitative imaging of cellular conditions as well as the circumstances around the fluorescent dyes.

In the present study, a FLIM measurement system was constructed and applied to *Halobacterium salinarum* (*Hb. salinarum*) loaded with 2′,7′-bis-(carboxyethyl)-5(6)-carboxyfluorescein (BCECF) to obtain information on the intracellular environment as well as the intracellular pH in each of the cells [9–12]. *Hb. salinarum* belongs to the family of extreme halophilic archaebacteria, and considerable attention has been paid to this bacterium in relation to proton transport, phototaxis or the adaptation of an organism to extreme environments [13–15]. Intracellular pH is an essential parameter for *Hb. salinarum* in the regulation of intracellular processes [14, 16, 17], and fluorescence intensity ratio methods have been used to measure the intracellular pH [18, 19]. The

Molecular Nano Dynamics, Volume II: Active Surfaces, Single Crystals and Single Biocells
Edited by H. Fukumura, M. Irie, Y. Iwasawa, H. Masuhara, and K. Uosaki
Copyright © 2009 WILEY-VCH Verlag GmbH & Co. KGaA, Weinheim
ISBN: 978-3-527-32017-2

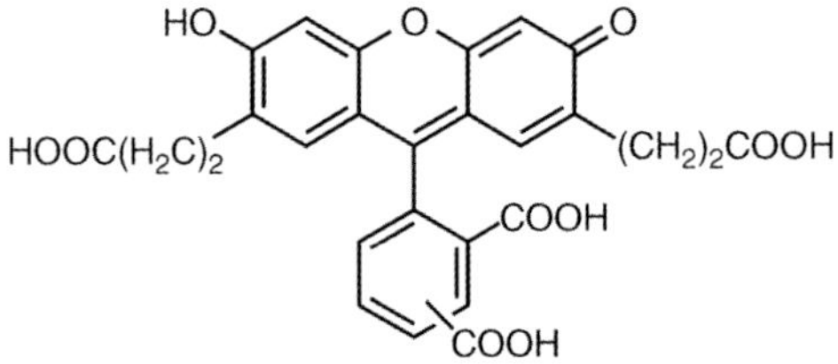

Figure 31.1 Chemical structure of BCECF.

fluorescence probe, BCECF, whose chemical structure is shown in Figure 31.1, is one of the most widely used fluorescent dyes for evaluation of pH because of its highly pH-dependent fluorescence intensity [4, 6, 18–22]. It is shown that the intracellular pH of *Hb. salinarum* can be evaluated from the fluorescence lifetime of BCECF retained in the cells, so that the intracellular pH as well as other intracellular properties is expected to be obtained from FLIM measurements on the halobacteria.

Application of electric fields to cells has been widely used in chemical biology for gene transfection, drug delivery, and modulation of intracellular ion concentrations [23–25]. However, the detailed mechanisms of such electroperturbation processes remain unclear. Therefore, the effect of an externally applied electric field on living cells was also examined by taking halobacteria as an example. We measured the time course of the space-resolved images of both the fluorescence intensity and the fluorescence lifetime of BCECF loaded in *Hb. salinarum* in the presence of external electric fields, which allow modulations of membrane permeabilization and intracellular environments [12]. Then, the field-induced changes in the shape and in the intracellular environment of the halobacteria were measured using the fluorescence of BCECF inside the cell. Charged and polar groups within protein structures have been reported to produce an electric field of 1–80 MV cm^{-1} for embedded molecules [26–28]. The fluorescence chromophore, which may be embedded in a protein cavity or attached to a membrane, is surrounded by both apolar and polar functional groups [29–31]. These groups produce a strong electric field inside the protein cavity or toward the chromophore. As a result, the excitation dynamics of the fluorescence chromophore may be affected by such a field [32, 33]. It is conceivable that the local electric field is one of the vital factors that control the rate of the non-radiative process of the fluorescent chromophore in cells. To analyze the fluorescence lifetime, this kind of effect may have to be considered in living cells. An external electric field effect on the fluorescence spectrum has also been observed for the fluorescent chromophore of BCECF in a polymer film.

31.2
Experimental

31.2.1
FLIM Measurement System

FLIM measurements were carried out using a four-channel time-gated detection system [9, 11, 34]. The experimental system is shown in Figure 31.2. A mode-locked

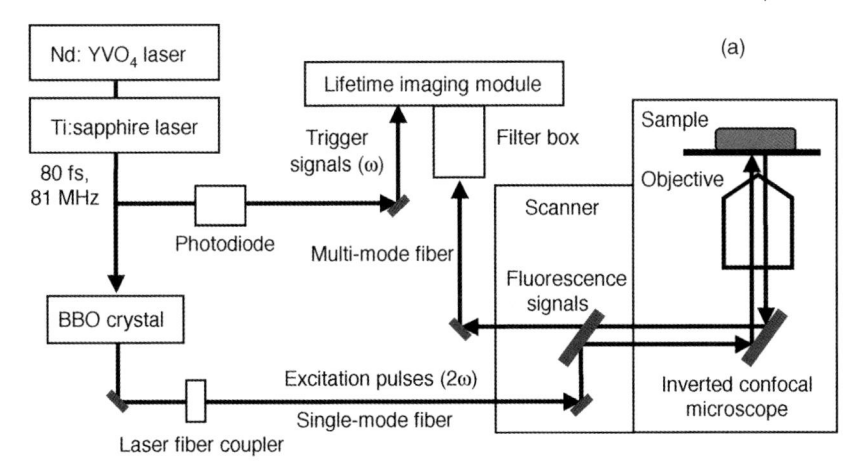

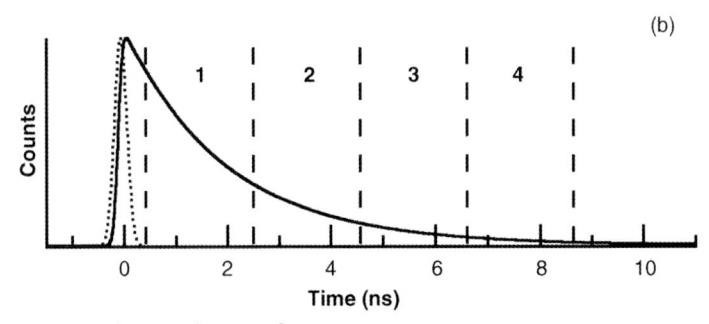

Figure 31.2 (a) Schematic diagram of a FLIM system.
(b) Schematic illustration of the time-gating method.
Fluorescence decay profile (solid line) and excitation pulse (dotted line). LIMO captures the fluorescence in the four time-windows.

Ti:sapphire laser (Spectra-Physics, Tsunami) pumped by a diode laser (Spectra-Physics, Millennia Xs) was used as the excitation light source. The pulse duration and the repetition rate of the laser pulse were 80 fs and 81 MHz, respectively. The second harmonic of an ultrafast harmonic system (Inrad) was used for excitation. The excitation beam was coupled to a single-mode optical fiber by a fiber coupler (Five Lab) and was introduced into the scanner head (Nikon, C1) of a confocal microscope (Nikon, TE2000-E). The intensity of the excitation pulse was attenuated to be <40 pJ pulse^{-1}. The excitation beam was focused onto the sample with a 40× or 60× oil microscope objective, and the fluorescence was collected by the same objective and was passed into two filters (Nikon, BA520 and EX510-560) to remove the scattered light. Fluorescence emission was detected by a pulse counting photomultiplier in a high-speed lifetime imaging module (Nikon Europe BV, LIMO). Fluorescence decay was measured for each pixel of the confocal microscope image. To minimize the amount of data generated, the lifetime imaging module captures the fluorescence decay trace into four time windows using the time-gating electronics (see Figure 31.2). Each reference trigger of the laser-pulse train enables four accumulation registers sequentially, and the detected fluorescence photons are counted and accumulated by one of the four accumulation registers. Each fluorescence lifetime was evaluated by analysis of the four time-window signals by assuming a single exponential decay, and the fluorescence lifetime image was obtained. The size of the image was 256×256 pixels. The background was evaluated by the counts at the area where fluorescent cells were not observed. Each measurement of the fluorescence lifetime image required ∼10 min. The compensation for the delay of the fluorescence photon signals was adjusted by measuring the fluorescence lifetime images of a standard slide (Molecular Probes) or dye molecules in a polymer film.

A time-to-amplitude converter (TAC) system was also employed to measure fluorescence decays without the microscope. Then, the fluorescence decay and the fluorescence lifetime were obtained precisely with the microchannel-plate photomultiplier (MCP-PM) as detection. The time resolution of the lifetime was determined, using a convolution method, to be ∼10 ps.

31.2.2
Preparation of *Hb. salinarum* Loaded with BCECF

The acetoxymethyl (AM) ester loading technique was used for loading BCECF into *Hb. salinarum*. The AM ester derivative of BCECF (BCECF/AM) is membrane-permeable and can enter into cells without disrupting their membranes. The AM ester is then cleaved by intracellular esterase to form BCECF, which has a very low membrane permeability, effectively trapping it inside the cell.

The strains of *Hb. salinarum*, S9, were cultured in peptone medium at 37 °C at pH 7.0 for six days [18, 35]. The cells were washed and resuspended in the basal salt solution (4 mol dm^{-3} NaCl containing 2.5×10^{-2} mol dm^{-3} HEPES) at pH 6.8. A 10 mm^3 of dimethylsulfoxide solution of BCECF/AM at 1.0×10^{-2} mol dm^{-3} was added to the 10 cm^3 cell suspension. The cells were incubated in the dark at 18 °C for three days to load the dye [18, 19]. The cell suspension was then centrifuged and the obtained cells were repeatedly washed with the basal salt solution at pH 6.8 until the supernatant showed no fluorescence. The dye-loaded cells were resuspended in the basal salt solution at pH 7.7.

31.2.3
Measurements of External Electric Field Effects

In order to measure fluorescence intensity images and fluorescence lifetime images of BCECF-loaded halobacteria in the presence of external electric fields, two stainless steel electrodes were inserted into the cell suspension. A dc field was applied between the electrodes, whose distance was 3.2 cm, and the halobacteria in the middle of the electrodes were observed.

Electric-field-induced changes in the fluorescence spectrum of BCECF itself were also measured using electric field modulation spectroscopy [36, 37]. For that purpose, a 0.4 ml portion of the aqueous solution (5 ml total) containing polyvinyl alcohol

(PVA, 88 mg) was mixed with 0.2 ml of HEPES buffer solution (1×10^{-2} mol dm^{-3} HEPES, 10 ml total) at pH 7.0 containing BCECF (5 mg) and stirred for several minutes at room temperature. The mixture was then cast on an ITO-coated quartz substrate by a spin-coating method, and water was removed by evaporation. A semitransparent aluminum (Al) film was deposited on the dried polymer film by a vacuum vapor deposition method. The ITO and Al films were used as electrodes. A sinusoidal ac voltage with a modulation frequency of 40 Hz was applied to a sample polymer, and the field-induced change in fluorescence intensity was detected with a lock-in amplifier at the second harmonic of the modulation frequency. A dc component of the fluorescence intensity was simultaneously observed.

31.3
Results and Discussion

31.3.1
FLIM of *Hb. salinarum*

Figure 31.3a–d show the time-resolved fluorescence intensity images of *Hb. salinarum* loaded with BCECF. These were observed using the experimental system shown in Figure 31.2. The excitation wavelength was 450 nm, and the detected fluorescence was in the region 515–560 nm where BCECF shows strong fluorescence intensity [10, 22]. Each of four time-windows was set to 2 ns. The fluorescence intensities of the halobacteria are different from each other. However, all the halobacteria exhibit a decrease in fluorescence intensity with time. The fluorescence lifetime image obtained from these intensity changes is shown in Figure 31.3e. It is clearly seen that some halobacteria give fluorescence lifetimes shorter than others. This suggests that at least two species with different fluorescence lifetimes exist in the cell suspension.

Figure 31.4a shows another example of FLIM of the halobacteria loaded with BCECF. The corresponding histograms of the fluorescence lifetime are shown in Figure 31.4c. The histogram of the fluorescence lifetime over the whole cells in the image shows a peak at around 2.4 ns. The histogram was also obtained for typical cells that seem to show long and short fluorescence lifetimes, respectively. The results suggest that a small amount of the halobacteria exhibits a lifetime as short as ∼1.9 ns, while most of the halobacteria have a lifetime of ∼2.4 ns. Figure 31.4b shows the changes in fluorescence intensity with time of the cells exhibiting long and short lifetimes. The intensity scale is normalized for intensities in the 0–2 ns region. In the present experiments, only four time-windows were used to detect the fluorescence intensity. However, it is clearly seen that the two cells exhibit different time dependences of the fluorescence intensity. These results indicate that there are (at least) two different populations of halobacteria with respect to the fluorescence lifetime, suggesting different environments in different halobacteria.

Because of the use of only four time-windows and the assumption of single exponential decay, an non-negligible error in the obtained lifetime values may be

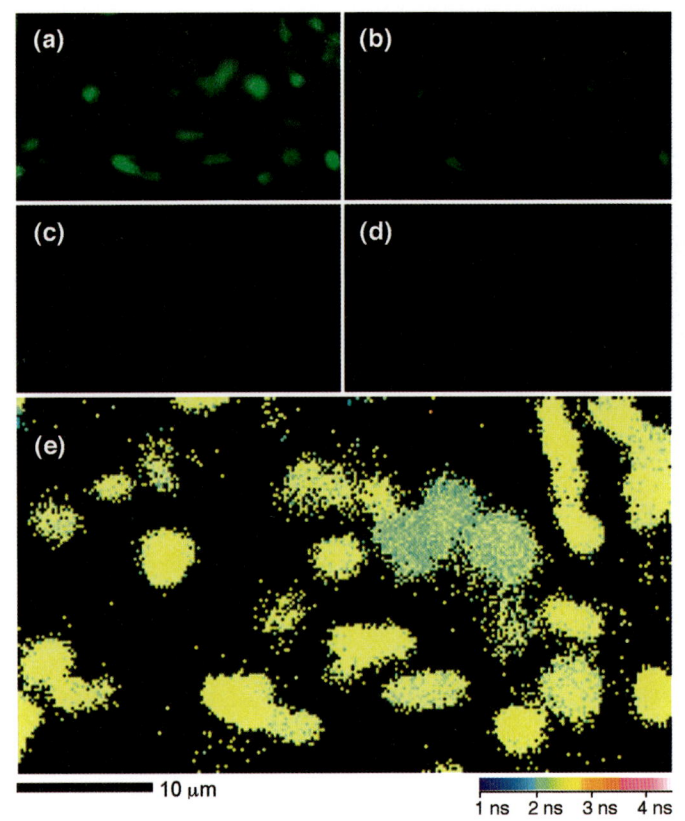

Figure 31.3 Time-resolved image of fluorescence intensity with a time interval of 0–2 (a), 2–4 (b), 4–6 (c), and 6–8 ns (d). (e) The corresponding fluorescence lifetime image of BCECF-loaded *Hb. salinarum*. Excitation wavelength was 450 nm. Fluorescence was detected in the region 515–560 nm.

inevitable. Actually, the fluorescence of BCECF in *Hb. salinarum* shows nonexponential decay and its lifetime as an averaged value over the spatial distribution was precisely determined to be 2.76 ns by using the TAC system, as described in Section 31.2.1. This value is larger by ∼0.4 ns than that obtained for most of the halobacteria (∼2.4 ns) implying that the lifetime values evaluated in the experiments may be smaller than the actual values by ∼0.4 ns. Thus, the cells exhibiting long and short lifetimes may give lifetimes of 2.8 and 2.3 ns, respectively.

As shown in the next section, the intracellular pH can be evaluated from the fluorescence lifetime of BCECF inside cells without any ratio methods [10]. The relation between the intracellular pH and the lifetime of BCECF in *Hb. salinarum* indicates that the lifetime decreases with decreasing intracellular pH. Based on the correlation function between the intracellular pH and the fluorescence lifetime, the average value of the intracellular pH of *Hb. salinarum* was estimated to be ∼7.1, which is roughly the same as that obtained with the intensity ratio method [18].

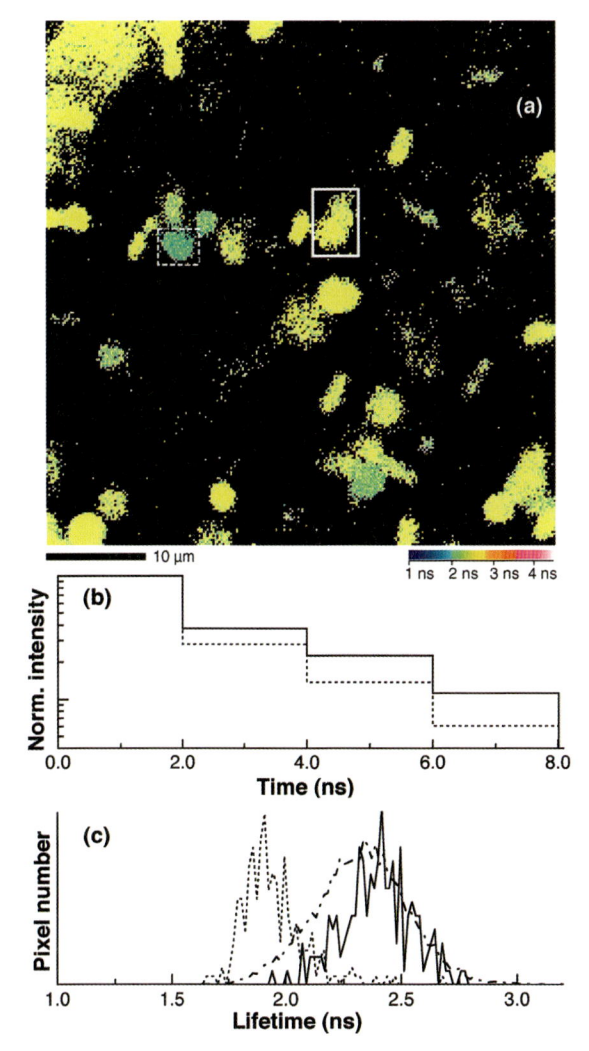

Figure 31.4 (a) Fluorescence lifetime image of BCECF-loaded *Hb. salinarum*. (b) Plots of the fluorescence intensities of the cells indicated by solid square (solid line) and dotted square (dotted line) against four time intervals of 0–2, 2–4, 4–6, and 6–8 ns on a logarithmic scale. The intensity scale is normalized for the intensities in the 0–2 ns region. (c) The histograms of the fluorescence lifetime for the cells indicated by solid squares and dotted squares and for the whole cells are shown by solid, dotted, and chain lines, respectively. Excitation wavelength was 450 nm. Fluorescence was detected in the region 515–560 nm.

However, if the observed difference in the fluorescence lifetime is only attributed to the difference in the intracellular pH, a pH value of 5~6 has to be assumed for the cells having such a short fluorescence lifetime. It is unlikely that the cellular pH is much lower than the outside pH because the intracellular pH of halobacteria has been reported to be 7~8, even when the extracellular pH is 5.5–7.7 [18, 19, 38].

Therefore, it is concluded that the observation of the short fluorescence lifetime cannot be explained only by the intracellular pH, and other factors should be considered as the origins of the short fluorescence lifetime. The fluorescence lifetime of BCECF can also be influenced by cellular environmental factors other than the pH dependence, for example, ion concentration or electric field inside a cell [10, 11]. The observation of two halobacteria species with short and long fluorescence lifetimes arises from the fact that the cells have different intracellular environments. It is suggested from the electrofluorescence spectrum of BCECF shown in Section 31.3.3 that the observed difference in the fluorescence lifetime is ascribed to the difference in electric fields inside a cell. In fact, intra- and inter-molecular electron transfer, excimer formation and intramolecular intersystem crossing processes have been shown to be significantly influenced by electric fields with a strength of the order of $1\,\mathrm{MV\,cm^{-1}}$ [32, 36, 37, 39]. Thus the short lifetime of BCECF in *Hb. salinarum* may result from the effect of the strong electric fields surrounding the BCECF located inside halobacteria, which depends on the cell condition. These results may suggest that fluorescence lifetime measurement of a fluorescent probe is applicable to evaluate the activity of each cell.

31.3.2
pH Dependence of the Fluorescence Lifetime in Solution and in Living Cells

The pH dependence of the fluorescence lifetime of BCECF has been examined in aqueous solution using the TAC system of time-resolved fluorescence spectroscopy. The fluorescence decay observed in solution is shown in Figure 31.5. The decay curves are fitted by assuming bi-exponential decay, that is, $\Sigma A_i \exp(-t/\tau_i)$, where A_i and τ_i denote the pre-exponential factor and the lifetime of component i ($=1, 2$), respectively. Plots of the solution pH against the average fluorescence lifetime (τ_f) given by $\Sigma A_i \tau_i$ are shown in Figure 31.6. The correlation between pH and the fluorescence lifetime in solution is almost the same as that obtained with the frequency-domain method [6, 40]. The correlation is well fitted by the following

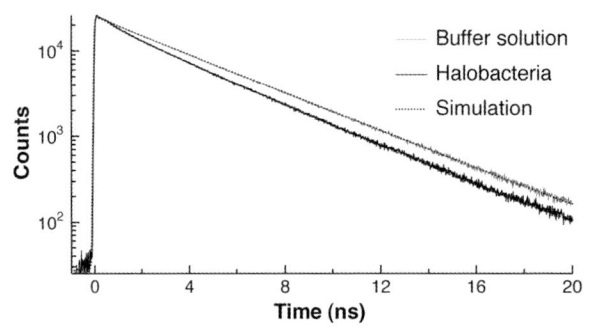

Figure 31.5 Fluorescence decay of BCECF in HEPES buffer at pH 8.5 and in *Hb. salinarum*. The excitation and fluorescence wavelengths were 450 and 530 nm, respectively.

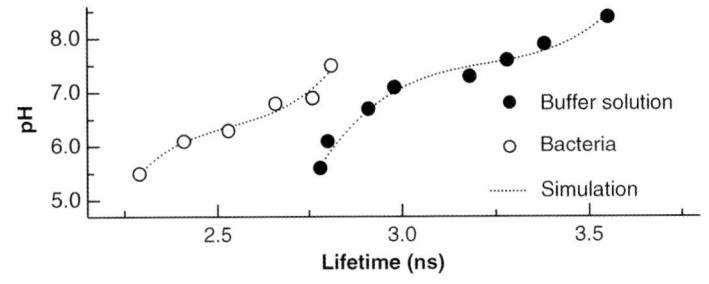

Figure 31.6 Plots of pH against the fluorescence lifetime of BCECF in HEPES buffer (●) and in *Hb. salinarum* (○). The excitation and fluorescence wavelengths were 450 and 530 nm, respectively.

polynomial function of pH in the range from 5.6 to 8.4 (Figure 31.6):

$$pH = -514.51 + 483.09 \times \tau_f - 149.38 \times \tau_f^2 + 15.434 \times \tau_f^3 \qquad (31.1)$$

The polynomial behavior of the correlation function may result from the pH dependence of the molar ratio between the monoanionic and dianionic species of BCECF.

The fluorescence decay of BCECF observed in *Hb. salinarum* is also shown in Figure 31.5. It was necessary to assume a tri-exponential decay to reproduce the decay observed *in vivo*. The pH dependence of the fluorescence lifetime in *Hb. salinarum* could be measured using monensin, which is a kind of Na^+/H^+ ionophore and forms an equilibrium between intracellular and extracellular pH [18, 19]. Thus the cell suspension was mixed with 2.5 mm^3 of DMSO solution of monensin at 1.0×10^{-2} mol dm^{-3}. After 10 min, the pH of the suspension was adjusted to give different values of the pH, at each of which the fluorescence decay was measured.

Plots of the intracellular pH against τ_f of BCECF *in vivo* are also shown in Figure 31.6. The correlation function between the intracellular pH and τ_f is different from that in solution, indicating that substantial consideration must be paid to calibration of intracellular pH using solution data [6]. The average fluorescence lifetime is shorter *in vivo* than in solution, even at the same pH. The correlation function of the intracellular pH in the range from 5.5 to 7.5 is given as follows:

$$pH = -440.42 + 523.91 \times \tau_f - 205.49 \times \tau_f^2 + 26.961 \times \tau_f^3 \qquad (31.2)$$

From Eq. (31.2), we can evaluate the pH of *Hb. salinarum* without ratio methods. The fluorescence lifetime in *Hb. salinarum* without monensin is evaluated to be 2.76 ns. The intracellular pH is then calculated to be 7.1, which is in reasonable agreement with that obtained from the excitation ratio method [18]. The reason why the fluorescence lifetime *in vivo* is smaller than that *in vitro* may be ascribed to the local field produced by some proteins and membranes that affects the chromophore *in vivo* [32, 41]. In fact, the fluorescence lifetime of BCECF is considered to become shorter in the presence of electric fields, as described in the next section.

31.3.3
External Electric Field Effect on Fluorescence of BCECF

Charged and polar groups of peptides or proteins cause electric fields in chromophores [26–28]. The gradient of pH across membranes also induces electric fields inside cells [42, 43]. Such electric fields influence the photoexcitation dynamics of the chromophore [32]. It is therefore conceivable that the short fluorescence lifetime observed in some halobacteria results from the effects of electric fields within the cell. To explore the possibility that the difference in the fluorescence lifetime of halobacteria is due to the field effects, the electric field effects on the fluorescence spectra of BCECF in a PVA polymer film have been investigated. Electrofluorescence spectra, that is, plots of the field-induced change in fluorescence intensity of BCECF as a function of fluorescence wavelength, have been observed. The results are shown in Figure 31.7. The excitation wavelength was 424 nm, where the field-induced change in absorption intensity was negligible. It was found that the fluorescence of BCECF was quenched by an applied electric field, suggesting that the fluorescence lifetime of BCECF becomes shorter in the presence of electric fields. The electrofluorescence spectrum can be reproduced by a linear combination of the fluorescence spectrum and its first derivative spectrum, indicating not only field-induced quenching but also the Stark shift induced by a change in molecular polarizability between the fluorescent state and the ground state. The magnitude of the quenching of the fluorescence of BCECF in PVA was ~0.4% with a field strength of 1.0 MV cm^{-1}. The lifetime of the minor parts observed in *Hb. salinarum* is shorter than that of most cells by ~20%. If a field strength of 7 MV cm^{-1} is applied to BCECF, the observed difference

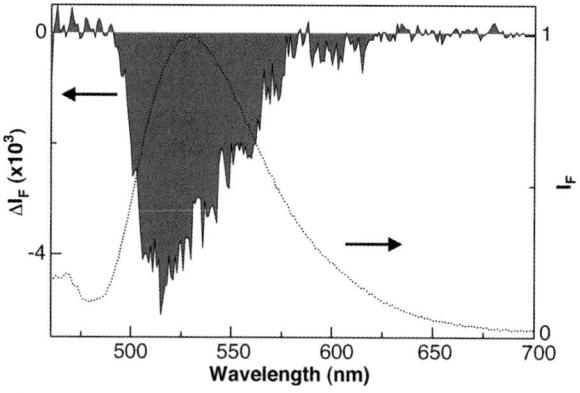

Figure 31.7 Fluorescence spectrum (dotted line) and electrofluorescence spectrum (shaded line) of BCECF at 0.1 mol% in a PVA film. Applied field strength was 1.0 MV cm^{-1}. Excitation wavelength was 424 nm.

in the fluorescence lifetime can be reproduced since the field effect is proportional to the square of the applied field strength in a random distribution system [32]. The field strength estimated in *Hb. salinarum* seems to be in the region expected for electric fields produced by proteins and membranes.

31.3.4
Electric-Field-Induced Aggregate Formation in *Hb. salinarum*

External electric field effects on *Hb. salinarum* were measured by using a dc field. Figure 31.8 shows the time course of the fluorescence intensity and the corresponding fluorescence lifetime images of *Hb. salinarum* loaded with BCECF. The applied field was 0.25 V cm^{-1}. Fluorescence in the 515–560 nm region was detected. The distribution of the fluorescence lifetime over the whole cells in each image is also shown at the right of the figure. As already described in Section 31.3.1 and as shown in Figure 31.8a, the halobacteria were homogeneously dispersed in the basal salt solution, and the fluorescence lifetime of each bacterium could be identified in the lifetime image before application of an electric field. The halobacteria have a cylindrical shape with a long axis of 1–5 μm. The distribution of the fluorescence lifetime in Figure 31.8a exhibits a peak at ~2.4 ns. However, it is found that aggregates of halobacteria are effectively formed after exposure to an electric field. As shown in Figure 31.8b–d, the aggregate becomes larger with exposure time, and an aggregate larger than 100 μm is formed at 100 min after exposure (see Figure 31.8d). The halobacteria cannot clearly be seen at electric fields larger than 1.0 V cm^{-1}, which is probably due to destruction of the halobacteria in the presence of high electric fields. It is noted that not all of the halobacteria are the same in panels Figure 31.8a–d because of movement of the halobacteria, but the aggregates are observed in any regions around the middle of the electrodes. The cylindrical shape of the halobacteria seems to remain unchanged, even after the formation of aggregates.

The field-induced morphological change is clearly observed in the fluorescence intensity image, but the peak position of the fluorescence lifetime distribution remains constant at around 2.4 ns at 100 min after exposure (see Figure 31.8). This result indicates that the halobacteria mostly exhibit a fluorescence lifetime of ~2.4 ns, regardless of exposure time; the pH in the halobacteria is essentially the same, irrespective of aggregate formation.

As shown in Section 31.3.1, at least two halobacteria species exhibiting different fluorescence lifetimes of BCECF exist in the cell suspension: small amounts of halobacteria exhibit a lifetime as short as 1.9 ns, while most of the halobacteria exhibit a lifetime of around 2.4 ns [11]. Halobacteria having a short fluorescence lifetime of ~1.9 ns are observed in the lifetime image at 0 min in Figure 31.8a, as presented by the blue color in the image, and still exist at 100 min after exposure (see Figure 31.8d). Thus it is concluded that the effect of the field-induced aggregate formation on the fluorescence lifetime of BCECF is very small for both the short and long fluorescence lifetimes. The negligible effect of aggregate formation on the fluorescence lifetime therefore indicates that the intracellular environments of the halobacteria remain unchanged after formation of aggregates.

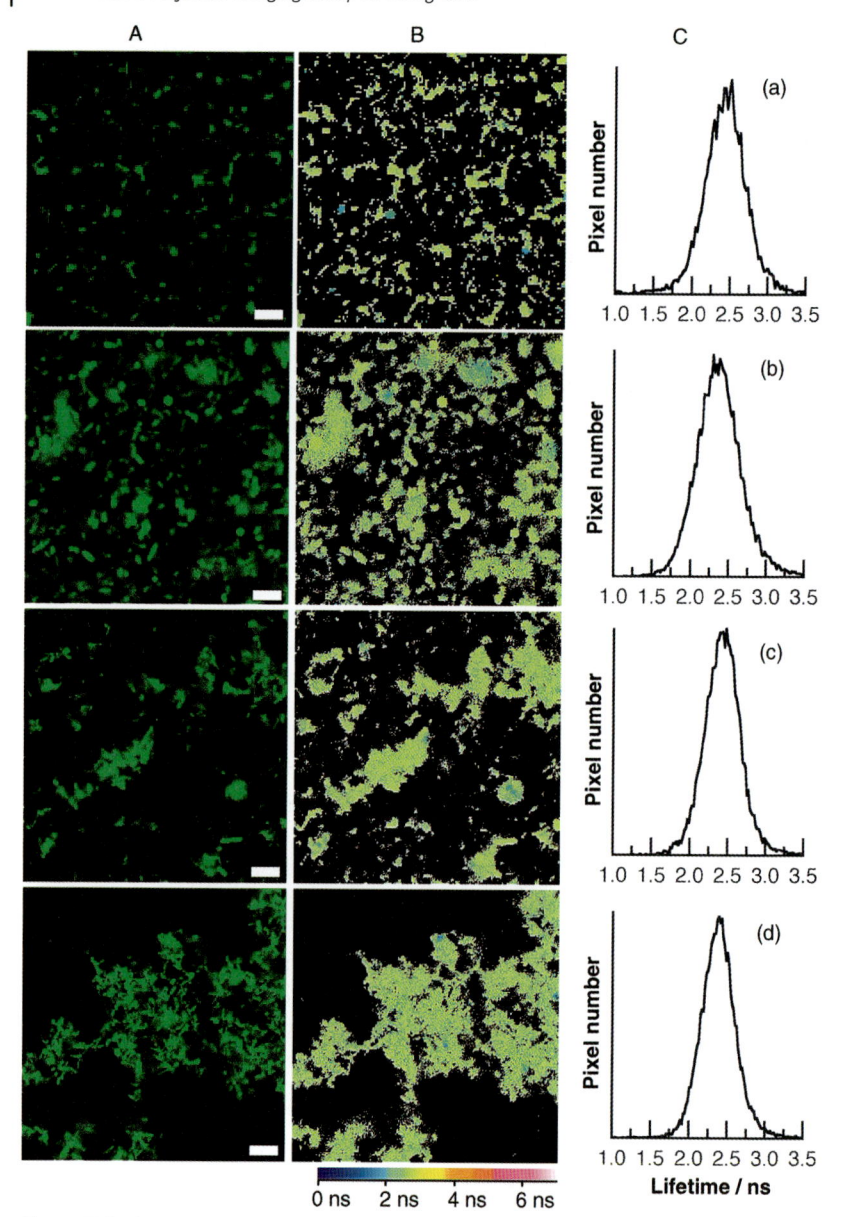

Figure 31.8 Fluorescence intensity images (A) of BCECF-loaded *Hb. salinarum* and the corresponding fluorescence lifetime images (B) and the distributions of the fluorescence lifetime (C) at 0 (a), 60 (b), 80 (c), and 100 min (d) after exposure to an external electric field. The applied voltage was 0.8 V. Scale bar is 10 μm. Excitation wavelength was 450 nm. Fluorescence in the 515–560 nm was detected.

31.4
Summary

FLIM measurements were applied to *Hb. salinarum* loaded with BCECF. At least two halobacteria species that exhibit different fluorescence lifetimes from each other are found to exist in the cell suspension, suggesting that the cells have different intracellular environments. The difference in the fluorescence lifetime may reflect the difference in activity of halobacteria. It is suggested that strong electric fields inside a cell play a significant role in the determination of the fluorescence lifetime of BCECF. The fluorescence intensity of BCECF in a polyvinyl alcohol film is quenched by an external electric field, indicating that the fluorescence lifetime becomes shorter in the presence of electric fields. The fact that the fluorescence lifetime in living cells is much shorter than that in buffer solution at the same pH also implies that fluorescence chromophores in living cells feel strong electric fields produced by proteins and membranes located near the fluorescence probe. The measurements of fluorescence lifetime images allow the study of the intracellular dynamics of a single living microorganism in response to changes in metabolism and environmental conditions. It is also shown that aggregates of the halobacteria are formed by an external electric field, even when the intracellular environment remains unchanged by exposure to an electric field. Electroporation seems to be important for aggregate formation; however, the generated pore should be small because of the preservation of the intracellular environment.

Acknowledgments

The authors thank Professor Hui-Ping Wang at Zhejiang University in China, Professor Kazuo Tsujimoto at the Japan Advanced Institute of Science and Technology (JAIST) in Ishikawa, and Professors Seiji Miyauchi and Naoki Kamo at Hokkaido University in Sapporo for their collaboration in this work. This work was supported by a Grant-in-Aid for Scientific Research on Priority Area (Area No. 432, No. 16072201) from the Ministry of Education, Culture, Sports, Science, and Technology (MEXT) of Japan.

References

1 Lichtman, J.W. and Conchello, J.-A. (2005) Fluorescence microscopy. *Nat. Methods*, **2**, 910–919.

2 Day, R.N. (2005) Imaging protein behavior inside the living cell. *Mol. Cell. Endocrinol.*, **230**, 1–6.

3 Tsien, R.Y. and Poenie, M. (1986) Fluorescence ratio imaging: a new window into intracellular ionic signaling. *Trends Biochem. Sci.*, **11**, 450–455.

4 Ritucci, N.A., Erlichman, J.S., Dean, J.B. and Putnam, R.W. (1996) A fluorescence technique to measure intracellular pH of single neurons in brainstem slices. *J. Neurosci. Methods*, **68**, 149–163.

5 Lakowicz, J.R. and Berndt, K.W. (1991) Lifetime-selective fluorescence imaging using an rf phase-sensitive camera. *Rev. Sci. Instrum.*, **62**, 1727–1734.
6 Hanson, K.M., Behne, M.J., Barry, N.P., Mauro, T.M., Gratton, E. and Clegg, R.M. (2002) Two-photon fluorescence lifetime imaging of the skin stratum corneum pH gradient. *Biophys. J.*, **83**, 1682–1690.
7 Wallrabe, H. and Periasamy, A. (2005) Imaging protein molecules using FRET and FLIM microscopy. *Curr. Opin. Biotechnol.*, **16**, 19–27.
8 Suhling, K., French, P.M.W. and Phillips, D. (2005) Time-resolved fluorescence microscopy. *Photochem. Photobiol. Sci.*, **4**, 13–22.
9 Nakabayashi, T., Iimori, T., Kinjo, M. and Ohta, N. (2006) Construction of a fluorescence lifetime imaging system and its application to biological systems and polymer materials. *J. Spectrosc. Soc. Jpn.*, **55**, 31–39 (in Japanese).
10 Nakabayashi, T., Wang, H.-P., Tsujimoto, K., Miyauchi, S., Kamo, N. and Ohta, N. (2007) A Correlation between pH and fluorescence lifetime of 2′,7′-bis(2-carboxyethyl)-5(6)-carboxyfluorescein (BCECF) *in vivo* and *in vitro*. *Chem. Lett.*, **36** (2007), 206–207.
11 Wang, H.-P., Nakabayashi, T., Tsujimoto, K., Miyauchi, S., Kamo, N. and Ohta, N. (2007) Fluorescence lifetime image of a single halobacterium. *Chem. Phys. Lett.*, **442**, 441–444.
12 Nakabayashi, T., Wang, H.-P., Tsujimoto, K., Miyauchi, S., Kamo, N. and Ohta, N. (2008) Studies on effects of external electric fields on halobacteria with fluorescence intensity and fluorescence lifetime imaging microscopy. *Chem. Lett.*, **37**, 522–523.
13 Spudich, J.L. (1998) Variations on a molecular switch: transport and sensory signalling by archaeal rhodopsins. *Mol. Microbiol.*, **28**, 1051–1058.
14 Subramaniam, S. and Henderson, R. (2000) Molecular mechanism of vectorial proton translocation by bacteriorhodopsin. *Nature*, **406**, 653–657.
15 Soppa, J. (2006) From genomes to function: haloarchaea as model organisms. *Microbiology*, **152**, 585–590.
16 Oesterhelt, D. and Stoeckenius, W. (1973) Functions of a new photoreceptor membrane. *Proc. Natl. Acad. Sci. USA*, **70**, 2853–2857.
17 Lanyi, J.K. and Weber, H.J. (1980) Spectrophotometric identification of the pigment associated with light-driven primary sodium translocation in *Halobacterium halobium*. *J. Biol. Chem.*, **255**, 243–250.
18 Tsujimoto, K., Semadeni, M., Huflejt, M. and Packer, L. (1988) Intracellular pH of halobacteria can be determined by the fluorescent dye 2′,7′-bis(carboxyethyl)-5 (6)-carboxyfluorescein. *Biochem. Biophys. Res. Commun.*, **155**, 123–129.
19 Urano, H., Mizukami, T. and Tsujimoto, K. (1997) Chemical tools for detection of intracellular pH-change in halobacteria. *Chem. Lett.*, **26**, 1217–1218.
20 Rink, T.J., Tsien, R.Y. and Pozzan, T. (1982) Cytoplasmic pH and free Mg^{2+} in lymphocytes. *J. Cell Biol.*, **95**, 189–196.
21 Paradiso, A.M., Tsien, R.Y. and Machen, T.E. (1984) Na^+—H^+ exchange in gastric glands as measured with a cytoplasmic-trapped fluorescent pH indicator. *Proc. Natl. Acad. Sci. USA*, **81**, 7436–7440.
22 Boens, N., Qin, W., Basaric, N., Orte, A., Talavera, E.M. and Alvarez-Pez, J.M. (2006) Photophysics of the fluorescent pH indicator BCECF. *J. Phys. Chem. A*, **110**, 9334–9343.
23 Sale, A.J.H. and Hamilton, W.A. (1967) Effects of high electric fields on microorganisms–I. Killing of bacteria and yeasts. *Biochim. Biophys. Acta*, **148**, 781–788.
24 White, J.A., Blackmore, P.F., Schoenbach, K.H. and Beebe, S.J. (2004) Stimulation of capacitative calcium entry in HL-60 cells by nanosecond pulsed electric fields. *J. Biol. Chem.*, **279**, 22964–22972.
25 Sun, Y., Vernier, P.T., Behrend, M., Wang, J., Thu, M.M., Gundersen, M. and Marcu,

L. (2006) Fluorescence microscopy imaging of electroperturbation in mammalian cells. *J. Biomed. Opt.*, **11**, 024010 1–8.

26. Callis, P.R. and Burgess, B.K. (1997) Tryptophan fluorescence shifts in proteins from hybrid simulations: an electrostatic approach. *J. Phys. Chem. B*, **101**, 9429–9432.

27. Park, E.S., Andrews, S.S., Hu, R.B. and Boxer, S.G. (1999) Vibrational Stark spectroscopy in proteins: a probe and calibration for electrostatic fields. *J. Phys. Chem. B*, **103**, 9813–9817.

28. Kriegl, J.M., Nienhaus, K., Deng, P., Fuchs, J. and Nienhaus, G.U. (2003) Ligand dynamics in a protein internal cavity. *Proc. Natl. Acad. Sci. USA*, **100**, 7069–7074.

29. Brejc, K., Sixma, T.K., Kitts, P.A., Kain, S.R., Tsien, R.Y., Ormö, M. and Remington, S.J. (1997) Structural basis for dual excitation and photoisomerization of the *Aequorea victoria* green fluorescent protein. *Proc. Natl. Acad. Sci. USA*, **94**, 2306–2311.

30. Ormö M., Cubitt, A.B., Kallio, K., Gross, L.A., Tsien, R.Y. and Remington, S.J. (1996) Crystal structure of the *Aequorea victoria* green fluorescent protein. *Science*, **273**, 1392–1395.

31. Yang, F., Moss, L.G. and Phillips, G.N. Jr (1996) The molecular structure of green fluorescent protein. *Nat. Biotechnol.*, **14**, 1246–1251.

32. Ohta, N. (2002) Electric field effects on photochemical dynamics in solid films. *Bull. Chem. Soc. Jpn.*, **75**, 1637–1655.

33. Ogrodnik, A., Eberl, U., Heckmann, R., Kappl, M., Feick, R. and Michel-Beyerle, M.E. (1991) Excitation dichroism of electric field modulated fluorescence yield for the identification of primary electron acceptor in photosynthetic reaction center. *J. Phys. Chem.*, **95**, 2036–2041.

34. Gerritsen, H.C., Asselbergs, M.A.H., Agronskaia, A.V. and van Sark, W.G.J.H.M. (2002) Fluorescence lifetime imaging in scanning microscopes: acquisition speed, photon economy and lifetime resolution. *J. Microsc.*, **206**, 218–224.

35. Lanyi, J.K. and MacDonald, R.E. (1979) Light-induced transport in *Halobacterium halobium*. *Methods Enzymol.*, **56**, 398–407.

36. Ohta, N., Koizumi, M., Umeuchi, S., Nishimura, Y. and Yamazaki, I. (1996) External electric field effects on fluorescence in an electron donor and acceptor system: ethylcarbazole and dimethyl terephthalate in PMMA polymer films. *J. Phys. Chem.*, **100**, 16466–16471.

37. Ohta, N., Umeuchi, S., Nishimura, Y. and Yamazaki, I. (1998) Electric-field-induced quenching of exciplex fluorescence and photocurrent generation in a mixture of ethylcarbazole and dimethyl terephthalate doped in a PMMA polymer film. *J. Phys. Chem. B*, **102**, 3784–3790.

38. Wagner, G. and Hope, A.B. (1976) Proton transport in *Halobacterium halobium*. *Aust. J. Plant Physiol.*, **3**, 665–676.

39. Nakabayashi, T., Wu, B., Morikawa, T., Iimori, T., Rubin, M.B., Speiser, S. and Ohta, N. (2006) External electric field effects on absorption and fluorescence of anthracene–$(CH_2)_n$–naphthalene bichromophoric molecules doped in a polymer film. *J. Photochem. Photobiol. A*, **178**, 236–241.

40. Szmacinski, H. and Lakowicz, J.R. (1993) Optical measurements of pH using fluorescence lifetimes and phase-modulation fluorometry. *Anal. Chem.*, **65**, 1668–1674.

41. Tsushima, M., Ushizaka, T. and Ohta, N. (2004) Time-resolved measurement system of electrofluorescence spectra. *Rev. Sci. Instrum.*, **75**, 479–485.

42. Hendler, R.W., Drachev, L.A., Bose, S. and Joshi, M.K. (2000) On the kinetics of voltage formation in purple membranes of *Halobacterium salinarium*. *Eur. J. Biochem.*, **267**, 5879–5890.

43. Koch, M.K. and Oesterhelt, D. (2005) MpcT is the transducer for membrane potential changes in *Halobacterium salinarum*. *Mol. Microbiol.*, **55**, 1681–1694.

32
Multidimensional Fluorescence Imaging for Non-Invasive Tracking of Cell Responses

Ryosuke Nakamura and Yasuo Kanematsu

32.1
Introduction

Fluorescence microscopy has become a powerful technique for the examination of fixed or living biological specimens. Recent progress in the development of laser-scanning microscopy [1, 2] and various fluorescent probes [3–5] has demonstrated that a fluorescence-based method allows the selective and sensitive detection of an object of interest labeled by a fluorescent probe with a good signal-to-background ratio. However, when the wavelength of the excitation light is tuned to the near-ultraviolet (UV) region, native molecules existing in cells and tissues also emit fluorescence because many molecules have photoabsorption in the near-UV region. These autofluorescence background signals significantly reduce the signal-to-background ratio in the conventional use of fluorescence microscopy.

Typical native molecules that show photoabsorption spectra in the near-UV region are aromatic amino acids (tryptophan, phenylalanine, tyrosine), the extracellular matrix (collagen, elastin), coenzymes relating to electron transfer systems [nicotinamide adenine dinucleotide (NADH), flavin adenine dinucleotide (FAD)], and many kinds of secondary metabolites. Since these molecules are, in general, physiologically important, autofluorescence has been used to monitor the metabolic state of living cells and applied to tissue diagnostics [6–16]. Physiological alterations in tissues are detected as changes in the fluorescence properties of associated molecules, including an increase/decrease in the fluorescence intensity and a peak shift of the fluorescence spectrum.

Fluorescence microscopy based on autofluorescence is a promising non-invasive and versatile method because it does not require labeling of the target molecules. However, it is essential to develop the following techniques to overcome possible challenges:

(1) A procedure for decomposing a mixture of unknown fluorescent components and tracking the spectral change of a specific fluorescent component, because most cells and tissues contain several autofluorescent molecules with broad and overlapping fluorescence spectra in the near-UV region.

(2) A method of reducing the acquisition time to avoid photo-damage of living cells during irradiation with the near-UV light and tracking cell responses in real time.

Utilization of a multidimensional fluorescence data set that includes, for example, a fluorescence spectrum, an excitation spectrum, a time profile, anisotropy, and spatial localization, is a straightforward and effective approach. Obviously, increasing a dimension (experimental variable) of the data improves the selectivity of the measurement and its subsequent analysis. It has been noticed that 3D or higher dimensional data is inherently different from 2D data, because the decomposition of 3D data is often unique while that of 2D data never is [17]. Therefore, 3D or higher dimensional fluorescence data can be decomposed into individual fluorescent components without any prior knowledge of the autofluorescent molecules.

The parallel factor analysis (PARAFAC) model [18–20] is based on a multilinear model, and is one of several decomposition methods for a multidimensional data set. A major advantage of this model is that data can be uniquely decomposed into individual contributions. Because of this, the PARAFAC model has been widely applied to 3D and also higher dimensional data in the field of chemometrics. It is known that fluorescence data is one example that corresponds well with the PARAFAC model [21].

Collecting a multidimensional fluorescence data set will require a long acquisition time and is time-consuming. The design of an optical configuration that achieves efficient and rapid acquisition of a whole data set is essential for the tracking of cell responses in real time without photo-damage of living cells.

We developed two kinds of multidimensional fluorescence spectroscopic systems: the time-gated excitation–emission matrix spectroscopic system and the time- and spectrally resolved fluorescence microscopic system. The former acquires the fluorescence intensities as a function of excitation wavelength (Ex), emission wavelength (Em), and delay time (τ) after impulsive photoexcitation, while the latter acquires the fluorescence intensities as a function of Em, τ, and spatial localization (x-, y-positions). In both methods, efficient acquisition of a whole data set is achieved based on line illumination by the laser beam and detection of the fluorescence image by a 2D image sensor, that is, a charge-coupled device (CCD) camera.

In this study, we propose an approach based on unique optical configuration, efficient acquisition of a multidimensional data set, and decomposition of unknown fluorescent components by using the PARAFAC model. Further, we demonstrate that our approach is powerful and effective enough to track complicated responses in living cells by analyzing the autofluorescence of native molecules.

To present our methodology, we describe the time-gated excitation–emission spectroscopic system in Section 32.3. 2D fluorescence spectroscopy acquiring excitation and fluorescence spectra has been widely used at research and diagnostic levels because of the high selectivity and simple configuration of the measurement system [12–16]. Here, we extended it to the 3D (Ex, Em, and τ) system with a time-resolution of 200 ps, by a combination of a spatially dispersed super continuum as the excitation light source and a CCD camera equipped with a 200-ps-gated intensifier. To evaluate our method, we measured the 3D fluorescence of a mixed solution composed of a number of fluorescent dyes, and analyzed the 3D data set by the PARAFAC model.

In Section 32.4, we describe the time- and spectrally resolved fluorescence imaging system for tracking autofluorescence spectral changes of target molecules in living cells. Conventional laser-scanning confocal microscopy requires a long time to obtain a whole data set of multidimensional fluorescence intensities. In addition, the detector requires a relatively long exposure time because autofluorescence is generally very weak. Therefore, we developed a line-scanning technique, which is based on line illumination of the laser beam and detection of the fluorescence image through a slit instead of a pinhole [22–25]. In this optical arrangement, the fluorescence image was obtained by scanning only one axis perpendicular to the excitation line, and the acquisition time was significantly reduced compared with conventional laser-scanning confocal microscopy.

The performance of line-scanning microscopy, in which time-gated and spectral-resolved fluorescence images are obtained and decomposed based on the PARAFAC model, was examined by applying it to the analysis of one of the induced plant defense responses: the accumulation of antimicrobial compounds, generally known as phytoalexins, in oat (*Avena sativa*). Oat leaves produce avenanthramides as phytoalexins when attacked by pathogens or treated with an elicitor [26–28]. Avenanthramides are substituted hydroxycinnamic acid conjugates, which demonstrate photoabsorption in the near-UV region. By using line-scanning microscopy, we have measured the autofluorescence of oat leaves and analyzed the accumulation of avenanthramides in response to the elicitor. The results demonstrate that our approach is powerful and effective for the analysis of complicated responses in living cells.

32.2
Materials and Methods

32.2.1
Time-Gated Excitation–Emission Matrix Spectroscopy

A schematic illustration of the time-gated excitation–emission spectroscopy is shown in Figure 32.1a. The output pulses (repetition rate of 200 kHz, pulse duration of 150 fs) from an amplified mode-locked Ti:sapphire laser (Coherent, RegA9000) were focused in a sapphire plate to generate a white light continuum, which was dispersed by a grating (300 grooves mm^{-1}) and focused on a sample cell. Therefore, different positions within the line illumination correspond to different excitation wavelengths. The fluorescence image of the line illumination on the sample was relayed to the end of the optical fiber bundle, which consisted of a linear array of 20 fiber cores, attached to a polychromator (Acton, SpectraPro-150, 300 grooves mm^{-1} grating). Fluorescence from the other end of the optical fiber bundle was spectrally dispersed by the grating

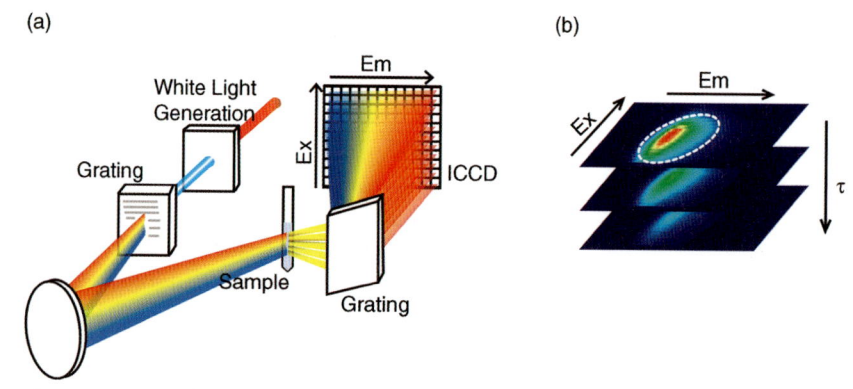

Figure 32.1 (a) A schematic illustration of time-gated excitation–emission matrix spectroscopy. (b) A typical example of the 3D fluorescence data measured for Rhodamine 590 in ethanol (10^{-5} mol l^{-1}).

and detected by a CCD camera (480 × 640 pixels) equipped with a 200-ps-gated intensifier (LaVision, PicoStar HR). The time delay between a laser pulse and a gating electronic pulse was changed by an electronic delay generator (Becker & Hickl GmbH, DEL-150). As a result, a single frame of the CCD contained information on the time-gated excitation–emission spectral map (Ex–Em map) with 20 excitation and 640 emission wavelengths, as shown in Figure 32.1b. In this study, the optical configuration was adjusted so that the excitation and emission wavelength ranges were 455–575 and 468–731 nm, respectively. The exposure time of the detector was typically set at 50 ms. Under these conditions, it takes only a few seconds to acquire a data set consisting of, for example, 20 (Ex) × 640 (Em) × 24 (τ).

32.2.2
Time- and Spectrally-Resolved Fluorescence Imaging

The experimental set-up for the time- and spectrally-resolved fluorescence imaging system is shown in Figure 32.2. The excitation laser source and the detection system are the same as those in the time-gated excitation–emission matrix spectroscopy. An amplified mode-locked Ti:sapphire laser operated at a wavelength of 780 nm and a repetition rate of 200 kHz. The second harmonics (center wavelength of 390 nm, pulse duration of 150 fs) generated in a thin BBO crystal was used as an excitation light. A line illumination pattern (in parallel with the y-direction in Figure 32.2) was created by a cylindrical lens ($f = 150$ mm) and was focused on a sample with a 10× objective lens (Olympus: numerical aperture of 0.30). The excitation intensity was reduced to 10 pJ, which was measured in front of the objective lens.

The fluorescence image of the line illumination on the sample was relayed to the entrance slit of a polychromator. The slit width was set at 70 μm, corresponding to 7 μm on the sample in this configuration. Fluorescence passing through the entrance slit was spectrally dispersed by the grating and detected by a CCD (480 × 640 pixels)

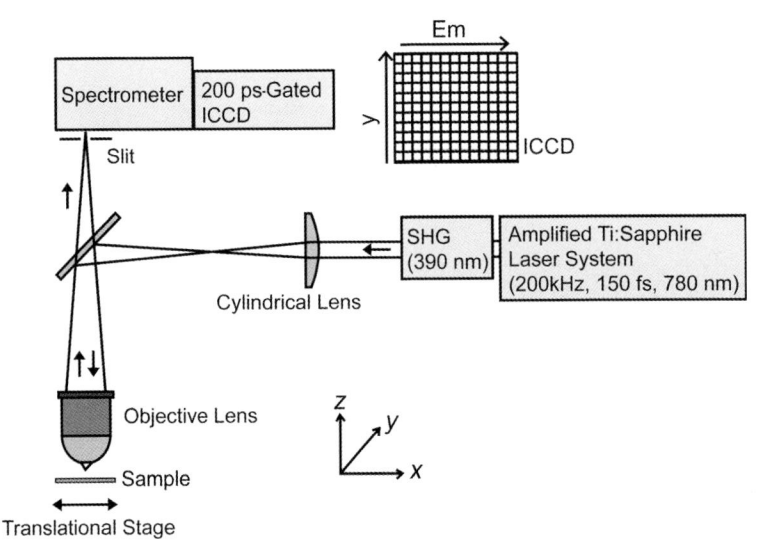

Figure 32.2 Experimental set-up for time- and spectrally-resolved fluorescence imaging based on the line illumination.

equipped with a 200-ps-gated intensifier. As a result, a single frame of the CCD provided information on the two-dimensional time-gated fluorescence data: The fluorescence image of the line illumination (y-direction) on the sample was vertically aligned on the 2D data while the fluorescence spectrum at each y-position in the excitation line was horizontally allocated. Hereafter, we refer to the 2D time-gated data as a "y-Em map." A y-Em map consists of 480-μm height (y) in length on the sample and a 265-nm spectral bandwidth (Em). We obtained the sample images, which we call "x–y images" by shifting the sample position along the x-direction (perpendicular to the excitation line) and reconstructing the data set of y-Em maps acquired at each x position with the time-gated fluorescence spectra. Furthermore, varying the gate timing of the intensifier, we finally obtained multiple fluorescence images as a function of Em and τ. Since the scanning dimensions of a sample position can be reduced by combining line excitation and multi-channel detection, this method acquires a whole data set remarkably faster than conventional confocal microscopy. In this study, the typical exposure time of the detector was set at 200 ms. Under this condition, it takes about 10 min to acquire a data set consisting of, for example, 640 (x) × 480 (y) × 640 (Em) × 2 (τ).

We obtained y-Em maps by scanning the x-position (1 μm step, 640 positions in total). Typically, two frames of different delay times ($\tau = 0.0$ and 3.0 ns) were obtained at each x-position. As a result, a fluorescence data set consisted of 640 (x) × 480 (y) × 640 (Em) × 2 (τ). For the PARAFAC calculations, this data set was binned with 25-nm steps along the fluorescence wavelength dimension to reduce data size. In addition, the spatial dimensions of 640 (x) × 480 (y) were reshaped to the one-dimensional array (size of 307 200), and then reshaped again to the spatial dimensions of 640 (x) × 480 (y) after calculations. Therefore, the data set, which consisted of 10 (Em) × 2 (τ) × 307 200 (xy), was fitted by the PARAFAC model.

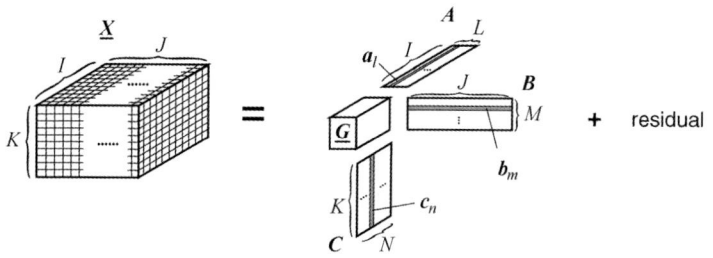

Figure 32.3 The Tucker3 model for a 3D data set of **X** with $I \times J \times K$. **G** is the 3D core array with $L \times M \times N$. a_{il}, b_{jm}, and c_{kn} in Eq. (32.1) are elements of the matrices of **A**$(I \times L)$, **B**$(J \times M)$, and **C** $(K \times N)$, respectively, or, in other words, are elements of the vectors of a_l, b_m, and c_n, respectively.

32.2.3
PARAFAC Model

There are several methods for decomposing the 3D data set **X** with $I \times J \times K$. The two major methods are PARAFAC and Tucker3. Since the PARAFAC model can be considered a constrained version of the Tucker3 model, we first describe the Tucker3 model and then give a description of the PARAFAC model.

The formulation of the 3D Tucker3 model (Figure 32.3) can be given as

$$x_{ijk} = \sum_{l=1}^{L} \sum_{m=1}^{M} \sum_{n=1}^{N} a_{il} b_{jm} c_{kn} g_{lmn} + e_{ijk}, \quad i = 1, \ldots, I; j = 1, \ldots, J; k = 1, \ldots, K, \quad (32.1)$$

where x_{ijk} is an element of the 3D data set **X** $(I \times J \times K)$. a_{il}, b_{jm}, and c_{kn} are elements of the matrices of **A**$(I \times L)$, **B**$(J \times M)$, and **C**$(L \times N)$, respectively, or, in other words, are elements of the vectors of a_l, b_m, and c_n, respectively (see Figure 32.3). e_{ijk} is a residual term and g_{lmn} is an element of the 3D core array **G**$(L \times M \times N)$, where L, M, N represent the numbers of components of the first, second and third modes, respectively. The off-superdiagonal elements of the core array represent the interactions between the different modes. Therefore, due to the degree of freedom of the rotation, the Tucker3 model does not have a unique solution.

The 3D PARAFAC model with F components is depicted in Figure 32.4 and can be formulated as follows:

$$x_{ijk} = \sum_{f=1}^{F} a_{if} b_{jf} c_{kf} + e_{ijk}, \quad i = 1, \ldots, I; j = 1, \ldots, J; k = 1, \ldots, K, \quad (32.2)$$

where x_{ijk} and e_{ijk} have the same meaning as in Eq. (32.1). a_{if}, b_{jf}, and c_{kf} are elements of the matrices of **A**$(I \times F)$, **B**$(J \times F)$, and **C**$(L \times F)$, respectively, or, in other words, are elements of the vectors of a_f, b_f, and c_f, respectively (see Figure 32.4). The PARAFAC model is recognized as a simplification of the Tucker3 model: The numbers of components of all dimensions are equal to F and there is no interaction between the

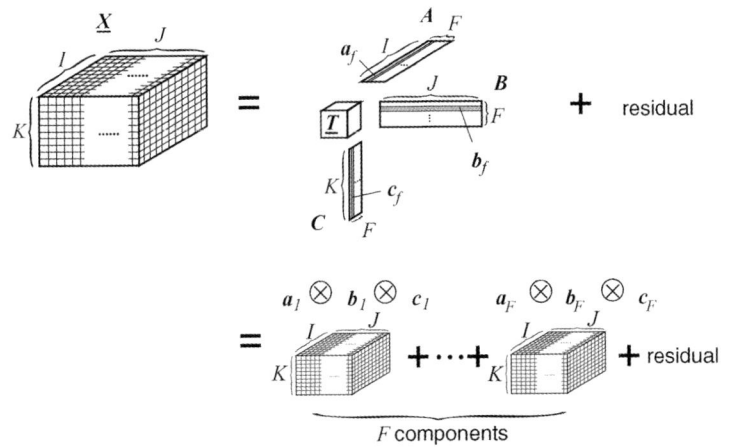

Figure 32.4 The PARAFAC model for a 3D data set of **X** with $I \times J \times K$. **T** is the 3D superdiagonal array with $F \times F \times F$. $a_{if}, b_{jf},$ and $c_{kf},$ are elements of the matrices of $\mathbf{A}(I \times F)$, $\mathbf{B}(J \times F)$, and $\mathbf{C}(K \times F)$, respectively, or, in other words, are elements of the vectors of $\mathbf{a}_f, \mathbf{b}_f,$ and $\mathbf{c}_f,$ respectively. $\mathbf{a}_f \otimes \mathbf{b}_f \otimes \mathbf{c}_f$ represents the contribution of the fth component.

different dimensions. In other words, when $\mathbf{G}(L \times M \times N) = \mathbf{T}(F \times F \times F)$, the Tucker3 model in Eq. (32.1) is in accordance with the PARAFAC model in Eq. (32.2), where **T** is a superdiagonal array with zeros in all places except for the superdiagonal which contains only one.

As shown in Figure 32.4, the 3D PARAFAC model in Eq. (32.2) can also be written as

$$X = \sum_{f=1}^{F} \mathbf{a}_f \otimes \mathbf{b}_f \otimes \mathbf{c}_f + E, \qquad (32.3)$$

where $E (I \times J \times K)$ is the 3D array of residual terms, and $\otimes$ is the Kronecker product. This formulation clearly represents the decomposition of **X** into each contribution of the fth component.

In this study, x_{ijk} was regarded as the fluorescence intensity element of the 3D fluorescence data set $X (I \times J \times K)$. In the case of time-gated excitation–emission spectroscopy, $\mathbf{a}_f (a_{1f}, \ldots, a_{If})$, $\mathbf{b}_f (b_{1f}, \ldots, b_{Jf})$, and $\mathbf{c}_f (c_{1f}, \ldots, c_{Kf})$ correspond to an excitation spectrum, a fluorescence spectrum, and a time profile of the fth component, respectively. On the other hand, in the case of time- and spectrally-resolved imaging, they correspond to a fluorescence spectrum, a time profile and an xy-image of the fth component, respectively. The PARAFAC model assumes that the fluorescence spectrum of each component is independent of the excitation wavelength, the delay time and the position, whereas only the relative contribution of each fluorescent component changes at each fluorescence variable. This assumption is reasonable in the sub-nanosecond time resolution of our system. It should be noted that scattered light (Rayleigh and Raman scattering) appearing as a diagonal line pattern on the Ex–Em map is inadequate for PARAFAC modeling. By using a time gate with 200-ps resolution, we removed the scattering pattern from a 3D data set, and then analyzed the set based on the PARAFAC model.

The number of components F was determined by the core consistency diagnostic, as proposed by Bro and Kiers [29]. This is based on evaluating the appropriateness of the PARAFAC model by comparing the core arrays of the Tucker3 and PARAFAC models, because the PARAFAC model is a constrained version of the Tucker3 model. Therefore, the core consistency can be defined as

$$\text{Consistency}(\%) = 100 \left(\frac{\sum_{l=1}^{F}\sum_{m=1}^{F}\sum_{n=1}^{F} (g_{lmn} - t_{lmn})^2}{F} \right) \quad (32.4)$$

where t_{lmn} is an element of the 3D superdiagonal array $T(F \times F \times F)$. The consistency (%) is always less than or equal to 100% and may also be negative. It is considered that an appropriate number of components has been obtained when the consistency drops from a high value (60–100%) to a low value with an increase in the number of components.

In PARAFAC modeling, non-negativity constraints were applied to all three dimensions. All analyses were performed with the N-way toolbox for MATLAB [30], which is a set of MATLAB routines designed to perform multi-way data analysis.

32.2.4
Sample Preparation

Oat seeds (*Avena sativa* L., cv. Shokan 1) were soaked in distilled water for 24 h to facilitate germination and then sown in wet vermiculite. They were maintained at 20 °C for 7 days under exposure to continuous artificial light at a photosynthetic photon flux density (PPFD) of 50 mol m^{-2} s^{-1} in the growth chamber.

Leaf segments were prepared from the primary leaves of 7-day-old oat seedlings. The lower epidermis was peeled away and the mesophyll cells were floated on 3 ml of the elicitor solution or distilled water in a petri dish with the peeled surface in contact with the solution [31]. Penta-*N*-acethylchitopentaose solution (concentration 1 mM) was used as an elicitor while distilled water was used as a control. All experiments were performed on leaf segments after incubation for 48 h at 20 °C. Penta-*N*-acetylchitopentaose was purchased from Seikagaku Kogyo, Tokyo. Avenanthramide A (*N*-(4-hydroxycinnamoyl)-5-hydroxyanthranilic acid) was a gift from Dr. A. Ishihara (Kyoto University, Kyoto).

32.3
Time-Gated Excitation–Emission Matrix Spectroscopy

32.3.1
The 3D Fluorescence Properties of Dye Solutions

The 3D fluorescence data set, consisting of 20 (Ex) × 640 (Em) × 24 (τ) data points was measured for four kinds of fluorescent dyes: Coumarin 540 (C540), DCM,

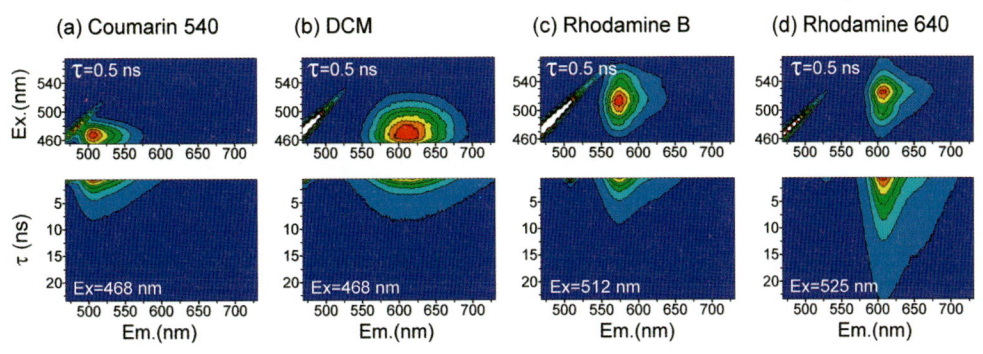

Figure 32.5 The 3D fluorescence data of each fluorescence component: (a) C540, (b) DCM, (c) RhB, (d) Rh640. The upper panels show Ex–Em maps sliced at 0.5 ns while the bottom panels show τ–Em maps sliced around the peaks of the excitation spectra.

Rhodamine B (RhB), and Rhodamine 640 (Rh640) in ethanol. The contour representations for the Ex–Em maps and the τ–Em maps are shown in Figure 32.5a–d. To avoid strong scattering of the excitation laser, τ was scanned from 0.5 ns. Nevertheless, small scattering of the excitation laser light appears as a diagonal pattern in each Ex–Em map. The fluorescence properties of these fluorescent dyes are clearly characterized in the 3D space of Ex–Em–τ. For example, C540 and DCM appear at similar positions in Ex but at different positions in Em, reflecting different Stokes shift energies. It is seen from the τ–Em maps that the fluorescence decay time of Rh640 is much longer than the others.

32.3.2
The 3D Fluorescence Property of a Mixed Solution

Next, we examined a mixed solution consisting of the fluorescent dyes: C540 (2.3×10^{-5} mol l^{-1}), DCM (1.5×10^{-5} mol l^{-1}), RhB (6.0×10^{-5} mol l^{-1}), and Rh640 (4.0×10^{-5} mol l^{-1}). The absorption spectra of the mixture and the constituents are shown in Figure 32.6. The spectral information of each constituent is buried in the featureless absorption spectrum of the mixture.

The 3D fluorescence data set consisting of 20 (Ex: 455–575 nm) × 640 (Em: 468–731 nm) × 24 (τ: 0.5–23.5 ns) data points was measured for the mixed solution. The acquisition time was just a few seconds. The Ex–Em maps at various τ are presented in the contour representation in Figure 32.7. The dotted white line is the scattering of the excitation light. It can be seen that the fluorescence pattern of the mixture varies drastically with τ. The small fluorescence pattern located around 470 nm (Ex) and 500 nm (Em) disappears with τ, and the broad fluorescence pattern located around the center of the map changes and becomes narrower with τ.

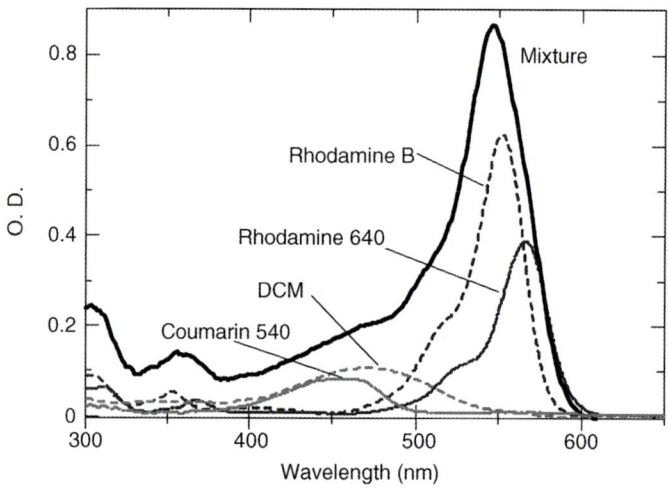

Figure 32.6 Absorption spectra of the mixture (thick solid line) and the constituents.

For the PARAFAC modeling, the Ex–Em map at 0.5 ns was removed, since the scattered light pattern was inadequate for PARAFAC modeling. Therefore, the data set consisting of 20 (Ex: 455–575 nm) × 640 (Em: 468–731 nm) × 23 (τ: 1.5–23.5 ns) was fitted by the PARAFAC model.

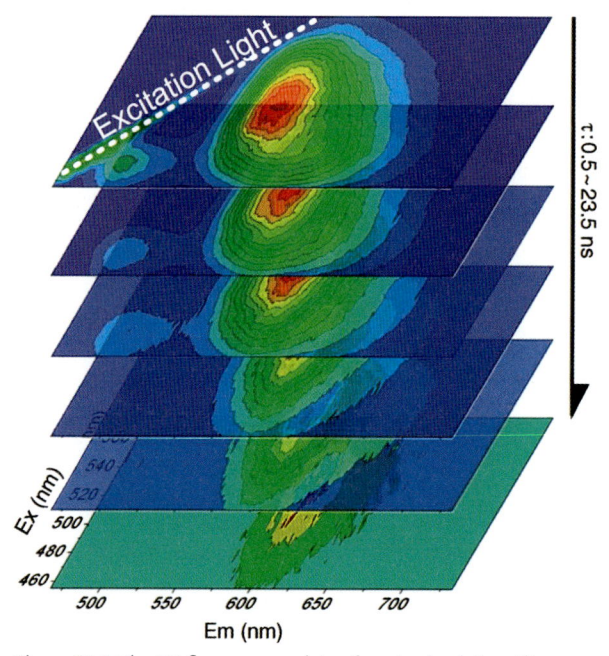

Figure 32.7 The 3D fluorescence data of a mixed solution. The Ex–Em maps sliced at $\tau = 0.5$, 2.5, 4.5, 9.5, 14.5 and 23.5 ns are shown.

32.3.3
PARAFAC Decomposition Without any Prior Knowledge of Constituents

First, we examined a number of fluorescent components from the 3D data for the mixture. Based on the formulation given by Eq. (32.4), we calculated the consistency of the PARAFAC model with various numbers of components. The consistency for the 1- and 2-component models was almost 100% in both cases, while that for the 3-component model decreased to 70% (Figure 32.8). In 4- or more component models, the consistencies were almost 0%, indicating that the PARAFAC model was no longer adequate in these cases. Therefore, we determined that 3 components was an appropriate number.

We fitted the 3D fluorescence data of the mixture solution by the 3-component PARAFAC model, and then obtained the vectors of a_f with elements of a_{if} ($I = 1, \ldots, 20$), b_f with elements of b_{jf} ($j = 1, \ldots, 640$), and c_f with elements of c_{kf} ($k = 1, \ldots, 23$) of the fth component. These corresponded to the excitation spectrum, the fluorescence spectrum, and the time profile of the fth component, respectively. For a comparison against Figure 32.5, we calculated the Kronecker products of $a_f \otimes b_f$ and $c_f \otimes b_f$, corresponding to Ex–Em and τ–Em maps of the fth component, respectively, and plotted them in Figure 32.9. Here, we labeled the components extracted by PARAFAC as Components I, II, and III.

Component I exhibited two peaks at 505 and 625 nm in Em (Figure 32.9a). These two peaks can be ascribed to C540 and DCM by comparison with Figure 32.5. In addition, Components II and III are naturally ascribed to RhB and Rh640, respectively. These results indicate that the PARAFAC method successfully decomposed the 3D data into individual components without any prior knowledge, except that C540 and DCM were regarded as a single component with a two-peak structure of fluorescence.

To evaluate dissimilarity among fluorescence properties of the fluorescent dyes examined in this study, we calculated orthogonality (sinθ) between the fluorescence

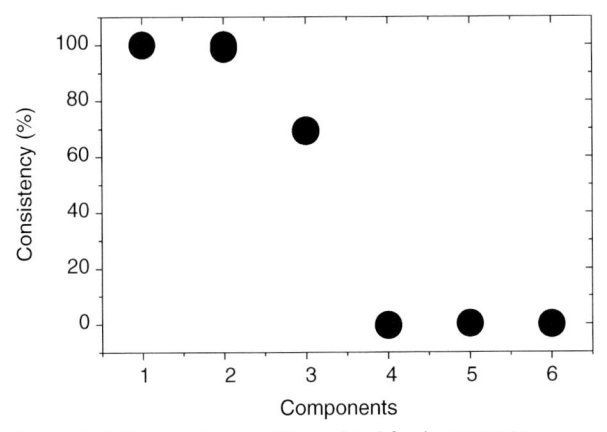

Figure 32.8 The consistency (%) calculated for the PARAFAC model with different numbers of components, using Eq. (32.4).

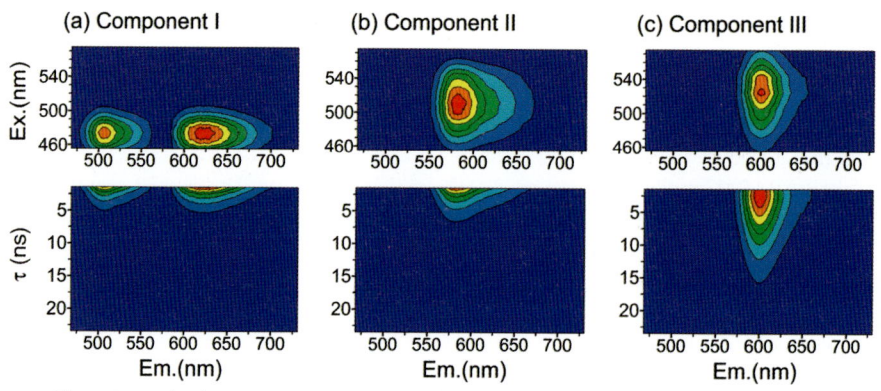

Figure 32.9 The three components extracted from the 3D fluorescence data of the mixture in Figure 32.7 using the PARAFAC model. The upper panels are Ex–Em maps constructed by $a_f \otimes b_f$, while the bottom panels are τ–Em maps constructed by $c_f \otimes b_f$.

variables (Ex, Em, τ) of each pair of two different fluorescent dyes (listed in Table 32.1). That is, if $\sin\theta$ between the excitation spectra of different dyes is zero for example, their excitation spectra are identical, while $\sin\theta = 1$ means that their excitation spectra are completely different. Table 32.1 shows that the pair with the smallest $\sin\theta$ value for Ex was RhB–Rh640 ($\sin\theta = 0.20$), while this pair had relatively large $\sin\theta$ values for Em and τ. Similarly, the pair with the smallest $\sin\theta$ value for Em was DCM–Rh640 ($\sin\theta = 0.40$), but it had relatively large $\sin\theta$ values for Ex and τ. On the other hand, a C540–DCM pair, which was regarded as a single component in PARAFAC decomposition, had relatively small $\sin\theta$ values for both Ex and τ. As expected, these results indicate that sufficient dissimilarities are required for at least two variables in the case of 3D data. Otherwise other dimensions (additional fluorescence variables) should be introduced for further decomposition. Such improvement in the fluorescence-based method is relatively easy due to the simple optical configuration of our system.

Summarizing this section, we developed the time-gated excitation–emission matrix spectroscopic system and applied it to the decomposition of a mixed solution of a number of fluorescent dyes. We demonstrated that our approach, which was based on unique optical configuration, efficient acquisition of a multidimensional data set, and decomposition of unknown fluorescent components by using the PARAFAC model, was effective for the analysis of unknown multi-component targets.

Table 32.1 Orthogonality ($\sin\theta$) calculated between fluorescence variables (Ex, Em, τ) of each pair of two different fluorescent dyes.

	(a) Ex				(b) Em				(c) τ		
	DCM	RhB	Rh640		DCM	RhB	Rh640		DCM	RhB	Rh640
C540	0.46	0.97	0.97	C540	0.99	0.97	1.00	C540	0.09	0.09	0.49
DCM		0.82	0.85	DCM		0.71	0.40	DCM		0.17	0.55
RhB			0.20	RhB			0.87	RhB			0.42

32.4
Time- and Spectrally-Resolved Fluorescence Imaging

32.4.1
Characterization of y–Em Maps

In this section, based on the methodology presented in the previous section, we describe multidimensional fluorescence imaging and its application to tracking cell responses. We developed the time- and spectrally-resolved fluorescence imaging system based on line illumination, which is capable of rapid acquisition of fluorescence intensities as a function of Em, τ, and xy-positions. We applied it to the analysis of an induced plant defense response, that is, the accumulation of antimicrobial compounds or phytoalexins, in oat (*Avena sativa*).

A transmission image of mesophyll cells of an oat leaf treated with an elicitor is shown in Figure 32.10a. Figure 32.10b shows a y–Em map (τ = 0.0 ns) observed at the position of x_1 indicated by a dotted line in Figure 32.10a. This kind of information on the time-gated fluorescence spectrum at each y-position can be obtained as a single frame of the CCD, demonstrating the unique and effective configuration of the line-scanning method. In Figure 32.10b, in addition to the strong fluorescent component at a wavelength longer than 650 nm, weakly fluorescent components can be recognized in the region of 450–650 nm: At least two components of a short-wavelength

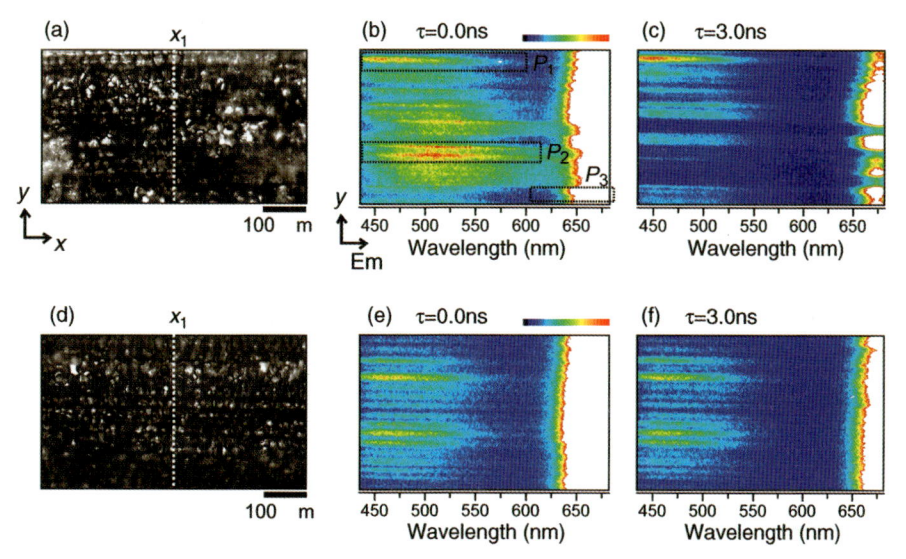

Figure 32.10 (a) Transmission image of mesophyll cells of an oat leaf treated with an elicitor. (b) Time-gated y-Em map observed at the x_1 position indicated by a dotted line in (a). Delay time τ is 0.0 ns. (c) The same as (b) but τ is 3.0 ns. (d)–(f) The same as (a)–(c), respectively, but the oat leaf was treated with distilled water as a control. All the time-gated y-Em maps are normalized by the fluorescence intensity of the short-wavelength components centered around 450 nm to focus on the weak fluorescence in 450–650 nm. Therefore, the intensity at a wavelength region longer than 650 nm is above the scale and is shown as a white area.

component centered at ~450 nm and a mid-wavelength component centered at ~510 nm. As a result, three components with different fluorescence spectra were observed in the elicitor-treated cells.

Figure 32.10c shows a y–Em map observed at the same position of x_1, but at a delay time of 3.0 ns, where fluorescence intensity was normalized by the short-wavelength component. The fluorescence pattern of the mid-wavelength component disappears and the fluorescence intensity of the long-wavelength component decreases, indicating that the fluorescence lifetimes of these components are shorter than that of the short-wavelength component.

Figure 32.10d is a transmission image of mesophyll cells of an oat leaf treated with distilled water as a control. In contrast to the three components observed in the elicitor-treated cells, only two components with different fluorescence spectra can be recognized in the y–Em map (Figure 32.10e): A short-wavelength component centered at ~450 nm and a long-wavelength component centered at a wavelength longer than 650 nm. In a y–Em map at a delay time of 3.0 ns (Figure 32.10f), where the fluorescence intensity was normalized by the short-wavelength component, the intensity of the long-wavelength component decreased. This indicates that the fluorescence lifetime of the long-wavelength component is shorter than that of the short-wavelength component, as is the case with the elicitor-treated cells.

The time-integrated fluorescence spectra and time profiles of the fluorescence intensities of these components, which were obtained by varying the delay time τ in steps of 0.02 ns and then averaging in each area of $P_1 - P_3$ (see Figure 32.10b), were plotted in Figure 32.11a and b, respectively. It should be mentioned that the fluorescence properties of the short- and long-wavelength components in elicitor-

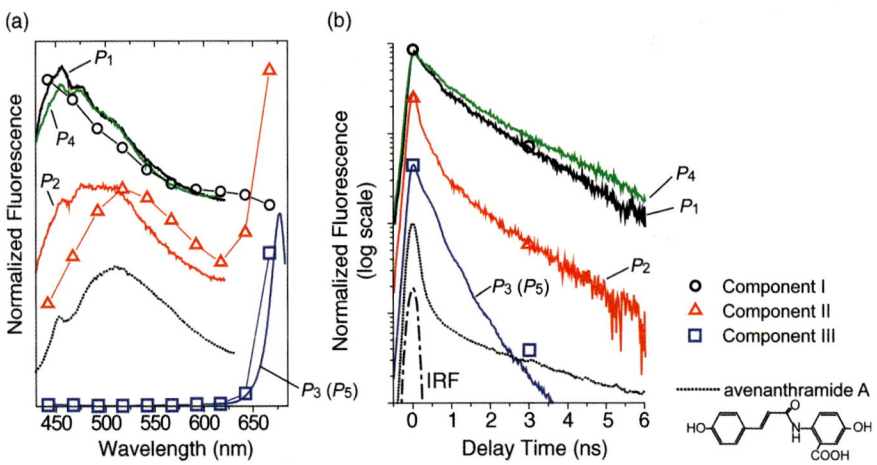

Figure 32.11 (a) Time-integrated fluorescence spectra and (b) time profiles of the fluorescence intensities. The lines are results experimentally observed and the symbols are the components extracted by PARAFAC. The fluorescence spectra and the time profiles were drawn with appropriate shifts in a vertical direction for comparison.

treated cells were similar to those in water-treated cells, respectively. The mid-wavelength component (P_2), which was observed only in the elicitor-treated cells, had a broad fluorescence spectrum, as shown in Figure 32.11a. The time profiles of these components are shown in Figure 32.11b. They can be fitted by two or three exponential decays convoluted with an instrumental response function (IRF). The parameters obtained by the fitting are listed in Table 32.2. As expected from the time-gated y–Em maps in Figure 32.10, the decay time of the short-wavelength component is longer than that of the others.

32.4.2
Spatial Localization of Fluorescent Components

To study the spatial localization of these fluorescent components, x–y images at different wavelengths are shown in Figure 32.12. Here, x–y images of the elicitor-treated cells were constructed with fluorescence at $\tau = 0.0$ ns and averaged in a 25-nm bin size centered at Em = 442, 567 and 667 nm (Figure 32.12a–c, respectively). Fluorescence at these selected wavelengths is mainly derived from the short-, mid-, and long-wavelength components, respectively. The minor contributions from other components exist because of spectral overlap among different components.

The short- and long-wavelength components have a similar spatial localization at this spatial resolution. Namely, the fluorescence of these components was observed in the same cell. On the other hand, the x–y image of the mid-wavelength component shows a complementary pattern to the other components. That is, the fluorescence intensity of the mid-wavelength component is strong in the cell, whereas the fluorescence of the long-wavelength component is weak.

As in the case of the elicitor-treated cells, the x–y images corresponding to the short- and long-wavelength components of the water-treated cells (control) were similar to each other (data not shown).

32.4.3
PARAFAC Decomposition

To extract the major fluorescent components from the multidimensional data set obtained in the elicitor-treated cells, we performed PARAFAC with three components, which was determined from the core consistency diagnostic (see

Table 32.2 Decay times and their relative amplitudes obtained by fitting to the fluorescence time profiles in Figure 32.11b.

Sample	Time constants in ns (amplitudes in %)		
P_1	<0.2 (45.5)	1.2 (49.5)	3.0 (5.0)
P_2	<0.2 (74.6)	0.6 (23.3)	2.0 (2.1)
P_3	<0.2 (56.5)	0.6 (43.5)	—
Avenanthiramide A	<0.2 (94.3)	0.5 (5.4)	3.5 (0.3)

(a) Em = 442 nm (b) Em = 567 nm (c) Em = 667 nm

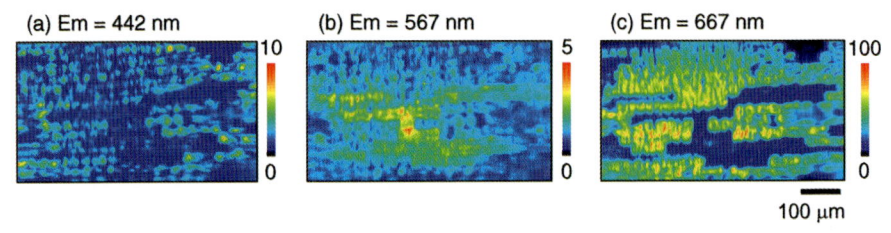

100 μm

Figure 32.12 The x-y images of the elicitor-treated cells constructed with the fluorescence at τ = 0.0 ns and averaged in a 25-nm binning size centered at Em = (a) 442, (b) 567, (c) 667 nm, respectively. The images are scaled so that the highest intensity in the long wavelength components becomes 100.

Section 32.2.3). The size of the data set used for the calculation was 10 (Em) × 2 (τ) × 307 200 (xy) as described in Section 32.2.2. We named the three components extracted by PARAFAC as Components I, II and III, which corresponded to the short-, mid-, and long-wavelength components, respectively. The fluorescence spectra of these components are plotted in Figure 32.11a. Component I is peak shifted to a shorter wavelength than the short-wavelength component (P_1) while Component II is peak shifted to a higher wavelength than the mid-wavelength component (P_2). It is considered that the overlapping region between the two components was successfully separated by PARAFAC. Component III agrees well with the long-wavelength component (P_3). The relative intensities of Components I, II, and III at 0.0 and 3.0 ns are plotted in Figure 32.11b. These were obtained in agreement with the time profiles of the short-, mid- and long-wavelength components, respectively.

Figure 32.13a–c shows the spatial localizations of Components I, II, and III extracted by PARAFAC, respectively. Components I and III agree well with those of the short- and long-wavelength components, respectively. On the other hand, the spatial localization of Component II is slightly different from that of the mid-wavelength component. The complementary relationship between Components II and III is not clear compared with that between the mid- and long-wavelength components. In the fluorescence spectrum of Component II (see Figure 32.11a), a rise in the component toward a longer wavelength exists at around 650 nm. This contribution is probably due to the cross talk of the long-wavelength component, which comes from very different fluorescence intensities between the mid- and long-wavelength components. It is suggested that the cross talk of the long wavelength component results in different spatial patterns between the mid-wavelength component and Component II.

To examine the spatial localizations of the short- and mid-wavelength components without the effect of cross talk, we performed PARAFAC with two components for a data set at a wavelength range up to 605 nm. We named the two components extracted by PARAFAC as Components I′ and II′, respectively. The fluorescence spectra of Components I′ and II′ are in agreement with those of Components I and II in a wavelength region lower than 605 nm. Further, the time profiles of Components I′ and II′ correspond well to those of Components I and II (data not shown). The spatial localizations of Components I′ and II′ are shown in Figure 32.13d and e. The

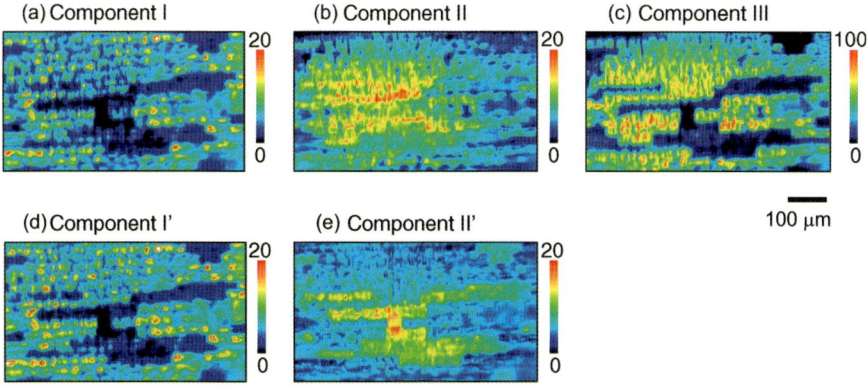

Figure 32.13 (a)–(c) The x-y images of Components I, II, and III, respectively, extracted by PARAFAC for the elicitor-treated cells. (d)–(e) The same as (a)–(b), respectively, but PARAFAC was performed for a data set in a wavelength range up to 605 nm (see the text). The images are scaled so that the highest intensity in the long wavelength components is 100.

former was found at exactly the same spatial localization as Component I (Figure 32.13a) and was similar to that of the short-wavelength component (Figure 32.12a). On the other hand, the latter was different from the spatial localization of Component II (Figure 32.13b) but very similar to that of the mid-wavelength component (Figure 32.12b). The results indicate that the different spatial pattern of Component II compared with the others (the mid-wavelength component and Component II′) was due to the cross talk of the long-wavelength component. Further, it is shown that cross talk can be partially avoided by appropriately limiting the data used for PARAFAC.

32.4.4
Possible Assignments of Fluorescent Components

We observed the short-, mid-, and long-wavelength components in the elicitor-treated cells, and the short- and long-wavelength components in the water-treated cells. First we focused our attention on the long-wavelength components, which were observed in both samples and found to have the same fluorescence properties. From its fluorescence properties, this component is naturally assigned to chlorophyll.

Second, we considered the mid-wavelength component, which was only observed in the elicitor-treated cells. This suggests the possibility that the component is associated with avenanthramides. It was reported that avenanthramide A is a major component of induced avenanthramides and reaches a maximum at 36–48 h after treatment with the elicitor [32]. We measured the time-resolved fluorescence spectrum of avenanthramide A in aqueous solution (pH. 7.0) *in vitro*. The time-integrated fluorescence spectrum is broad and centered at 510 nm (Figure 32.11a). The spectral shape is, in part, similar to that of P_2 and very similar to that of Component II. The comparison of the time-integrated fluorescence spectra supports the suggestion that the mid-wavelength component (or Component II) originates from avenanthramide A. However, the time profiles of the fluorescence intensities are very different from each other (Figure 32.11b). This presumably indicates the different environment of avenanthramide A *in vivo* and *in vitro* because the fluorescence decay (or fluorescence quantum yield) is generally very sensitive to the surrounding environment of the molecules. Viscosity is known to be one of the parameters which affects fluorescence quantum yield. The relationship between fluorescence quantum yield and solvent viscosity has been studied for various dye molecules with a flexible structure, such as diphenyl-methane and polymethine dyes, and discussed in terms of conformational changes induced by internal rotation [33, 34]. We measured the viscosity dependence of fluorescence for avenanthramide A in an aqueous solution of saccharides and observed an increase in fluorescence quantum efficiency with increasing solvent viscosity. Therefore, it seems that the discrepancy between the fluorescence decay profiles of the mid-wavelength component (or Component II) and avenanthramide A in aqueous solution is due to the different environments of avenanthramide A *in vivo* and *in vitro*. To identify the mid-wavelength component (or Component II) observed in this study more reliably, a combination of other analytical methods is essential and now in progress.

Finally, we discuss the short-wavelength components observed both in the elicitor- and in water-treated samples. These two components were similar in their fluorescence properties, implying that they have a common origin. Although a reasonable assignment for this component is difficult at present, it is considered that nicotinic coenzyme, NAD(P)H, is a possible candidate for the short-wavelength component, partly because the fluorescence spectra are similar [10, 14, 16].

Summarizing this section, we developed the time- and spectrally-resolved fluorescence imaging system based on line illumination, which is capable of rapid acquisition of fluorescence intensities as a function of Em, τ, and *xy*-positions. We applied this method to the analysis of a plant defense response, accumulation of antimicrobial compounds of phytoalexin in oat leaves, induced by the elicitor. In addition to the strong fluorescence from chlorophyll molecules, weakly fluorescent components, one of which possibly originated from avenanthramide A as phytoalexin, were observed in oat leaves treated with an elicitor.

32.5
Concluding Remarks

In this chapter, we have presented the methodology for non-invasive tracking of cell responses, which is based on the following:

(1) Use of autofluorescence signals of native molecules in cells or tissues.
(2) Detection of multidimensional fluorescence data such as Ex, Em, τ, *xy*, and so on.
(3) Efficient and rapid acquisition of a multidimensional data set based on a unique optical configuration.

(4) Analysis and decomposition of a multidimensional data set by using the PARAFAC model.

First, to demonstrate the effectiveness of our approach, we developed the time-gated excitation–emission matrix spectroscopic system and applied it to the decomposition of a mixed solution of a number of fluorescent dyes. The combination of a spatially dispersed super continuum as the excitation light source and a CCD camera equipped with a 200-ps-gated intensifier achieved rapid acquisition of a 3D fluorescence data set (Ex, Em, τ). In fact, it takes only a few seconds to obtain a whole data set consisting of 20 (Ex) × 640 (Em) × 24 (τ). A captured 3D data set was successfully decomposed into individual contributions without any prior knowledge of the constituents. This method is effective not only for multi-component analysis of a mixed sample but also for dynamical analysis of a more complicated system with time-varying processes such as enzyme reactions, photo-induced reactions, and other chemical reactions. In addition, it is relatively easy to improve this system by increasing the fluorescence variables acquired or by introducing additional laser pulses, because the experimental set-up is based on a simple optical configuration.

In addition, the methodology was applied to fluorescence imaging based on the autofluorescence signals of native molecules in cells or tissues. The imaging system is capable of the rapid acquisition of fluorescence intensities as a function of Em, τ, and xy-positions, which is achieved by line illumination of the excitation laser beam. We applied this system to the analysis of a plant defense response, accumulation of phytoalexin in oat leaves, induced by elicitor treatment. In oat leaves treated with an elicitor, we successfully observed weakly fluorescent components, one of which possibly originated from avenanthramide A as a phytoalexin, in addition to the strong fluorescence from chlorophyll molecules.

We presented the application of this method for the detection and analysis of autofluorescent molecules in living cells. In addition to autofluorescent molecules, fluorescence indicators for Ca^{2+}, pH, and so on may be unique targets for this method. In general, the fluorescence quantum efficiency of autofluorescence is much lower than that of the fluorescence of the indicators. Therefore, a technique for separating unknown fluorescent components with very different quantum efficiencies is essential. Simultaneous analysis of the spatiotemporal dynamics of autofluorescent molecules and fluorescence indicators would be a powerful approach for revealing complicated responses in living cells.

Acknowledgments

The work described in Section 32.4 is in collaboration with Professor Akio Kobayashi, Dr Shin'ichiro Kajiyama, and Mr Yoshihiro Izumi of Osaka University. This work was supported in part by a Grant-in-Aid for Scientific Research on Priority Areas (No.432) from the Ministry of Education, Culture, Sports, Science and Technology (No.17034033) and by a grant from CREST of Japan Science and Technology Agency (JST).

References

1 Webb, R.H. (1996) Confocal optical microscopy. *Rep. Prog. Phys.*, **59**, 427–471.
2 Halbhuber, K.-J. and Konig, K. (2003) Modern laser scanning microscopy in biology, biotechnology, and medicine. *Ann. Anat.*, **185**, 1–20.
3 Chudakov, D.M., Lukyanov, S. and Lukyanov, K.A. (2005) Fluorescent proteins as a toolkit for *in vivo* imaging. *Trends Biotechnol.*, **23**, 605–613.
4 Shaner, N.C., Steinbach, P.A. and Tsien, R.Y. (2005) A guide to choosing fluorescent proteins. *Nat. Methods*, **2**, 905–909.
5 Gao, X., Yang, L., Petros, J.A., Marshall, F.F., Simons, J.W. and Nie, S. (2005) *In vivo* molecular and cellular imaging with quantum dots. *Curr. Opin. Biotechnol.*, **16**, 63–72.
6 Richards-Kortum, R. and Sevick-Muraca, E. (1996) Quantitative optical spectroscopy for tissue diagnosis. *Annu. Rev. Phys. Chem.*, **47**, 555–606.
7 Andersson-Engels, S., af Klinteberg, C., Svanberg, K. and Svanberg, S. (1997) *In vivo* fluorescence imaging for tissue diagnostics. *Phys. Med. Biol.*, **42**, 815–824.
8 Andersson, H., Baechi, T., Hoechl, M. and Richter, C. (1998) Autofluorescence of living cells. *J. Microsc.*, **191**, 1–7.
9 Rigacci, L., Alterini, R., Bernabei, P.A., Ferrini, P.R., Agati, G., Fusi, F. and Monici, M. (2000) Multispectral imaging autofluorescence microscopy for the analysis of lymph-node tissues. *Photochem. Photobiol.*, **71**, 737–742.
10 Huang, S., Heikal, A.A. and Webb, W.W. (2002) Two-photon fluorescence spectroscopy and microscopy of NAD(P)H and flavoprotein. *Biophys. J.*, **82**, 2811–2825.
11 Ashjian, P., Elbarbary, A., Zuk, P., DeUgrate, D.A., Benhaim, P., Marcu, L. and Hedrick, M.H. (2004) Noninvasive *in situ* evaluation of osteogenic differentiation by time-resolved laser-induced fluorescence spectroscopy. *Tissue Eng.*, **10**, 411–420.
12 Zangaro, R.A., Silveira, L., Manoharan, R., Zonios, G., Itzkan, I., Dasari, R.R., Van Dam, J. and Feld, M.S. (1996) Rapid multiexcitation fluorescence spectroscopy system for *in vivo* tissue diagnosis. *Appl. Opt.*, **35**, 5211–5219.
13 Zuluaga, A.F., Utzinger, U., Durkin, A., Fuchs, H., Gillenwater, A., Jacob, R., Kemp, B., Fan, J. and Richards-Kortum, R. (1999) Fluorescence excitation emission matrices of human tissue: a system for *in vivo* measurement and method of data analysis. *Appl. Spectrosc.*, **53**, 302–311.
14 Coghlan, L., Utzinger, U., Drezek, R., Heintzelman, D., Zuluaga, A., Brookner, C., Richards-Kortum, R., Gimenez-Conti, I. and Follen, M. (2000) Optimal fluorescence excitation wavelengths for detection of squamous intra-epithelial neoplasia: results from an animal model. *Opt. Express*, **7**, 436–446.
15 Zellweger, M., Grosjean, P., Goujon, D., Monnier, P., van den Bergh, H. and Wagnieres, G. (2001) *In vivo* autofluorescence spectroscopy of human bronchial tissue to optimize the detection and imaging of early cancers. *J. Biomed. Opt.*, **6**, 41–51.
16 Shirakawa, H. and Miyazaki, H.S. (2004) Blind spectral decomposition of single-cell fluorescence by parallel factor analysis. *Biophys. J.*, **86**, 1739–1752.
17 Leurgans, S. and Ross, R.T. (1992) Multilinear models: Applications in spectroscopy. *Statist. Sci.*, **7**, 289–310.
18 Harshman, R.A. and Lundy, M.E. (1994) PARAFAC: Parallel factor analysis. *Comp. Stat. Data Anal.*, **18**, 39–72.
19 Bro, R. (1997) PARAFAC. Tutorial and applications. *Chemom. Intell. Lab. Syst.*, **38**, 149–171.
20 Bro, R. (2006) Review on multiway analysis in chemistry – 2000–2005. *Crit. Rev. Anal. Chem.*, **36**, 279–293.
21 Anderson, C.M. and Bro, R. (2003) Practical aspects of PARAFAC modeling of

fluorescence excitation-emission data. *J. Chemom.*, **17**, 200–215.

22 Veirs, D.K., Ager, J.W., Loucks, E.T. and Rosenblat, G.M. (1990) Mapping materials properties with Raman spectroscopy utilizing a 2-D detector. *Appl. Opt.*, **29**, 4969–4980.

23 Brakenhoff, G.J. and Visscher, K. (1991) Confocal imaging with bilateral scanning and array detectors. *J. Microsc.*, **165**, 139–146.

24 Brakenhoff, G.J., Squier, J., Norris, T., Bliton, A.C., Wade, M.H. and Athey, B. (1996) Real-time two-photon confocal microscopy using a femtosecond, amplified Ti:sapphire system. *J. Microsc.*, **181**, 253–259.

25 Stimson, M.J., Haralampus-Grynaviski, N. and Simon, J.D. (1999) A unique optical arrangement for obtaining spectrally resolved confocal images. *Rev. Sci. Instrum.*, **70**, 3351–3554.

26 Mayama, S., Tani, T., Matsuura, Y., Ueno, T. and Fukami, H. (1981) The production of phytoalexins by oat in response to crown rust, *Puccina-coronata* f. sp. *avena*. *Physiol. Plant Pathol.*, **19**, 217–226.

27 Mayama, S., Matsuura, Y., Iida, H. and Tani, T. (1982) The role of avenalumin in the resistance of oat to crown rust, *Puccina-coronata* f. sp. *avena*. *Physiol. Plant Pathol.*, **20**, 189–199.

28 Bordin, A.P.A., Mayama, S. and Tani, T. (1991) Potential elicitors for avenalumin accumulation in oat leaves. *Ann. Phytopathol. Soc. Jpn.*, **57**, 688–695.

29 Bro, R. and Kiers, H.A.L. (2003) A new efficient method for determining the number of components in PARAFAC models. *J. Chemom.*, **17**, 274–286.

30 Andersson, C.A. and Bro, R. Chemom. (2000) The N-way toolbox for MATLAB. *Intell. Lab. Syst.*, **52**, 1–4.

31 Ishihara, A., Miyagawa, H., Matsukawa, T., Ueno, T., Mayama, S. and Iwamura, H. (1998) Induction of hydroxyanthranilate hydroxycinnamoyl transferase activity by oligo-*N*-acetylchitooligosaccharides in oats. *Phytochemistry*, **47**, 969–974.

32 Ishihara, A., Ohtsu, Y. and Iwamura, H. (1999) Induction of biosynthetic enzymes for avenanthramides in elicitor-treated oat leaves. *Planta*, **208**, 512–518.

33 Oster, G. and Nishijima, Y. (1956) Fluorescence and internal rotation: Their dependence on viscosity of the medium. *J. Am. Chem. Soc.*, **78**, 1581–1584.

34 Sharafy, S. and Muszkat, K.A. (1971) Viscosity dependence of fluorescence quantum yields. *J. Am. Chem. Soc.*, **93**, 4119–4125.

33
Fluorescence Correlation Spectroscopy on Molecular Diffusion Inside and Outside a Single Living Cell

Kiminori Ushida and Masataka Kinjo

33.1
Introduction

33.1.1
Investigation on Biological System Based on Molecular Identification and Visualization

Recent progress in biological science, including molecular biology, structural biology, chemical biology, molecular genetics, and others, provides us with a chance to solve "life" as a combination of a huge number of complex but understandable mechanisms. The most important approaches are analyses of each biological tissue as an assembly of chemical substances using the accumulated knowledge of material science. Investigation focusing on molecules at the nanometer scale (or lower) involves a variety of techniques which enables us to reveal another aspect of life as a dynamic engine where various materials transport between organs, generating, storing, and losing chemical and thermodynamic energies. Among various techniques in nanoscience, two important strategies which bring large contributions to the understanding of life exist, that is, molecular identification and its visualization.

Molecular identification studies on various substances involved in biological systems have been extensively performed and have progressed remarkably in the last hundred years. Now these approaches can treat the world of nanometer scale and, at the same time, the individual behavior of single molecules [1–4]. Analytical methods for this kind of approach, such as various purification, spectroscopic, chromatographic, immunological and generic methods have been and still are being developed, Today, materials in our scope involve nucleotides, proteins, carbohydrates, and other organic and inorganic compounds. These materials are also classified according to their function, such as generic compounds, enzymes, immunological compounds, molecular recognizers, signaling molecules, framework or shaping materials, membranes, gases, nourishment and so on. As a result, the dynamic aspects of biological systems (or "life") can be considered as a sequence of many chemical reactions and material transportations: A and B react in organ 1 to generate C, C transports to organ 2 to react with D, D is stimulated and activated to generate E, E goes ... and so on. Researchers like to write down a scenario or a recipe which is composed of a list of reaction equations which have accumulated to be a huge number during the history of biological study.

Visualization techniques in biological systems have also been showing widespread progress. This started with the invention of the optical microscope several centuries ago, and nowadays, electron microscopes, laser scanning microscopes (LSM), near-field microscopes, and atomic force microscopes (AFM), are utilized in biological studies because visualization brings a great impact to many researchers – "Seeing is believing". The ability of these visualization techniques reaches both the nanometer scale and single molecular level and can be called "molecular visualization". Support from relating chemical and biological inventions is also significant such as labeling techniques, secretion of fluorescing proteins such as green fluorescence protein (GFP), polymerase chain reaction (PCR), immunological and generic methods.

If we want an image to illustrate a biological system, such as a single cell, we can consider a map of our real community, such as one district or country as shown in Figure 33.1 because the activity of one biological system is similar to the activity (or economy) of one society which can be estimated from material transportations. The above-mentioned two approaches, that is, identification and visualization, also contribute to drawing a correct map of the country. Visualization techniques provide

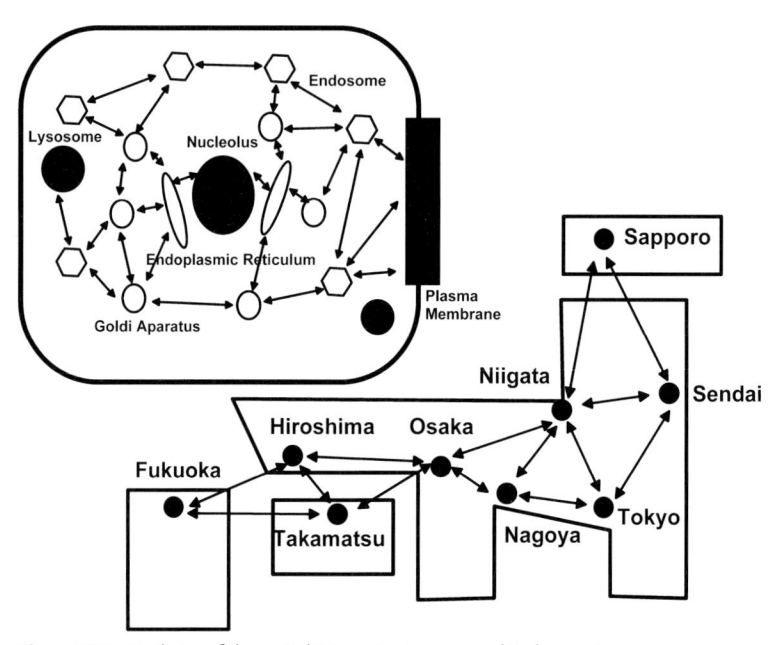

Figure 33.1 Similarity of dynamic bioimaging to geographical mapping.

a map of the country with a fine network composed of railroads or highways. The map tells us which lines connect the others and takes us to our destination and which stations are convenient for transit. On the other hand, identification methods provide us with information about what kind of trains or automobiles run on these lines: cars, trucks, express, urban and local trains, and freight trains. The map of each line partly indicates the condition of daily activity of this country; however, the information is insufficient to know the extent of social activity. For example, we need to know when the trains start, what is their speed, when do they reach a station, and whether we can change trains there. We need timetables of all trains and, if possible, a fine motion picture of all activities is very informative. Quantitative statistics of transporting people and materials is also helpful to know the real-time activity of the total society.

Returning to biological investigations described here in parallel, the investigation process proceeds as follows. Visualization studies make a map of each organ. Next, researchers identify substances existing there which may be key materials in some bioactivity. Then, they visualize the distribution map of these substances to obtain a roughly drawn picture of material traffic inside the biological organ. However, in a similar manner as with the geographic maps, we need to know the timetable of this material traffic or the statistics of transporting materials. Therefore, needless to say, real-time observation of material transport is the next target of implementation in biological science which has now started to be developed by many researchers.

In biological systems, most of these transports are diffusion processes driven by thermal energy involved in the surrounding medium. It should be noted that this transport itself is a random motion and the most important question is what controls the course and the destination of each molecule in such stochastic processes.

33.1.2
Technical Restrictions and Regulations in Real-Time Visualization of Material Transport in Biological System [2]

Real-time observation of movements of identified substances attracts much attention from researchers in biological science. However, we are confronting various physical limits or regulations which must be removed before the realization of desired experiments. These limits mainly concern spatial resolution, time resolution, and sensitivity in detection.

33.1.2.1 Spatial Resolution
The limit of spatial resolution depends significantly on the probing method. If we perform optical observation with an alignment forming a microscope of image optics, the spatial resolution limit is around 50% of the wavelength of light. In a similar manner, the resolution of an electron microscope depends on its de Broglie wavelength. For example, a 300 kV electron microscope, with a de Broglie wavelength of a few picometers, can provide resolution finer than a few angstroms (0.1 nm).

A scanning method for each observation may improve the spatial resolution. A LSM with confocal optics and a scanning near-field optical microscope (SNOM) can provide finer spatial resolution with a limit of a few nanometers. The highest resolution of AFM also reaches the subnanometer scale. However, these methods are inappropriate to record the movement of materials because of the scanning period needed to obtain a full view of a sizable area, during which time the objective must be immobilized. For the electron microscope, vacuum conditions are preferred to obtain a higher resolution.

Since the typical size of molecules involved in biological activities is less than a few nanometers, this spatial limit is a serious problem for real-time observation of molecular transportation. For organs, the size of a ribosome is about 20 nm, which is also a critical size for optical observation. For example, a large particle (> 20 nmϕ) such as a nanocolloid is connected to small molecules to visualize their real movement by dark-field observation of light scattering [3, 4]. This method changes the total size of the objective molecule and, accordingly, its diffusion coefficient. The real-time visualization of the molecule without labeling is still impossible. (See the discussion in Section 33.1.3.)

33.1.2.2 Time Resolution

If we use an optical sensor, the time resolution of detection depends on both the physical mechanism of generation of electric signals from photons and the duration of signal processing and recording through electronics and the software. A typical limit of responding speed is 10–100 ps for a single photon on a single channel. However, this kind of pulse detection brings a sizable (~10 ns or longer) dead time after the single pulse. The hardware for collecting photons, measurement (AD conversion), and storage of each signal also regulate the number of signals which can be processed in a short period. For two-dimensional detectors, the wiring architectures for read-out also restrict the time resolution. Ordinal charged-coupled device (CCD) elements emit stored signals for each pixel as sequential data and the resulting video-frame speed depends strongly on the total architecture.

33.1.2.3 Sensitivity

The performance of real-time observation of molecular transportation depends strongly on the sensitivity of the detectors, especially in single molecular detection (SMD). The sensitivity of SMD by fluorescence has been improved by the emergence of appropriate dye molecules with high quantum efficiency. The optical apparatus must be carefully designed to minimize the background from stray light and/or autonomous fluorescence. Although the higher the sensitivity the better, its performance is a trade-off between the spatial and time resolution of the measurement. For example, position sensitive detectors (PSDs) are less sensitive than single detectors and fast video detectors have relatively low sensitivity.

33.1.3
Time and Space Resolution Required to Observe Anomalous Diffusion of a Single Molecule in Biological Tissues

With any of the super detecting techniques available today, we cannot observe the real-time movement of single molecules in sufficient resolution because the objective molecule is too small (<10 nm) and moves too fast (<10 ps) to be detected. Instead of this, we discuss only statistical aspects of movements, for example, we use the diffusion coefficient D as a statistical parameter derived from measurements on many molecules.

Various new techniques suitable for estimating D are now available: fluorescence correlation spectroscopy (FCS) [5], fluorescence recovery after photobleaching (FRAP) [6], pulsed field gradient nuclear magnetic resonance (PFG-NMR) [7], diffusion ordered NMR spectroscopy (DOSY) [8], and others. Among these, FCS and FRAP are popular in biological studies because they are often installed on a commercial LSM system and conveniently coupled with it.

Normally, the movement of a single molecule is a kind of diffusion process which is expressed as a sequence of random positions, that is, trajectory vectors X_i as a function of stepwise time, $t_0, t_1, \ldots, t_i, \ldots$,

$$X_i = X(t = t_i) \tag{33.1}$$

as shown in Figure 33.2 [9]. This expression is very similar to the recording principle of a motion picture that is a sequence of static photographs. We can express the same phenomena using a summation of stepwise displacements [9] as

$$X_i = \sum_{j=1}^{i}(X_j - X_{j-1}) = \sum_{j=1}^{i} \Delta x_j \tag{33.2}$$

When we use a time step of constant interval Δt ($t_i = i\,\Delta t$), the diffusion process can be expressed as a simple sequence of Δx_i. This random walk model is convenient to express the normal Brownian motion [9, 10]. The definition of Brownian motion,

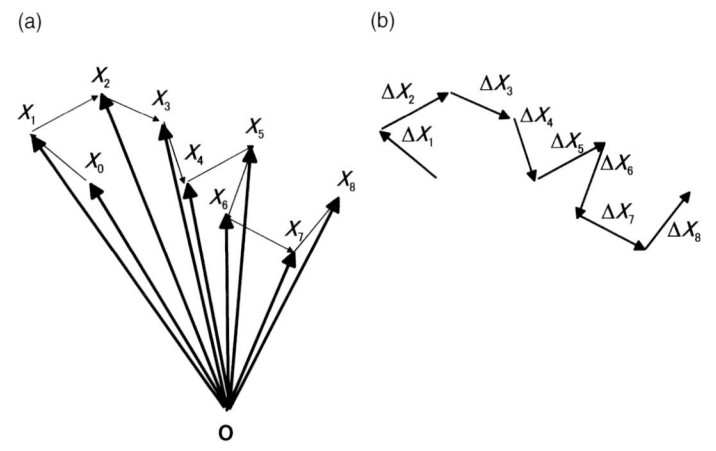

Figure 33.2 Two vector models expressing the random walk (trajectory) of diffusion. (a) Sequence of position vectors on each time step, $t_1, t_2, \ldots, t_n, \ldots$ (b) Sequence of displacement vectors on each time step. Although the two expressions are equivalent, (b) is invariant for origin.

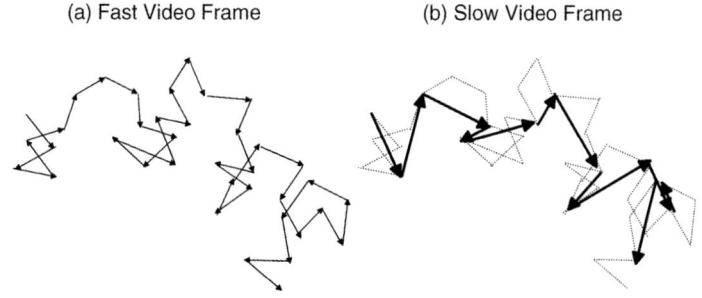

Figure 33.3 Different results of motion picture with different video frame speeds for the identical trajectory of random walk. In contrast to the case of normal diffusion, observed mean square displacements (MSD) significantly depend on the frame speed in the case of anomalous diffusion.

that is, a Wiener process, leads us to a proposition that both distributions of X_i and Δx_i are Gaussian line shaped and their correlation functions are

$$\langle X_i, X_j \rangle = \delta(t_i - t_j) \tag{33.3}$$

and

$$\langle \Delta x_i, \Delta x_j \rangle = \delta(t_i - t_j) \tag{33.4}$$

where $\delta(t)$ is a delta function. For Δx_i, its Gaussian distribution $W(\Delta x, \Delta t)$ is expressed using the dispersion $\sigma(\Delta t)$ as

$$W(\Delta x, \Delta t) = \frac{1}{\sqrt{2\pi\sigma(\Delta t)}} \exp\left(-\frac{\Delta x^2}{2\sigma(\Delta t)}\right) \tag{33.5}$$

The definition of diffusion coefficient is given as

$$\sigma(\Delta t) = 2D\Delta t \tag{33.6}$$

As long as we use this Brownian motion model, the diffusion coefficient D is independent of Δt. This proves that no high speed observation is necessary to obtain the simple diffusion coefficient D of Brownian motion. Moreover, observation over a long period is more valuable because no high spatial resolution is needed to resolve relatively large X_i and Δx_i. This principle has been employed to measure D of visible particles in multiple-particle-tracking microrheology (MPTM) techniques [11]. The area of application of MPTM is still spreading because of recent improvements in video frame speed [12] Figure 33.3.

However, this proposition is not valid for all biological systems because the diffusion occurs in an inhomogeneous space where various kinds of structures are interfering with the diffusing molecules. In a precise definition, the diffusion in a biological system is not true Brownian motion.

This aspect of diffusion in inhomogeneous space is called "anomalous diffusion" [13–18] and ordinal diffusion which can be expressed as a Brownian motion is called "normal diffusion" or "Euclid diffusion" [14]. The definition of normal and anomalous diffusion is that the mean-square displacement (MSD) increases in proportion to the time evolution or not, that is,

$$\langle \Delta x_i^2 \rangle \propto \Delta t \text{ or } \langle X_i^2 \rangle \propto t \text{ (Normal diffusion)} \quad (33.7)$$

$$\langle \Delta x_i^2 \rangle \not\propto \Delta t \text{ or } \langle X_i^2 \rangle \not\propto t \text{ (Anomalous diffusion)} \quad (33.8)$$

Most of the cases found in a biological system involve anomalous diffusion. If we use a time-dependent diffusion coefficient $D(t)$, the MSD of d-dimensional diffusion is expressed as

$$\langle \Delta x_i^2 \rangle = 2dD(t)t \quad (33.9)$$

This equation tells us that both short and long time measurements are important to acquire a full lineshape of $D(t)$ where continuous variation of t is ideal. If our goal is a full-recording of $D(t)$, the best approach is to calculate MSD from the statistical accumulation of X_i data, real-time movements (trajectory) of molecules. Therefore, fast detection is better because the MSD in a long period can be obtained from the summation of Δx_i in any range. However, when Δx_i becomes short with fast observation, finer spatial resolution is also required. This is the reason why the dilemma between time resolution and spatial resolution still remains a problem in handling the real-time movements of substances in any biological system.

In practice, poor time and space resolution lead to a wrong evaluation of the diffusion coefficient in a case including anomalous diffusion. One typical example is the use of video cameras with different frame speeds. With a vector model as shown in Figure 33.4, the results of two different detections for the same Brownian motion

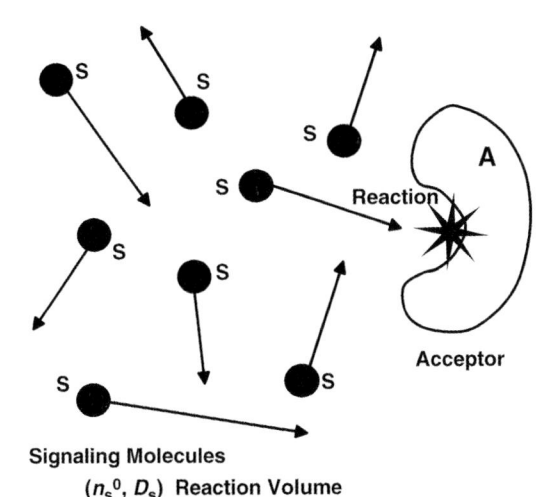

Figure 33.4 A model illustration of the signaling reaction in a small reaction volume. The signaling molecules (S) with population n_S^0 are diffusing (diffusion coefficient D_S) to reach an acceptor (A). A signal is recognized when single S reacts with A to trigger another activity.

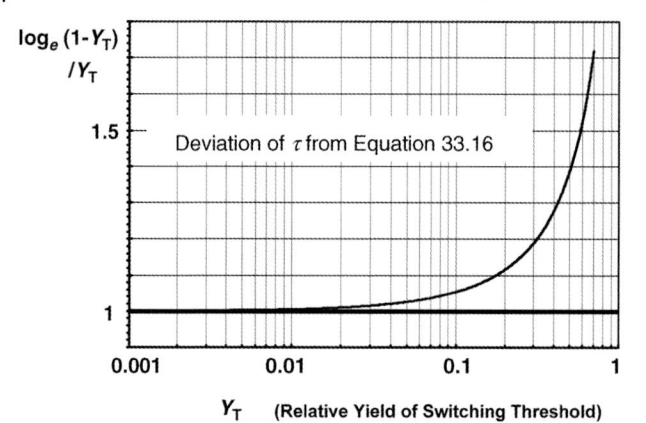

Figure 33.5 Deviation of Eq. (33.16) from the approximate Eq. (33.17) against relative yield of switching threshold Y_T.

are depicted in Figure 33.5. In the case of normal diffusion, the obtained values of the diffusion coefficient that are calculated from MSD are invariant with different time resolution. However, in the case of anomalous diffusion, the calculated value depends on the frame speed of each video monitoring. It should be noted that one to two order faster detection is required to record the random walk of a particle in anomalous diffusion than the time range of the time-dependent diffusion coefficient $D(t)$. For example, if one would like to know the behavior of D in microseconds, random walks in nanoseconds or faster must be monitored.

Instead of this, we can use the typical scale of structures which may induce anomalous diffusion in biological systems: The size of a small apparatus such as a ribosome of Goldi apparatus, the size of polymers involved in cytoplasm, and the size of membranes, mesh structures involved in extracellular matrices. The size of these inhomogeneous structures are of the order of 10–100 nm and the change in $D(t)$ occurs around the diffusion distance one order larger than the structural scale (10 nm^{-1} µm). The D value of Rhodamine 6G (a typical small molecule) in water is 2.8×10^{-10} (m^2 s^{-1}) [19], the diffusion distance of 10 nm^{-1} µm correlates with a 100 ns^{-1} ms diffusion time. Since larger molecules such as proteins or DNAs diffuse more slowly, the requirement of time resolution for the D measurement is 100 ns or longer.

33.1.4
General Importance of Anomalous Diffusion in a Signaling Reaction

Now we present a kinetic consideration of a type of chemical reaction which can be regarded as a signaling process in a biological system, as shown in Figure 33.4.

$$S + A \rightarrow \text{(products: acknowledgement of signal)} \quad (33.10)$$

A signaling molecule S reacts with an acceptor A, and then, the system recognizes that a signaling process (communication) is completed. We define a very tiny reaction space (we refer to "reaction volume" here, probably nm^3–μm^3 scale) of 10^{-24}–10^{-15} l. When a sufficient number of molecules are involved in the reaction volume and their Brownian motions are sufficiently high, the reaction can be treated as a stochastic process and ordinal rate constants and diffusion coefficients can be used for description of the reaction. When only a few molecules are involved, the reaction probability must be evaluated with statistics based on a Poisson process and the uncertainty of signaling will increase.

The rate of second-order diffusion-controlled chemical reaction is expressed as

$$-\frac{dn_S}{dt} = 4\pi(D_S + D_A)(r_S + r_A)n_S n_A \tag{33.11}$$

where D_S (or D_A), r_S (or r_A) and n_S (or n_A) are the diffusion coefficient [20], the reaction radius, and the population of the signaling molecule S (or the acceptor A) in the reaction volume, respectively. Here we ignore the effect of molecular volumes. Diffusion coefficients may also depend on the reaction area in an inhomogeneous medium because of anomalous diffusion.

If the acceptor is almost immobile ($D_S \gg D_A$) such as those fixed on membranes, Eq. (33.11) is reduced to

$$-\frac{dn_S}{dt} = 4\pi D_S (r_S + r_A) n_S n_A \tag{33.12}$$

In biological signaling systems, the reaction period until the signal has "reached" the acceptor as destination is important, rather than the rate or yield of reaction. We assume the existence of a threshold of reaction yield n_T which is defined as the minimum number of reactant molecules to switch the acceptor to "on".

Some acceptors can be recovered to the initial state by other repairing processes and the entire reaction may be regarded as "cathartic". In both cases, the second-order reaction can be simplified to be pseudo-first order and the decay curve of A is obtained as

$$n_S = n_S^0 \exp[-4\pi D_S (r_S + r_A) n_A t] \tag{33.13}$$

with the initial concentration n_S^0. The relative reaction yield Y is

$$Y = \frac{n_S^0 - n_S}{n_S^0} = 1 - \exp[-4\pi D_S (r_S + r_A) n_A t] \tag{33.14}$$

The reaction period τ is a function of Y_T that is the relative yield for $n_T = n_S^0 - n_S$

$$\tau = -\frac{\log_e(1 - Y_T)}{4\pi n_A D_S (r_S + r_A)} \tag{33.15}$$

where $Y_T = n_T/n_S^0$ or $n_T = n_S^0 Y_T$. This is the speed or the efficiency of communication in a biological system. If $Y_T \ll 1$ where a huge number of A molecules are provided as reactant, Eq. (33.15) is reduced to

$$\tau = \frac{Y_T}{4\pi n_A D_S (r_S + r_A)} = \frac{n_T}{4\pi n_A D_S (r_S + r_A) n_S^0} \tag{33.16}$$

This primitive equation is often used in chemical engineering analysis. If we would like to decrease the reaction period τ until we obtain the required number of products, n_T, we can increase the value of n_S^0, in other words, the only thing we need to do is provide more S.

Although this strategy is effective in mass production reactions such as occur in an industrial plant, the situation in a biological system, that is a small and closed system, is completely different. Provision of excess S is not valuable because it introduces unwanted burdens, including the synthesizing process for S and the clean-up process of excess S molecules, on total biological activities. For example, if only one molecule of S is needed to switch A, secretion of 10 000 S molecules should accelerate the signaling process more effectively than the secretion of 10 molecules. However, the system would have to synthesize 10 000 S molecules, 9999 of which are in vain and must be cleaned up after completion of the signaling process. The increase in extra activity in these two operations (synthesis and cleaning) exhausts a large amount of biological energy causing damage to the total biological system.

Therefore, in a signaling system with a quick response, that is, with a shorter τ required to sustain the life system, the most serious problem is the dilemma between following the two requirements emerging in Eq. (33.15). (i) The amount of reactants to be synthesized must be as small as possible. (ii) The maximum speed of reaction is needed to establish molecular signaling. Under requirement (i), the relative yield of threshold Y_T becomes large and we must use Eq. (33.15) rather than Eq. (33.16). In Figure 33.5, $\log_e(1 - Y_T)/Y_T$ is plotted against Y_T and the deviation is large in the area of $Y_T > 0.01$. In this area, the response time τ is additionally increased by the factor shown in Figure 33.5.

On the other hand, the effectiveness of the signaling reactions also depends on the diffusion coefficient, as shown in Eq. (33.11). Although other parameters in Eq. (33.11) (r_S, r_A, and n_T) are not variable as determined for each reaction (33.10), only the diffusion coefficients (D_S and D_A) can be controlled by the existence of the surrounding media. Moreover, as mentioned in the previous section, the diffusion coefficient of anomalous diffusion depends on the diffusion time and the dimensions of the reaction space. In such a situation, the diffusion coefficient observed by one method (e.g., FCS, FRAP) is only a local value, depending on the time constant and the spatial size of a proper experiment. As mentioned for Figure 33.4 in the beginning of this section, the size of the reaction volume for signaling reaction is of the order of pL–fL and measurement of the diffusion coefficient in such a microspace is important.

This model for a signaling reaction can be applied to various activities occurring inside and outside cells, such as neurotransmission and drug delivery. In all biological systems, swift, efficient and errorless material transport is required. To establish a good network, it should be noted that negotiation of parameters in Eq. (33.11), that is, n_S^0 and D_S is essential. Another characteristic of a biological reaction is the extremely small reaction volume (pL–fL) and a countable number of molecules are involved in the total reaction. Normally the signaling molecules travel only a short distance (of the order of μm) after being synthesized at the secretion point. Therefore, FCS has great merit in that it can evaluate the diffusion coefficient involved in a small volume. In such a small space, the obtained value of

33.2 Use of Fluorescence Correlation Spectroscopy (FCS) for Investigation of Biological Systems

D_S may be different from that obtained for a larger volume due to anomalous diffusion.

In the next section, we present a simple description of an FCS experiment especially for a biological system as an inhomogeneous medium. We also describe our recent challenge to observe anomalous diffusion by a modified FCS system and its application to polymer solutions, which is are model media for extracellular matrices. In the third section, we introduce various applications of FCS to the observation of a single cell and its surroundings.

33.2 Use of Fluorescence Correlation Spectroscopy (FCS) for Investigation of Biological Systems

33.2.1 Use of FCS for Biological Systems

FCS [5] is a powerful tool to analyze the real-time observation of molecular diffusion. Fluorescence from continuously photo-irradiated molecules is detected by a confocal microscope. Since the small confocal volume (CV) less than fL involves only a small number of molecules, for example 10 or less, the intensity of the fluorescence fluctuates due to the population change within the CV. Figure 33.6 shows a typical geometry of the CV and a cylindrical shape approximation was used with a radius from several hundreds of nm to several μm of which the volume is 0.1–10 fL, depending on the optics. Solutions below 10^{-7} M, which can be obtained by an ordinal diluting operation, supply only 10 or less molecules inside the CV on average. Under such conditions, the photon signal fluctuates depending on the number of molecules which varies dynamically due to diffusion.

An autocorrelation function of signal intensity is analyzed by a fitting method. One typical equation is

$$G(\tau) = \frac{\langle I(t)I(t+\tau)\rangle}{\langle I(t)\rangle^2} = 1 + \frac{1}{N}\left(\frac{1}{1+4D\tau/w_{xy}^2}\right)\left(\frac{1}{1+4D\tau/w_z^2}\right)^{1/2} \quad (33.17)$$

where, τ, $I(t)$, N, D, w_{xy}, and w_z are correlation time, fluorescence intensity function, population of emitting molecule, its diffusion coefficient, horizontal dimension (radius) of the CV, and vertical dimension of the CV, respectively. It should be noted that the equation is based on the use of a constant D. Therefore the equation assumes normal diffusion implicitly.

This fitting method is partly valid for the analysis of anomalous diffusion, however, careful consideration is needed in the use of Eq. (33.17). For severe anomalous diffusion, the lineshape of $G(\tau)$ shows apparent deformation. In such a case, fractal expansion of Eq. (33.17) can be applied, as shown in the literature. In moderate anomalous diffusion, deformation of $G(\tau)$ is not clear but only the fitting result for

(a) Large Molecules **(b) Small Molecules**

Confocal Volume

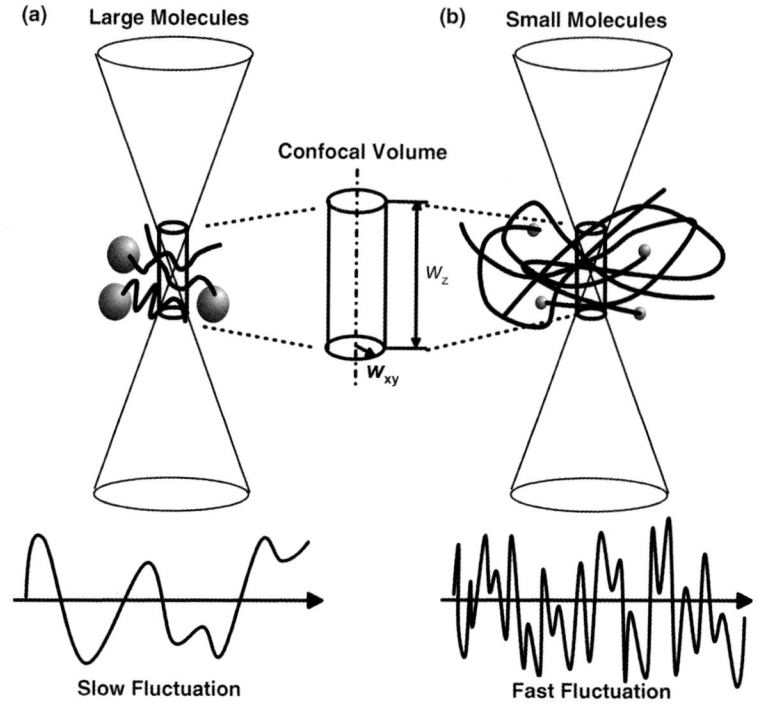

w_z

w_{xy}

Slow Fluctuation   Fast Fluctuation

Figure 33.6 Illustration of confocal volume in fluorescence correlation spectroscopy (FCS) describing the experimental principle for evaluation of diffusion coefficients from the fluctuation of photon signals. (a) Fluctuation due to large and less mobile molecules is slow and small D values are obtained. (b) Fluctuation due to small and mobile molecules is fast and large D values are obtained. The figure also indicates the shape of the confocal volume which is approximated as a cylindrical space with parameters w_{xy} and w_z in Eq. (33.17) See Ref. [5].

D may be changed with a different size of CV. A theoretical study provides a proof with analytical solutions.

33.2.2
Experimental Example of Anomalous Diffusion Observed in a Model System for Extracellular Matrices

Recently, we developed a new type of FCS measurement called "sampling volume controlled (SVC)" FCS where we can change the size of the confocal volume (CV) continuously [21–26]. The rough alignment of SVC-FCS instruments is indicated in Figure 33.7. The radius of the CV was changed in the range 200–700 nm. The dimension of the reaction volume of a biological reaction is comparable to the smallest volume size (200 nm in radius) or smaller. A similar approach using an iris to change the size of the laser illumination aperture has been presented by another group independently. The difference in our approach from that of the other group is the conservation of a Gaussian distribution of laser intensity in the CV.

33.2 Use of Fluorescence Correlation Spectroscopy (FCS) for Investigation of Biological Systems

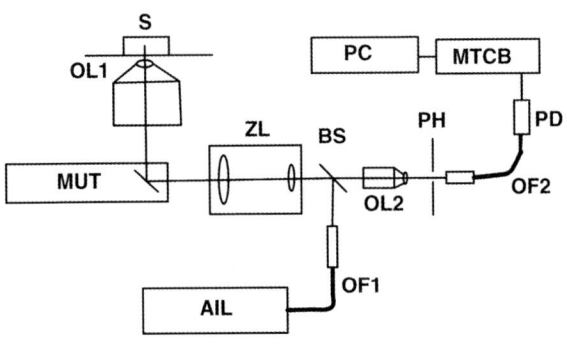

Figure 33.7 A diagram of sa ampling volume controlled FCS measurement system. The laser light from an argon ion laser (AIL) is introduced to a beam splitter (BS) through an optical fiber (OF1). The diameter of the laser beam is expanded continuously by a motorized zoom lens (ZL). The laser beam is reflected onto a mirror in a mirror unit turret (MUT) and focused on sample (S) through an objective lens (OL1). Emission is corrected by OL1, and again focused on a pinhole (PH) by another objective lens (OL2), The light is introduced to a photodetector (PD) through another optical fiber (OF2) and the photon signals are analyzed by multiple tau correlation board (MTCB) coupled with a personal computer (PC) The system was modified after Ref. [21].

We changed the size of CV, that is, the parameter w_{xy} in Eq. (33.17). This changed the averaged diffusion distance L and we obtained

$$L = \sqrt{\frac{3}{2}} w_{xy} \tag{33.18}$$

L can be converted to diffusion time τ_{obs} by

$$\tau_{obs} = \frac{L^2}{6 D_{obs}} \tag{33.19}$$

with the observed value of D_{obs}. The change in D_{obs} can be plotted against both L (distance dependence) and τ_{obs} (time dependence).

The medium used was an aqueous solution of hyaluronan (HA) [27, 28] which is a model system for an extracellular matrix (ECM) such as cartilage. In cartilage, as shown in Figure 33.8, no blood vessels are found and signaling molecules, nutrient, and water are provided directly through the space between the meshwork constructed from collagens, HA and other glycoproteins [29]. Therefore, HA solution can be regarded as a model system for ECM. HA is a long polymer which entangles with itself forming a meshwork space without any interchain affinity. The space structure is similar to a gel but the spacing of the polymer (mesh size) is larger (10–100 nm).

The results of the distance dependence of the diffusion coefficient (DDDC) and the time dependence of the diffusion coefficient (TDDC) of Alexa488 (Alexa) in an aqueous solution of HA measured by SVC-FCS are quoted from Ref. [23] in Figure 33.9. The molecular diameter of Alexa is about 1 nm and the estimated mesh sizes were 33, 15, 7 nm for 0.1, 0.9, 1.5 wt% of HA. When the diffusion distance was increased, the diffusion coefficient seemed to be converted to a value smaller than that obtained in a solution without HA. This is direct evidence of anomalous

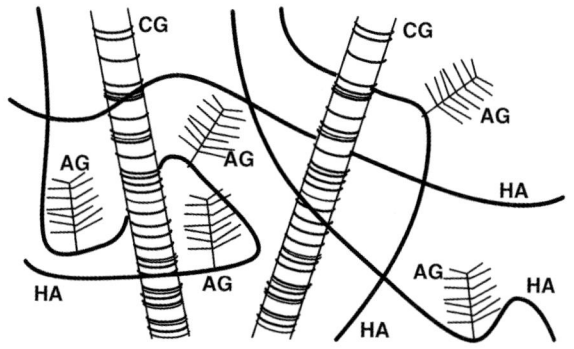

Figure 33.8 Typical composition of cartilage: Stiff chains of collagens (CG), soft chains of hyaluronan (HA), and a smaller molecular group composed of protein and glycosaminoglucans (Agrican: AG) form a hybrid mesh structure. See Ref. [29].

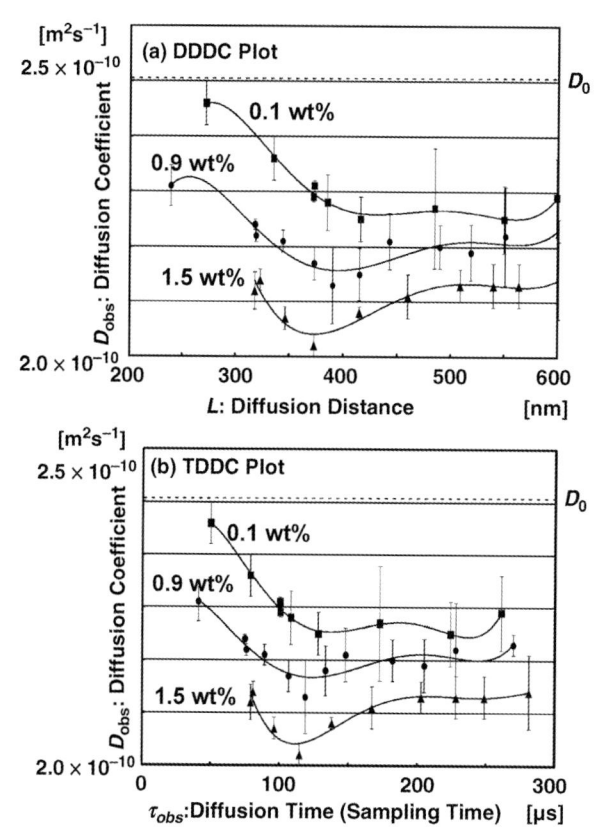

Figure 33.9 (a) Distance dependence and (b) time dependence of diffusion coefficient of Alexa 488 in HA aqueous solution (0.1, 0.9, 1.5 wt%) obtained by SVC-FCS. The vertical scale is the absolute value of D. The figures are reproduced from Refs. [23] and [25].

33.2 Use of Fluorescence Correlation Spectroscopy (FCS) for Investigation of Biological Systems

diffusion. The transient region appears in the several hundred nm region sandwiched by two plateaus (short distance limit and long distance limit).

On increasing the HA concentration from 0.1 to 1.5 wt%, this transient region appears to shift to a smaller diffusion distance and the D value of the right plateau (long distance value) is lowered. In the long distance region, where the interaction between the mesh and the molecules lowers the D value, the relative change is expressed by Ogston's concentration law [30] as

$$D/D_0 = \exp(-\alpha[\text{HA}]^{0.5}) \tag{33.20}$$

where D_0 is the diffusion coefficient without HA. The HA concentration changes the mesh size and the magnitude of depression of D depends on the relative size of the diffusing molecules in reference to the mesh size through the parameter α.

The results for cytochrome c (cytc) are also reproduced from Refs. [23] and [25] in Figure 33.10. In addition to the results of SVC-FCS, the results of photochemical

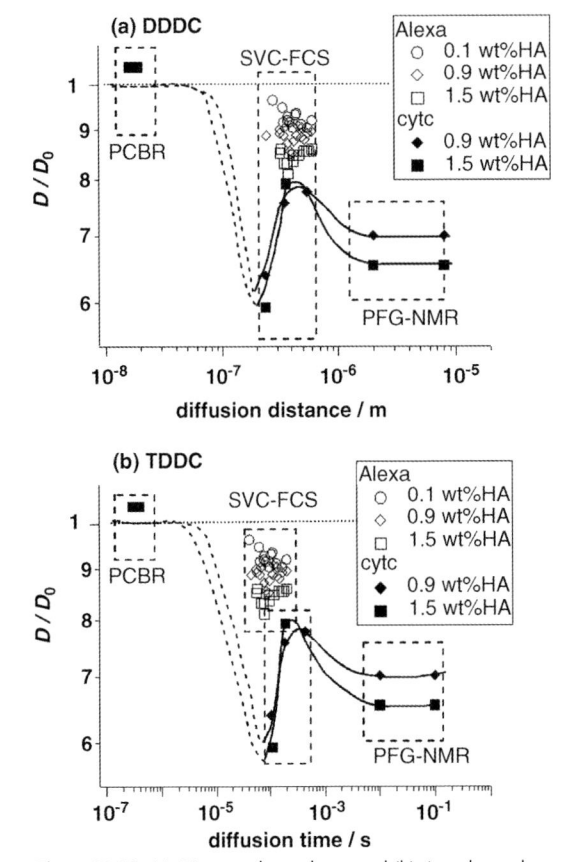

Figure 33.10 (a) Distance dependence and (b) time dependence of diffusion coefficient of Alexa 488 and cytc in HA aqueous solution (0.1, 0.9, 1.5 wt%) obtained by PCBR, SVC-FCS, PFG-NMR methods. The values are normalized with those obtained without HA (D_0). The figures are reproduced from Refs. [23] and [25].

biomolecular reaction (PCBR) [31] and PFG-NMR [31, 32] are also indicated. The molecular diameter of cytc is 3.5 nm. It was also found that the transient area was shifted to the shorter diffusion distance area and that the D value for long distance was lowered. Two results in Figures 33.9 and 33.10 are consistent and provide a qualitative explanation of anomalous diffusion induced by mesh space. Both the position of the transient area and the magnitude of depression of D are controlled by the relative relationship between the mesh and diffusing molecules. The mechanism of this simple mesh model can be summarized as follows with Figure 3.11.

1. When observed D values are plotted against L or τ_{obs}, the lineshape is divided into three regions.

2. In the short distance (time) limit, the D plot makes a plateau at the value almost equal to D_0 and is invariant on addition of mesh materials. In this area only a minority of diffusing particles interact with polymer chains during their short travel.

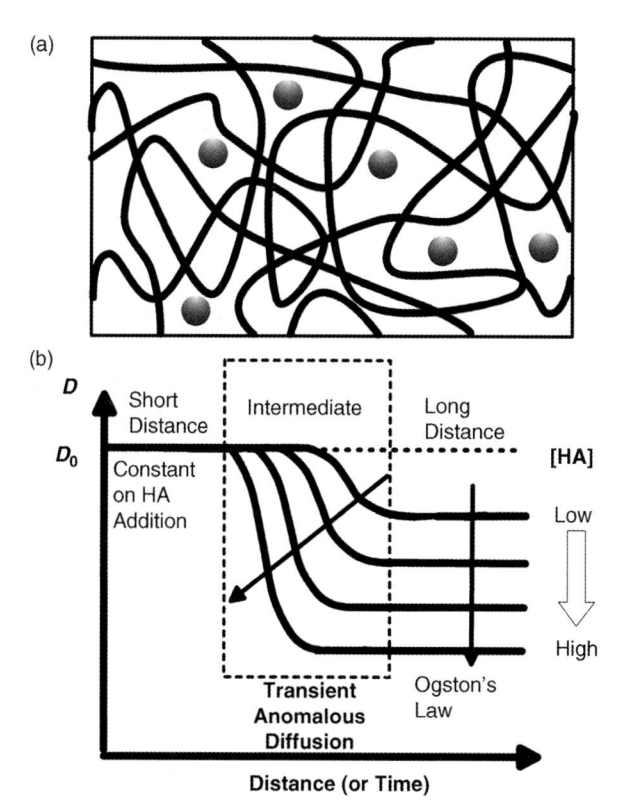

Figure 33.11 (a) An illustration of molecular diffusion in an aqueous solution of polymer materials containing random meshwork structures. (b) General behavior of anomalous diffusion in inhomogeneous solution with simple mesh structure.

3. In the long distance (time) limit, the D plot makes another plateau whose value decreases on addition of HA. The decrease is described by Ogston's law (Eq. (33.20)) and this behavior is explained with a mechanism of retardation by the mesh which act as a continuous medium with friction. Therefore, the distance or time dependence is negligible to form a plateau.

4. In the intermediate region a transient lineshape emerges whose position is 1–2 order larger than the mesh size. This area is sometimes called a transient anomalous diffusion area [18]. The position of the transient is shifted to smaller L (or τ_{obs}) on increase in the HA concentration, that is, decrease in mesh size.

For small molecules, the decrease in D occurs at L slightly larger than the mesh size. The small molecule is gradually decelerated passing through several numbers of mesh units. This mechanism is frequently referred to as the "Ant in the Labyrinth" mechanism in percolation theory [33, 34]. Application of percolation to anomalous diffusion in the mesh space suggests the possible appearance of low-dimensional (tube-like) transportation inside the space. This phenomenon will be detected by another type of anomalous behavior in DDDC or TDDC.

The results of the study in HA solution suggest a simple but secure strategy to control the reaction by anomalous diffusion occurring in biological systems. The simple polymer solution used here is not only a model for ECM but also one to be extended to other biological space such as cytoplasm and membranes.

33.2.3
Quantitative Estimation of Reaction Volume in Signaling Reaction

As pointed out in Sections 33.2 and 33.3, the dimension of the reaction space where the magnitude of D is the key factor for completeness of signaling reaction, is smaller than 10–1000 nm. In this small space, the diffusion time for a small molecule such as Rhodamine 6 G [19] is shorter than 10 μs. If we assume a diffusion controlled reaction in Eq. (33.16) and the reaction radius $(r_S + r_A)$ of 1 nm. The second-order rate constant is $2.1 \times 10^9 \, M^{-1} \, s^{-1}$. If only one acceptor exist in the reaction volume, the switching time until the acceptor reacts with one signaling molecule is estimated from Eq. (33.16) as shown in Table 33.1. The number of molecules involved in the reaction space is very critical. If the system requires 3.6 μs switching, both a 100 nm cubic reaction space with 1000 signaling molecules or a 1 μm cubic reaction space with 1 000 000 signaling molecules are allowed. However, the latter case needs 10^3 times greater labor for synthesis and cleaning up 999 999 signaling molecules in proportion to the size of the reaction volume. Therefore, the reaction space of a 10–100 nm cube, which involves both the acceptor and provider of signaling molecules, is the most appropriate size biological reaction.

These values are consistent with our previous results concerning anomalous diffusion in HA solution in which the location of the transient anomalous diffusion area is 10–100 nm. The existence of HA never disturbs a reaction space smaller than the 10–100 nm scale where the observed value of D is unchanged from the value without HA (D_0). If the signaling molecules are provided within this small volume, in

Table 33.1 Switching time of signaling reaction with various reaction volume estimated from Eq. (33.16): One acceptor (A) is involved in one reaction volume and reacts one signaling molecule (S) to be "On."

Size of reaction space		10 nm Cube	100 nm Cube	1 μm Cube
Volume (l)		10^{-21}	10^{-18}	10^{-15}
Number of Signaling Molecules	10	1.81 μs	1.8 s	180 000 s
	10^2	17.2 ns	17.2 ms	17 200 s
	10^3	—	171 μs	171 s
	10^4	—	1.71 μs	1.71 s
	10^5	—	17.1 ns	17.1 ms
	10^6	—	171 ps	171 μs

other words, if the signaling molecules are secreted by some organs inside the volume, they react smoothly with an acceptor and tend to stay inside because of the depression of the diffusion coefficient at a long distance. Therefore, the required number of signaling molecules can be the minimum necessary for quick (less than 1 ms) and error-free switching. At the same time, the probability that the acceptor reacts with other molecules coming from outside the volume (this is also a kind of error communication) is decreased. The line shape of transient anomalous diffusion in Figures 33.9, 33.10 and 33.11 aids the correctness and swiftness of signaling reactions which can be described with the model shown in Figure 33.4.

The characteristics of FCS in the detection of molecular transport in biological tissues are summarized in the following two points. (i) Instead of giving up visualizing the real-time movements of single molecules, we handle a statistical value of movement, that is, a diffusion coefficient (D) with an assumption that all molecules have Brownian motion. We need careful consideration in treating anomalous diffusion because its results deviate from Brownian motion. (ii) The size of the detecting area (i.e., the size of the CV) is of the same order as the reaction spaces of a biological reaction such as the signaling process described in Figure 33.4. Therefore, the most valuable values of the diffusion coefficient are automatically obtained.

33.3
A Short Review of Recent Literature Concerning FCS Inside and Outside a Single Cell

33.3.1
FCS Measurement Inside Single Cells

From the early stages of the history of FCS measurement using confocal microscopes, the observation of single cells has been extensively performed. A huge number of studies have been published all of which cannot be introduced here thoroughly. FCS is now an established protocol to study the dynamics of intracellular molecules.

The majority of this kind of FCS studies treat intermolecular interactions such as molecular association to clarify the function of biomolecules in the target. In addition to the concentration of molecules, the binding constants of the two interacting molecules are easily obtained by FCS. A number of researchers have found it convenient to obtain a cellular map indicating where the objective molecular interaction occurs. This approach requires only qualitative values of D from FCS to distinguish the on–off of the molecular interaction. Therefore, only a small number of studies have stated the existence of anomalous diffusion in cells [35–37] or discussed the strangeness of the absolute values of D obtained from intracellular FCS measurements [38].

However, the existence of serious anomalous diffusion provides a problem in each FCS analysis. One example of a study by Banks et al. [37] showed a small deviation of the auto-correlation curve from the theoretical lineshape, one of which is presented as Eq. (33.17). The results showed that the lineshape is an overlap of many theoretical lineshapes in continuous distribution or with continuous change of D depending on τ, similar to $D(t)$ in Eq. (33.9) as explained by Schwille et al. [35, 36]. These two extreme models cannot be distinguished from a poor auto-correlation curve with a logarithmic time scale. Conventional analyses for cases with anomalous diffusion use alternative types of equation with fractal consideration with an exponent on D or t as

$$\langle \Delta x^2 \rangle = 6 D^\alpha t \text{ (exponent } \alpha \text{ on } D) \tag{33.21}$$

$$\langle \Delta x^2 \rangle = 6 D t^\alpha \text{ (exponent } \alpha \text{ on } D) \tag{33.22}$$

Both equations are useful to obtain well-defined D values in each experiment based on a fitting method. Although we understand that the form in Eq. (33.9) is more general, the numerical data from FCS measurement is not sufficient to obtain the full lineshape of $D(t)$ in Eq. (33.9). Seki et al. obtained an analytical solution of auto-correlation curves for $D(L)$ in a step function [39]. They proved that the solution lineshape is different from that of normal diffusion with a non-linear least square algorithm if the deviation from Eq. (33.17) is too small. Even in this case of moderate anomalous diffusion, the observed value of D changes sensitively, depending on t or L.

Viscoelastic measurements of a living cell, that also show remarkable progress nowadays, provide the macroscopic viscosity of cytoplasm. The diffusion coefficient of a spherical particle in cytoplasm is approximately obtained by the Stokes–Einstein relation. FRAP measurement also provide information about diffusion at a long distance. However, recent FCS studies pointed out the strangeness of D values in living cells, which are unexpectedly large indicating high mobility of the interacting molecule [38]. This contradiction can be partly explained by the step function model we presented in Section 33.2.2.

Now the tide of FCS imaging is turning towards the observation of real-time movements in a number of groups [40]. When multichannel (two-dimensional) detectors are employed, ordinal confocal imaging optics is not sustained and the vertical resolution is lost. Several studies use a pair of spinning disks [41] and illumination by the optics of total internal reflection [42].

33.3.2
FCS Measurement Outside Cells

Our study in HA solution in Section 33.2.2 suggests an appropriate model of anomalous diffusion in ECM and will be a guide for other systems. A recently suggested diagnosis method using MPTM for synovial fluid is to clarify their viscoelastic behavior [12]. We also performed FCS measurement of a dye molecule diffusing inside sectioned cartilages [43]. The diffusion measurement could be an efficient tool for medical diagnoses in the near future. The advantage of FCS is its suitability to observe the dynamical conversion of extracellular organs at a molecular level. One typical application is the detection of amyloid-β aggregations [44].

FCS study is now gradually expanding from singular cells or organs to the examination of individual animal. The dynamic behavior of nuage protein in medaka was investigated [45] and the flow in blood vessels of zebrafish and microchannels [46] has been monitored by FCS. These techniques will be extended to medical diagnosis. The technical improvement in FCS for extracellular materials and individuals will realize many fruitful investigations in medical, zoological, and agricultural science in the next decade.

33.4
Summary

FCS study inside and outside of cells will be a powerful tool to investigate the dynamic behavior of molecules involved in biological systems. The visualization and characterization of transporting molecules reflect the activity of life itself. An appropriate treatment of diffusion in inhomogeneous media, that is, anomalous diffusion, will provide a break-through to the wide application of FCS to medical, zoological, and plant science where researchers would like to operate on individuals. The space size of FCS, typically a 10–100 nm cubic space, is comparable with that of ordinal chemical reaction in biological systems. Different from FRAP or other methods which need a visible (large) space size, FCS can obtain the most appropriate D value to discuss ordinal biological reaction, such as signaling.

Acknowledgments

The authors are grateful to Professor Hiroshi Masuhara and Professor Hiroshi Fukumura for their encouragement of our FCS studies. One of us (KU) expresses his thanks to many coworkers in his study: Dr Akiko Masuda, Dr Takayuki Okamoto, Dr Hiroyuki Koshino, Dr Goro Nishimura, and Professor Mamoru Tamura. A part of our study, described in Refs. [21–25, 31] is supported by Grants-In-Aid for Scientific Research (Kakenhi) No. 17034067 in the Priority Area "Molecular Nano Dynamics" and No.17300166 from the Ministry of Education, Culture, Sports, Science and Technology (MEXT) of Japan.

References

1 Rigler, R. and Vogel, H. (eds) (2008) *Single Molecules and Nanotechnology*, Springer Series in Biophysics, vol. 12, Springer.

2 Sauer, M., Hofkens, J. and Enderlein, J. (eds) (2007) *Handbook of Fluorescence Spectroscopy and Imaging: From Ensemble to Single Molecules*, Wiley-VCH, Weinheim.

3 Kusumi, A., Nakada, C., Ritchie, K., Murase, K., Suzuki, K., Murakoshi, H., Kasai, R.S., Kondo, J. and Fujiwara, T. (2005) Paradigm shift of the plasma membrane concept from the two-dimensional continuum fluid to the partitioned fluid: high-speed single-molecule tracking of membrane molecules. *Annu. Rev. Biophys. Biomol. Struct.*, **34**, 351–378.

4 Kusumi, A., Ike, H., Nakada, C., Murase, K. and Fujiwara, T. (2005) Single-molecule tracking of membrane molecules: plasma membrane compartmentalization and dynamic assembly of raft-philic signaling molecules. *Sem. Immunol.*, **17**, 3–21.

5 Rigler, R. and Elson, E.S. (eds) (2001) *Fluorescence Correlation Spectroscopy: Theory and Applications*, Springer.

6 Sprague, B.L., Pego, R.L., Stavreva, D.A. and McNally, J.G. (2004) Analysis of binding reactions by fluorescence recovery after photobleaching. *J. Biophys.*, **86**, 3473–3495.

7 Stejskal, E.O. and Tanner, J.E. (1965) Spin diffusion measurements: spin echoes in the presence of a time-dependent field gradient. *Chem. Phys. J*, **42**, 288–292.

8 Barjat, H., Morris, G.A., Smart, S., Swanson, A.G. and Williams, S.C.R. (1995) High-resolution diffusion-ordered 2D spectroscopy (HR-DOSY) – a new tool for the analysis of complex mixtures. *J.Magn. Reson. Ser. B*, **108**, 170–172.

9 Krafter, J., Shlesinger, M.F. and Zumofen, G. (1996) Beyond brownian motion. *Phys. Today*, 33–39.

10 Borodin, A.N. and Salminen, P. (2002) *Handbook of Brownian Motion, Facts and Formulae (Probability and Its Applications)*, 2nd edn, Birkhäuser.

11 Apgar, J., Tseng, Y., Fedorov, E., Herwig, M.B., Almo, S.C. and Wirtz, D. (2000) Multiple-particle tracking measurements of heterogeneities in solutions of actin filaments and actin bundles. *Biophys. J.*, **79**, 1095–1106.

12 Jay, G.D., Torres, J.R., Warman, M.L., Laderer, M.C. and Breuer, K.S. (2007) The role of lubricin in the mechanical behavior of synovial fluid. *Proc. Nat. Acad. Sci*, **104**, 6194–6199.

13 Klafter, J. and Sokolov, I.M. (2005) Anomalous diffusion spreads its wings. *Phys. World*, **2005-8**, 29–32.

14 Netz, P.A. and Dorfmüller, T. (1997) Computer Simulation Studies of Diffusion in Gels: Model Structures. *J Chem. Phys.*, **107**, 9221–9233.

15 Metzler, R. and Klafter, J. (2003) When Translocation Dynamics Becomes Anomalous. *Biophys. J.*, **85**, 2776–2779.

16 Saxton, M.J. (1994) Anomalous Diffusion Due to Obstacles: a Monte Carlo study. *Biophys. J.*, **66**, 394–401.

17 Saxton, M.J. (2005) Fluorescence corralation spectroscopy. *Biophys. J.*, **89**, 3678–3679.

18 Saxton, M.J. (2007) A biological interpretation of transient anomalous subdiffusion I. Qualitative model. *Biophys. J.*, **92**, 1178–1191.

19 Magde, D., Elson, E.L. and Webb, W. (1974) Fluorescence correlation spectroscopy II. An experimental realization. *Biopolymers*, **13**, 29–61.

20 Keizer, J. (1987) Diffusion effects on rapid bimolecular chemical reactions. *Chem. Rev.*, **87**, 167–180.

21 Masuda, A., Ushida, K. and Okamoto, T. (2005) New fluorescence correlation spectroscopy enabling direct observation of spatiotemporal dependence of diffusion constants as an evidence of anomalous transport in extracellular matrices. *Biophys. J.*, **88**, 3584–3591.

22 Masuda, A., Ushida, K. and Okamoto, T. (2005) Direct observation of spatiotemporal dependence of anomalous diffusion in inhomogeneous fluid by sampling-volume-controlled fluorescence correlation spectroscopy. *Phys. Rev. E*, **72**, 060101.

23 Masuda, A., Ushida, K. and Okamoto, T. (2006) New fluorescence correlation spectroscopy (FCS) suitable for the observation of anomalous diffusion in polymer solution: time and space dependences of diffusion coefficients. *J. Photochem. Photobiol. A*, **183**, 304–308.

24 Ushida, K. and Masuda, A. (2007) General importance of anomalous diffusion in biological inhomogeneous systems, in *Nano Biophotonics Science and Technology Handai Nanaophotonics*, vol. 3 (eds H. Masuhara, S. Kawata and F. Tokunaga), Elsevier, pp. 175–188, Chapter 11.

25 Ushida, K. (2008) Anomalous diffusion in polymer solution as probed by fluorescence correlation spectroscopy and its universal importance in biological systems. AIP Conference Proceedings, 982, pp. 464–469.

26 Fletcher, K.A., Fakayode, S.O., Lowery, M., Tucker, S.A., Neal, S.L., Kimaru, I.W., McCarroll, M.E., Patonay, G., Oldham, P.B., Rusin, O., Strongin, R.M. and Warner, I.M. (2006) Molecular fluorescence, phosphorescence, and chemiluminescence spectrometry. *Anal. Chem.*, **78**, 4047–4066.

27 Lapčík, L. Jr, Lapčík, L., De Smedt, S., Demeester, J. and Chabreek, P. (1998) Hyaluronan: preparation, structure, properties, and applications. *Chem. Rev.*, **98**, 2663–2684.

28 Laurent, T.C. and Balazs, E.A. (eds) (1998) *Chemistry, Biology and Medical Applications of Hyaluronan and Its Derivatives*, Portland Press.

29 Poole, A.R. (2001) *Cartilage Structure and Function in Arthritis and Allied Conditions*, 14th edn (ed. W.J. Koopman), Willams & Wilkins, pp. 226–284.

30 Ogston, A.G., Preston, B.N., Wells, J.D. and Snowden, J.M. (1973) On the transport of compact particles through solutions of chain-polymers. *Proc. R. Soc. London Ser. A*, **333**, 297–316.

31 Masuda, A., Ushida, K., Nishimura, G., Kinjo, M., Tamura, M., Koshino, H., Yamashita, K. and Kluge, T. (2004) Experimental evidence of distance-dependent diffusion coefficients of a globular protein observed in polymer aqueous solution forming a network structure on nanometer scale. *J Chem. Phys.*, **121**, 10787–10793.

32 Masuda, A., Ushida, K., Koshino, H., Yamashita, K. and Kluge, T. (2001) Novel distance dependence of diffusion constants in hyaluronan aqueous solution resulting from its characteristic nano-microstructure. *J Am. Chem. Soc.*, **123**, 11468–11471.

33 Aharony, A. and Stauffer, D. (1984) Possible breakdown of the alexander-orbach rule at low dimensionalities. *Phys Rev. Lett.*, **52**, 2368–2730.

34 Stauffer, D. and Aharony, A. (1994) *Introduction to Percolation Theory*, Taylor & Francis.

35 Kim, S.A., Heinze, K.G. and Schwille, P. (2007) Fluorescence correlation spectroscopy in living cells. *Nature Methods*, **4**, 963–973.

36 Bacia, K. and Schwille, P. (2007) Practical guidelines for dual-color fluorescence cross-correlation spectroscopy. *Nature Protocols*, **2**, 2842–2856.

37 Banks, D.S. and Fradin, C. (2005) Anomalous diffusion of proteins due to molecular crowding. *Biophys. J.*, **89**, 2960–2971.

38 Mikuni, S., Tamura, M. and Kinjo, M. (2007) Analysis of intranuclear binding process of glucocorticoid receptor using fluorescence correlation spectroscopy. *FEBS Lett.*, **581**, 389–393.

39 Seki, K., Masuda, A., Ushida, K. and Tachiya, M. (2005) A theoretical method to analyze diffusion of probe molecules in nanostructured fluids by fluorescence

correlation spectroscopy. *J. Phys. Chem. A*, **109**, 2421–2427.

40 Gryczynski, Z.K. (2008) FCS imaging—a way to look at cellular processes. *Biophys. J.*, **94**, 1943–1944.

41 Saisan, D.R., Arevalo, R., Graves, C., McAllister, R. and Urbach, J.S. (2006) Spatially resolved fluorescence correlation spectroscopy using a spinning disk confocal microscope. *Biophys. J.*, **91**, 4241–4252.

42 Ohsugi, Y., Saito, K., Tamura, M. and Kinjo, M. (2006) Lateral mobility of membrane-binding proteins in living cells measured by total internal reflection fluorescence correlation spectroscopy. *Biophys. J.*, **91**, 3456–3464.

43 Lee, J.I., Koike, R., Morito, T., Ushida, K. and Sato, M., unpublished results.

44 Garai, K., Sureka, R. and Maiti, S. (2007) A pathway and genetic factors contributing to elevated gene expression noise in stationary phase. *Biophys. J.*, **92**, L55–L57.

45 Nagao, I., Aoki, Y., Tanaka, M. and Kinjo, M. (2008) Analysis of the molecular dynamics of medaka nuage proteins by fluorescence correlation spectroscopy and fluorescence recovery after photobleaching. *FEBS J.*, **275**, 341–349.

46 Pan, X., Yu, H., Shi, X., Korzh, V. and Wahland, T. (2007) Characterization of flow direction in microchannels and zebrafish blood vessels by scanning fluorescence correlation spectroscopy. *J. Biomed. Opt.*, **12**, 014034.

34
Spectroscopy and Photoreactions of Gold Nanorods in Living Cells and Organisms
Yasuro Niidome and Takuro Niidome

34.1
Introduction

34.1.1
Spectroscopic Properties of Gold Nanorods

Gold nanorods are rod-shaped anisotropic gold nanoparticles with unique optical properties [1–5]. They show two surface plasmon bands corresponding to the transverse and longitudinal surface plasmon bands in the visible (~520 nm) and the near-IR regions, respectively (Figure 34.1). The longitudinal band has a substantially larger extinction coefficient than the transverse band.

With reference to spherical gold nanoparticles, Maxwell's equations for the optical response to an electromagnetic field of light wave were solved analytically by Mie [6]. For gold nanorods, the Mie theory has been readily used for quantitative studies of the optical properties of nanorods of different size, shape, and aggregation conditions. Link *et al.* applied the Mie theory to gold nanorods using an ellipsoidal model [7, 8], and found that the intensities and wavelength positions of the two plasmon bands depended on the aspect ratio of the gold nanorods and the dielectric constant of the media. Using the same ellipsoidal model, Gluodenis and Foss discussed the effects of mutual orientation on the spectra of metal nanoparticle pairs (rod–rod and rod–sphere) [9]. They indicated that interactions between the two gold nanorods drastically change the longitudinal surface plasmon bands. These theoretical treatments are useful for a qualitative estimation of the peak positions of the longitudinal surface plasmon bands depending on the shapes, sizes, and aggregation states of gold nanorods. Recently, the discrete dipole approximation (DDA) method [10, 11], which is a kind of finite element method, has been applied to gold nanorods [12–14]. Figure 34.2 shows the surface plasmon bands of gold nanorods theoretically elucidated using the DDA methods. These spectra indicate that the DDA method is useful for prediction of the surface plasmon bands depending on the shapes at the ends of the gold nanorods. It has also been shown that the DDA method is useful

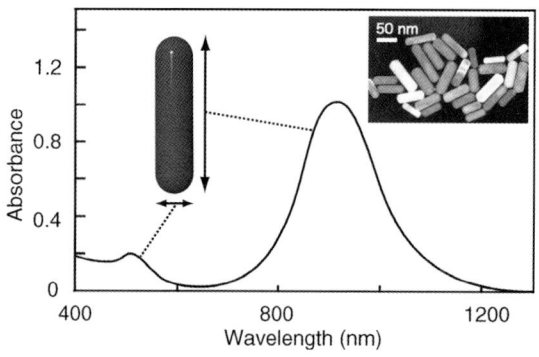

Figure 34.1 Extinction (absorption) spectrum of gold nanorods and their transmitted electron microscopic image (inset). Gold nanorods (10 × 60 nm) showed peaks at 520 and 900 nm corresponding to the transverse and longitudinal surface plasmon bands, respectively.

for discussing the correlations between the surface plasmon bands and the size and shape of the gold nanoparticles.

As explained by the Mie theory and the DDA calculations, the surface plasmon bands of metallic nanoparticles consist of absorption and scattering. In the case of particles larger than ∼50 nm, the contribution of light scattering is not negligible [14]. Thus, gold particles larger than ∼50 nm can act as probe materials for light scattering observations. In the case of gold nanorods, surface plasmon bands can be observed in the near-IR region, indicating that the nanorods act as light-scattering probe particles in this region [13]. Thus, gold nanorods are unusual materials with an intense surface plasmon band that affords light scattering observation, as well as near-IR absorption [15]. By taking advantage of such unique optical characteristics, many biological and medical applications for gold nanorods are possible.

34.1.2
Biocompatible Gold Nanorods

A cationic detergent, hexadecyltrimethylammonium bromide (CTAB), is indispensable as the stabilizing agent during the preparation of gold nanorods [2]. However, because CTAB is highly cytotoxic, these gold nanorods cannot be used in biological fields despite their unique optical characteristics. To obtain a CTAB-free nanorod solution, gold nanoparticles can be washed using centrifugation [2]. However, CTAB bilayers usually remain on the surface of the gold nanorods, and can also be adsorbed non-covalently onto the surface. Therefore, further removal of the CTAB will result in aggregation of the nanorods. To reduce the cytotoxicity of the gold nanorods and stabilize them under biological conditions, Liao et al. prepared nanorods conjugated to poly(ethylene glycol) (PEG) [16]. The PEG-conjugated nanorods could be dispersed in buffered solutions, and antibodies could be modified by using a bifunctional cross-linker molecule that had a disulfide group and a succinimidyl group at the ends of an aliphatic chain. This work showed the possibility that functional gold nanorods

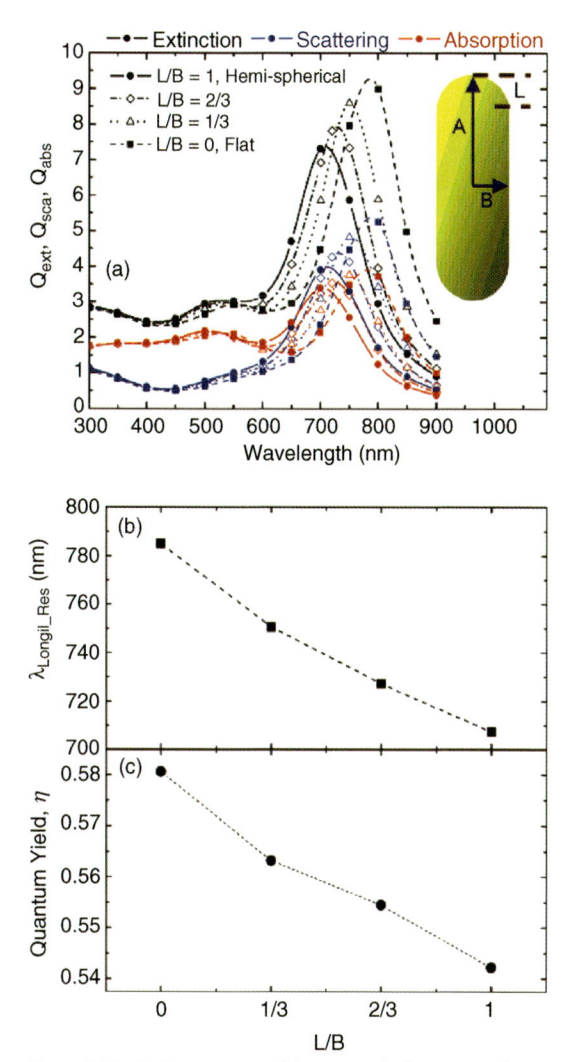

Figure 34.2 (a) Comparison of the DDA calculations as a function of end-cap shape factor, L/B, for the Au nanorods with $R = 3$. The inset of (a) illustrates the end-cap shape definition. (b) Longitudinal surface plasmon resonance peak position and (c) radiative quantum yield η of an Au nanorod as a function of end-cap shape factor, L/B [13]. (The figures were reproduced from Ref. [13] with permission.).

could be used as bioprobes *in vitro* and *in vivo*. We have also developed a technique to remove CTAB by extracting it from the nanorod solution into a chloroform phase containing phosphatidylcholine (PC) (Figure 34.3) [17–19]. After this modification, the PC molecules on the gold surface were identified with transmission electron microscopy and energy-dispersive X-ray analysis. The PC-modification on the gold nanorods decreased the zeta-potential of the nanorods from $+67$ to $+21$ mV,

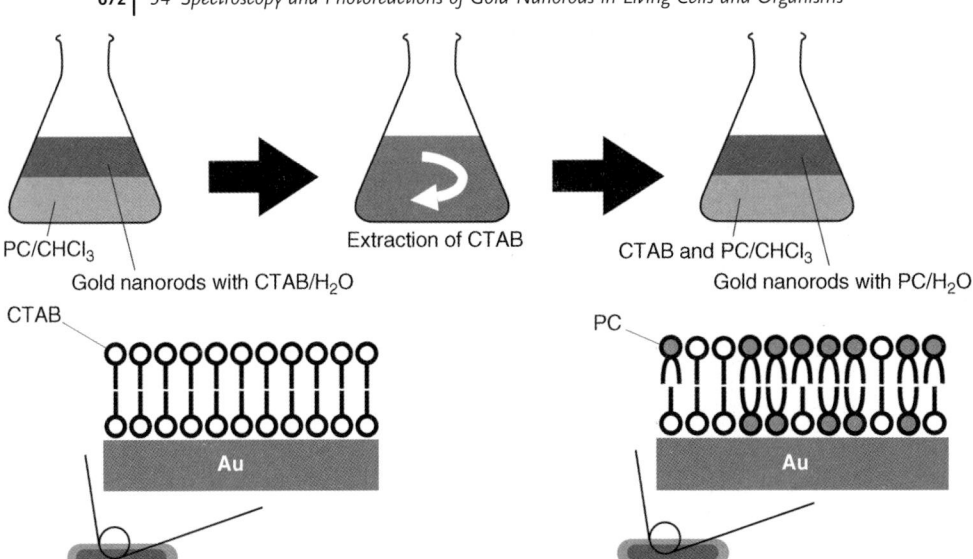

Figure 34.3 Schematic illustration of the preparation of PC-modified gold nanorods. CTAB in the original gold nanorods solution was extracted into chloroform solution of PC. After performing three extraction procedures, the aqueous solutions containing the PC-modified gold nanorods were centrifuged, and then dispersed again in water.

indicating that cationic CTAB was removed from the surface, and the cytotoxicity was markedly reduced (Figure 34.4).

For application of gold nanorods in biological and medical fields, including tumor imaging, photothermal therapy, gene and drug delivery, the targeted delivery of the nanorods after systemic injection must be achievable [20–23]. For the targeted delivery *in vivo*, a stealth character in blood circulation is certainly required. Providing the stealth character to nanorods will enable efficient delivery to the specific target sites, higher contrast images of the sites, and more effective photothermal therapy compared to the current techniques. To test whether the PC-modified nanorods can be useful in applications *in vivo*, biodistribution of the gold nanorods in mice after intravenous injection was evaluated (Figure 34.5).

Unfortunately, most of the injected gold was detected in the liver, indicating that the PC-modified gold nanorods were cleared from the blood circulation within 30 min, and then trapped in the liver, probably in the Kupffer cells which are adept at taking in such cationic particles. To overcome the instability of the gold nanorods in the blood circulation, we also modified them with a PEG chain, and then evaluated the cytotoxicity *in vitro* and the biodistribution after intravenous injection into mice [24]. The PEG-modification was achieved by adding thiol-terminated PEG (mPEG$_{5,000}$-SH, NOK Corp.) to a nanorod solution stabilized with CTAB. The mPEG$_{5,000}$-SH was adsorbed onto the nanorod surfaces, and then excess CTAB was removed by dialysis. The PEG-modified gold nanoparticles showed neutral

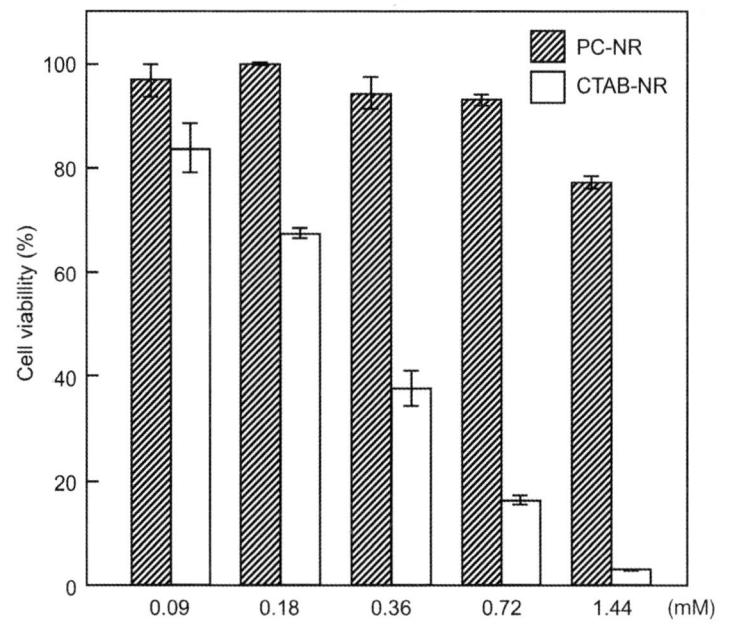

Figure 34.4 Viabilities of HeLa cells after contacting the PC-modified gold nanorods (gray bars) and the original CTAB-stabilized gold nanorods (open bars). The gold nanorod solutions (0.09, 0.18, 0.36, 0.72 and 1.44 mM as Au atom at final concentrations) were added to the cells. The cells were incubated for 24 h. Viabilities of the cells were evaluated by the MTT assay.

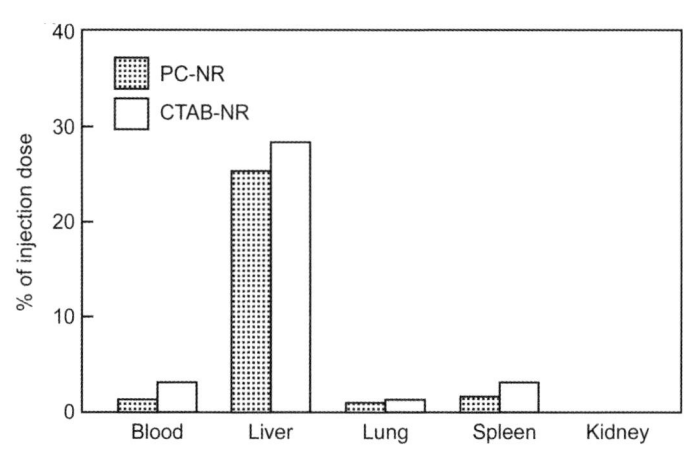

Figure 34.5 Biodistribution of gold nanorods in mice after intravenous injection. Biodistributions of PC-modified and CTAB-stabilized gold nanorods are indicated with gray bars and open bars, respectively. At 30 min after the injection, organs were collected and lysed in aqua regia. Quantities of gold ion in the lysates were quantified by ICP mass spectrometry.

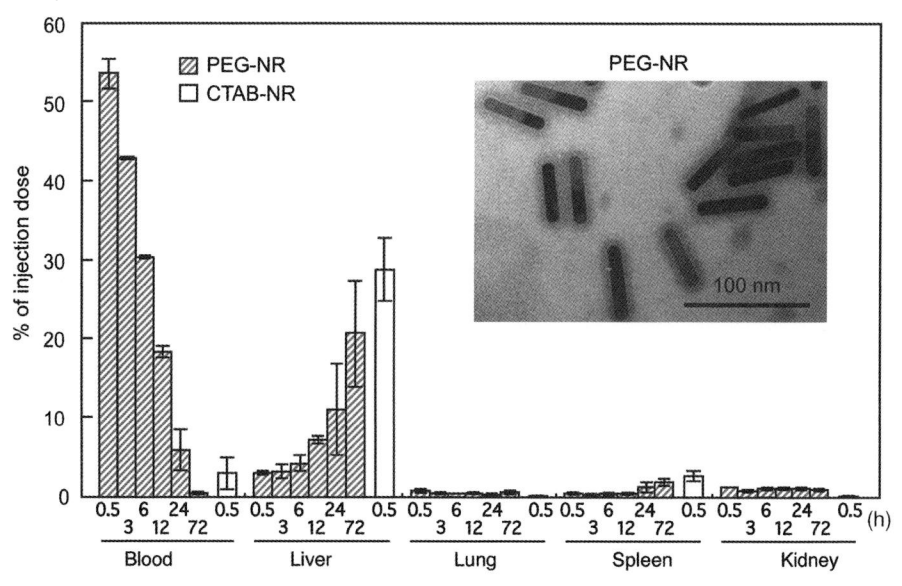

Figure 34.6 Biodistribution of PEG-modified gold nanorods in mice after intravenous injection. At several time points after injection, the quantities of gold in the tissue samples were evaluated by ICP mass spectrometry. Closed bars show biodistribution of PEG-modified gold nanorods at 0.5, 3, 6, 12, 24 and 72 h after injection. Open bars show that of CTAB-stabilized gold nanorods at 0.5 h after injection. Inset shows the electron micrograph of the PEG$_{5,000}$-modified gold nanorods.

zeta-potential (−0.5 mV), and had little cytotoxicity *in vitro*. Following intravenous injection into mice, 54% of the injected PEG-modified gold nanoparticles were found in blood 0.5 h after intravenous injection, whereas most of the gold from the nanorods stabilized with CTAB was detected in the liver (Figure 34.6). Thus, due to the formation of a stealth character, modification of ligands such as RGD peptides and antibodies with gold nanorods will allow us to develop targeted delivery systems [25–30].

34.2
Spectroscopy of Gold Nanorods in Living Cells

34.2.1
Gold Nanorods Targeting Tumor Cells

Methods to stain/label specific cells and tissues have been intensively studied for the purposes of developing diagnostic and investigational techniques in biology and therapeutic application. A variety of techniques using absorption and fluorescent dyes, such as Malachite Green and Rhodamine-6G, have been reported as contrast agents to distinguish between normal and diseased cells and tissues [31]. However, organic molecules are not stable in living systems; they tend to degrade over the long term or under light irradiation. In order to overcome this problem, semiconductor quantum dots, which show intense and stable photoluminescence, have been used recently for biological and cell imaging [32–35]. Because the peak wavelength of the photoluminescence can be tuned using their sizes, the quantum dots are very useful contrast agents for multi-color fluorescence staining. The potential cytotoxicity of the semiconductor material, however, limits the *in vivo* use of quantum dots.

Colloidal gold nanoparticles are expected to be a new class of contrast agents and imaging labels due to their strong surface plasmon bands located in the visible to near-IR regions [36–38]. In earlier works, colloidal gold nanoparticles have also been used for biological labeling for transmission electron microscopic observations [39]. The surface-modified gold nanoparticles can be prepared via the strong interactions between gold and thiol, disulfide, and amine groups. For example, single-stranded DNA [40–43], sugars [44], RGD peptides [45, 46], antibodies [38, 43], and folate [47, 48] have been reported as functional capping-agents of gold nanoparticles, and the latter three were expected to be effective in delivering the gold nanoparticles to tumor cells.

Surface modification of gold nanorods is similar to modifying spherical gold nanoparticles; however, as described above, the surface of the gold nanorods were capped with cationic CTAB layers. Nonspecific interactions between the cationic gold nanorods and anionic biomaterials should be taken into account. El-Sayed and coworkers wrapped the positive gold nanorods with polyanionic poly(styrenesulfonate) (PSS). The PSS-wrapped gold nanorods were dispersed in a buffered solution, and anti-epidermal growth factor receptor (anti-EGFR) monoclonal antibodies were fixed on the surface of the nanorods [49]. Figure 34.7 shows light scattering images of cell cultures incubated with the anti-EGFR antibody-conjugated spherical gold nanoparticles and nanorods. The spectra indicated typical surface plasmon bands that were assignable to well isolated gold nanoparticles and nanorods, and thus the peak intensities of the surface plasmon bands reflected the amount of anti-EGFR antibody-conjugated nanorods bound on the cells. The malignant-type cells (HOC 313 clone 8 and HSC 3) showed intense surface plasmon bands, because the antibody-conjugated gold nanorods had a high affinity due to the overexpressed EGFR on the cytoplasmic membrane of the malignant cells. This indicated that the antibody-conjugated nanorods can be functional nanoparticles to target tumor cells, and the accumulation of the nanorods can be seen through light scattering observations.

Recently, the surface-modified gold nanorods were used as probe materials for surface-enhanced Raman scattering (SERS) [20, 21] and photoacoustic spectroscopy [50]. The SERS can give information on the vibrational modes of organic molecules on or near the gold nanorods, while photoacoustic spectroscopy is a sensitive way to detect photothermal conversion. These methods will open up new applications of gold nanorods as bioprobes.

34.2.2
Spectroscopy of Gold Nanorods *In Vivo*

As described in papers using gold nanoshells that have also shown surface plasmon bands in the near-IR region [36, 37], the targeted delivery of nanoparticles allows us to

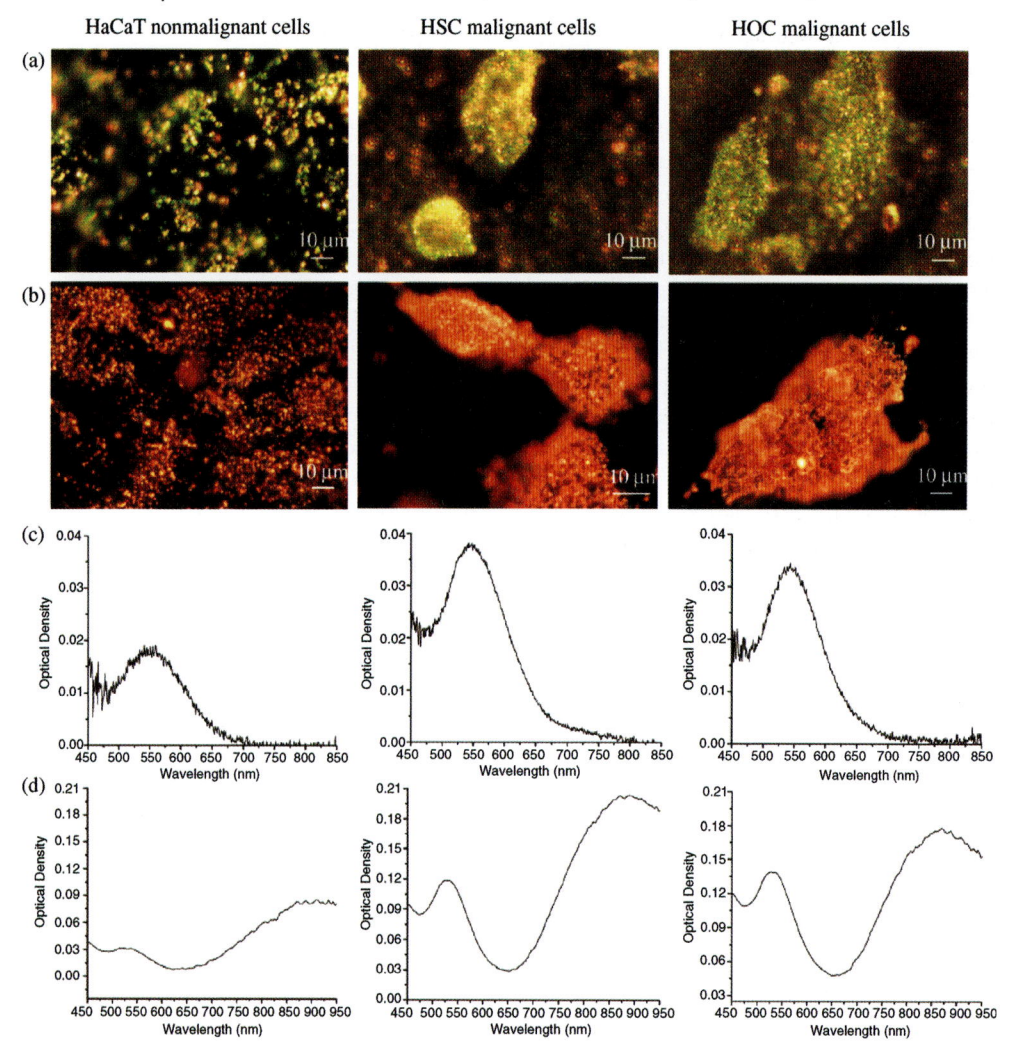

Figure 34.7 (a) Light scattering images of anti-EGFR/Au nanospheres after incubation with cells for 30 min at room temperature. (b) Light scattering images of anti-EGFR/Au nanorods after incubation with cells for 30 min at room temperature. (c) Average extinction spectra of anti-EGFR/Au nanospheres from 20 different single cells for each kind. (d) Average extinction spectra of anti-EGFR/Au nanorods from 20 different single cells for each kind. From gold nanospheres, the green to yellow color is most dominant, corresponding to the surface plasmonic enhancement of scattering light in the visible region, and from gold nanorods, the orange to red color is the most dominant, corresponding to the surface plasmonic enhancement of the longitudinal oscillation in the near-infrared region. (The figures were reproduced by permission of reference [49].).

identify the target site, and induce tissue damage by the photothermal effects of near-IR laser light. The targeted delivery and bioimaging of gold nanorods are important topics of interest for the development of diagnostic and therapeutic systems using near-IR light. As well as conventional staining using gold nanorods as near-IR contrast agents, El-Sayed and coworkers reported efficient two-photon luminescence of gold nanorods [51, 52]. They remarked that their gold nanorods were "lightning" [52]. The luminescent gold nanorods were used for two-photon luminescence imaging [53–55]. This method revealed the distribution of the gold nanorod that had been taken up in the living cells *in vitro*, and showed luminescent images of gold nanorods that were circulating in the blood flow of a mouse using a microscopic technique [54]. This was pioneering work in applying the gold nanorods as an *in vivo* bioprobe.

We also attempted to detect the surface plasmon bands of gold nanorods in a mouse using a conventional spectrophotometer [56]. In order to record the absorption spectra and investigate the dynamics of the surface plasmon bands of the gold nanorods in a mouse, we used an integral sphere (Figure 34.8). Monochromatic light from the spectrophotometer was introduced into the abdomen of an anesthetized mouse that was put on a port of the integral sphere. Because the integral sphere collected and normalized scattered light from the abdomen, the absorption spectra from the abdomen could be obtained.

Figure 34.9 shows an absorption spectrum from the abdomen of the mouse. There was intense absorption of hemoglobin in the visible region, and the onset of absorption of water in the near-IR region (>900 nm). The integral sphere was found to be useful in order to obtain the absorption spectra of the abdomen of a mouse.

Figure 34.10 shows the difference spectra for the abdomen after intravenous injections using the spectrum of the abdomen before the injections as the baseline. The lines (a) and (b) indicate the spectra immediately after and 30 min after the injection, respectively. Following the injection of PEG-modified gold nanorods (Figure 34.10a), characteristic peaks at around 900 nm were observed. These peaks

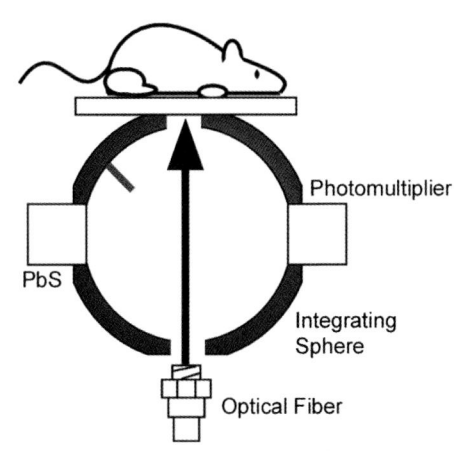

Figure 34.8 Schematic illustration of the spectroscopic analysis of gold nanorods in a mouse using an integral sphere. The anesthetized mouse was placed on a port of an integrating sphere. Monochromatic light from a spectrophotometer was introduced into the abdomen of the mouse through optical fibers.

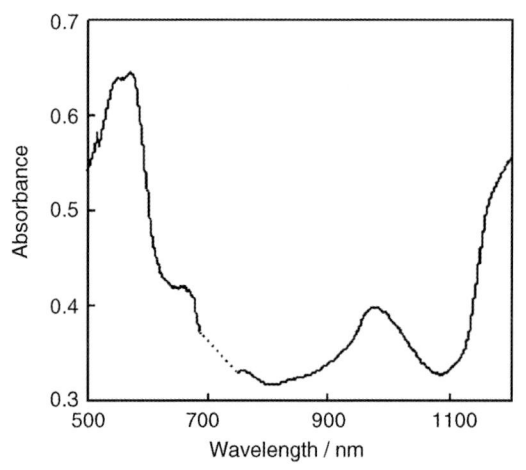

Figure 34.9 Absorption spectra of the abdomen of a mouse. The absorption peaks at around 550 and 980 nm can be attributed to hemoglobin in blood and water, respectively. At around 700 nm, no reasonable absorption signals were obtained due to lack of sensitivity of the photodetectors.

were assigned to the gold nanorods in the abdomen which did not form aggregates. It was reported that the aggregation of the gold nanorods dramatically changed the longitudinal surface plasmon bands in the near-IR region [57]. The results from Figure 34.10d which show the 5% glucose injection did not have these peaks in the near-IR region (Figure 34.10d). Thus, the results from Figure 34.10a indicated that the PEG-modified gold nanorods circulated for at least 30 min without forming aggregates. In the case of the gold nanorods modified with CTAB (Figure 34.10b), the surface plasmon band of the gold nanorods was detected immediately after the injection. However, 30 min after the injection, the peak had diminished in intensity. The injection of PC-modified gold nanorods also produced a surface plasmon band

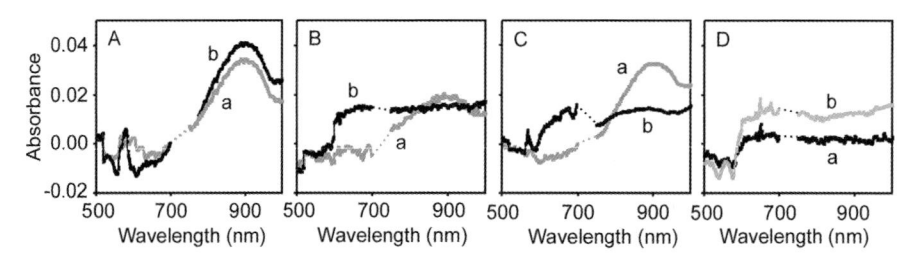

Figure 34.10 Absorption spectra of gold nanorods in the abdomen of mice after the intravenous injection of gold nanorods (300 μl of 2 mM Au atoms). The dashed lines correspond to the region where accurate values could not be obtained due to the lower sensitivity of the spectrophotometer. The spectrum before the injection was used as baseline data. Panels A–D show spectra immediately after injection (a) and 30 min after injection (b) of PEG- (A), CTAB- (B), and PC-modified (C) gold nanorods (300 μl of 2 mM Au atoms), and 5% glucose solution as control (D).

resulting from the gold nanorods immediately after the injection (Figure 34.10c), but no absorption peak was observed in the spectrum (Figure 34.10c) 30 min after the injection of the PC-modified gold nanorods. With both the CTAB- and PC-modified nanorods, the circulation of the gold nanorods in mice was not as good as that with the PEG-modified gold nanorods. The CTAB- and PC-modified gold nanorods were trapped outside the monitoring area of the abdomen.

The inductively coupled plasma mass spectrometry (ICP-MS) measurements of gold in organs indicated that the CTAB- and PC-modified gold nanorods accumulated in the liver 30 min after intravenous injection [24]. The accumulated nanorods in the liver could not be detected spectroscopically using a conventional spectrophotometer. In contrast, surface modification with PEG was very effective in improving the circulation of gold nanorods in the blood stream, and these could be detected in the abdomen for at least 30 min (Figure 34.10a). We have shown that using a combination of a spectrophotometer and an integral sphere was useful for monitoring the spectroscopic properties of gold nanorods that are circulating in the blood stream of a mouse.

Absorbance at 900 nm was continuously monitored after intravenous injection. Figure 34.11 shows the time courses of the absorption changes induced by the intravenous injection of PEG- (Figure 34.11a), PC- (Figure 34.11b), and CTAB-modified nanorods (Figure 34.11c). The injection of the PEG-modified gold nanorods resulted in an increase in absorption in the abdomen, which then reached a plateau (Figure 34.11). At 60 min after the first injection, the same solution was injected again. The second injection also resulted in a stepwise increase in absorption. On the other hand, the injections of the PC- (Figure 34.11b) and the CTAB-modified nanorods (Figure 34.11c) also resulted in an increase in absorption, but the absorption intensities immediately decreased. When the PC- and the CTAB-modified

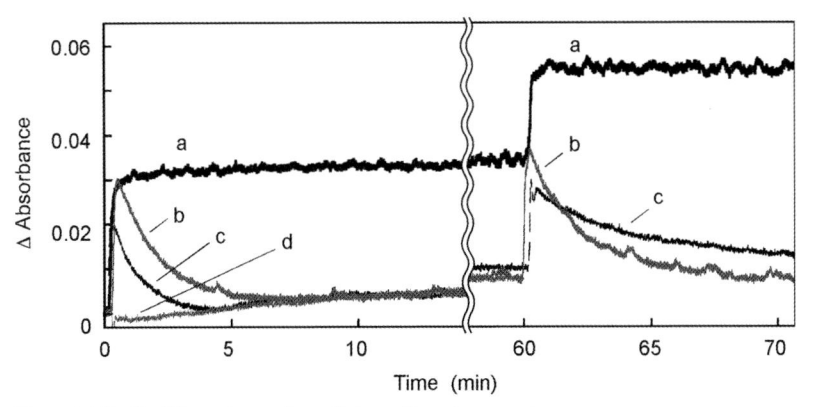

Figure 34.11 Real time observation of absorption changes at 900 nm in mice. (a) PEG-modified gold nanorods, (b) PC-modified gold nanorods, (c) CTAB-modified gold nanorods, (d) 5% glucose solution as control. After 60 min the same amount of gold nanorods was injected again.

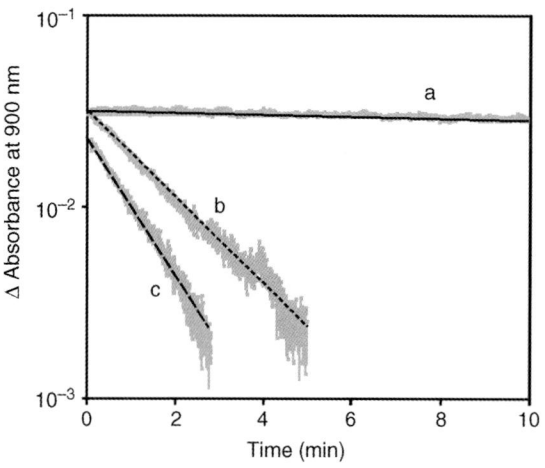

Figure 34.12 Logarithm plots of absorption changes in the abdomen after intravenous injection. (a) PEG-modified gold nanorods, (b) PC-modified gold nanorods, (c) CTAB-modified gold nanorods.

gold nanorods were injected into the mice again, the same changes were observed in both cases.

From Figure 34.11, the first absorption decay signals after the injections were re-plotted in Figure 34.12 using a logarithmic scale. The absorption changes of the control experiment (5% glucose injection) were then subtracted from those changes. In the case of the PEG-modified gold nanorods, absorption at 900 nm slowly decreased. This indicated that the PEG-modified gold nanorods could circulate stably in blood. On the other hand, the absorption spectra of the PC- (Figure 34.12b) and CTAB-modified gold nanorods (Figure 34.12c) showed linear decays, indicating that the decreased longitudinal surface plasmon bands were due to single exponential decays. Thus, the decays mainly originated from the accumulation of gold nanorods in the liver; that is, the spectral changes induced by the aggregation of the gold nanorods were negligible in the present experiments. According to the single-exponential fitting method, the half-lives of the surface plasmon bands of the PEG-, PC-, and CTAB-modified gold nanorods were estimated to be 231, 1.3, and 0.8 min, respectively. The half-lives of the PEG-modified gold nanorods obtained from our present system were essentially consistent with those of the results of the ICP-MS measurements [24]. It was shown that the real-time monitoring at 900 nm using an integral sphere revealed the dynamics of gold nanorods in mice.

34.3
Photoreactions of Gold Nanorods for Biochemical Applications

El-Sayed's group has reported that use of a continuous near-IR laser induced cell destruction in the presence of bioconjugated gold nanorods [49]. This is because the

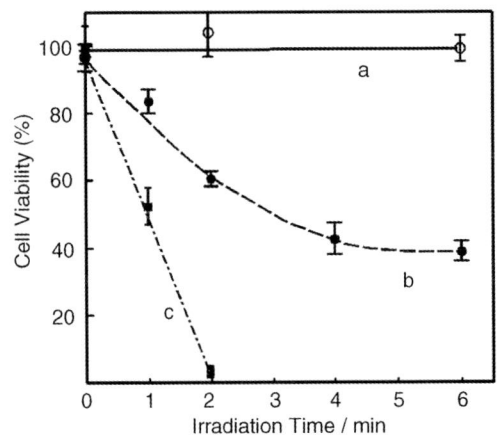

Figure 34.13 Viabilities of HeLa cells following pulsed laser irradiation (1064 nm, 10 Hz, 250 mJ pulse^{-1}) without (a) and with (b, c) PC-NRs in the medium. PC-NR concentrations in the medium: (a) 0, (b) 0.4, (c) 0.8 mM (Au atoms).

distinct surface plasmon bands of gold nanorods in the near-IR region allow efficient photothermal conversion. Using a combination of gold nanorods which can target tumors [20–22, 54, 55, 58, 59] and the hyperthermia treatment, the gold nanorod may be promising as a bifunctional material for probing and killing tumor cells.

The lipid-modified gold nanorods [17, 19] which showed no significant cytotoxicity have also been used as photothermal converters for photoinduced cell death of tumor cells [60, 61]. Because pulsed near-IR light introduces excessive heat in the immediate area surrounding the gold nanorods in a very short period, pulsed near-IR laser irradiation is useful for destruction of single cells [61]. We demonstrated that this unique photoreaction (photothermal conversion and reshaping) can be used to prevent unwanted cell damage [60].

Figure 34.13 shows the relationships between the time of laser irradiation and cell viabilities following laser irradiation. In the absence of PC-nanorods, laser irradiation triggered no cell damage (Figure 34.13a). On the other hand, in the presence of PC-nanorods, cell viabilities decreased with increasing laser irradiation time. In the presence of 0.8 mM PC-nanorods (Figure 34.13c), a 2 min laser irradiation damaged almost all the cells. In the presence of 0.4 mM PC-nanorods (Figure 34.13b), cell viabilities after 2 and 4 min of laser irradiation decreased to about 60 and 40%, respectively. Thus, it was obvious that the photothermal reaction of PC-nanorods induced cell death, depending on the PC-NR concentration and laser irradiation time.

Figure 34.14 shows the absorption spectra of PC-nanorods without (Figure 34.14a) and with (Figure 34.14b–d) laser irradiation. The spectrum of PC-nanorods without laser irradiation (Figure 34.14a) showed a broad longitudinal surface plasmon band in the near-IR region. This indicated that PC-NRs formed small aggregates in PBS buffer [3, 24]. The spectrum showed intense absorption of the longitudinal surface plasmon band at 1064 nm. At 2 min after laser irradiation (Figure 34.14b), the surface plasmon band in the near-IR region decreased remarkably. This indicated

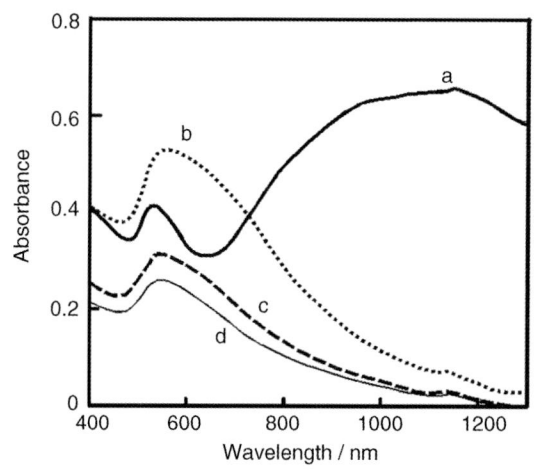

Figure 34.14 Absorption spectra of PC-nanorods without (a) and with (b) laser irradiation (250 mJ pulse^{-1}) in PBS buffer (Laser irradiation times: (b) 2, (c) 4, (d) 6 min). PC-NR concentration: 0.4 mM (Au atoms).

that PC-nanorods were reshaped into spherical nanoparticles by the pulsed laser irradiation. Laser irradiation for a 4 min period (Figure 34.14c) decreased the intensity of the shoulder peak in the near-IR region, and resulted in a single peak at around 580 nm that corresponded to spherical particle aggregates. Since reshaped PC-nanorods hardly absorbed laser light at 1064 nm, a further 2 min of laser irradiation (total irradiation = 6 min) caused no additional spectral changes (Figure 34.14d). It should be noted that the decreased rate of cell viability was suppressed when the irradiation time was longer than 4 min. As absorbance at 1064 nm disappeared, photoinduced cell death was suppressed. This means that PC-nanorods do not damage other cells after the photoreaction. This will allow the achievement of selective cell death without unwanted damage to neighboring cells.

34.4
Conclusions and Future Outlook

Gold nanorods are an attractive research target because of their unique optical and photothermal properties. Because the surface plasmon bands show strong scattering and absorption properties, it makes them a highly effective class of contrast agents for *in vitro* and *in vivo* imaging of tumor cells. Many ideas for their use in practical diagnosis and therapy of diseases have been proposed [62–64]. Specific imaging and therapy of tumor cells have been achieved using bi- or multi-functional gold nanorods that have been modified with antibodies, folates, and peptides. Nevertheless, even with the progress of recent research, many factors need to be optimized. For example, the sizes and shapes of the nanorods that are optimal for tumor targeting should be studied further. An effective method for delivering near-IR light to the tumor cells also needs more research. Targeting an *in vivo* system will also need additional effort so that a practical amount of gold nanorods can be localized to diseased cells or tissues. By employing optimized targeting strategies, an imaging/therapy regimen using gold nanorods should become a practical medical treatment.

Acknowledgments

We acknowledge the contribution of Dr Hironobu Takahashi of Utah State University, Dr Koji Nishioka from Sumitoto Chemical Co. Ltd., and Dr Takahiro Kawano of the Institute for Materials Chemistry and Engineering, Kyushu University for work included in this review. We also thank the following people for use of their facilities and support: Kanako Honda, Yukichi Horiguchi, Yasuyuki Akiyama, Kohei Shimoda, Keisuke Higashimoto, and Yoshifumi Okuno from the Department of Applied Chemistry, Kyushu University. We also acknowledge the financial support of the Grant-in-Aid for Scientific Research (No. 15350085), KAKENHI (Grant-in-Aid for Scientific Research) on Priority Area "Molecular Nano Dynamics (No. 432)" and "Strong Photon-Molecule Coupling Fields (No. 470)", and a Grant-in-Aid for the Global COE Program, "Science for Future Molecular Systems" from the Ministry of Education, Culture, Sports, Science and Technology (MEXT) of the Japanese Government.

References

1 Esumi, K., Matsuhisa, K. and Torigoe, K. (1995) Preparation of rodlike gold particles by UV irradiation using cationic micelles as a template. *Langmuir*, **11**, 3285–3287.

2 Yu, Y.-Y., Chang, S.-S., Lee, C.-L. and Wang, C.R.C. (1997) Gold nanorods: electrochemical synthesis and optical properties. *J. Phys. Chem. B*, **101**, 6661–6664.

3 van der Zande, B.M.I., Böhmer, M.R., Fokkink, L.G.J. and Schöneberger, C. (1997) Aqueous gold sols of rod-shaped pa. *J. Phys. Chem. B*, **101**, 852–854.

4 van der Zande, B.M.I., Koper, G.J.M. and Lekkerkerker, H.N.W. (1999) Optical properties of aligned rod-shaped gold particles dispersed in poly(vinyl alcohol) films. *J. Phys. Chem. B*, **103**, 5754–5760.

5 van der Zande, B.M.I., Pages, L., Hikmet, R.A.M. and van Blaaderen, A. (1999) Optical properties of aligned rod-shaped gold particles dispersed in poly(vinyl alcohol) films. *J. Phys. Chem. B*, **103**, 5761–5767.

6 Mie, G. (1908) Beiträge zur optik trüber medien, speziell kolloidaler metallösungen. *Ann. Phys. (Lipzig)*, **25**, 377.

7 Link, S., Mohamed, M.B. and El-Sayed, M.A. (1999) Simulation of the optical absorption spectra of gold nanorods as a function of their aspect ratio and the effect of the medium dielectric constant. *J. Phys. Chem. B*, **103**, 3073–3077.

8 Link, S. and El-Sayed, M.A. (2005) Simulation of the optical absorption spectra of gold nanorods as a function of their aspect ratio and the effect of the medium dielectric constant. *J. Phys. Chem. B*, **109**, 10531–10532.

9 Gloudenis, M., Colby, J. and Foss, A. (2002) The effect of mutual orienttion on the

spectra of metal nanoparticles rod-rod and rod-sphere pairs. *J. Phys. Chem. B*, **106**, 9484–9489.
10 Purcell, E.M. and Pennypacker, C.R. (1973) Scattering and absorption of light by nonspherical dielectric grains. *Astrophys. J.*, **186**, 705–714.
11 Draine, B.T. and Flatau, P. j. (2004) User Guide for the Discrete Dipole Approximation Code DDSCAT 6.1.
12 Brioude, A., Jiang, X.C. and Pileni, M.P. (2005) Optical properties of gold nanorods: DDA simulations supported by experiments. *J. Phys. Chem. B*, **109**, 13138–13142.
13 Lee, K.-S. and El-Sayed, M.A. (2005) Dependence of the enhanced optical scattering efficiency relative to that of absorption for gold metal nanorods on aspect ratio, size, end-cap, shape, and medium refractive index. *J. Phys. Chem. B*, **109**, 20331–20338.
14 Jain, P.K., Lee, K.S., El-Sayed, I.H. and El-Sayed, M.A. (2006) Calculated absorption and scattering properties of gold nanoparticles of different size, shape, and composition: application in biological imaging and biomedicine. *J. Phys. Chem. B*, **110**, 7238–7248.
15 Ni, W., Kou, X., Yang, Z. and Wang, J. (2008) Tailoring longitudinal surface plasmon wavelength, scattering and absorption cross sections of gold nanorods. *ACS Nano.*, **2**, 677–686.
16 Liao, H. and Hafner, J.H. (2005) Gold nanorods bioconjugates. *Chem. Mater.*, **17**, 4636–4641.
17 Takahashi, H., Niidome, Y., Niidome, T., Kaneko, K., Kawasaki, H. and Yamada, S. (2006) Modification of gold nanorods using phosphatidylcholine to reduce cytotoxicity. *Langmuir*, **22**, 2–5.
18 Honda, K., Kawazumi, H., Yamada, S., Nakashima, N. and Niidome, Y. (2007) Extraction of hexadecytrimethylammonium bromide from gold nanorod solutions: adsorption of gold nanorods on anionic glass surfaces. *Trans. Mater. Res. Soc. Jpn.*, **32**, 421–424.
19 Niidome, Y., Honda, K., Higashimoto, K., Kawazumi, H., Yamada, S., Nakashima, N., Sasaki, Y., Ishida, Y. and Kikuchi, J.-i. (2007) Surface modification of gold nanorods with synthetic cationic lipids. *Chem. Commun.*, 3777–3779.
20 Huang, X., El-Sayed, I.H., Qian, W. and El-Sayed, M.A. (2007) Analysis of adenosine triphosphate and glutathione through gold nanoparticles assisted laser desorption/ionization mass spectroscopy. *Nano. Lett.*, **7**, 1591–1597.
21 Oyelere, A.K., Chen, P.C., Huang, X., El-Sayed, I.H. and El-Sayed, M.A. (2007) Peptide-conjugated gold nanorods for nuclear targeting. *Bioconj. Chem.*, **18**, 1490–1497.
22 Pissuwan, D., Valenzuela, S.M., Miller, C.M. and Cortie, M.B. (2007) A golden bullet? Selective targeting of toxoplasma gondii tachyzoites using antibody-functionalized gold nanorods. *Nano. Lett.*, **7**, 3808–3812.
23 Tong, L., Zhao, Y., Huff, T.B., Hansen, M.N., Wei, A. and Cheng, J.-X. (2007) Gold nanorods mediated tumor cell death by compromising membrane integrity. *Adv. Mater.*, **19**, 3136–3141.
24 Niidome, T., Yamagata, M., Okamoto, Y., Akiyama, Y., Takahashi, H., Kawano, T., Katayama, Y. and Niidome, Y. (2006) PEG-modified gold nanorods with a stealth character for *in vivo* application. *J. Controlled Release*, **114**, 343–347.
25 Arap, W., Pasqualini, R. and Rouslahti, E. (1998) Cancer treatment by targeted drug delivery to tumor vasculature in a mouse model. *Science*, **279**, 377–380.
26 Shadidi, M. and Sioud, M. (2003) Selective targeting of cancer cells using synthetic peptides. *Drug. Resist. Update*, **6**, 363–371.
27 Dharap, S.S., Wang, Y., Chandna, P., Khandare, J.J., Qiu, B., Gunaseelan, S., Sinko, P.J., Stein, S., Farmanfarmaian, A. and Minko, T. (2005) Tumor-specific targeting of an anticancer drug delivery system by LHRH peptide. *Proc. Natl. Acad. Sci. USA*, **102**, 12962–12967.

28 Schrama, D., Reisfeld, R.A. and Becker, J.C. (2006) Antibody targeted drugs as cancer therapeutics. *Nat. Rev. Drug Discov.*, **5**, 147–159.

29 Shukla, R., Thomas, T.P., Peters, J.L., Desai, A.M., Kukowska-Latallo, J., Patri, A.K., Kotlyar, A. and Baker, J.J.R. (2006) HER2 specific tumor targeting with dendrimer conjugated anti-HER2 mAb. *Bioconjug. Chem.*, **17**, 1109–1115.

30 Myc, A., Majoros, I.J., Thomas, T.P. and Baker, J.J.R. (2007) Dendrimer-based targeted delivery of an apoptotic sensor in cancer cells. *Biomacromolecules*, **8**, 13–18.

31 Sevick-Muraca, E.M., Houston, J.P. and Gurfinkel, M. (2002) Fluorescence-enhanced, near infrared diagnostic imaging with contrast agents. *Curr. Opin. Chem. Biol.*, **6**, 642–650.

32 Chan, W.C.W. and Nie, S. (1998) Quantum dot bioconjugates for ultra sensitive nonisotopic detection. *Science*, **281**, 2016–2018.

33 Bruchez, M. Jr, Moronne, M., Gin, P., Weiss, S. and Alvisatos, A.P. (1998) Semiconductor nanocrystals as fluorescent biological labels. *Science*, **281**, 2013–2014.

34 Åkerman, M.E., Chan, W.C.W., Laakkonen, P., Nhatia, S.N. and Ruoslahti, E. (2002) Nanocrystal targeting *in vivo*. *Proc. Nat. Acad. Sci. USA*, **99**, 12617–12621.

35 Wang, X., Yang, L., Chen, Z.G. and Shin, D.M. (2008) Application of nanotechnology in cancer therapy and imaging. *CA Cancer J. Clin.*, **58**, 97–110.

36 Hirsch, L.R., Jackson, J.B., Lee, A., Halas, N.J. and West, J.L. (2003) A whole blood immunoassay using gold nanoshells. *Anal. Chem.*, **75**, 2377–2381.

37 Hirsch, L.R., Stafford, R.J., Bankson, J.A., Sershen, S.R., Rivera, B., Price, R.e., Hazle, J.D., Halas, N.J. and West, J.L. (2003) Nanoshell-mediated near-infrared thermal therapy of tumors under magnetic resonance guidance. *Proc. Nat. Acad. Sci. USA*, **100**, 13549–13554.

38 Sokolov, K., Follen, M., Aaron, J., Pavlova, I., Malpica, A., Lotan, R. and Richards-Kortum, R. (2003) Real-time vital optical imagined of precancer using anti-epidermal growth factor receptor antibodies conjugated to gold nanoparticles. *Cancer Res.*, **63**, 1999–2004.

39 Hayat, M.A. (1989) *Colloidal Gold*, Academic Press, New York.

40 Mirkin, C.A., Letsinger, R.L., Mucic, R.C. and Storhoff, J.J. (1996) A DNA-based method for ratinally assembling nanoparticles into macroscopic materials. *Nature*, **382**, 607–609.

41 Elghanian, R., Storhoff, J.J., Mucic, R.C., Letsinger, R.L. and Mirkin, C.A. (1997) Selective colorimetric detection of polynucleotides base on the distance-dependent optical properties of gold nanoparticles. *Science*, **277**, 1078–1081.

42 Storhoff, J.J., Elghanian, R., Mucic, R.C., Mirkin, C.A. and Letsinger, R.L. (1998) One-pot colorimetric differentiation of polynucleotides with single base imperfections using gold nanoparticles. *J. Am. Chem. Soc.*, **120**, 1959–1964.

43 El-Sayed, I.H., Huang, X. and El-Sayed, M.A. (2005) Surface plasmon resonance scattering and absorption of anti-EGFR antibody conjugated gold nanoparticles in cancer diagnostics: applications in oral cancer. *Nano. Lett.*, **5**, 829–834.

44 Reynolds, A.J., Haines, A.H. and Russel, D.A. (2006) Gold glyconanoparticles for mimics and measurement of metal ion-mediated carbohydrate-carbohydrate interactions. *Langmuir*, **22**, 1156–1163.

45 Tkachenko, A.G., Xie, H., Coleman, D., Glomm, W., Ryan, J., Anderson, M.F., Franzen, S. and Feldheim, D.L. (2003) Multifunctional gold nanoparticle-peptide complexes for nuclear targeting. *J. Am. Chem. Soc.*, **125**, 4700–4701.

46 Tkachenko, A.G., Xie, H., Liu, Y., Coleman, D., Ryan, J., Glomm, W.R.,

Shipton, M.K., Franzen, S. and Feldheim, D.L. (2004) Cellular trajectroies of peptide-modified gold particle complexes: comparison of nuclear localiztion signals andpeptide transductiondomains. *Bioconj. Chem.*, **15**, 482–490.

47 Quintana, A., Raczka, E., Piehler, L., Lee, I., Myc, A., Majoros, I., Patri, A.K., Thomas, T., Mulé, J. and Baker, J.J.R. (2002) Design and function of a dendrimer-based therapeutic nanodevice targeted to tumor cells through the folate receptor. *Pharm. Res.*, **19**, 1310–1316.

48 Bhattacharya, R., Patra, C.R., Earl, A., Wang, S., katarya, A., Lu, L., Kizhakkedathu, J.N., Yaszemski, M.J., Greipp, P.R., Mukhopadhyay, D. and Mukherjee, P. (2007) Attaching folic acid on gold nanoparticles using noncovalent interaction via different polyehtylene glycol backbones and targeting of cancer cells. *Nanomedicine*, **3**, 224–238.

49 Huang, X., El-Sayed, I.H., Qian, W. and El-Sayed, M.A. (2006) Cancer cell imaging and photothermal therapy in the near-infrared region by using gold nanorods. *J. Am. Chem. Soc.*, **128**, 2115–2120.

50 Eghtedari, M., Oraevsky, A., Copland, J.A., Kotov, N.A., Conjusteau, A. and Motamedi, M. (2007) High sensitivity of *in vivo* detection of gold naonrods using a laser optoacoustic imaging system. *Nano. Lett.*, **7**, 1914–1918.

51 Link, S. and El-Sayed, M.A. (2000) Shape and size dependent of radiative, non-radiative and photothermal properties of gold nanocrystals. *Int. Rev. Phys. Chem.*, **19**, 409–453.

52 Mohamed, M.B., Volkov, V., Link, S. and El-Sayed, M.A. (2000) The 'lighting' gold nanorods: fluorescence enhancement of over a million compared to the gold metal. *Chem. Phys. Lett.*, **317**, 517–523.

53 Wang, H., Huff, T.B., Zweifel, D.A., He, W., Low, P.S., Wei, A. and Cheng, J.-X. (2005) *In vitro* and *in vivo* two-photon luminescence imaging of single gold nanorods. *Proc. Nat. Acad. Sci. USA*, **102**, 15752–15756.

54 Huff, T.B., Hansen, M.N., Zhao, Y., Cheng, J.-X. and Wei, A. (2007) Controlling the cellular uptake of gold nanorods. *Langmuir*, **23**, 1596–1599.

55 Huff, T.B., Tong, L., Zhao, Y., Hansen, M.N., Cheng, J.-X. and Wei, A. (2007) Hyperthermic effects of gold nanorods on tumor cells. *Nanomedicine*, **2**, 125–132.

56 Niidome, T., Akiyama, Y., Shimoda, K., Kawano, T., Mori, T., Katayama, Y. and Niidome, Y. (2008) *In vivo* monitoring of intravenously injected gold nanorods using near-infrared light. *Small*, **4**, 1001–1007.

57 Foss, C.A. Jr, Hornyak, G.L., Stochkert, J.A. and Martin, C.R. (1994) Template-synthesis nanoscopic gold particles: optical spectra and the effects of particle size and shape. *J. Phys. Chem.*, **98**, 2963–2971.

58 Ding, H., Yong, K.-T., Roy, I., Pudavar, H.E., Law, W.C., Bergy, E.J. and Prasad, P.N. (2007) Gold nanorods coated with mulilayer polyelectrolyte as contrast agents for multimodal imaging. *J. Phys. Chem. C*, **111**, 12552–12557.

59 Stone, J.W., Sisco, P.N., Goldsmith, E.C., Baxter, S.C. and Murphy, C.J. (2007) Using gold nanorods to probe cell-induced collagen deformation. *Nano Lett.*, **7**, 116–119.

60 Takahashi, H., Niidome, T., Nariai, A., Niidome, Y. and Yamada, S. (2006) Photothermal reshaping of gold nanorods prevents further cell death. *Nanotechnology*, **17**, 4431–4435.

61 Takahashi, H., Niidome, T., Nariai, A., Niidome, Y. and Yamada, S. (2006) Gold nanorod-sensitized cell death: microscopic observation of single living cells irradiated by pulsed near-infrared laser light in the presence of gold nanorods. *Chem. Lett.*, **35**, 500–501.

62 Pérez-Juste, J., Pastoriza-Santos, I., Liz-Marzán, L.M. and Mulvaney, P. (2005)

Gold nanorods: synthesis, characterizaiton and application. *Coord. Chem. Rev.*, **249**, 1870–1901.

63 Jain, P.K., El-Sayed, I.H. and El-Sayed, M.A. (2007) Au nanoparticles target cancer. *Nanotoday*, **2**, 18–29.

64 Murphy, C.J., Gole, A.M., Hunyadi, S.E., Stone, J.W., Sisco, P.N., Alkilany, A., Kinard, B.E. and Hankins, P. (2008) Chemical sensing and imagind with metalic nanorods. *Chem. Commun.*, 544–557.

35
Dynamic Motion of Single Cells and its Relation to Cellular Properties

Hideki Matsune, Daisuke Sakurai, Akitomo Hirukawa,
Sakae Takenaka, and Masahiro Kishida

35.1
Introduction

35.1.1
Single Cell Analysis

Cell behavior is induced by external stimuli surrounding the cell. Some of the external factors involve chemical stimuli from cytokines, growth factors, hormones, and nutrients; and biological stimuli derived from cell–cell and cell–matrix contacts. The external factors affect the cell simultaneously and induce internal cascade reactions consisting of many elementary processes. The reactions propagate three-dimensionally inside the cell. As a result, the cell displays some recognizable temporal behavior including pseudopod formation, cell motion and morphological transformation. Further repetition of the stimuli from the external factors to the cell sometimes induces dynamic responses, including cell proliferation, cell differentiation, and apoptosis. (Figure 35.1) Biological researchers have a great interest in formulating an overall picture of the programmed cellular system on the basis of the relationship between individual external factors and cellular responses. The understanding of the system is connected to the design of tissues and organisms. However, even if a uniform culture condition is prepared for the cell assembly, microenvironments for individual cells may be heterogeneous. For ensemble cultivation, a broad distribution is expected for the concentration of oxygen, nutrient, cytokines, and also for the coordination number of contacting cells. The individual cells must receive various localized reactions and input signals. As a result, ensemble measurements yield a broad distribution of the cellular behavior. One must often investigate input–output relationships from an average value of the broad distribution. To reveal an effect of the external factors on individual cell behavior, it is important to investigate the input–output relationship one by one precisely, after eliminating the other input factors as much as possible (Figure 35.2). Single-cell analysis is a simple

Molecular Nano Dynamics, Volume II: Active Surfaces, Single Crystals and Single Biocells
Edited by H. Fukumura, M. Irie, Y. Iwasawa, H. Masuhara, and K. Uosaki
Copyright © 2009 WILEY-VCH Verlag GmbH & Co. KGaA, Weinheim
ISBN: 978-3-527-32017-2

External Stimuli

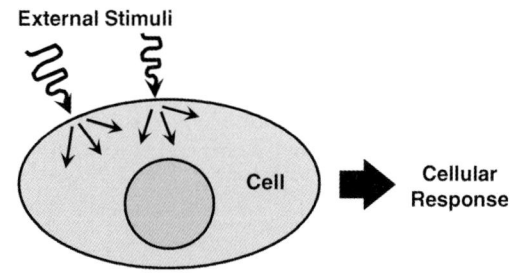

Figure 35.1 Cellular behavior originating from the external stimuli surrounding the cell.

analytical system compared to ensemble measurement because individual cells are placed in a situation affected by restricted external factors, where the biological signals originating from cell–cell contact are negligible. A uniform cellular microenvironment can be provided to an isolated cell more easily than to the ensemble. Nowadays, single-cell techniques have become an emerging area in the field of biotechnology [1, 2]. Optical tweezers have a limited number of trapping objects, but provide high resolution for trapping single cells [3]. They are suitable for a precise analysis of a single cell. More details on the application of optical tweezers for biological cells will be given later.

35.1.2
Dynamic Motion of Murine Embryonic Stem Cell

Murine embryonic stem (mES) cells are pluripotent cell lines established directly from the early embryo [31, 32]. Since the early 1980s, numerous mES cell lines from several strains have been isolated and maintained *in vitro* according to

External factors surrounding cell
1. **Chemical factors;** growth factors, cytokines, nutrients, hormones
2. **Physical factors;** mechanical force, interaction with a substrate, distribution of soluble factors
3. **Biological factors;** cell-cell interaction, cell-matrix (ECM) interaction and
   oxygen, temperature, pH •••

For investigation of interaction between cell and matrix

comparison

floating single cell   adhering cell

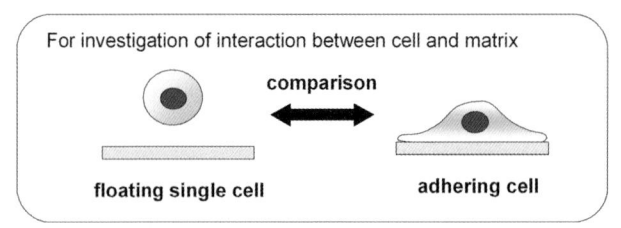

Figure 35.2 A number of external factors affecting cellular behavior simultaneously.

various protocols [4, 5]. *In vitro*, mES showed the capacity to reproduce the various somatic cell types and develop into the germ lines. When maintained in the presence of leukemia inhibitory factor (LIF) or in coculture with fibroblastic feeder cells, mES cells retain their pluripotency and are capable of self-renewal [6, 7]. In addition to the self-renewal and pluripotency, mES cells are characterized by a high frequency of morphological change. mES cells placed on a dish show remarkable cellular motion such as extending and spreading of the cellular surface. We have great interest in how the real-time motion of the cell is related to the cellular ability, including proliferative and differentiation capacities. The relationship between the dynamic motion and the properties or functions of biological cells allows us to predict the ability of the cells, if any, before cultivation. This is important knowledge in the field of cell biology. However, investigation of the relationship needs precise observation and long-term monitoring of individual single cells in the foregoing culture. Recent progress in techniques for culture and observation of cells makes it easier to carry out the real-time observation and long-term monitoring of individual cells.

35.2
Laser Trapping of Biological Cells

35.2.1
Optical Tweezers

A single gradient laser beam traps a particle near its focus, which is sometimes referred to as "optical tweezers". Since the first single-beam trapping of a micrometer-sized bead by Ashkin *et al.* in 1986, it has emerged as a powerful tool for micromanipulation in biology [8–10]. In a typical experiment, an IR continuous-wave laser beam is focused close to the diffraction limit using a high numerical aperture microscope objective lens. A light gradient near the focal region of a near-IR laser beam gives rise to the forces of radiation pressure that make possible a stable trap of micron-sized particles, including living biological cells and organelles within cells. Trapped biological particles have included *E. coli* bacterium, yeast cell, lymphocytes, red blood cells, and macrophages [12, 13]. Trapping with IR laser light causes less optical damage to living cells than does that with visible laser light [11]. The yeast cell and bacteria even in the IR traps showed cell division [11]. Multiple optical traps created by scanning one single beam trap along a variable number of positions were applied for the rotation and bending of a filamentous *E. coli* bacterium [14]. Recently, holographic optical tweezers have been used to provide a number of trapping sites easily and rapidly, and allowed precise manipulation and controlled rotation of biological objects [15]. The optical tweezers have been applied for the manipulation and sorting of cells [16, 17]. The combination of the optical traps with a pulsed UV laser microbeam has been used for precise cell fusions of pairs of selected cells. [2.4] The optical trapping of biological cells has been combined with Raman imaging or spectroscopy [18–22]. In order to apply the trapping techniques to living mammalian cells, the required condition for a set-up is a combination of the optical trap with a cultivation chamber utilizable on a microscope, where the culture conditions such as temperature and atmosphere can be controlled. We have produced a cultivation chamber combined with optical tweezers and applied it to the cultivation of a single mES cell trapped with optical tweezers.

35.2.2
Set-up for Optical Trapping of a Living Cell

Figure 35.3 shows a schematic representation of our experimental optical set-up, which is a dual single-beam gradient optical tweezers similar to that reported by Fallman and Axner [23]. The expanded and parallel polarized laser beam (in TEM_{00} mode) from a Nd:YAG laser (wavelength = 1064 nm) (Compass 1064–1500N, Coherent Japan, Inc.) was directed into an invert microscope (Olympus IX 71) from the side port and was focused by a high numerical aperture (NA) microscope objective (UplanApo 100X/NA 1.30). Figure 35.1b shows a home-made chamber for the cultivation of the cells on the microscope. The combination of the optical tweezers and the chamber allows us to culture the trapped cell for long time. Our chamber consisted of a cover glass plate (30 mm × 60 mm × 0.2 mm), silicon rubber (30 mm × 60 mm × 5 mm) with a flow channel (0.01 mm × 35 mm × 5 mm) and a plastic cover. The chamber was maintained under 5% CO_2 atmosphere at 37 °C. (Figure 35.4) The medium exchange chamber is connected to a peristaltic pump that continuously circulates a fresh medium, keeping the nutrient conditions around the cells constant. The trapped single cells were observed morphologically, and cell conformation was continuously recorded using a charge-coupled device (CCD) camera. Time-sequential images were obtained at regular time intervals of 5 min and fed into a personal computer. To avoid unexpected heating of the chamber, the mechanical shutter was positioned after the halogen lamp for bright-field observation, and was open only while capturing the bright field images.

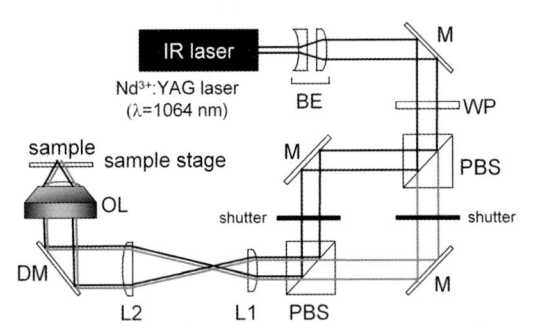

Figure 35.3 Schematic layout for a dual-trap optical tweezers system. BE, beam expander; WP, half-wave plate; PBS, polarizing beam-splitting cube; L1, L2, lenses 1, 2; M, Mirror; OL, objective lens; DM, dichromic mirror.

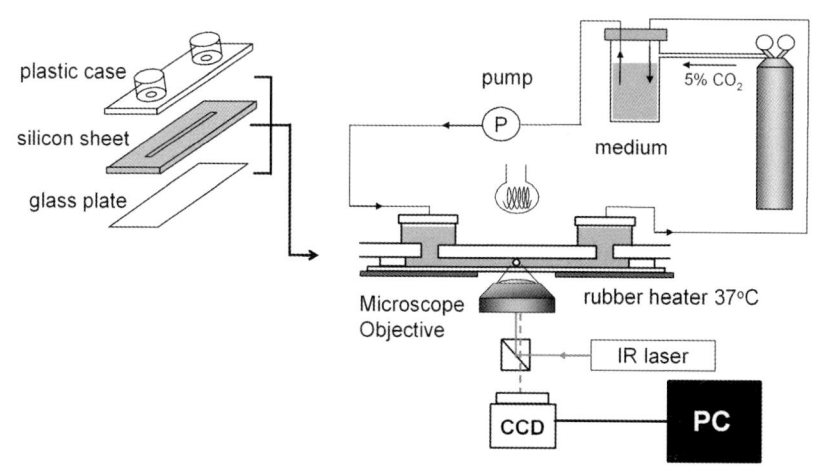

Figure 35.4 Set-up of flow chamber for the cultivation of a mammalian cell on a microscope, which maintains the culture condition such as temperature, atmosphere and nutrient concentration of a medium. The cell trapped with optical tweezers can be cultured within the chamber.

35.2.3
Murine Embryonic Stem Cell Trapped with Optical Tweezers

A single mES cell trapped with optical tweezers was manipulated into the small compartment prepared beside a channel of the flow chamber to reduce the possibility of it escaping from the laser focal spot by the pressure of a flowing medium. We demonstrated the single-cell observation trapped with the optical tweezers, and continuously monitored the cell behavior in the chamber. After trapping, mES cells showed dynamic conformational changes in a few minutes. The trapped cell at the laser focus repeatedly extended and shortened the pseudopods in a medium. The dynamic conformational change of the cell stopped a few minutes after the trapping, but there remained a small fluctuation of the surface for more than 10 h. Half of cells in the culture dish showed the dynamic shape change, which was clearly different from thermal Brownian motion. The Brownian variation of 10 μm-sized polystyrene is calculated as less than 30 nm. This suggests that the trapped ES cell survived more than 10 h of trapping. After 12 h of the trapping, the ES cell started a dynamic conformational change again, and began to divide into two cells. (Figure 35.5). Finally, the two cells aggregated into one. Generally, ES cells tend to make aggregates and to form a colony or embryonic body. The soft surfaces of the ES cells make the cell–cell interfaces unclear. However, this demonstrates that optical trapping is a promising technique for the precise investigation of single biological cells suspended in a medium. This technique makes possible the real-time monitoring of cellular behavior for individual single cells. Here, we also have an interest in the real-time motion of the mES cell. The individual mES cells floated with optical tweezers showed remarkable dynamic motion of their morphologies, but the degrees and

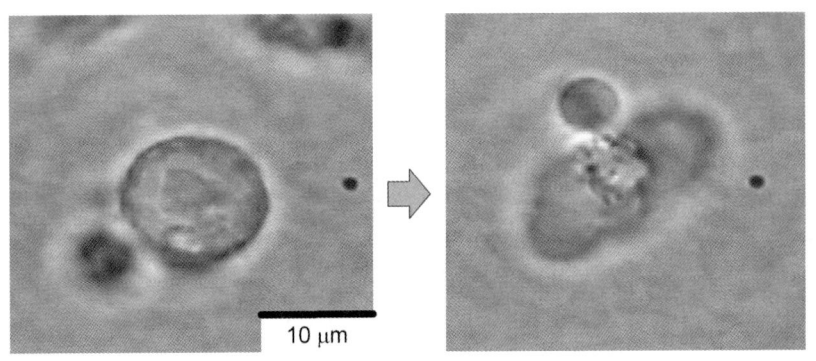

Figure 35.5 The bright-field images of morphological change in the mES cell after 12 h of optical trapping.

frequencies of them were different from each other. We considered that this short-term behavior of individual mES cells originated from an intrinsic factor, and that the difference in the motility was caused by cellular activity. If a relationship between the dynamic motion and cellular activity of mES exists, it is important to reveal it quantitatively. In the next section we describe an investigation of the relationship between the dynamic motion that the floating cell showed and the cellular properties including proliferation and differentiation ability.

35.3
Relationship Between Cellular Motion and Proliferation

It is ideal to reach an understanding of cellular properties only by precise analysis of a particular single cell. Unfortunately, the cell often shows a broad distribution of behavior even if a uniform condition is provided, probably because of the heterogeneous cellular state and microenvironment around cells. A large number of single cell analyses are needed to investigate cellular properties. From the viewpoint of measurement efficiency, it is preferable to perform single-cell analysis and ensemble measurement in parallel. Below, we investigate the relationship between cellular motion and proliferation by a combination of single-cell analysis and ensemble measurement.

35.3.1
Dynamic Motion of a Murine Embryonic Stem Cell [28]

The mES cells placed on a dish show remarkable cellular motion such as extending and spreading of the cellular surface (Figure 35.6). It is considered that the cellular motions offer physiological information on the state and potential of cells. Kino-oka et al. reported the relation between individual cellular motions and growth potentials for various cells through time-lapse observation [24–27]. The state of the cell could be estimated without staining. However, a relatively long-time observation of several

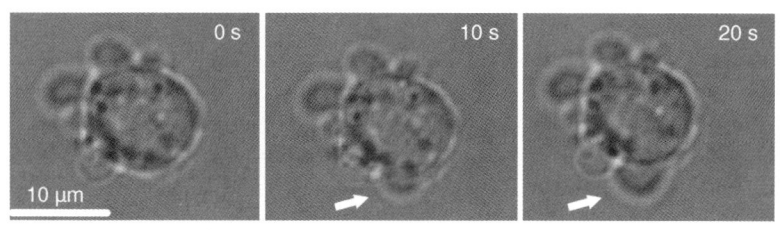

Figure 35.6 Example of the ES cell exhibiting morphological changes within 1 min. The surface (arrowed) of the mES cells changes. A cell exhibiting morphological changes in 1 min was defined as a cell with high activity. This figure was reproduced by permission of the Society for Biotechnology, Japan.

hours was required to determine the state of the cells. In this study, we propose a new parameter to evaluate the state of individual mES cells. The real-time morphological changes in a mES cell at the early stage of a subculture are related to the foregoing cellular proliferation. This reveals that a one minute observation of an individual mES cell is sufficient to estimate the proliferative potential.

35.3.2
Experimental Procedure

Our experimental procedure consisted of two parts, that is, microscopic observation for the estimation of cellular activity and cultivation for the evaluation of growth rate. (Figure 35.7) The frozen mES cells in the vial were thawed and then suspended in a medium. An aliquot of the mES cells was replaced on the glass-bottom dish in Cell Saver W1 (Waken Electronics, Japan), an instrument to maintain the cells at 37 °C in

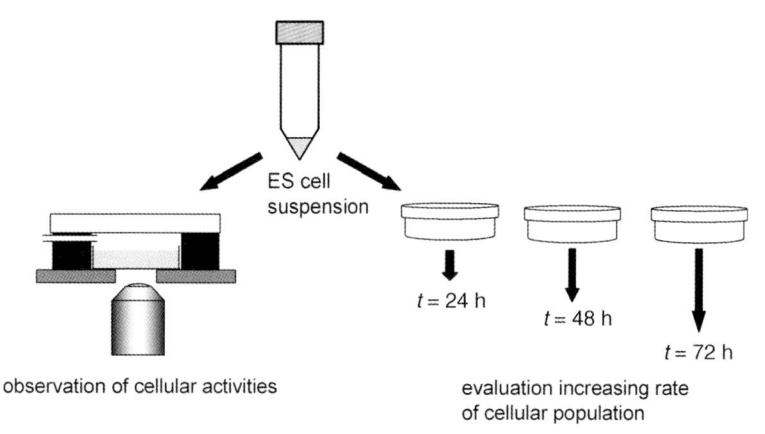

Figure 35.7 Schematic drawing of experimental procedure. The suspended ES cells were divided into two, for observation of cellular morphology and estimation of an increasing ratio of cellular population in the subsequent subculture. This figure was reproduced by permission of the Society for Biotechnology, Japan.

humidified air with 5% CO_2 on an inverted microscope. We microscopically monitored the real-time motion of the mES cells with a ×100 objective. The cellular motility was estimated by one minute observation for each cell. Here we define a cell exhibiting morphological changes within 1 min as a cell with high activity. On the contrary, a cell showing no change in morphology in 1 min was defined as a cell with low activity. We monitored the dynamic behavior of 40–50 individual cells in randomly picked areas. The frequency of the cell with high activity (α) in the suspension was evaluated from Eq. (35.1).

$$\alpha = \frac{\text{number of cells with high activity}}{\text{total number of cells examined}} \qquad (35.1)$$

This evaluation was conducted at the beginning of each subculture in the same manner.

Next, in order to estimate the increasing ratio of the mES cells in the subculture, they were seeded into three gelatin-coated dishes, 35 mm in diameter, at 2×10^5 cells per 0.85 mL of the culture medium containing LIF. The dishes were maintained in a 5% CO_2-containing humidified incubator to estimate the increasing number of cells. The whole medium was renewed every 24 h. The number of cells, N_t at a given culture time, t, was counted by the trypan-blue exclusion test on a hemocytometer under an optical microscope. The dishes were used for estimation of the increasing number of cells at culture times of 24, 48 and 72 h, respectively. An aliquot of the cells after 72 h culture was subjected to the next experiment consisting of observation and subsequent subculture as outlined above. The experiments were repeated 15 to 18 times for investigation of the relationship between the cellular activity and the proliferation potential. Alkaline phosphatase activity was demonstrated to confirm the retention of undifferentiation of the examined mES cells.

We first examined the activities of the mES cells from each suspension placed on a glass-bottom dish. The mES was characterized by the high frequency of extending the projections (Figure 35.6). Here, we defined the mES cells extending the projections, recognizable by direct microscopic observation with a ×100 objective, as cells with high activity, and then the α value was evaluated from Eq. (35.1) for each suspension before a subculture. The α values ranged from 0.35 to 0.85. At the same time as the estimation of the α values, an aliquot of the suspension before the subculture was conducted to the trypan-blue (TB) exclusion test that was generally used to distinguish between living cells and dead cells in suspension. Figure 35.8 shows the plots of the frequency of TB-negative (living) cells against the α values of the suspensions. The α values were less dependent on the frequency of TB-negative cells, which was relatively constant among the suspensions.

Figure 35.9 shows the plots of increasing ratios of cellular population in the subculture, N_{24}/N_0 (from 0 to 24 h) and N_{72}/N_0 (from 0 to 72 h) against α. This demonstrates that N_{24}/N_0 and N_{72}/N_0 were proportional to the α values, as shown by the dashed line and solid line ($N_{24}/N_0 = 1.9\alpha$ and $N_{72}/N_0 = 9.8\alpha$). The relationship shown in Figure 35.9 is considered to reflect that the cellular activity is related to the cell proliferation, that is, the proliferative potential of a cell exhibiting morphological changes in 1 min was higher in the subculture than that of a cell exhibiting no changes. Kino-oka et al. reported the relation between the mean value of the rotation

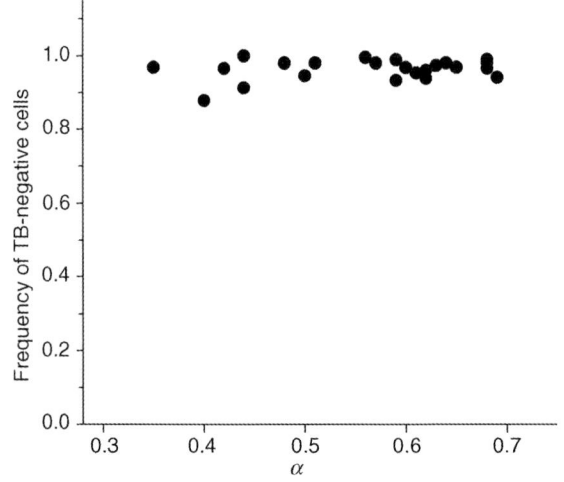

Figure 35.8 Frequency of trypan-blue (TB) negative cells against the α values before subculture. The α values were evaluated by direct observation according to Eq. (35.1). The TB-stained cells were counted twice by trypan-blue exclusion test among about 100 cells. This figure was reproduced by permission of the Society for Biotechnology, Japan.

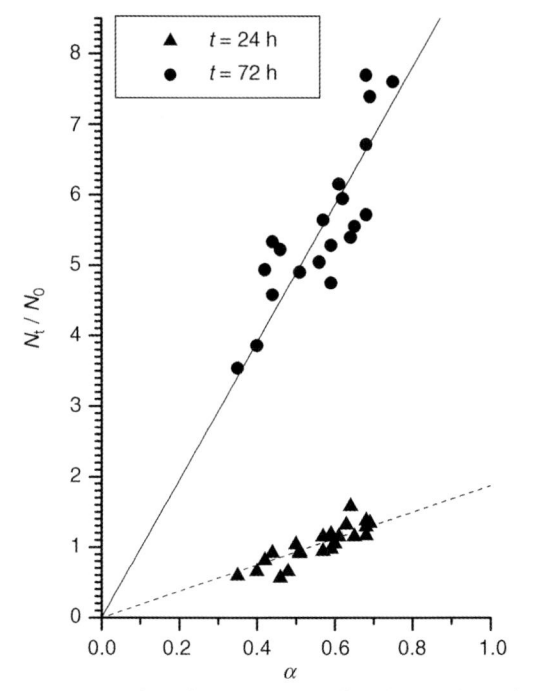

Figure 35.9 Plots of increasing ratio of cellular population from 0 to t h, N_t/N_0 against the α values closed circles; $t = 72$ h, closed triangle; $t = 24$ h. The line was drawn using the equation ($\beta = 0.89$ and $t_D = 16$ h). Adapted from Ref. [28].

rate of human keratinocyte and the doubling time. (8) The mean value of the rotation rate decreased with an increase in doubling time caused by the progress of cellular age. In our case, the passage numbers (P) of the examined mES cell were between $P=15$ and $P=18$. The α values seem to be independent of the passage number.

Here, we propose a kinetic model of the relationship between α and N_t/N_0 in Figure 35.9. Assuming that only the mES cell with high activity is capable of cell proliferation and divides into two daughter cells once in 25 h and that the mES cell with a low activity, on the other hand, is going to die without cell division, the increasing ratio of cellular population from 0 to 25 h is given as follows:

$$N_{25}/N_0 = 2\alpha \tag{35.2}$$

The Eq. (35.2) is in accord with the experimental result, shown in Figure 35.4. Furthermore, the data of N_{72}/N_0 were proportional to the α values, as were those of N_{24}/N_0. This implies that the doubling time and the population ratio of the viable cells were constant after 24 h. Accordingly, the increasing ratio of cellular population (N_t/N_0) from 0 to an overall culture time, t (h), is given as follows:

$$N_t/N_0 = 2\alpha \times (2\beta)^{(t-24)/t_D} \tag{35.3}$$

where t_D (h) is a doubling time after $t=24$ h; β is a population ratio of the cells capable of next cell division after $t=24$ h, which was estimated from the number of dead cells floating in the medium exchanged for a fresh one at 48 and 72 h, as shown by Eq. (35.4).

$$\beta = \frac{\text{number of dead cells floating in medium}}{\text{total number of cells seeded}} \tag{35.4}$$

The amount of dead cells was constant at 11% every 24 h, which corresponds to $\beta=0.89$. The data of N_{72}/N_0 against α in Figure 35.4 were fitted by Eq. (35.3) using the nonlinear least squares method. The fitting result showed $t_D=17.5$ h, which was different from the value of 24 h for the doubling time for the first cell division in the model mentioned above. This might be derived from the lag time, characteristic of a first division.

The cell behavior offers important information concerning cellular activity. One-minute observation of morphological change in the mES cell before seeding is a nondestructive and noninvasive method, but sufficient to predict the foregoing proliferation without staining. In general, viable cells were estimated by a trypan-blue staining. However, it is not sufficient to predict the proliferation rate of the cell. The frequency of morphological change of the mES cell is proposed as a new parameter to estimate the viable cells. In our experimental results, the increasing ratio of cellular population of the mES cells was proportional to the α values. In addition, the α values were mainly concerned with the potential for the first cell division in 24 h, but not with the proliferation potential after the second cell division.

35.4 Cell Separation by Specific Gravity

35.4.1 Cell Separation

The cell separation process plays an important role in many biological applications. However, the current cell separation technique, including FACS and MACS, needs expensive instruments. In addition, the antibodies are consumed for the labeling of the cells, whenever the separation instruments are used. On the other hand, centrifugation is a classical technique, but separates cells more easily on the basis of specific gravity. We have new insight into the relation between the specific gravity of the cells and the differentiation capacity. Two populations of murine ES cells divided by centrifugation were individually cultured for further *in vitro* differentiation. Figure 35.10 shows the experimental procedure. In the first step of the separation procedure, a small amount of cell dispersion was placed on top of the medium filled in a centrifuge tube. Immediately after that, the cells were collected using centrifugation at 500 rpm for 5 min to obtain the cells with high specific gravity. Next, the cells were collected from the supernatant by centrifugation at 1000 rpm for 5 min to obtain the cells with low specific gravity. The cells from the two cellular populations were injected into 96 round-bottom well plates at 500 cells/well to form embryonic bodies (EBs). The 5-day-old EBs were transferred onto gelatin-coated dishes, which is one of major procedures to differentiate into cardiac cells [30]. A regular occurrence in ES differentiation cultures is the development of foci of cells within EBs that begin rhythmic contractions (beating), which are an indication of cardiac muscle development [7, 29]. The EBs transferred to the dish were periodically observed to detect the onset of cardiac myogenesis.

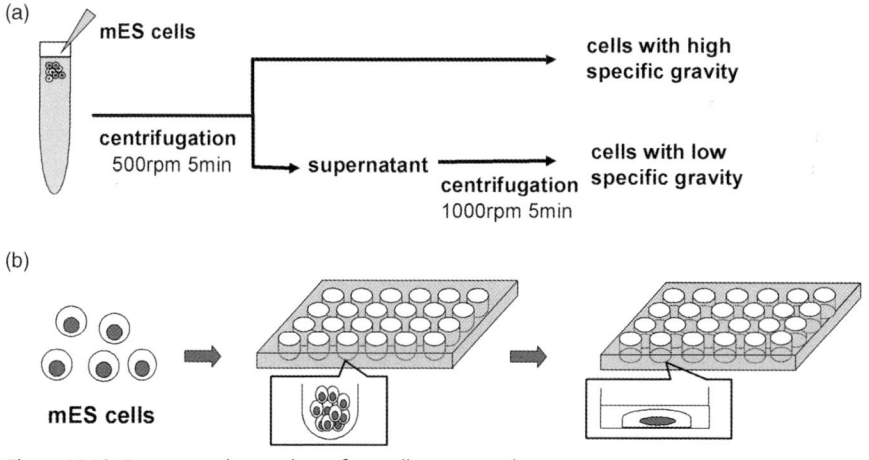

Figure 35.10 Experimental procedure of (a) cell separation by centrifugation into two parts and (b) further culture process for differentiation into cardiac muscle.

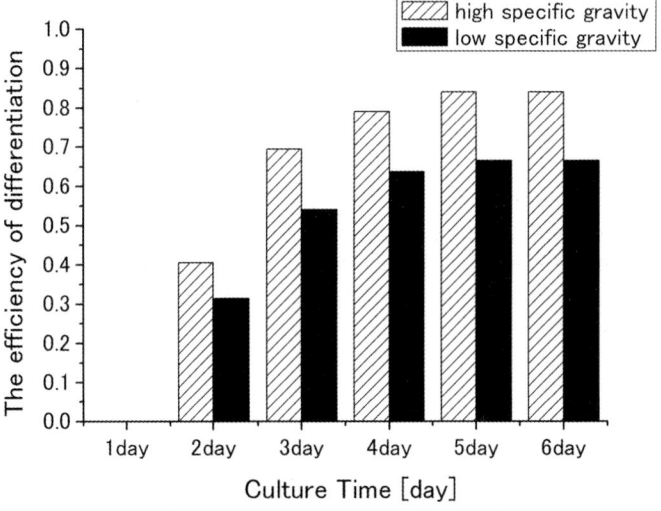

Figure 35.11 The efficiency of generation of beating muscle against culture time.

Figure 35.11 shows the efficiency of generation of beating EBs against culture time. The beating EBs first appeared at 2 d of attachment culture for the EBs derived from the ES cells with both high and low specific gravity. However, the efficiency of production of beating EBs from the cells with high specific gravity finally reached 84%, which is higher than the 67% for the EBs with low specific gravity. This suggests that the differentiation ability of ES with high specific gravity is higher than that of ES with low specific gravity. Here, the differentiation ability into cardiac muscle generally depends on the size of the EBs. In our experiment, the average size of five day old EBs was 472 μm for those with high specific gravity, which is similar to the average size of 464 μm for EBs with low specific gravity. These findings indicate that the centrifugation might allow us to separate the ES cell ensemble into two populations with higher and lower differentiation ability, in spite of it being a very simple process. A further precise investigation and alignment of the procedure is necessary for separation of cells at a high level, we propose it as a new technique to divide cells into groups with different differentiation abilities.

35.5
Summary

The optical tweezers technique has been a powerful tool for biological application. We believe that precise control of the cellular microenvironment and single-cell analysis provide opportunities to predict the effects of external stimuli including cell–cell, cell–ECM and cell–soluble factor interaction on the cell behavior and fate, which are link to revealing the internal cellular signaling system. There still exists a broad distribution of cell responses even by single-cell analysis. Researchers need to improve and develop the technique to one utilizable for a precise analysis. The understanding of specific effects of external factors has application for the optimization of culture conditions for tissue engineering.

Acknowledgments

The authors thank Associate Professor Mizumoto and coworkers of Kyushu University for technical advice. This study was partially supported by a Grand-in-Aid for Scientific Research (KAKENHI) on Priority Areas "Molecular Nano Dynamics" (No. 16350012) from the Ministry of Education, Culture, Sports, Science, and Technology (MEXT) of Japan.

References

1 Khademhosseini, A., Langer, R., Borenstein, J. and Vacanti, J.P. (2006) Microscale technologies for tissue engineering. *Proc. Natl. Acad. Sci. USA*, **103**, 2480–2487.

2 Carlo, D.D., Aghdam, N. and Lee, L.P. (2006) Single-cell enzyme concentrations, kinetics, and inhibition analysis using high-density hydrodynamic cell isolation arrays. *Anal. Chem.*, **78**, 7918–7925.

3 Ashkin, A. (2000) History of optical trapping and manipulation of small-neutral particle, atoms, and molecules. *IEEE J. Sel. Top. Quant.*, **6**, 841–856.

4 Wobus, A.M. and Boheler, K.R. (2005) Embryonic stem cells: Prospects for developmental biology and cell therapy. *Physiol. Rev.*, **85**, 635–678.

5 Smith, A.G., Heath, J.K., Donaldson, D.D., Wong, G.G., Moreau, J., Stahl, M. and Rogers, D. (1988) Inhibition of pluripotential embryonic stem cell differentiation by purified polypeptides. *Nature*, **336**, 688–690.

6 Williams, R.L., Hilton, D.J., Pease, S., Willson, T.A., Stewart, C.L., Gearing, D.P., Wagner, E.F., Metcalf, D., Nicola, N.A. and Gough, N.M. (1988) Myeloid leukaemia inhibitory factor maintains the developmental potential of embryonic stem cells. *Nature*, **336**, 684–687.

7 Keller, G.M. (1995) *In vitro* differentiation of embryonic stem cells. *Curr. Opin. Cell Biol.*, **7**, 862–869.

8 Ashkin, A. (1970) Acceleration and trapping of particles by radiation pressure. *Phys. Rev. Lett.*, **24**, 156–159.

9 Ashkin, A., Dziedzic, J.M., Bjorkholm, J.E. and Chu, S. (1986) Observation of a single-beam gradient force optical trap for dielectric particles. *Opt. Lett.*, **11**, 288–290.

10 Ashkin, A. (1987) Optical trapping and manipulation of viruses and bacteria. *Science*, **235**, 1517–1520.

11 Ashkin, A., Dziedzic, J.M. and Yamane, T. (1987) Optical trapping and manipulation of single cells using infrared laser beams. *Nature*, **330**, 769–771.

12 Berns, M.W., Wright, W.H., Tromberg, B.J., Profeta, G.A., Andrews, J.J. and Walter, R.J. (1989) Use of a laser-induced optical force trap to study chromosome movement on the miotic spindle. *Proc. Natl. Acad. Sci. USA*, **86**, 4539–4543.

13 Ashkin, A. and Dziedzic, J.M. (1989) Internal cell manipulation using infrared-laser traps. *Proc. Natl. Acad. Sci. USA*, **86**, 7914–7918.

14 Visscher, K., Brakenhoff, G.J. and Krol, J.J. (1993) Micromanipulation by multiple optical traps created by a single fast scanning trap integrated with bilateral confocal scanning laser microscope. *Cytometry*, **14**, 105–114.

15 Dufresne, E.R., Spalding, G.C., Dearing, M.T., Sheets, S.A. and Grier, D.G. (2001) Computer-generated holographic optical tweezer arrays. *Rev. Sci. Instrum.*, **72**, 1810–1816.

16 Buican, T.N., Smyth, M.J., Crissman, H.A., Salzman, G.C., Stewart, C.C. and Martin, J.C. (1987) Automated single-cell manipulation and sorting by light trapping. *Appl. Opt.*, **26**, 5311–5316.

17 Grover, S.C., Skirtach, A.G., Gauthier, R.C. and Grover, C.P. (2001) Automated single-cell sorting system based on optical trapping. *J. Biomed. Opt.*, **6**, 23–30.

18 Xie, C., Mace, J., Dinno, M.A., Li, Y.Q., Tang, W., Newton, R.J. and Gemperline, P.J. (2005) Identification of single bacterial cells in aqueous solution using confocal laser tweezers Raman spectroscopy. *Anal. Chem.*, **77**, 4390–4397.

19 Creely, C.M., Volpe, G., Singh, G.P., Soler, M. and Petrov, D.V. (2005) Raman imaging of floating cells. *Opt. Express*, **13**, 6105–6110.

20 Creely, C.M., Singh, G.P. and Petrov, D. (2005) Dual wavelength optical tweezers for confocal Raman spectroscopy. *Opt. Commun.*, **245**, 465–470.

21 Ajito, K. and Torimitsu, K. (2002) Laser trapping and Raman spectroscopy of single cellular organelles in the nanometer range. *Lab on a Chip*, **2**, 11–14.

22 Uzunbajakava, N., Lenferink, A., Kraan, Y., Willekens, B., Vrensen, G., Greve, J. and Otto, C. (2003) Nonresonant Raman imaging of protein distribution in single human cells. *Biopolymers*, **72**, 1–9.

23 Fallman, E. and Axner, O. (1997) Design for fully steerable dual-trap optical tweezers. *Appl. Opt.*, **36**, 2107–2113.

24 Kino-oka, M., Agatahama, Y., Hata, N. and Taya, M. (2004) Evaluation of growth potential of human epithelial cells by motion analysis of pairwise rotation under glucose-limited condition. *Biochem. Eng. J.*, **19**, 109–117.

25 Umezaki, R., Murai, K., Kino-oka, M. and Taya, M. (2002) Correlation of cellular life span with growth parameters observed in successive cultures of human keratinocytes. *J. Biosci. Bioeng.*, **94**, 231–236.

26 Hirai, H., Umezaki, R., Kino-oka, M. and Taya, M. (2002) Characterization of cellular motions through direct observation of individual cells at early stage in anchorage-dependent culture. *J. Biosci. Bioeng.*, **94**, 351–356.

27 Hata, N., Agatahama, Y., Kino-oka, M. and Taya, M. (2005) Relations between individual cellular motions and proliferative potentials in successive cultures of human keratinocytes. *Cytotechnology*, **47**, 127–131.

28 Matsune, H., Sakurai, D., Niidome, Y., Takenaka, S. and Kishida, M. (2008) Relationship between degree of dynamic morphological change and proliferative potential of murine embryonic stem cells. *J. Biosci. Bioeng.*, **105**, 58–60.

29 Kurosawa, H. (2007) Methods for inducing embryoid body formation: *in vitro* differentiation system of embryonic stem cells. *J. Biosci. Bioeng.*, **103**, 389–398.

30 Kurosawa, H., Imamura, T., Koike, M., Sasaki, K. and Amano, Y. (2003) A simple method for forming embryoid body from mouse embryonic stem cells. *J. Biosci. Bioeng.*, **96**, 409–411.

31 Evans, M.J. and Kaufman, M.H. (1981) Establishment in culture of pluripotential cells from mouse embryos. *Nature*, **292**, 154–156.

32 Martin, G. (1981) Isolation of a pluripotent cell line from early mouse embryos cultured in medium conditioned by teratocarcinoma stem cells. *Proc. Natl. Acad. Sci. USA*, **78**, 7634–7638.

Index

a
ab initio methods   357ff.
– DFT calculations   370
absorbtion
– acceptor   6
– band-edges   136
– bleached   46
– bulk   350
– coefficient   14, 386, 388
– condensed media   103
– cross-section   14, 136
– cyclic-multiphoton absorption mechanism   212
– fluence   348
– four-photon absorption cross-section   136ff.
– induced   46
– IR beam   79
– mid-infrared   25
– multiphoton   136f.
– nonlinear   156f.
– one-photon   14, 138
– polarized absorption spectra   261ff.
– steady-state spectra   148
– three-photon absorption cross-section   136, 138
– transient absorption spectra   271, 549
– two-photon absorption   156, 591
– two-photon absorption cross-section   156ff.
– UV   176
activation
– barrier   319, 329, 331
– energy   319, 324, 329, 331, 411f.
adenine–silver complex model   31ff.
adsorbate
– conformation   327f.
– -induced electronic states   348
– –metal complex   10, 30
– surface   84
– surface-enhanced HRS   96
adsorption
– associative   327f.
– chemical   30
– CO   84f.
– dissociative   111, 319, 327ff.
– electrostatic   242
– energies   327f., 330
– ion   81
– metal surface   84
– molecular   34, 327
– monomer   390
– selective   177, 204
– structure   327
AEM (Auger electron microscopy)   242, 342
– current oscillation   242f.
– depth profile   248
– ionization   163, 306ff.
– process   156, 164
AFM (atomic force microscopy)   4, 23f., 26, 206ff.
– cantilever   24, 118
– contact mode   26, 33
– non-contact mode   320
– sensitivity   118
– surface morphology changes   449f., 498f., 527, 531, 534ff.
annihilation
– electron–electron   307
– exciton–exciton   205, 217, 220, 307
– one-electron   365
– operator   378
– sequential   217
– singlet–singlet   217
anti-Stokes Raman scattering   28
– coherent, see CARS
– spectra   16
Antoniewicz model   364
aperture
– closed   156f.

– NA (numerical aperture) 158
– open 156f.
– size 57
– sub-wavelength 56
Arrhenius plots 411f.
atomic orbitals (AOs) 358f., 368, 370
Auger electron microscopy, see AEM
Auger recombination 351
autocorrelation
– curves 141ff.
– interferometric 134, 137
– trace 343
autocorrelation function
– fluorescence 141, 143, 149f.
– second-order orientational 62
– temporal fluctuation of fluorescence intensity 140f.

b

background
– far-field 28
– free-detection 106
– scattering 379
band gap
– bulk 300
– energies 36, 158
– quantum confinement 300
– size-tunable 294
– structure of CdS quantum dots 298f.
– TiO_2 320
band theory 357ff.
Belousov–Zhabotonsky (BZ) reaction 178
blinking
– dynamics 156, 163
– intrinsic 306
– kinetics model 310f.
– modified 308ff.
– photoluminescence 162f., 306f.
– Raman spectrum 34
– suppression 310
Bloch basis 359, 368
block copolymer
– diblock copolymer 203f., 206ff.
– film 203ff.
– functional chromophore 203f.
– functionalization 203, 208
– polymer microspheres 205ff.
– site-selective doping 208ff.
– surface morphology 208ff.
Boltzmann
– constant 141, 360
– distribution 118, 122
– kinetics 357
bonding
– chemical 25
– covalent 25
– distance 32f.
– π-type 389
Born approximation, see SCAB
Bose–Einstein distribution 379
bottom-up method 239
boundary condition 145, 359
– Direchlet 366, 369
– nonperiodic 376
– periodic 375f.
boundary element method 6f.
bridge model 386, 389
bright field optical microscope 178
Brillouin zone 345
Brownian motion 118, 120, 149, 159, 161, 227, 552ff.
Brownian ratchet mechanism 229
Brønsted acid 428ff.
building block
– heterologous 187
– homologous 187

c

CAI (computer-assisted irradiation) 174, 177ff.
– designing polymers 177ff.
CARS (coherent anti-Stokes Raman scattering) 28, 547, 549
– nonlinear CARS process 29f.
– polarization 28f.
– tip-enhanced near-field CARS microscopy 29
catalyst
– acidity–basicity 317f., 332f.
– Ag/Al_2O_3 402ff.
– CuO/ZnO 427
– electrocatalyst 84
– heterogeneous oxide 317, 384, 427
– PdO/ZrO_2 427
– three-way 401
– zeolite-based 401f., 405, 407, 420
catalytic
– dehydration 317ff.
– dehydrogenation 317ff.
– dynamic catalytic reaction 327ff.
cavitation bubble 550ff.
CCD (charge coupled device), see detector
cell, see living cell
channel
– micro- 229f.
– nano- 204
charge carriers
– recombination process 164, 270

– transfer   4, 203
– transient trapping   311
– trapped   305f., 310
charge transfer/transport, *see* (CTs)
chemometrics   624
chirality
– bundle   517, 519f.
– helical   517, 519
– molecular   516ff.
– supramolecular   515, 517, 520f.
– three-axial   516ff.
– tilt   517f., 520
cluster
– ad-   363f.
– Ag   408ff.
– agglomeration   432, 435
– C_{60} clusters   259f., 264ff.
– carbonyl   428
– chiral   521
– clustering process   435ff.
– $C_{60}N^+$–MePH optically transparent nanoclusters   264f., 270f.
– dynamic coalescence   432ff.
– growth   438
– hydrogen bonded [4+4]   514ff.
– [4+4] ion-pair   514
– model   359, 368
– nano-   260
– Pd   427f., 431f., 434
– pseudo-cubic hydrogen bonding   505
CO
– adsorption   84ff.
– monolayer   88
– multibonded   96
– oxidation   85
– TR-SFG measurements   84ff.
– transient site migration   87ff.
computer-assisted irradiation, *see* CAI
conductance   144f., 361, 380f.
– -drop   379
– electrode–molecule–electrode (E–M–E)   361
– molecular   365
configurational interaction (CI) method   358f.
confocal light scattering microspectroscopy   561
confocal microscope   57, 133ff.
– confocal volume   141
conformation   279ff.
contact angle   283f., 288
coordination number (CN)
– Ag–Ag   408f.
– Pd–O   433ff.
– Pd–N   435f.

– Pd–Pd   430ff.
coupling   173
– constant   378
– dielectric   302
– dipole–dipole   302
– electron–nuclear (eN)   360, 375
– electron–phonon (eph)   340, 360, 377ff.
– exciton–phonon   299f.
– nondiabatic   360
– vibronic   6
crankshaft motion   461
cross-correlation analysis   120, 125ff.
CTs (charge transfer/transport)
– bulk level   357
– heterogeneous   357f., 382
– homogeneous   358
– inelastic   375ff.
– interfaces   357ff.
– molecular level   357
– semi-classical transport theory   384
– sequential   364
– static   358
current
– density   245, 247
– elastic correction   379
– in-flux   383
– oscillation   241ff.
– reduction   244
– STM tunneling   14
– –time curve   11
– tunneling   16, 317

d
decay
– emissive   6
– processes   3
– rate   13
decomposition
– DCOOH   332
– HCOOH   325f., 331
– parallel factor analysis (PARAFAC) model   633f., 637ff.
– reaction   317, 325f.
– thermal   213
– time–domain data   346
– unimolecular   329
degree of freedoms (DoFs)   363
– nuclear   364, 382
– rotation   628
dehydration
– DCOOD on TiO_2 318ff.
– HCOOH on TiO_2 317ff.
– unimolecular   329
dehydrogenation

- bimolecular 318, 329, 331f.
- HCOOH on TiO_2 317ff.
- pathway 329, 331
- temperature 330f.

density functional theory, see DFT
density matrix 359, 365f.
density of states (DOS)
- local 45
- projected (PDOS) 389f.

deposition
- chemical vapor deposition (CVD) 280f., 428
- dopant-induced laser ablation 204f., 211ff.
- electrodeposition, see oscillatory electrodeposition
- Langmuir–Blodgett method 58, 60
- layer-by-layer 268
- oscillatory electrodeposition 239ff.
- reprecipitation method 217
- spin-coating 49, 58, 208, 215, 218
- thin-film 204

desorption-induced by electronic transition (DIET) 364, 381ff.
desorption-induced by multiple electronic transition (DIMET) 364, 382f.

detection
- efficiency 73
- -interface-selective 104f.
- interferometric 137
- single chromophore 162

detector
- avalanche photodiode 134, 218
- CCD (charge coupled device) camera 27, 41, 43, 73, 112, 427, 624ff.
- closed aperture 156f.
- ICCD (image-intensified charge coupled device) 562
- microchannel-plate photomultiplier (MC-PMT) 610
- multi-channel spectral 41
- open aperture 156f.
- photomultipliers (PMTs) 62, 73, 75, 77, 105, 343
- quadrant photodiode (QPD) 119
- single-channel photodetector 41, 45f.

DFG (difference frequency generation) 74, 77f.
DFT (density functional theory) 34, 317, 358f., 366
- HCOOH on TiO_2 317ff.
- HF–DFT 358
- in situ 318
- KS–DFT 358
- time-dependent 393, 490

dielectric constant 19, 21, 74, 165f., 565
DIET, see desorption-induced by an electronic transition
difference frequency generation, see DFG
diffraction limit
- infrared 580f.
- visible light light 3, 22, 39, 56f., 94, 137, 573, 578

diffusion
- anomalous diffusion 648f., 656ff.
- coefficient 141f., 649ff.
- constant 142
- -controlled electron transfer processes 308
- DDDC (distance dependence of diffusion coefficient) 657ff.
- distance 160, 652, 657
- -limited aggregation (DLA) 241, 250
- normal 650ff.
- real-time visualization 647ff.
- rotational/translational 133
- second-order diffusion-controlled 653
- spectral 162
- TDDC (time dependence of diffusion coefficient) 657f., 661
- time 141, 151
- two-dimensional 226f.
- vector models 649
- velocity 141f.

DIMET, see desorption-induced by multiple electronic transition
dipped adcluster model (DAM) 359
discret dipole approximation (DDA) method 5, 12, 71f., 669ff.

dispersion
- curve 20
- group velocity 108
- Pd 432ff.
- relation 20

dissipative structures 189f.
- Bénard convection cells 190ff.
- droplet arrays 193ff.
- evaporating polymer solutions 191ff.
- hierarcially ordered structures 193
- large-scale 190f.
- polymer dewetting pattern 195f.
- polymer-rich finger structures 192
- tears-of-wine 190ff.

DSC (differential scanning calorimetry) 180
double layer, see electric double layer
droplet
- deformation 283, 286, 288
- manipulation 279, 281
- net transport 285f.
- rachet motion 284ff.

– spontaneous motion 281
dye exclusion test 554
Dyson equation 366, 378, 383
DZP, *see* polarized double zeta

e

eigenchannel 367
electric dipole moment 73, 94
– dynamic 7
– IR transition 76
– orientation 8
– Raman transition 76
electric dipole 75f.
electric double layer 80f.
electric field
– distribution 48ff.
– enhanced 48f.
– flourescence lifetime imaging microscopy (FLIM) 607f., 610f.
– -induced aggregate formation 617ff.
– -induced quenching 616
electric field modulation spectroscopy 610
electrochemical
– etching 10
– oscillations 240f.
electrochemical quartz crystal microbalance 248
electrochemical reaction systems 240f.
– Stark tuning 85
electrode
– asymmetric 284f.
– Au 80, 84
– counter 78, 253, 272
– crystalline disk 79
– ITO (indium–tin oxide) 264ff.
– photosensitive 260, 272
– potential 78, 81, 241
– Pt 80ff.
– Pt-poly 85
– reference 78, 244, 284, 286
– semi-infinte 362
– silver 5
– surface 81, 83f.
– working 78, 253, 272
electrolyte
– concentration 81
– solution 78f.
– ultra-thin layer 250ff.
electromagnetic
– distributions 9
– enhancement 4ff.
– field 94, 159
– simulations 48
– theory 46, 48

electron
– bath 382
– conduction 42
– density 329f., 338, 446
– distribution function 360
– –electron scattering 46
– –hole pairs 250, 350, 364, 382
– hot 347, 349, 363, 381, 384
– injection 338, 362, 364, 381
– lifetime 338, 344
– photoinduced electron transfer 260
– photoinduced hot 350
– –photon scattering 46
– primary 384
– secondary 384f., 389
– solvation 344ff.
electron transfer 6, 163, 260, 328
– inter-molecular 614
– intra-molecular 614
– ultrafast proton-coupled 345
electron
– transient trapping 307, 383
– tunneling 13, 163, 319
– wet 345
electronic
– absorption resonance 349
– excitation probability 14
– state 30, 348
electronic structure theory 359, 381
electronic transition state theory (eTST) 383
electrophoresis 228f.
emission
– angle 75
– count rate 218f
– fluorescence 3, 14, 512ff.
– intensity 220
– solid-state 513
enantiomeric
– crystals 528
– pseudo-cubic network 520
enantioresolution 511f.
endothermic reactions 104
energy
– acceptor 6
– binding 32f., 324
– confinement 299
– diagrams 94f., 327, 329
– level diagram 299
– profiles 360
– reorganization 360
energy–time uncertainty relation 104
energy transfer 6, 268
– non-radiative 310
– rate 310, 364

- ultra-fast non-radiative 310
enhanced green fluorescence protein (EGFP) 557
enhancement
- chemical 30
- efficiency 5
- electromagnetic 4f., 30
- excitation 13
enhancement factor 5
- field (FEF) 6ff.
- Raman 5, 49
- TERS 5
entropy 189
- conformational 191
- production 189
- transfer 189
EPR (electron paramagnetic resonance) spectroscopy 419f., 469
equal-pulse transmission correlation method 45
evaporation 22, 24
exchange-correlation (XC) functionals 358, 385
excitation
- de- 3
- density 603
- direct 362, 364
- electronic 361ff.
- incoherent 348
- indirect 362, 364, 381f.
- intensity 163f.
- intensity fluctuations 607
- inter-molecular 362
- intra-molecular 362
- laser power 219
- mechanisms 349ff.
- multiple pulse 351, 593
- one-electron 362
- out-of-focus 591, 593
- photon energy 338
- power dependence 218f.
- probability images 47
- pulsed 95
- Raman 105
- ratio method 615
- substrate-mediated 350
- surface adsorbate 350
- two-photon 591f.
- wavelength 50, 611ff.
exciton
- Bohr radius 293
- deep-trapped 302
- dynamics 220
- inter-band exciton recombination 298
- inter-exciton distances 220
- migration 218, 220
- non-radiative exciton recombination 303, 307
- self-trapped 218
exposure time 27, 33
extinction
- band 42
- coefficient 144f., 411
- optical 44
- peak 42f.
- peak shifts 42
- spectra 39, 43

f

fast braodband spectral acquisition 592
fcc model 386, 389
FCS (fluorescence correlation spectroscopy) 139ff.
- biological systems 655
- confocal volume (CV) 655, 656
- experimental set-up 139f.
- inside single cells 662f.
- local temperature measurement 140ff.
- molecular diffusion 228, 645ff.
- non-emissive relaxation dynamics in CdTe quantum dots 148ff.
- outside single cells 664
- sampling volume controlled (SVC-FCS) 656ff.
femtosecond transient absorption spectroscopy 549
Fermi
- energy 319
- functions 367, 379
- level 338, 345, 350f., 362, 367f.
figures of merit (FOM) 158
finite-differential time-domain (FDTD) method 5, 23
finite element electromagnetic simulations 8
first-order expansion 94
FLIM (flourescence lifetime imaging microscopy) 602, 607ff.
- electric field effects 607f., 610f.
- in vitro 615
- in vivo 615f.
- pH dependence 607f.
- system 608ff.
fluorescence
- anisotropy 61f.
- anisotropy decay 63ff.
- auto- 591, 623f.
- beads 574, 579f.
- chromophores 162, 208f., 607f.

– components 636f., 639ff.
– decay rate 12f., 602, 614f.
– delay time 624
– depletion effect 572
fluorescence depolarization method 61ff.
fluorescence
– donut-like enhancement pattern 6
– electrofluorescence spectrum 614, 616
– γ–Em maps 635ff.
– enhancement 12, 303ff.
– epi- 557
– intensity 6, 13, 58f., 140
– lifetime 62, 607f., 610ff.
– line illumination 624, 627, 635
fluorescence microscopy
– flourescence lifetime imaging microscopy, see FLIM 602, 607ff.
– time- and spectrally-resolved system 624f., 629
– super-resolution laser-scanning fluorescence microscope 571ff.
fluorescence
– multidimensional 623ff.
– polymer latex bead 558
– quenching 6
fluorescence recovery after photobleaching (FRAP) 14, 649, 654
fluorescence
– solid-state emission 512ff.
– solution 630ff.
fluorescence spectroscopy 3f., 6, 25
– confocal system 559, 590f.
– excitation–emission matrix system 624ff.
– fluorescence correlation spectroscopy, see FCS
– fluorescence dip spectroscopy 572
– line-scanning 591ff.
– point-by-point scanning 591, 597
– single (sub-) molecule STM spectroscopy 13f., 16
– three-dimensional 558, 630ff.
– time-resolved 614, 636
– total internal reflection fluorescence, see TIRF
– transient fluorescence detected IR, see TFD-IR
fluorescence
– spectral imaging 591ff.
– two-color 571f.
– two-photon 561, 602, 677
– up-conversion fluorescence depletion 571f.
Förster resonance energy transfer (FRET) 302
force

– adhesion 118
– atomic 33
– effect 35
– electric double layer 118
– electrostatic 119
– femto-newton order 120
– friction 89
– gradient 159, 162
– hydrodynamic interaction 120ff.
– Lorentz 259, 266
– magnetic 259, 264, 266
– photon 117ff.
– repulsive 33, 282
– scattering 159
– van der Waals 535
– weak 118f.
Fourier electron density map 446
Fourier transformation 104f.
– SHG spectrum 111f.
– time-domain response 105f., 108
– voltage oscillation 251
Fourier transform infrared spectroscopy, see FTIR
fourth-order coherent Raman scattering 103ff.
– buried interfaces 103ff.
– frequency domain detection 112
– frequency domain spectrum 112
– time-domain detection 112
Frank–Condon
– factor 361
– mechanism 95, 364
– state 448
FRAP (fluorescence recovery after photobleaching) 228
frequency
– angular 20f.
– -domain detection 112
– -domain filter 351
– -domain method 614
– -domain spectrum 112
– intrinsic frequency shift 25
– IR 76
– peak 30
– resonance 19, 21, 29
– second harmonic 104
– shifts 31f., 87
– transverse-mode 44
– vibrational 31
Fresnel
– coefficients 91, 340
– factor 75, 79
– formula 384
friction 89, 91

FWHM (full width at half maximum) 29, 86f., 95
– Gaussian pulse 105f.
– NIR laser microscope 134, 137
FTIR (Fourier transform infrared) spectroscopy 410ff.

g

Gaussian
– beam 157
– conformation 56
– distribution 261
– function 86, 111
– profile 10
– pulse 135
glass transition temperature 180
Gran–Taylor prism 77
Green dyadic formalism 45
Green's function theory 359, 361, 366ff.
– matrix (GFM) 366, 368
– NEGF (nonequilibrium) 365ff.
– NEGF–DFT 365, 367f., 373, 383f.
– NEGF–SCF 367ff.
– perturbation (PT-GF) 368
– retarded 379
– time-dependent 394
guest
– exchange reaction 522
– molecules 511
– volumes 510

h

Hamiltonians 365ff.
Hartree–Fock (HF) method 358f.
Heaviside function 10, 383
higher order multiphoton excitation, see NIR laser microscope
higher order multiphoton fluorescence 135ff.
highest occupied molecular orbital, see HOMO
highly oriented pyrolytic graphite (HOPG) 11f.
hole-trapping 308
holographic Super Notch filters 78
Holstein model 376, 381
HOMO (highest occupied molecular orbital) 373
homodyne scheme 106
hopping mechanism 308, 324, 360
host–guest
– assemblies 505
– chemistry 465, 505
– complexation 522
host frameworks 510f.

hot spot 49f.
HRS (hyper-Raman scattering) 72, 94ff.
Huygens–Fresnel integration 157
hydrocarbons
– oxidation 412f.
– oxidative activation 416ff.
hydrodynamic force measurement 122ff.
hydrogel 89
hydrogen-bonding 91, 512
– acceptor 515, 520
– donor 520
– multipoint 205
– networks 512, 515f.
– water 80, 83f., 91
hydrogen
– effect 404f., 409f., 415, 417f.
– switching 409, 416f.
– waves 80, 85
hydrophilic
– core 516
– glass beads 288ff.
– substrate 234
– surface 91, 229, 282f., 285, 287
hydrophobic
– beads 290
– phenyls 515
– surface 92, 230, 282, 287
hyper-Raman spectroscopy 94, 96

i

ICP-MS (inductively coupled plasma mass spectroscopy 679
IETS (inelastic electron tunneling spectroscopy) 4, 368, 375ff.
illumination 14, 27
– fluorescence line 624, 627, 635
– light field 25f.
– Raman scattering 25f.
– side 10
image
– CARS 29
– excitation probability images 47
– flourescence intensity 610f.
– fluorescence SNOM 57ff.
– optical transmission 137
– near-field Raman excitation 50f.
– near-field transient transmission 46f.
– near-field two-photon excitation 47ff.
– scanning fluorescence 138
– simultaneous STM–TERS 11ff.
– NIR laser microscope four-photon fluorescence 138
– NIR laser microscope three-photon fluorescence 138

– NIR laser microscope two-photon
 fluorescence 138
– transient fluorescence 577f.
– transmission 635f.
imaging
– bioimaging 646, 676
– fluorescence lifetime 607ff.
– high-speed 556
– light scattering spectroscopic 547ff.
– multidimensional fluorescence 623f.,
 635ff.
– multiphoton fluorescence 134, 137
– nano- 23
– near-field two-photon excitation 49f.
– nonlinear 42
– off-resonant 29
– on-resonant 29
– quantitative 607
– real-space 56
– single molecule 294
– spectral fluorescence imaging 591f., 598ff.
– three-dimensional biological 156
– time- and spectrally resolved
 fluorescence 626f., 635ff.
– total internal reflection fluorescence
 (TIRF) 557
inelastic electron tunneling spectroscopy, see
 IETS
instrumental response function (IRF) 637
intensity ratio method 612
interactions
– acid–base 430f.
– attractive 89
– bilayer–substrate 230ff.
– carrier–carrier 164
– chemical 30
– coefficient 123f., 128
– Coulomb 20, 266, 318
– dimer–dimer 385
– dipole–dipole 300
– electric double layer 231
– electrostatic 89, 231, 261
– force 118, 120, 122ff.
– homologous pair–pair 188
– hydration 231
– hydrophobic 264
– inter-chain 63f.
– intermolecular 3, 133, 218, 512f., 535
– intra-chain 64
– lattice–electron 300
– metal–support 427f.
– molecular–molecular 279
– molecule–substrate 76, 279
– repulsive 61, 89

– van der Waals 117f., 230f.
interface
– aqueous solution/nitrobenzene 283
– Au thin film electrode/electrolyte
 solution 80
– buried 103f.
– concentric pattern 241
– disk/water 79
– electrode/electrolyte 71, 80
– fourth-order coherent Raman
 scattering 108ff.
– gel/solid 89
– hexadecane/solution 108
– ice/air 92
– liquid/air 241, 250, 252ff.
– liquid/liquid 108ff.
– molecule/bulk 360
– OTS/quarz 92
– protic/solvent metal-oxide 345
– PVA gel/hydrophobic 91
– PVA gel/OTS-modified quartz 92, 96
– PVA gel/quartz 89, 92, 97
– Pt thin film electrode/electrolyte
 solution 80ff.
– quartz/electrolyte 79
– quartz/water 79, 81
– -selective detection 104f., 113
– semiconductor/solution 250
– solid/liquid 71f., 84, 96, 260, 270, 340
– solid/liquid/liquid 289
– solid/solid 89
interfacial
– charge transport 365
– force 283
– frictional behavior 89, 92
– molecule/bulk CTs 357f.
– molecular structure 89
– structures 71
– tension 283f.
– water 71, 79, 89ff.
intersystem crossing 6
ion beam etching 239, 243
ion-exchange 429
IPNs (interpenetrating polymer
 networks) 175
IR (infrared) absorption spectroscopy 25, 85,
 94
IR microspectroscopy
– application 573f.
– single cells 571
– super-resolution, see TDF-IR
IR–visible
– double resonance signal 577
– SFG measurement 72f.

isomer
– closed-ring 444, 446f., 451f.
– open-ring 446f., 451
isomerization dynamics 340

j

Joule heating 375
junction
– electrode–molecule–electrode 365ff.
– metal–molecule–metal 360

k

Kadanoff–Baym ansatz 367, 383, 393
Keldysh formalism 365f.
Keldysh–Kadanoff–Baym (KKB) equation 366
Kerr effect 156f.
kinetic potential analysis 123ff.
Kohn–Sham (KS) framework 358
Kretchmann configuration 21
Kronecker product 629
Kubelka–Munk 409

l

Lambert–Beer law 176f.
Landauer formula 360ff.
Landauer–Buttliker 367
Langevin equations 123
Langmuir–-Blodgett transfer process 26
Laplace equation 6
Laplacian field 250
laser ablation dynamics 548ff.
– etching 550
– femtosecond 550
– phthalocyanine film 549
laser
– coherent beam 527
– continuous 15
– femtosecond 41f., 45, 47, 338ff.
– fluence 349, 555f.
– focal point 11, 558, 560
– free-electron 104
– nanosecond 338
– near-infrared laser microscope, see NIR
– non-destructive removal 555
– power 14, 27
– profile 86f.
– pulsed 14, 29, 134
– pump–pulse irradiation 86ff.
– repetition-rate 14, 159, 162
laser-scanning confocal microscope, see LSCM
laser
– short pulse 84
– spot size 10
– tabletop 104
laser trapping 119f., 122f., 158f.
– biological cells 691ff.
laser tsunami processing 547ff.
– in situ micro-patterning 556
– injection of nanoparticles 558ff.
– removal technique 556
– single animal cell manipulation 554ff.
– single polymer bead manipulation 551ff.
laser
– tweezer 554f., 691
– ultrashort pulse 134, 337, 340, 352
layered nanostructure, see oscillatory electrodeposition
LEED (low-energy electron diffraction) optics 318
light field 20ff.
– collection 26
– confinement 23
– evanescent 21, 119
light
– nano-light-source 23f., 26
– non-propagating 56
light scattering microspectroscopy 561ff.
– confocal 561ff.
– imaging system 561f.
– spectra 563
– supercontinuum 561
linear synchronous transit (LST) 319
lipid bilayer 225
– artificial 225ff.
– composition 233
– fluidic 227f.
– molecular manipulation 233ff.
– patterned 226
– properties 225f.
– self-spreading 229ff.
– two-dimensional diffusion 226f.
lithography
– dip-pen nanolithography (DPN) 226f.
– electron-beam lithography (EBL) 233f., 239
– nano-sphere lithography (NSL) 233f.
– photo- 239
living cell
– dynamic motion 689f.
– extracellular matrix (ECM) 657ff.
– extracellular pH 613, 615
– gold nanorods 674ff.
– injection of nanoparticles 558ff.
– intracellular pH 607f., 612ff.
– intracellular spectral gradient 596ff.
– laser trapping 691ff.
– laser tsunami 554ff.
– manipulation of single animal 554, 556

– membrane, see photosynthetic membranes
– molecular diffusion   645
– patterning methods   555f.
– photochemical damage   556, 560, 602, 624
– photothermal damage   556, 602, 624, 676
– separation   699f.
– suspension   615, 617, 695ff.
– TDF-IR imaging   581ff.
local field effect   158
local plasmon, see plasmon polaritons
location model   511f.
lock-in amplifier   46, 134, 611
longitudinal optical (LO) phonon   300
Lorenz–Mie regime   159
low-energy electron diffraction, see LEED
lower critical solution temperature (LCST)   178
lowest order expansion (LOE)   378f., 381, 393
lowest unoccupied molecular orbital, see LUMO
LSCM (laser-scanning confocal microscope)   177, 561, 590f., 599, 604, 625
LUMO (lowest unoccupied molecular orbital)   322, 373, 389

m
Mach–Zehnder interferometry (MZI)   182
magnetic dipole   96
magnetic field effects (MFEs)   259f., 268
magnetic
– gradient   259, 263, 266
– orientation   259ff.
– processing   264ff.
magnetohydrodynamic mechanism   259
mapping
– high-speed   8
– optical   22
– spatial   578
– spectral   11, 27
– STM–TERS intensity   11ff.
– tip-enhanced near-field Raman spectral   28
Marcus theory   357, 360ff.
Maxwell's equation   5, 669
mean-field level theory   358
mean residence time   141
mean-square displacement (MSD)   227f., 650ff.
membranes, see photosynthetic membranes
Menzel–Gomer–Redhead (MGR) model   364
metal
– electroactive ion   244f.
– –molecule complexes   30
– nanostructure   4f.
– periodic nano-architechtures   225

– subsurface   5
microfluidic system   230, 233, 281, 285
– channel   281f.
– three-dimensional   230
microdomain   204, 208, 213
micro-spectroscopic systems   133ff.
Mie theory   5, 669
migration
– exciton   217
– nanoparticle   564
mode
– alkali–substrate stretching   346f.
– alternative bond length (ABL)   379
– assembly   519
– axial   36
– bulk phonon   111
– C–C stretching   27
– C–N stretching   27
– Cs–Pt stretching   341, 345, 350f.
– dipolar   43f.
– frustrated rotation   88, 104
– frustrated translation   88, 104
– hyper-Raman active   94ff.
– IR-active vibrational   72
– lateral   347ff.
– longitudinal plasmon   42ff.
– molecular diffusion   225
– packing   517
– phonon   345, 379
– Raman active   94ff.
– ring breathing   27, 30, 35
– ring stretching   29f.
– selective excitation   351f.
– silent   72
– stacking   519
– surface normal   341, 348
– surface phonon   111, 341, 347, 351f.
– transverse plasmon   42ff.
– vibrational   16, 34
modulators   513f.
molecular
– aggregation   188
– architectures   522f.
– arrangements   530f., 538
– bond length   30
– conformation   531
– diffusion dynamics   225f., 228, 648ff.
– dynamics   133ff.
– fingerprint   571
– fluorescence   6
– information   516ff.
– manipulation   225. 228, 233ff.
– nano-identification   19
molecular orbital (MO) theory   26, 358ff.

- renormalized (RMO)   373
- restricted   374
molecular
- orientation   35
- segregation   225
- single molecular detection (SMD)   648
- separation   226, 228
- structure distortion   6
- transportation   229f.
- vibration   3, 13, 76
- vibration energy   25
- vibration frequencies   29
- weight   58ff.
molecule
- amphiphilic lipid   225
- dimer   49f.
- lipid   225
- steroidal   522f.
- transport   104
- vibrational excitation   13f.
moletronics   357, 360, 365
momentum conservation   73, 75
monolayer
- OTS (octadecyltrichlorosilane)   92f.
- polar   75
- self-spreading   229f.
morphology changes
- bulk crystals   531f., 539f.
- laser ablation dynamics   550
- microcrystals   532ff.
- organic crystals   527ff.
- photochromic single crystals   498f.
- triisobenzophenone   537ff.
morphosynthesis of polymeric systems   173ff.
Møller–Plesset perturbation (MP) method   358
Mulliken charges   327f., 331, 373f.
multichromophoric systems   217, 219
multiphoton microscopy   602
multiple multipole method (MMA)   12
multiple-particle-tracking microrheology (MPTM)   650
multi-reference self-consistent field (MSCF)   358f.

n

nano-analysis   19f., 31
nanocrystals
- adenine   27f., 30ff.
- monodispersed   148
- organic dye   217, 221
- size   217f.
- surface   33

nanodevices   268, 272
nanoparticle
- aggregates   301f.
- CdS   268, 293ff.
- CdTe   148ff.
- colloidal   293, 295
- crystalline   42
- diameter   42
- gap   48
- gold   42f.
- metal   19, 21f., 39
- Mn^{2+}-doped ZnS   268ff.
- noble metal   48ff.
- organic   205
- patterning   203
- polarization   21
- self-defocusing materials   157
- -spheres   42f., 48ff.
- synthesis   295ff.
nanorods   42f.
- aggregation   670
- aspect ratio   42f.
- biocompatible   670ff.
- biodistribution   673f.
- CTAB-stabilized gold nanorods   670ff.
- gold   42f., 669ff.
- in vivo   675ff.
- in vitro   677
- lightning rod effect   47
- living cells   674f.
- metal   45
- PC-modified gold nanorods   671f., 678ff.
- PEG-modified gold nanorods   672, 674, 677ff.
- photoreaction   680ff.
- plasmon-mode wavefunctions   42ff.
- spectroscopic properties   669f.
- two-photoninduced photoluminescence, see TPI-PL
- ultrafast near-field imaging   45ff.
nanostructure
- construction by magnetic fields   259ff.
- construction by spin chemistry   259
- metal filament   250ff.
- modification   203ff.
near-field optical spectroscopy   22, 40f.
near-field transmission images   42f., 46f.
near-field transmission spectra   42ff.
near-field two-photon excitation images   47ff.
NEGF, see Green's function theory
NIR (near-infrared) laser microscope   133ff.
- femtoseond   133ff.
- higher order multiphoton excitation   133ff.

– higher order multiphoton fluorescence from organic crystals   135ff.
– multiphoton fluorescence imaging   134, 137
– time-resolved   134
NMR (nuclear magnetic resonance)   516
– diffusion-ordered NMR spectroscopy (DOSY)   649
– pulsed field gradient (PFG-NMR)   649, 659f.
noncolinear optical parametric amplification (NOPA) system   340ff.
nonlinear optical (NLO)
– phenomena   27
– properties of CdTe quantum dots   155ff.
– second-order process   71f., 94
– third-order effect   156
NO
– consumption rate   415
– oxidation   411, 415
– selective reduction   413, 428f.
NOx
– conversion   402ff.
– NSR (NOx storage-reduction)   401
– reduction technologies   401ff.
– storage materials   401
NSR, see NOx
nuclear wavepacket
– coherent   347, 349, 351
– displacement   341, 350
– motion   104, 343ff.
nuclear wavepacket dynamics   337ff.

o

OH
– oscillators   91
– -stretching region   79, 81f., 90, 92f., 96
one-bond-flip (OBF) motion   461
OPA path   77
open density matrix   364
optical fiber probe   40ff.
optical
– field   47ff.
– fourth-order coherent Raman scattering transitions   104ff.
– manipulation   159
– subwavelength scale measurements   49
– switching   158
– trapping   158f., 552
organic field effect transistor (OFET)   197
organic fluorophores   512f.
organic inclusion crystals   505ff.
– dynamics of steroidal   506ff.
– guest-responsive structures   506f.
– intercalation in bilayer crystals   508f.
– solid-state fluorescence emission   512ff.
ORTEP drawings   446, 463, 477f., 491, 530, 538
oscillation
– amplitude   43
– collective electronic   42f.
– current   241
– electrochemical   241
– frequency   43, 349
oscillatory electrodeposition   241ff.
– Au   252f.
– Cu/Cu_2O 247ff.
– Cu–Sn alloy   242f.
– growth   252f.
– iron-group alloys   246f.
– layered nanostructure   242ff.
– ordered architectures   241
Ostwald ripening   295
OTS (octadecyltrichlorosilane)   92f.
oxygen
– bridging   319ff.
– defect   319ff.
– selective transfer   495
– vacancies   318, 324, 326

p

packing coefficient   510f.
packing diagram
– Cp* rings   496f.
– inclusion crystals   508
parallel factor analysis (PARAFAC) model   624f., 627ff.
– decomoposition   633f., 637ff.
particle-in-a-sphere model   293
particulate materials (PM)   401
Pauli principle   382
penetration   103, 138
perturbation
– expansion   378, 381
– order   364, 378
– theory   378
pH
– electrolyte   250
– in situ measurements   248
– SFG intensity   81
phase separation   178, 180ff.
– microphase separation structure   203, 206, 208f., 215
– nanoscale networklike   208f., 216
– reversible   181
– sea–island   206, 208f., 215f.
– worm-like   208, 210f., 217
phase transition

– benzyl muconate crystal   476
– thermal   449, 527f.
phonon
– bath   379
– coherent surface   341, 345, 350ff.
– dynamics   340f.
– frozen-phonon approximation   378
– longitudinal optical (LO)   340f.
photoactivation   303, 305ff.
photoactive species   390f.
photoautotrophic   594ff.
photobleaching   25, 163, 305, 596, 607
photocatalysis   357
photochemical reactions   173ff.
photochemistry   381ff.
photochromic reaction   490ff.
photochromism
– crystalline-state   487ff.
– rhodium dithionite complexes   487ff.
– single crystals   443ff.
– transition-metal based   488
photocurrent   267, 272
photocyclization reaction   446ff.
– asymmetric   528f.
– enantiospecific   528f.
– microcrystals   532ff.
– Norish type II   528
– quantum yield   447f.
– single-crystal-to-single-crystal   527ff.
photodesorption   364
– metal   364
– nitric oxide on Ag surface   384ff.
photodimerization   174, 178, 527
– [2+2]   465ff.
– anthracene   181f.
– benzyl muconates   465ff.
– cyclodimerization   465
– 1,3 dienes   465
– reversible   181f.
photodiode, see detector
photodissociation   174, 181f.
photoelectrochemical reaction   259f.
photoexcitation   46f., 267, 273, 338, 616
photoinduced
– charge separation   270
– electron-transfer reactions   270f.
– intermolecular electron-transfer   271f.
– reverse electron-transfer process   272
– surface dynamics of CO   84ff.
photoinhibition   596
photoirradiation   267, 273, 443, 448ff.
photoisomerization   282, 443, 449, 453, 460ff.
– benzyl (Z,Z) muconates   460ff.

– bicycle-pedal model   461
– EZ-photoisomerization   460ff.
– hula-twist process   461
– solid-state   460f.
photoluminescence
– CdS quantum dots   161f., 293ff.
– color   294, 298
– electro-   197
– enhancement   303ff.
– intensity   161, 164f.
– intensity time-trajectories   163f.
– intermittency, see blinking
– lifetime   304, 310
– monolayer films   268ff.
– on–off, see blinking
– polarization degree (p-value)   269f.
– polarized spectra   269
– quantum efficiency   303f.
– size-dependent   294, 298
– spectral shifts   299ff.
– time-dependent   161
– two-photon absorption-induced   161f.
photomechanical effect   448f., 550
photomultipliers (PMTs), see detector
photon
– antibunching   205, 217, 219ff.
– bunching   163
photon correlation   218, 220f.
– Hanbury–Brown set-up   218
– Twiss set-up   218
photon force measurement   117, 121f.
photon
– -in/photon out/photon method   73
– interphoton times   219
– local density of states   45
– multiphoton   133ff.
– pulse   75
– pressure potential   159ff.
– Raman-shifted   27
– single-photon sources   217, 220
photonics dewetted structures in organic   196f.
photoreversible shape changes   450ff.
photosensitization   178f.
photosynthetic membranes
– fluorescence microscopy   590ff.
– heterotrophic conditions   595f.
– photoautotrophic conditions   594ff.
– pigment-protein complexes   580f., 592
– spectral fluorescence imaging   591ff.
– thylakoid membranes in chloroplasts   590f., 598ff.
– thylakoid membranes of cyanobacteria   590, 594

Index | 717

plasmon polaritons 19ff.
– local 19, 21, 23, 30, 39
plasmon resonance
– localized 19ff.
– wavelength 13
point-spread function 597
Poisson process 653
polarizability
– dielectric particles 159
– first-hyper 73
– hyper- 75
– linear 73
– molecular 74
– second-hyper 73
polarization 23, 41
– anisotropy 61
– coherent 343
– excitation 61
– incident polarization directions 41, 48ff.
– in-plane 34
– IR beam 77
– out-of-plane 34
– p- 23, 81, 106, 110, 350
– parallel incident 48, 50, 61
– perpendicular incident 49, 61
– s- 23, 119, 350
– static 74
polarized double zeta (DZP) 385
polymer
– amorphous 189
– bio- 522
polymer blend 178ff.
– recycling 181ff.
– scattering profiles 182f.
polymer brush 61ff.
– dynamics 63ff.
– graft density 63ff.
polymer chain
– confined 55f.
– conformations 55f., 58ff.
– dynamics 61ff.
– grafting 55, 61, 63
– mobility 63f.
– PMMA (polymethyl methacrylate) 58ff.
– three-dimensional 60
– two-dimensional 60
polymer
– conjugated 261
– copolymer, *see* block copolymer
– crosslink 174, 181, 214, 216
– diblock copolymer 203ff.
– droplet 193ff.
– graded co-continuous morphology 176
– interpenetrating polymer networks, *see* IPNs

– microspheres 205ff.
– multicomponent 173f.
– solution, *see* dissipative structures
– spatially graded structures 175ff.
– syndiotactic 476
– wrapping 260f
polymerization 61, 175, 466f.
– benzyl muconates 465f., 471ff.
– electro- 259
– expanding 472ff.
– heterogeneous 476ff.
– homogeneous 476ff.
– rate 472ff.
– reactivity 470ff.
– shrinking 472ff.
– solid-state 469, 471f.
– strain 474ff.
– surface-initiated atom transfer radical 63
– thermally-induced 475
– topochemical 469
– UV- induced 473ff.
– X-ray-induced 474
potential
– cycling 78
– embedding 371
– energy slope 359
potential energy surfaces (PESs) 359f., 363f.
– Helmholtz double layer 244
– kinetic, *see* kinetic potential analysis
– oscillation 242, 247f., 250
– profile 81, 118, 120f.
– thermal 161
– three-dimensional profile 118, 120, 122f.
– trapping, *see* trapping potential analysis
potential
– induced 371f.
– mean-field 371f.
promotion effect 247f.
protrusion 320
– formic acid on TiO_2 321ff.
– gold 22
– nano- 22f.
PSTM (photon scanning tunneling microscopy) 4
pump–probe
– delay times 46, 338, 340, 343
– measurement 73
– pulses 339
– spectroscopy 337
– transient absorption scheme 45f.
– transitions 105
PVA (polyvinyl alcohol) gel 90f.
– pressure 91ff.

q

quantum chemical calculations 31
quantum confinement effect 293, 300, 360
quantum dots, see nanoparticle
quantum
– information science and technology 162
– intensity trajectory 310
– single quantum systems 205
– size effect 268
– yield 12, 512, 598
quartz surface 90ff.
quenching 25, 176

r

radiation pressure 119f.
radical
– chain mechanism 469
– decay of radical pair 270ff.
– pair mechanism 272f.
– photogenerated triplet biradicals 259
radius of gyration 56, 59f.
radio-frequency identification (RFIDs) tags 197
Raman active material 49
Raman bands 26ff.
– intensity 27, 50
– Rhodamine molecules 50
– shift 30
Raman microscope 26
Raman scattering, see Stokes Raman scattering
Raman spectroscopy
– confocal 248
– ex situ 252
– in situ 248, 252
– intensity 5, 8f., 11
– intensity distribution 28
– micro- 25, 27, 252ff.
– near-field nano-Raman spectroscopy 25f.
– near-field Raman excitation images 49f.
– nonlinear 25, 28, 41, 71ff.
– oscillatory electrodeposition 252ff.
– resonance Raman scattering, see RRS
– signal enhancement factor (EF) 11
– single-molecule 19
– STM–Raman spectroscopy 4
– surface-enhanced Raman scattering, see SERS
– third-order 105
– tip effect 30ff.
– tip-enhanced Raman scattering, see TERS
Raman spectrum
– adenine 31ff.
– far-field 28, 30
– near-infrared (NIR) 30f.
– peak shift 33f.
– temporally fluctuating 35
– tip-enhanced near-field 33f.
Rayleigh light scattering spectroscopy 548, 561ff.
Rayleigh
– particles 160
– scattering 3, 26
reaction 357ff.
– cathartic 653
– cavities in crystalline-state reaction 495ff.
– cross-linking reaction 214
– crystalline-state photochromism 490ff.
– photochemical biomolecular reaction (PVBR) 659f.
– signaling 652ff.
reaction path
– HC–SCR 414f.
– lowest-energy 494
– switchover 331
reaction propability 382ff.
recognition mechanism 511f.
recrystallization method 510, 528
refractive index 20f., 75
– complex 384
– dispersion 134f.
– imaginary part 120
– nonlinear 157
– relative 22
relaxation
– anisotropy 62
– charge-induced non-radiative 163
– non-radiative 302
– non-radiative Auger 298
– phonon-assisted non-radiative 298
– radiationless 364
– radiative 298, 302
– tension 558
– time 76, 165, 338, 343
resolution
– atomic scale 4, 14
– lateral 8
– nanometric spatial 19
– SNOM 57
– spatial 3f., 14, 22, 28, 39f., 57, 134, 563, 580, 647
– sub-nanosecond time 629
– temporal 134, 648
– vertical 103
resonance
– condition 21, 24, 47
– transition 343

S

SAM (self-assembled monolayer) 188f., 268f., 272, 279f.
– partial decomposition 281
SANS (small angle neutron scattering) 56, 60
scanning electron microscopy, see SEM
scanning near-field optical microscopy, see SNOM
SCBA (self-consistent Born approximation) 378
scatter-in (out) function of electrons 367
scattering
– electron–electron 384
– electron–phonon 378
– inelastic 3, 13, 25, 72, 94, 347
– inelastic nonlinear 94
– Rayleigh 3, 26
second harmonic generation, see SHG
second-order rate equation 329
SEHRS (surface-enhanced hyper-Raman spectroscopy) 72
SEIRAS (surface-enhanced infrared spectroscopy) 71
selective catalytic reduction (SCR) 401
– decane–SCR reaction 419
– HC (hydrocarbons)–SCR 401ff.
– n-hexane–SCR 411, 414
– promotion effect 415, 417, 420f.
– urea-SCR 401
self-assembly 49f., 187ff.
– bilayer vesicles 225
– blockcopolymer microphase seoaration 188
– bottom-up 197
– chiral salt crystals 528
– lipid molecules 558
– micelles 188
– nanoparticle 49f.
– self-assembled monolayer, see (SAM)
– snow crystals 188
– top-down 197
– vesicle fusion process 226
self-energy 366f., 369, 381
– matrices 368, 370
self-organization 49f., 522
– Beouzoc–Zhabotinsky reaction 188
– biological cell 188
– dynamic 239f.
– oscillatory electrodeposition 239ff.
– static 239
self-spreading, see lipid bilayer
SEM (scanning electron microscopy) 24, 194, 196
– current oscillation 242f.

semiconductor 43
– magnetic 268
– n-type 320
– p-type 250
– quantum dots 155, 159, 162, 260, 268, 299
– wide-band gap 110
sensor
– bio- 226
– chemical 226
– immunosensor 21
SERS (surface-enhanced Raman scattering) 4f., 30, 48ff.
– in situ 253ff.
– oscillatory electrodeposition 252ff.
– SERS–TERS 8ff.
– spectrum 30f.
SFG (sum frequency generation) spectroscopy 71ff.
– conformational changes 282
– electrochemical measurements 78f.
– experimental arrangement 77
– in situ measurements 79
– intensity 76, 81, 86, 96
– interfacial water structures at PVA gel/quartz interfaces 89ff.
– photoinduced surface dynamics of adsorbed CO 84ff.
– potential-depending structure of water 80ff.
– second-order 104
– spectrum 76, 78, 80f., 86f.
– time-resolved (TR-SFG) 71, 84, 87
– vibrational 74, 103, 113
SHG (second harmonic generation) 72, 74, 77, 105, 134f.
– experimental set-upff.
– intensity 111f., 340f.
– interferometric autocorrelation 134f.
– nuclear wavepacket motion at surfaces 345ff.
– time-resolved (TR-SHG) 105, 338, 340ff.
ship-in-bottle method 428
signal-to-noise-ratio (S/N) 140, 623
single crystals
– absorption intensity 452
– diarylethene 444ff.
– molecular structural chnages 490ff.
– photochromic 443ff.
– photomechanical effect 448f.
– photoreversible shape changes 450ff.
– surface morphology changes 449f., 498f.
single molecule microscopy 308
single molecule spectroscopy 117, 163, 217
site-selective doping 205, 208ff.

– dye-doped copolymer films 211ff.
– photoactive chromophore 208ff.
– nanoscale surface morphology 208ff.
slab model 359, 368
small angle neutron scattering, see SANS
SNOM (scanning near-field optical microscopy) 3, 39ff.
– aperture-less 3, 40
– conformation of single polymer chains 56f.
– device 57
– scattering-type 40
– shear-force feedback method 41
solvation dynamics 344ff.
spatiotemporal 239f.
– non-uniform conditions 181
– patterns 241
– synchronization 239, 241f.
spectroelectrochemical cell 78f.
spectrophotometer 26
spin chemistry 259f., 272
spin–lattice relaxation 272f.
spin–orbit coupling (SOC) 272
SPM (scanning probe microscopy) 4, 13, 39
Stark shift 616
states
– antibonding unoccupied 343
– deep-trap 302
– delocalized 344
– donor/acceptor 360
– electron image potential 344, 350
– electronic density 317, 338
– excited 3, 6, 13f., 16, 76
– ground 6, 76, 105, 108, 272, 349, 364, 549
– inclusion 510
– intermediate 338
– intrinsic molecular 6
– liquid 510
– localized 344
– one-electron 358f., 361f.
– photostationary 452f.
– quasi-stable 329
STM (scanning tunneling microscopy) 4, 23, 239
– chemistry 363, 376
– HCOOH on TiO_2 317ff.
– in situ 318
STS (scanning tunneling spectroscopy) 357
Stokes–Einstein model 141
Stokes law 159
Stokes Raman scattering 25
– cross-section 25
– hyper-Raman scattering, see HRS
– probability 25
– RRS (resonance Raman scattering) 10
Stokes shift 300, 631

sum frequency generation, see SFG
superconducting magnet 259, 261, 263f.
super-resolution laser-scanning fluorescence, see fluorescence microscopy
supramolecular synthons 522
surface
– acid–base property 317
– adsorbed species 412, 414f.
– charge 282, 299
– coverage 280f.
surface defect 287, 294, 298, 305, 319, 322, 328f.
surface-enhanced hyper-Raman spectroscopy, see SEHRS
surface-enhanced infrared spectroscopy, see SEIRAS
surface-enhanced optical microscopy 22
surface-enhanced Raman scattering, see SERS
surface fourth-order coherent Raman scattering 107f.
surface
– gold 6, 8ff.
– gradient 280f.
– Hirshfeld 511
– liquid 107f.
– metal subsurface 5
– morphology 208ff.
– morphology changes 527ff.
– passivation of defects 303, 306
– photochemical reactions 337
surface plasmon polaritons 20ff.
surface plasmon resonance (SPR)
– longitudinal SPR peak of nanorods 671, 681
surface
– polarization 301
– stoichiometric 327, 329f.
– switching 282ff.
– tension 281
– TMA-covered TiO_2 110f.
– -to-volume ratio 293
surface wetting 279ff.
– boundary 285, 287
– distribution 285
– gradient 283, 286ff.
surfactant 242f., 281f., 242f., 245
super-exchange mechanism 361
susceptibility
– anisotropic 263
– first-order 74
– fourth-order 105
– second-order 74, 76, 103
– second-order nonlinear 340, 347
– third-order 74
– third-order nonlinear 155f.

SWCNT (single wall carbon nanotube) 11, 35, 260
– -based composites 260ff.
– bundle 35, 263
– radial deformation 36
– semiconducting 262
– shortened 262f.
symmetry
– centro- 75, 94f.
– inversion 71f., 103
– molecular 72
– spatial 173

t

tempertaure
– deviation 145
– elevation coefficient 144ff.
temperature programmed desorption (TPD) 324
TERS (tip-enhanced Raman scattering) 4f., 8, 10f., 24, 26ff.
– BCB (brilliant cresyl blue) 8f., 13
– intensity mapping 11ff.
– SERS–TERS 8ff.
– signal intensities 10
– SWNT bundle spectrum 30f., 35
TFD-IR (transient fluorescence detected IR) 572ff.
– application to cells 581ff.
– fluorescence detection system 574f.
– optical layout for biological samples 575f.
– picosecond laser system 574
– picosecond time-resolved 578f.
– super-resolution 572ff.
– time-profiles 583
thermodynamic
– analysis of three-dimensional potentials 118, 120, 122f.
– equilibrium 187f., 239
– instability 191, 196
THG (third harmonic generation) 77
thin film
– homopolymer 209, 213
– metallic 21, 24
– passivating 250
– polymer 55f., 208ff.
– polymer ultra- 55f., 58ff.
– Pt electrode 80
– thickness 56, 59f., 208, 241f., 249
– transparent polymer 217
third harmonic generation, see THG
third-order nonlinearity 28
tight-binding-layer (TBL) 369
time-of-flight (TOF) analyzer 339f.

tip
– aspect ratio 6f.
– design 24
– forces 31, 33, 35f.
– geometry 6, 8, 11f.
– glass fiber 23
– metal 4, 6, 22
– nano- 19, 22ff.
– optical fiber 3
– optimization 24
– –sample distance 8
– –sample gaps 6, 13, 23
– –sample separation 317
– silver 6, 8ff.
– –substrate separation 8
– tunneling 4ff.
TIRF (total internal reflection fluorescence) 557
TOF, see time-of-flight
top-down method 239
topography 30, 51, 162
total internal reflection microscopy 119
TPI-PL (two-photon induced photoluminescence) 47f.
transition
– dipole 62
– electronic moments 61f.
– electronic states 25, 573
– inter/intra band 388
– one-photon downward 95
– optical 104f., 362
– probability 104
– resonant electronic 350
– state (TS) 318, 329f.
– surface interband 350
– two-photon upward 95, 105
– vibrational 76, 573
transmission 597
– coefficient 361, 377
– matrix 367
trapping potential analysis 118, 124f.
Tucker3 model 628, 630
tunneling
– bias repetition rate 16
– coherent 361
– current 4, 375
– efficiency 163
– inelastic electron 16
– position 8
– probe 4
– quantum mechanical 300
– resonant 308
two-color double resonant spectroscopy 572
two-photon photomession (2PPE) spectroscopy
– excitation probability 49

– experimental set-up 339f.
– femtosecond laser pulses 338ff.
– interferometric time-resolved 339, 343
– time-resolved 338
two-state model dynamics 364∧, 384
two-temperature model 382

u

UHV (ultrahigh vacuum) 318, 320, 339f.
– SHG 342f.
ultrafast
– electron dynamics 338
– laser 134f.
ultrafast spectroscopy 41f., 45f., 337
– near-field imaging of gold nanorods 45ff.
– time-resolved images 45
ultrafast surface dynamics 84
ultrashort NIR pulse 134
UV–VISs
– band height 409, 417
– in situ 405f., 410, 416
– –NIR spectrometer 262
– spectra 406f., 409

v

vacuum
– high 13
– ultrahigh 13
velocity
– fluctuations 126
– translational diffusion 141
vibration
– CO-stretching 85, 87
– frequency 3, 324
– low-frequency 104
– mode 25
– OH oscillators 80, 84
– OH-stretching 79, 81, 83
vibrational anharmonicity 87
vibrational coherence 104f.
– alkali-covered metal surfaces 345ff.
– diphasing 347ff.
– TMA-covered TiO_2 surface 110ff.
vibrational
– dynamics 337
– eigenstates 337
– nanospectroscopy 19, 84
– overtone 161
– relaxation dynamics 582ff.
viscosity 174
– -induced drag force 159
– solution 141
– temperature dependences 142
– water 232f.

voltammetry
– cyclic (CV) 79f., 80f., 85
– differntial pulse (DPV) 266

w

water
– cluster 80, 91, 233
– ice-like 80, 84, 91f., 96f.
– liquid-like 80, 84, 91f., 96f.
– molecule 80f., 83
– ordered layer 81
– potential-depending structure 80
wavefunction
– images 42
– longitudinal plasmon modes 43f.
– plasmon-wavefunction amplitudes 44
– steady-state 47
– total electron 360
waveguide effect 20, 368, 370ff.
wavelength
– component 636f.
– nonresonant 156
– observation 3, 16
– photon 19
– resonance peak 44
– sub- 49
wavenumber 20f., 87, 407
– bands 110f.
– -dependent sensitivity 107
– depth-dependent 111
– IR 103
wavepacket dynamics, see nuclear wavepacket
   dynamics
wide-band-limit (WBL) approximation 367

x

X-ray absorption fine structure (XAFS)
– dispersive (DXAFS) 427f., 430ff.
– extended (EXAFS) 407, 410, 416f., 427, 429
– in situ 410, 416f., 427, 430ff.
– quick (QXAFS) 427, 432ff.
– time-resolved EXAFS 435ff.
X-ray absorption near-edge structure
   (XANES) 436
X-ray diffraction (XRD) 80, 217
– diarylethene single crystal 444ff.
– organic inclusion crystals 507, 510
– polymerization 474, 476
– solid-state fluorescence spectra 513f.
– XRD–EXAFS 427
X-ray photoelectron diffraction (XPD) 329, 342

z

Zeeman splitting 272